Lecture Notes in Networks and Systems

Volume 167

The series "Lecture Notes in Networks and Systems" publishes the latest developments in Networks and Systems—quickly, informally and with high quality. Original research reported in proceedings and post-proceedings represents the core of LNNS.

Volumes published in LNNS embrace all aspects and subfields of, as well as new challenges in, Networks and Systems.

The series contains proceedings and edited volumes in systems and networks, spanning the areas of Cyber-Physical Systems, Autonomous Systems, Sensor Networks, Control Systems, Energy Systems, Automotive Systems, Biological Systems, Vehicular Networking and Connected Vehicles, Aerospace Systems, Automation, Manufacturing, Smart Grids, Nonlinear Systems, Power Systems, Robotics, Social Systems, Economic Systems and other. Of particular value to both the contributors and the readership are the short publication timeframe and the world-wide distribution and exposure which enable both a wide and rapid dissemination of research output.

The series covers the theory, applications, and perspectives on the state of the art and future developments relevant to systems and networks, decision making, control, complex processes and related areas, as embedded in the fields of interdisciplinary and applied sciences, engineering, computer science, physics, economics, social, and life sciences, as well as the paradigms and methodologies behind them.

Indexed by SCOPUS, INSPEC, WTI Frankfurt eG, zbMATH, SCImago.

All books published in the series are submitted for consideration in Web of Science.

More information about this series at http://www.springer.com/series/15179

Ajith Abraham · Oscar Castillo ·
Deepali Virmani
Editors

Proceedings of 3rd International Conference on Computing Informatics and Networks

ICCIN 2020

Springer

Editors
Ajith Abraham
Scientific Network for Innovation
and Research Excellence
Machine Intelligence Research Labs
(MIR Labs)
Auburn, WA, USA

Oscar Castillo
Tijuana Institute of Technology
Tijuana, Mexico

Deepali Virmani
Bhagwan Parshuram Institute of Technology
New Delhi, India

ISSN 2367-3370 ISSN 2367-3389 (electronic)
Lecture Notes in Networks and Systems
ISBN 978-981-15-9711-4 ISBN 978-981-15-9712-1 (eBook)
https://doi.org/10.1007/978-981-15-9712-1

This Springer imprint is published by the registered company Springer Nature Singapore Pte Ltd.
The registered company address is: 152 Beach Road, #21-01/04 Gateway East, Singapore 189721, Singapore

Organization

ICCIN-2020 was organized (as e-conference, due to COVID-19 pandemic) at Bhagwan Parshuram Institute of Technology, PSP-4, Sector-17, Rohini, New Delhi, India, during 29–30 July 2020.

Chief Patron

Sh. Atam Prakash Kaushik, Chairman, Bhartiya Brahmin Charitable Trust, New Delhi

Patrons

Padma Shri Sh. Surender Sharma, Sr. Vice President, Bhartiya Brahmin Charitable Trust, New Delhi
Sh. Vinod Vats, General Secretary, Bhartiya Brahmin Charitable Trust, New Delhi
Sh. Ram Babu Sharma, Secretary, Bhartiya Brahmin Charitable Trust, New Delhi
Sh. B. N. Sharma, Secretary, Bhartiya Brahmin Charitable Trust, New Delhi
Dr. Payal Pahwa, Principal, Bhagwan Parshuram Institute of Technology, New Delhi
Dr. A. K. Tandon, Founder Director, Bhagwan Parshuram Institute of Technology, New Delhi
Dr. Y. D. Gaur, Executive Director, Bhagwan Parshuram Institute of Technology, New Delhi

Convenor

Dr. Deepali Virmani, Professor and Head, Department of Computer Science and Engineering, Bhagwan Parshuram Institute of Technology, New Delhi

Keynote Speakers

Prof. Ajith Abraham, Director, Machine Intelligence Research Labs, USA
Prof. Oscar Castillo, Professor and Research Chair of Graduate Studies, Tijuana Institute of Technology, Tijuana, Mexico

Invited Speaker

Prof. (Retd.) Narasimman Sundararajan, School of Electrical and Electronics Engineering, Nanyang Technological University, Singapore

Steering Committee

Dr. C. R. Jagga, Deputy Director, Bhagwan Parshuram Institute of Technology, New Delhi
Sh. M. S. Vats, OSD, Bhagwan Parshuram Institute of Technology, New Delhi
Dr. S. D. Chauhan, Professor and Head, Department of Applied Mathematics, Bhagwan Parshuram Institute of Technology, New Delhi
Dr. Abhijit Nayak, Professor and Dean (Admin), Bhagwan Parshuram Institute of Technology, New Delhi
Dr. Anuradha Bhasin, Professor and Dean (Academics), Bhagwan Parshuram Institute of Technology, New Delhi
Dr. Sundaram Priyadarshini, Professor and Head, School of Business Administration, Bhagwan Parshuram Institute of Technology, New Delhi
Dr. Abhishek Swaroop, Professor and Head, Department of Information Technology, Bhagwan Parshuram Institute of Technology, New Delhi
Dr. Nidhi Sharma, Professor and Head, Examination Department, Bhagwan Parshuram Institute of Technology, New Delhi
Dr. Niranjan Bhattacharya, Professor and Head, Department of Electronics and Communications Engineering, Bhagwan Parshuram Institute of Technology, New Delhi
Dr. Peeyush Pant, Associate Professor, Head of Department of Electrical and Electronics Engineering, Bhagwan Parshuram Institute of Technology, New Delhi

Organizing Committee

Dr. Bhawna Suri, Associate Professor, Department of Computer Science and Engineering, Bhagwan Parshuram Institute of Technology, New Delhi
Dr. Shweta Taneja, Associate Professor, Department of Computer Science and Engineering, Bhagwan Parshuram Institute of Technology, New Delhi
Ms. Charu Gupta, Assistant Professor, Department of Computer Science and Engineering, Bhagwan Parshuram Institute of Technology, New Delhi
Ms. Palak Girdhar, Assistant Professor, Department of Computer Science and Engineering, Bhagwan Parshuram Institute of Technology, New Delhi
Mr. Dinesh Bhardwaj, Assistant Professor, Department of Computer Science and Engineering, Bhagwan Parshuram Institute of Technology, New Delhi

Core Committee Members

Dr. Achal Kaushik, Professor, Department of Computer Science and Engineering, Bhagwan Parshuram Institute of Technology, New Delhi
Mr. Shailendra S. Gaur, Assistant Professor, Department of Information Technology, Bhagwan Parshuram Institute of Technology, New Delhi
Ms. Deepti Jain, Assistant Professor, Department of Computer Science and Engineering, Bhagwan Parshuram Institute of Technology, New Delhi
Ms. Preeti Arora, Assistant Professor, Department of Computer Science and Engineering, Bhagwan Parshuram Institute of Technology, New Delhi
Ms. Himani Sharma, Assistant Professor, Department of Computer Science and Engineering, Bhagwan Parshuram Institute of Technology, New Delhi
Ms. Kritika Vashisht, Assistant Professor, Department of Computer Science and Engineering, Bhagwan Parshuram Institute of Technology, New Delhi
Ms. Minakshi Dhamija, Assistant Professor, Department of Computer Science and Engineering, Bhagwan Parshuram Institute of Technology, New Delhi
Ms. Mugdha Sharma, Assistant Professor, Department of Computer Science and Engineering, Bhagwan Parshuram Institute of Technology, New Delhi
Ms. Nikita Nijhawan, Assistant Professor, Department of Computer Science and Engineering, Bhagwan Parshuram Institute of Technology, New Delhi
Ms. Akanksha Dhamija, Assistant Professor, Department of Computer Science and Engineering, Bhagwan Parshuram Institute of Technology, New Delhi
Ms. Tanisha Madan, Assistant Professor, Department of Computer Science and Engineering, Bhagwan Parshuram Institute of Technology, New Delhi
Ms. Richa Sharma, Assistant Professor, Department of Computer Science and Engineering, Bhagwan Parshuram Institute of Technology, New Delhi

Technical Programme Committee

Prof. Ajith Abraham, Director of Machine Intelligence Research Labs (MIR Labs), USA
Prof. Oscar Castillo, Professor and Research Chair of Graduate Studies, Tijuana Institute of Technology, Tijuana, Mexico
Dr. Aws Zuheer Yonis, Senior Lecturer, Department of Communications Engineering, Ninevah University, Mosul, Iraq
Dr. Dimitris Kanellopoulos, Member of Educational Software Development Laboratory (ESD Lab), Department of Mathematics, University of Patras, Greece
Prof. Muhammad Sarfraz, Professor and Director of MSIT, Department of Information Science, Kuwait University, Kuwait
Dr. Vania Vieira Estrela, Visiting Faculty, Department of Computer Engineering and Industrial Automation (DCA), University of Campinas, Brazil
Dr. Froilan Mobo, Professor, Columban College Graduate School, Philippines
Prof. Jaekwang Kim, Professor, Department of Statistics, Iowa State University (ISU), USA
Prof. Alexander Gelbukh, Research Professor and Head of Natural Language Laboratory, Centre of Computer Research (CIC), National Polytechnic Institute (IPN), Mexico
Dr. Cheng-Yuan Chang, Professor, Department of Electrical Engineering, Chung Yuan Christian University, Taiwan
Dr. Ching-Hao Lai, Deep Learning Algorithm Researcher, Kneron, San Diego, USA
Prof. B. K. Dass, Professor, Department of Mathematics, University of Delhi, India
Prof. Amit Prakash Singh, Professor, University School of Information and Communication Technology, GGSIPU, New Delhi, India
Dr. I. Joe Louis Paul, Associate Professor, Department of Information Technology, SSN College of Engineering, Chennai, India
Dr. Jue-Sam Chou, Professor, Department of Business Administration, Nanhua University, Taiwan
Dr. Binh P. Nguyen, Senior Lecturer in Data Science, School of Mathematics and Statistics, Victoria University of Wellington (VUW), New Zealand
Dr. Trang Thi Thu Do, National University of Singapore, Singapore
Dr. Lesia Baranovska, Associate Professor and Senior Scientific Researcher, Institute for Applied System Analysis, Igor Sikorsky Kyiv Polytechnic Institute, Ukraine
Dr. P. S. Jagadeesh Kumar, Research Professor at Stanford Research Park, Stanford University, USA
Dr. Apostolos Gkamas, Associate Professor, University Ecclesiastical Academy of Vella, Ioannina, Greece
Dr. Mark Burgin, Department of Mathematics, University of California, Los Angeles, California, USA

Ms. Neha Singh, Assistant Professor, Department of Information Technology, Maharaja Agrasen Institute of Technology, New Delhi, India
Mr. Puneet Singh Lamba, Assistant Professor, Department of Computer Science and Engineering, Bharati Vidyapeeth's College of Engineering, New Delhi, India
Ms. Gunjan Ansari, Assistant Professor, Department of Information Technology, JSS Academy of Technical Education, Noida, Uttar Pradesh, India
Dr. Pinaki Chakraborty, Assistant Professor, Netaji Subhas University of Technology, New Delhi, India
Dr. Alka Singhal, Assistant Professor, Department of Computer Science and Engineering and Information Technology, Jaypee Institute of Information Technology (JIIT), Noida, Uttar Pradesh, India
Dr. K. Satyanarayan Reddy, Professor and HOD, Information Science and Engineering, Cambridge Institute of Technology, Bangalore, India
Prof. Ritu Sibal, Professor, Division of Computer Engineering, Netaji Subhas University of Technology, New Delhi, India
Dr. Malaya Dutta Borah, Assistant Professor, Department of Computer Science and Engineering, NIT Silchar, Assam, India
Mr. Puneet Misra, Assistant Professor, Department of Computer Science, University of Lucknow, India
Mr. R. Manoharan, Associate Professor, School of Mechanical Engineering, VIT University, India
Dr. Gopal Choudhary, Assistant Professor, Bharati Vidyapeeth's College of Engineering, Delhi, India
Dr. Vandana Niranjan, Associate Professor, Department of Electronics and Communication Engineering, IGDTUW, Delhi, India
Dr. Jothilakshmi Vishnu Prakash, Professor, Department of Electronics and Communication Engineering, Sri Venkateswara College of Engineering, Chennai, India
Dr. Froilan D. Mobo, Associate Professor, Philippine Merchant Marine Academy, Philippines
Prof. Jue-Sam Chou, Associate Professor, Department of Information Management, Nanhua University, Taiwan
Dr. Noor Elaiza Abdul Khalid, Associate Professor, Department of Computer Science and Engineering, University Teknologi MARA, Japan
Dr. Apostolos Gkamas, Associate Professor, Department of Computer Science and Technology, University Ecclesiastical Academy of Vella, Greece
Dr. Anuradha Chug, Assistant Professor, University School of Information and Communication Technology, GGSIPU, New Delhi, India
Dr. Prateek Agrawal, University of Klagenfurt, Austria, Associate Professor, Lovely Professional University, India
Ms. Vishu Madaan, Assistant Professor, Department of Computer Science, Lovely Professional University, India
Ms. Ishpreet Singh Virk, Assistant Professor, Department of Computer Science and Engineering, Baba Banda Singh Bahadur Engineering College, Fatehgarh Sahib, India

Preface

In today's time, computer science through its technologies and innovations is making a wide difference in various spheres of our life. The skills developed in the process are used to address global issues, versatile technology and building cost-effective solutions to various problems.

ICCIN-2020 provided a multidisciplinary platform, where substantive deliberations/interaction of thoughts and ideas on critical issues of computer science and information technology took place. This international collaboration of ICCIN-BPIT with academicians, scholars, educators fulfilled the need of moral, ethical and cultural connect among all. This event had brought a harmonious relationship between faculties and students as well. This platform facilitated an open and constructive dialogue on the research works of the best practices in computer science and information technology. The responses to the call for papers had been overwhelming—both from India and from overseas.

This book contains the research papers presented in the conference. Papers have been divided into the following tracks:

- Big Data Analytics and Business Intelligence
- Network and Information Security
- Machine Learning and Soft Computing
- High Performance Computing
- Internet-of-Things and Environmental Monitoring
- Computer Applications and Technological Innovations in Social Sciences Health care
- Computational Mathematics

We express our sincere gratitude to the eminent keynote speakers, invited speakers, authors and participants. Our diligent thanks to Chief Patron, Sh. Atam Prakash Kaushik, (Chairman, BBCT), Padma Shri Sh. Surender Sharma (Sr. Vice President, BBCT), Sh. Vinod Vats (Gen.Secretary, BBCT), Sh. B. N. Sharma (Secretary, BBCT) and other earnest members of the Bhartiya Brahmin Charitable Trust. We extend our heartfelt thanks to Principal, Prof. Payal Pahwa, Bhagwan Parshuram Institute of Technology, for giving us constant support and encouragement throughout the course of this endeavour. Encouragement from the directors/OSD/deans/HODs/faculty members of BPIT is highly appreciated.

We are grateful to Springer, especially to Mr. Aninda Bose (Senior Publishing Editor, Springer India Pvt. Ltd.), and the entire Springer team for the excellent collaboration, patience and help during the unrolling of this volume.

New Delhi, India

Organizing Committee
ICCIN 2020

Contents

Computer Applications and Technological Innovations In Social Sciences Health-Care

Computational Mathematics

Editors and Contributors

About the Editors

Prof. Ajith Abraham is Director of Machine Intelligence Research Labs (MIR Labs), which has members from more than 100 countries. Dr. Abraham's research and development experience includes more than 30 years in the industry and academia. He received the M.S. degree from Nanyang Technological University (NTU), Singapore, and the Ph.D. degree in Computer Science from Monash University, Melbourne, Australia. He works in a multi-disciplinary environment involving machine (network) intelligence, cyber security, sensor networks, web intelligence, scheduling, data mining and applied to various real-world problems. As an Investigator/Co-Investigator, he has won research grants worth over 100+ Million US$ from Australia, USA, EU, Italy, Czech Republic, France, Malaysia and China. In these areas, he has authored/co-authored more than 1400+ research publications out of which there are 100+ books covering various aspects of Computer Science. He has 1200+ co-authors originating from 40+ countries. Dr. Abraham has more than 38,000+ academic citations (h-index of 93 as per google scholar). He has given more than 100 plenary lectures and conference tutorials (in 20+ countries). He serves/has served the editorial board of over 50 international journals and has also guest edited 40 special issues on various topics.

Prof. Oscar Castillo holds the Doctor in Science degree (Doctor Habilitatus) in Computer Science from the Polish Academy of Sciences (with the Dissertation "Soft Computing and Fractal Theory for Intelligent Manufacturing"). He is a Professor of Computer Science in the Graduate Division, Tijuana Institute of Technology, Tijuana, Mexico. In addition, he is serving as Research Director of Computer Science and head of the research group on Hybrid Fuzzy Intelligent Systems. Currently, he is President of HAFSA (Hispanic American Fuzzy Systems Association) and Past President of IFSA (International Fuzzy Systems Association). Prof. Castillo is also Chair of the Mexican Chapter of the Computational Intelligence Society (IEEE). He also belongs to the Technical Committee on Fuzzy

Systems of IEEE and to the Task Force on "Extensions to Type-1 Fuzzy Systems". He is also a member of NAFIPS, IFSA and IEEE. He belongs to the Mexican Research System (SNI Level 3). His research interests are in Type-2 Fuzzy Logic, Fuzzy Control, Neuro-Fuzzy and Genetic-Fuzzy hybrid approaches. He has published over 300 journal papers, 10 authored books, 40 edited books, 200 papers in conference proceedings, and more than 300 chapters in edited books, in total 901 publications according to Scopus (H index=63, and more than 1050 publications according to Research Gate (H index=74 in Google Scholar). He has been Guest Editor of several successful Special Issues in the past, like in the following journals: Applied Soft Computing, Intelligent Systems, Information Sciences, Non-Linear Studies, Fuzzy Sets and Systems, JAMRIS and Engineering Letters. He is currently Associate Editor of the Information Sciences Journal, Applied Soft Computing Journal, Engineering Applications of Artificial Intelligence, Granular Computing Journal and the International Journal on Fuzzy Systems. Finally, he has been elected IFSA Fellow in 2015 and MICAI Fellow member in 2017. He has been recognized as Highly Cited Researcher in 2017 and 2018 by Clarivate Analytics because of having multiple highly cited papers in Web of Science.

Prof. Deepali Virmani is Head of the Department and Professor in the Department of Computer Science and Engineering at Bhagwan Parshuram Institute of Technology affiliated to GGSIPU, New Delhi. Dr. Deepali Virmani has received her B.Tech. degree in Computer Science from MDU, Rohtak, M.Tech. degree in Information Technology, from GGSIPU, and the Ph.D. degree in Computer Science from Delhi University,India. She has an innovative work experience of more than 19 years in both research and academics. She has published more than 90 research papers in International journals/National journals/International conferences of repute. She works in a multi-disciplinary environment involving sensor networks, web intelligence, data mining and intelligent information retrieval systems and machine learning applied to various real-world problems. Dr. Virmani has more than 500+ academic citations index as per google scholar .She has guided more than 70 B.Tech Projects. Presently, she is guiding many Ph.D. scholars registered with reputed universities like GGSIPU and UPTU. She is branch counselor of BPIT-IEEE student chapter and BPIT-CSI student branch. She is the associate editor of journal Open Computer Science De Grutyer and also on the reviewer board of various IEEE, Elsevier and springer journals. She has organized many professional activities like FDPs, workshops, expert lectures and conferences. She has been the session chair in national/international conferences.

Contributors

Surabhi Adhikari Department of Computer Science and Engineering, Delhi Technological University, New Delhi, India

Ashish Agarwal Time and Frequency Metrology,, National Physical Laboratory, New Delhi, India

Neha Agarwal Amity University, Noida, Uttar Pradesh, India

Isha Aggarwal Department of Computer Science and Engineering, Bhagwan Parshuram Institute of Technology, New Delhi, India

Priyanka Aggarwal Information Technology, BPIT, New Delhi, Delhi, India

Ahmad T. Al-Taani Department of Computer Science, Yarmouk University, Irbid, Jordan

Ayush Anand Unitedworld School of Computational Intelligence, Karnavati University, Gandhinagar, Gujarat, India

Joy Aneja Cadence Design Systems, Gurgaon, India

Monika Arora Information Technology, BPIT, New Delhi, India

Preeti Arora Bhagwan Parshuram Institute of Technology, GGSIPU, New Delhi, India

Amutha Arunachalam Time and Frequency Metrology,, National Physical Laboratory, New Delhi, India

Ashu MRIIRS, Faridabad, India

Neha Bagwari Ajay Kumar Garg Engineering College, Ghaziabad, UP, India

Sidharth Bansal Department of Computer Science and Engineering, Delhi Technological University, New Delhi, Delhi, India

Usha Batra G D Goenka University, Gurugram, Haryana, India

Piyush Bhardwaj Bhagwan Parshuram Institute of Technology (affiliated to Guru Gobind Singh Indraprastha University, Delhi 110078), Rohini, Delhi, India

Sameer Bhardwaj Bharati Vidyapeeth's College of Engineering, New Delhi, India

Tanish Bhola Bharati Vidyapeeth's College of Engineering, New Delhi, India

Kritarth Bisht Department of Computer Science and Engineering, Bhagwan Parshuram Institute of Technology, New Delhi, Delhi, India

Siddhartha Sankar Biswas Department of CSE, SEST, Jamia Hamdard, New Delhi, India

Samuel Jacob Chacko International Institute of Information Technology Bhubaneswar, Bhubaneswar, India

Mohit Chandna ThoughtWorks, Gurgaon, India

Sarthak Chandna Department of Computer Science and Engineering, Maharaja Agrasen Institute of Technology, GGSIPU, Dwarka, India

Pravin Chandra University School of Information, Communication and Technology, Guru Gobind Singh Indraprastha University, New Delhi, India

Pavithra Chatter Rao Bahadur Y Mahabaleswarappa Engineering College, Ballari, Karnataka, India

Sudipta Chattopadhyay Department of Electronics and Communication Engineering, Mizoram University, Tanhril, Aizawl, Mizoram, India

Harindri Chaudhary Department of Mathematics, Jamia Millia Islamia, New Delhi, Delhi, India

Tanvi Chaudhary Amity University, Noida, Uttar Pradesh, India

Monika Chauhan Division of Forensic Science, School of Basic and Applied Sciences, Galgotias University, Greater Noida, UP, India

Geetansh Chawla Department of Computer Science and Engineering, Bhagwan Parshuram Institute of Technology, New Delhi, Delhi, India

Garima Chhabra Department of Information Technology, Bhagwan Parshuram Institute of Technology, Delhi, India

Arpit Choudhary Department of Information Technology, Bhagwan Parshuram Institute of Technology, Delhi, India

Rohit Dev Bharati Vidyapeeth's College of Engineering, New Delhi, India

Arun Kumar Dubey USICT Guru Gobind Singh Indraprastha University, New Delhi, India

Rajneesh Dubey BPIT, GGSIPU, New Delhi, India

Abhinav Duggal Department of Computer Science and Engineering, Delhi Technological University, New Delhi, India

Rakhi Garg Department of Computer Science, MMV, BHU, Varanasi, UP, India

Shailendra Singh Gaur Bhagwan Parshuram Institute of Technology, New Delhi, Delhi, India

Vinay Gautam Chitkara University Institute of Engineering and Technology, Chitkara University, Rajpura, Punjab, India

Udayan Ghose University School of Information, Communication and Technology, Guru Gobind Singh Indraprastha University, New Delhi, India

Abhijyoti Ghosh Department of Electronics and Communication Engineering, Mizoram University, Tanhril, Aizawl, Mizoram, India

Harsh Goel Bhagwan Parshuram Institute of Technology (affiliated to Guru Gobind Singh Indraprastha University, Delhi 110078), Rohini, Delhi, India

Kartik Goel Bhagwan Parshuram Institute of Technology, GGSIPU, New Delhi, India

Lakshay Goyal Department of Computer Science and Engineering, Bhagwan Parshuram Institute of Technology, New Delhi, Delhi, India

Abhinav Gupta Bharati Vidyapeeth's College of Engineering, New Delhi, India

Charu Gupta Department of Computer Science and Engineering, Bhagwan Parshuram Institute of Technology, New Delhi, Delhi, India

Gautam Gupta Bharati Vidyapeeth's College of Engineering, Delhi, India

Jatin Gupta Bharati Vidyapeeth's College of Engineering, New Delhi, India

Megha Gupta Department of Computer Science and Engineering, Bhagwan Parshuram Institute of Technology, New Delhi, Delhi, India

Shelly Gupta ASET, Amity University, Noida, Uttar Pradesh, India

Shilpa Gupta Bharati Vidyapeeth's College of Engineering, New Delhi, India

Shubham Gupta Bharati Vidyapeeth's College of Engineering, New Delhi, India

Yatin Gupta Department of Computer Science and Engineering, Delhi Technological University, New Delhi, Delhi, India

Shreevidya Gurudath Ramaiah Institiute of Technology, Bangalore, India

Himanshi BPIT, GGSIPU, New Delhi, India

Nida Iftekhar Department of CSE, SEST, Jamia Hamdard, New Delhi, India

Achin Jain Communication and Technology, University School of Information, GGSIPU, Dwarka, Delhi, India

Harsh Jain Bharati Vidyapeeth's College of Engineering, New Delhi, India

Minni Jain Department of Computer Science and Engineering, Delhi Technological University, New Delhi, Delhi, India

Parsid Kumar Jain HCMS, Yamunanagar, Haryana, India

Rachna Jain Department of Computer Science and Engineering, Bharati Vidyapeeth's College of Engineering, New Delhi, India

Vanita Jain Bharati Vidyapeeth's College of Engineering, New Delhi, India

Khan Ghazala Javed International Institute of Information Technology Bhubaneswar, Bhubaneswar, India

Aarnav Jindal Department of Computer Science and Engineering, Delhi Technological University, New Delhi, India

Prashant Johri Department of Computer Science, Galgotias University, Greater Noida, India

Ashish Joshi University School of Information Communication and Technology, GGSIPU, Dwarka, Delhi, India

Dheeraj Joshi EE Department, DTU, Delhi, India

Harjas Kalsi BPIT, GGSIPU, New Delhi, India

Pranjal Kandhari Bharati Vidyapeeth's College of Engineering, New Delhi, India

Shreya Kapoor Dr. Akhilesh Das Gupta Institute of Technology and Management, New Delhi, India

Rahul Katarya Department of Computer Science and Engineering, Delhi Technological University, New Delhi, India

Satvik Kaul Bharati Vidyapeeth's College of Engineering, New Delhi, India

Prabhjot Kaur Chitkara University Institute of Engineering and Technology, Chitkara University, Rajpura, Punjab, India

Samia Khan Department of CSE, SEST, Jamia Hamdard, New Delhi, India

Taqseer Khan Department of Mathematics, Jamia Millia Islamia, New Delhi, Delhi, India

Zaifa Khan Department of Computer Science and Engineering, Ramaiah Institute of Technology, Bangalore, India

Sunil K. Khatri Amity University, Tashkent, Uzbekistan

Jalaj Khurana Department of Computer Science and Engineering, Bhagwan Parshuram Institute of Technology, New Delhi, India

Nishant Khurana Bharati Vidyapeeth's College of Engineering, New Delhi, India

Kritika BPIT, GGSIPU, New Delhi, India

Avneesh Kumar Department of Computer Science, Galgotias University, Greater Noida, India

Deepika Kumar G D Goenka University, Gurugram, Haryana, India

Nitin Kumar Department of Computer Science and Engineering, Bhagwan Parshuram Institute of Technology, Delhi, India

Somesh Kumar Moradabad Institute of Technology, Moradabad, UP, India

Puneet Singh Lamba University School of Information, Communication and Technology, GGSIPU, Sector 16 C, Dwarka, New Delhi, India;
Bharati Vidyapeeth's College of Engineering, Paschim Vihar, New Delhi, India

Rishab Lamba Bharati Vidyapeeth's College of Engineering University, New Delhi, India

Rakesh Kumar Lenka International Institute of Information Technology Bhubaneswar, Bhubaneswar, India

Lourembam Lolit Kumar Singh Department of Electronics and Communication Engineering, Mizoram University, Tanhril, Aizawl, Mizoram, India

Aarnav Madan Department of Computer Science and Engineering, Bhagwan Parshuram Institute of Technology, Delhi, India

Mamta Madan Vivekananda Institute of Professional Studies, Delhi, India

Rashima Mahajan MRIIRS, Faridabad, India

Dheeraj Malhotra Vivekananda Institute of Professional Studies, New Delhi, Delhi, India

Megha Malhotra Department of Computer Science and Engineering, Bhagwan Parshuram Institute of Technology, New Delhi, Delhi, India

Neha Malhotra Vivekananda Institute of Professional Studies, New Delhi, Delhi, India

Akshay Manchanda Department of Computer Science, Bhagwan Parshuram Institute of Technology, New Delhi, Delhi, India

Aditee Mattoo Noida Institute of Engineering and Technology, Greater Noida, UP, India

Akshat Minocha Department of Computer Science and Engineering, Bhagwan Parshuram Institute of Technology, New Delhi, Delhi, India

Akash Mishra University School of Information, Communication and Technology, Guru Gobind Singh Indraprastha University, New Delhi, India

Sushruti Mishra Department of Computer Science and Engineering, Delhi Technological University, New Delhi, India

Vrinda Mittal Information Technology, BPIT, New Delhi, Delhi, India

Yash Mittal BPIT, GGSIPU, New Delhi, India

Amar Kumar Mohapatra Indira Gandhi Delhi Technical University for Women, Kashmere Gate, Delhi, India

Pooja Mudgil Department of Information Technology, Bhagwan Parshuram Institute of Technology, Delhi, India

Rajat Mudgil BPIT, Rohini, I.P University, India

Ashish Varghese Muttumana Bhagwan Parshuram Institute of Technology (affiliated to Guru Gobind Singh Indraprastha University, Delhi 110078), Rohini, Delhi, India

Ashish Nagpal Dr. Akhilesh Das Gupta Institute of Technology and Management, New Delhi, India

Preeti Nagrath Department of Computer Science and Engineering, Bharati Vidyapeeth's College of Engineering, New Delhi, India

Pradyumn Nand Makemytrip Pvt Ltd, Gurgaon, India

Anoushka Panwar Vivekananda Institute of Professional Studies, New Delhi, Delhi, India

Manu S. Pillai Bharati Vidyapeeth's College of Engineering, New Delhi, India

Paras Prakash Department of Computer Science and Engineering, Bharati Vidyapeeth's College of Engineering, New Delhi, India

Lalji Prasad Sagar Institute of Research and Technology, SAGE University, Indore, India

Sai Tiger Raina Department of Computer Science and Engineering, Bharati Vidyapeeth's College of Engineering, New Delhi, India

Ravika Rajput International Institute of Information Technology Bhubaneswar, Bhubaneswar, India

Mamta Rani Maharaja Surajmal Institute of Technology (affiliated to Guru Gobind Singh IndraPrastha University), Janakpuri, New Delhi, India

Neeru Rathee Maharaja Surajmal Institute of Technology (affiliated to Guru Gobind Singh IndraPrastha University), Janakpuri, New Delhi, India

Ria Rawal Bhagwan Parshuram Institute of Technology, GGSIPU, New Delhi, India

Karan Rawlley Department of Computer Science and Engineering, Bhagwan Parshuram Institute of Technology, New Delhi, Delhi, India

Jimmy Rinsangzuala Department of Electronics and Communication Engineering,, Mizoram University, Tanhril, Aizawl, Mizoram, India

Munish Sabharwal Department of Computer Science, Galgotias University, Greater Noida, India

Abhishek Saini Department of Computer Science, Bhagwan Parshuram Institute of Technology, New Delhi, Delhi, India

Dharmender Saini Bharati Vidyapeeth's College of Engineering University, New Delhi, India

Deepika Sandil BPIT, Rohini, I.P University, India

B. Sathvik Department of Computer Science and Engineering, Ramaiah Institute of Technology, Bangalore, India

Ayush Kumar Saxena Department of Computer Science and Engineering, Bhagwan Parshuram Institute of Technology, New Delhi, India

Kumud Saxena Noida Institute of Engineering and Technology, Greater Noida, UP, India

S. Seema Department of Computer Science and Engineering, Ramaiah Institute of Technology, Bangalore, India

K. Seetharaman Department of Computer and Information Sciences, Annamalai University, Chidambaram, India

Bhavya Sethi Department of Information Technology, Bhagwan Parshuram Institute of Technology, Delhi, India

Abhishek Shah Department of Computer Science and Engineering, Delhi Technological University, New Delhi, India

Akshat Shakarwal Department of Computer Science and Engineering, Bhagwan Parshuram Institute of Technology, Delhi, India

Raju Shanmugam Unitedworld School of Computational Intelligence, Karnavati University, Gandhinagar, Gujarat, India

Mahak Sharma Department of Computer Science and Engineering, Bhagwan Parshuram Institute of Technology, New Delhi, Delhi, India

Mugdha Sharma Department of Computer Science and Engineering, Bhagwan Parshuram Institute of Technology, New Delhi, India

Richa Sharma ASET, Amity University, Noida, Uttar Pradesh, India

Simmi Sharma BPIT, Delhi, India

Arpit Shrotiya Bharati Vidyapeeth's College of Engineering, New Delhi, India

Abhay Pratap Singh Department of Computer Science and Engineering, Bhagwan Parshuram Institute of Technology, Delhi, India

Aniket Singh Department of Computer Science and Engineering, Ramaiah Institute of Technology, Bangalore, India

Dipika Singh Department of Computer Science, Institute of Science, BHU, Varanasi, UP, India

Neha Singh Maharaja Agrasen College (affiliated to Guru Gobind Singh Indraprastha University, Delhi 110078), Rohini, Delhi, India

Shailendra Narayan Singh ASET, Amity University, Noida, Uttar Pradesh, India

Tushar Singh Vellore Institute of Technology, Vellore, Tamil Nadu, India

Apoorvi Sood University School of Information, Communication & Technology, Guru Gobind Singh Indraprastha University, New Delhi, India

Shshank Sourabh Division of Forensic Science, School of Basic and Applied Sciences, Galgotias University, Greater Noida, UP, India

B. J. Sowmya Department of Computer Science and Engineering, Ramaiah Institute of Technology, Bangalore, India

K. G. Srinivasa Department of Information Management and Coordination, NITTTR, Chandigarh, India

P. V. Suma Latha Rao Bahadur Y Mahabaleswarappa Engineering College, Ballari, Karnataka, India

Bhawna Suri Department of Computer Science, Bhagwan Parshuram Institute of Technology, New Delhi, Delhi, India

Shakhzod Suvanov Faculty of Mathematics and Informatics, Samarkand State University, Samarkand, Uzbekistan

S. Suzain Rao Bahadur Y Mahabaleswarappa Engineering College, Ballari, Karnataka, India

D. V. Swetha Ramana Rao Bahadur Y Mahabaleswarappa Engineering College, Ballari, Karnataka, India

Shweta Taneja Department of Computer Science and Engineering, Bhagwan Parshuram Institute of Technology, GGSIPU, Dwarka, India

Yash Teotia Bhagwan Parshuram Institute of Technology (affiliated to Guru Gobind Singh Indraprastha University, Delhi 110078), Rohini, Delhi, India

Surendrabikram Thapa Department of Computer Science and Engineering, Delhi Technological University, New Delhi, India

Kanika Thapliyal Department of Computer Science, Bhagwan Parshuram Institute of Technology, New Delhi, Delhi, India

Anuj Thareja Department of Computer Science and Engineering, Bharati Vidyapeeth's College of Engineering, New Delhi, India

Pawar Umesh Ankush Indian Army, New Delhi, India

Abhishek Upadhyay International Institute of Information Technology Bhubaneswar, Bhubaneswar, India

Ritivik Vaidya Department of Computer Science and Engineering, Bhagwan Parshuram Institute of Technology, New Delhi, India

Moksh Verma Department of Computer Science, Bhagwan Parshuram Institute of Technology, New Delhi, Delhi, India

Deepali Virmani Department of Computer Science and Engineering, Bhagwan Parshuram Institute of Technology, New Delhi, Delhi, India

Nitin Vohra Bharati Vidyapeeth's College of Engineering, New Delhi, India

Niti Vore Sagar Institute of Research and Technology, SAGE University, Indore, India

Misbahur Rahman Warsi Aligarh Muslim University, Aligarh, India

Rashmi Yadav Department of Computer Science and Engineering, Acropolis Technical Campus, Indore, India

Sherin Zafar Department of CSE, SEST, Jamia Hamdard, New Delhi, India

Zonunmawii Department of Electronics and Communication Engineering,, Mizoram University, Tanhril, Aizawl, Mizoram, India

Abbreviations

2D and 3D hand signals	Two-dimensional and three-dimensional hand signals
3D models	Three-dimensional models
AAL	Ambient-assisted living
AC	Alternate current
ACT	Active control technique
ADHD	Attention-deficit/hyperactivity disorder
ADL	Activities of daily living
AI	Artificial intelligence
AIE	Adaptive interface ecosystems
AMH	Anti-Mullerian hormone
AmI	Ambient intelligence
ANC	Antenatal care
ANFIS	Adaptive neuro-fuzzy inference system
ANN	Artificial neural networks
ANTHOCNET	Ant-based hybrid routing algorithm for mobile and ad hoc networks
AODV	Ad hoc on-demand vector
API	Application programming interface
AR	Augmented reality
ASC	Acoustic scene classification
ASHAs	Accredited Social Health Activists
ASM	Active shape model
ASR	Automatic speech recognition
ATCS	Adaptive traffic control system
ATM	Automated teller machine
ATNs	Augmented transition networks
AUC	Area under curve
AV	Augmented virtuality
Avg. F	Average follicle
BAN	Body area network

BERT	Bidirectional encoder representations from transformers
BIPM	Bureau International des Poids et Mesures
BMI	Body mass index
BOINC	Berkeley Open Infrastructure for Network Computing
BOW	Bag of words
BP	Blood pressure
BSD	Berkeley Software Distribution
C	Capacitor
CA	Certifying authorities
CASME	Chinese Academy of Sciences Micro-Expression
CBIR	Content-based image retrieval
CBOW	Continuous bag of words
CCA	Controller of Certifying Authorities
CCTV	Closed-circuit television
CDS	Cooperative driving strategy
CEA	Central Electrical Authority
CED	Canny edge detection algorithm
CIPM	Certificate in Investment Performance Measurement
CK+	Cohn-Kanade+
CLAHE	Contrast limited adaptive histogram equalization algorithm
CMA	Circular microstrip antenna
CMHCSM	Comprehensive mental healthcare stepped-care model
CNN	Convolution neural network
CP	Co-polarization
CSC	Chaos synchronization and control
CSIR	Council of Scientific and Industrial Research
CSV	Comma-separated values
CV	Computer vision
D	Duty cycle
DA	Data analytics
DA	Data augmentation
DBN	Deep belief network
DC	Direct current
DCT	Discrete cosine transformation
DDoS	Distributed denial of service
DDYDAS	Driver drowsiness, yawn detection and alert system
DEAP	Database for Emotion Analysis using Physiological Signals
DGS	Defected ground structure
DL	Deep learning
DL	Description logic
DLF	Deep layered features
DLR	Deep layered representation

DMMH-SVM	Dual-margin multi-class hypersphere support vector machine
DNN	Deep neural network
DoS	Denial of service
DPS	Defected patch surface
DRC	Design rule check
DRMF	Discriminative response map fitting
DST	Daylight savings time
DT	Decision tree
EAGLE	Easily Applicable Graphical Layout Editor
EAR	Eye Aspect Ratio
ECG	Electrocardiography
EDA	Electrodermal activity
EEG	Electroencephalography
EFS	Evolution fuzzy systems
EHR	Electronic health records
ELU	Exponential linear unit
EM	Energy management
EM	Expectation maximization algorithm
EMG	Electromyography
ERC	Electrical rule check
ERS	Entity relationship schema
ESP8266	Espressif Systems
EV	Emergency vehicle
EVM	Eulerian video magnification
FBI	Federal Bureau of Investigation
FER	Facial expression recognition
FFANNs	Feed-forward artificial neural networks
Fig.	Figure
FLAC	Free lossless audio file
FLC	Fuzzy logic controller
FLV	Flash video
FN	False negative
FP	False positive
FPS	Frames per second
FSH	Follicle-stimulating hormone
FTK	File toolkit
GANs	Generative adversarial networks
GDP	Gross domestic product
GIF	Graphics Interchange Format
GKVK	Gandhi Krishi Vigyan Kendra
GLCM	Grey-level co-occurrence matrix
Glove	Global vector for word representation
GLV	Generalized Lotka–Volterra
GMT	Greenwich Mean Time

GNP	Genetic network programming
Govt.	Government
GPIO	General-purpose input/output
GPRS	General Packet Radio Service
GPS	Global Positioning System
GSM	Global System for Mobile Communication
GUI	Graphical user interface
H&E	Hematoxylin–eosin
HBA	Human behaviour analysis
HBU	Human behavioural awareness and understanding
HCR	Hybrid clustering routing
HCS	Hybrid combination synchronization
HDL	High-density lipoprotein
HMM	Hidden Markov model
HOG	Histogram of arranged angles
HOHA	Hollywood Human Action
HOOF	Histogram of optical-oriented flow
HSV	Hue, saturation, value
HTML	Hypertext Markup Language
HWMP	Hybrid Wireless Mesh Protocol
I	Integral
IBPE	Indexing-based peculiarity extrication from biomedical images
IBS	ID-based signature
IC	Integrated circuits
ICT	Information and Communication Technology
IEEE	Institute of Electrical and Electronics Engineers
ILD	Inductive loop detectors
INIS dataset	International Nuclear Information System data set
IoMT	Internet of Medical Things
IoT	Internet of Things
iRPROP	Improved resilient backpropagation algorithm
IST	Indian Standard Time
ITS	Indian Telecom Services
ITS	Intelligent transportation system
JIT	Just-in-time
JPG	Joint Photographic Expert Group
JSSK	Janani Shishu Suraksha Karyakaram
JSY	Janani Suraksha Yojna
KEEL	Knowledge extraction based on evolutionary analysis
KNN	K-nearest neighbours
L	Inductor
LBP	Local binary pattern
LDL	Low-density lipoprotein
LDR	Light-dependent resistor

LED	Light-emitting diode
LH	Luteinizing hormone
LLD	Late life depression
LM	Legal metrology
LoRa	Long-range radio system
LoRaWAN	Low-power wide area network protocol
LOSO	Leave one subject out
LOVO	Leave one video out
LReLU	Leaky rectified linear unit
LSB	Least-significant bit
LST	Lyapunov stability theory
LSTM	Long-short-term memory
LTE	Long-term evolution
LTE-M	Long-term evolution for machines
LUA	Programming language
LWM	Local weight mean
MAB	Multi-armed bandit
MaMSE	(ensemble) Maximum of mean squared errors
MAN	Metropolitan area networks
MANET	Mobile ad hoc networks
MAR	Mobile augmented reality
MAR	Mouth aspect ratio
MBCT	Mindfulness-based cognitive technology
MCU	Microcontroller unit
MDGs	Millennium Development Goals
MDT	Mobile data terminal
MeMSE	(ensemble) Median of mean squared errors
MFI	Multidimensional feature image
MiMSE	(ensemble) Minimum of mean squared errors
MIST	Montreal Imaging Stress Task
MIT	Massachusetts Institute of Technology
ML	Machine learning
MMR	Maternal mortality rate
MMSE	(ensemble) Mean of mean squared errors
MN	Minnesota
MPA	Microstrip patch antenna
MPP	Maximum power point
MPU	Message processing unit
MRA	Mutual recognition arrangement
MS	Master system
MSBR	Mindfulness-based stress reduction
MSU	Message steganography unit
MTCNN	Multitask cascaded neural network
MTG	Model Target Generator
MW	Mega Watt

NABL	National Accreditation Board of Laboratories
NB	Naïve Bayes
NER	Named-entity recognition
NIR	Near-infrared
NLP	Natural language processing
NMI	National Measurement Institute
NMS	Non-maximal suppression
NN	Neural network
NPD	Non-priority district
NPL	National Physical Laboratory
NRHM	National Rural Health Mission
NS-2	Network simulator-2
NTA	National Time Authority
NTCA	National Time Commerce Authority
NZMT	New Zealand unit of time
OASIS	Open Access Series of Imaging Studios
OBU	On-board unit
OC	Opto coupler
OGY	Ott Grebogi Yorke
OLSR	Optimized Link State Routing Protocol
OMG	Object Management Group
OpenCV	Open Source Computer Vision Library
ORS	Object relationship schema
OSF	Optical strain feature
OWL	Web Ontology Language
P	Proportional
PAD	Pleasure, Arousal and dominance
PC	Personal computer
PCB	Printed circuit board
PCOS	Polycystic ovarian syndrome
PD	Priority District
PDF	Portable document format
PEC	Perfect electric conducting
PERCLOS	Percentage of eyelid closure
PFC	Prefrontal cortex
PLR	Packet loss ratio
PMC	Perfect magnetic conducting
PNG	Portable network graphics
PNN	Probabilistic neural network
PRG	Progesterone
PRL	Prolactin
PSD	Power spectral density
PSO	Particle swarm optimization
PTSD	Post-traumatic stress disorder
PV	Photovoltaic

QoL	Quality of life
QoS	Quality of service
RBF	Radial basis function
RBS	Random blood sugar
RCAI	Root Certifying Authority of India
R-CNN	Region-based convolutional neural networks
RDBMS	Relational database management system
RDF	Resource Description Framework
ReLU	Rectified linear unit
RF	Random forest
RFID	Radio-frequency identification
RG	Rubenstein and Goodenough
RGB	Red, green and blue
RL	Reinforcement learning
RMA	Rectangular microstrip patch antenna
RNNs	Recurrent neural networks
ROC	Receiver operating characteristic
ROI	Region of interest
RR	Respiratory rate
RRSL	Regional Reference Standard Laboratories
Rs, R, R1, R2	Load resistors
RSUs	Roadside Units
RW	Rare words
S	Switch
SAMM	A spontaneous micro-facial movement
SAODV	Secured AODV
SBA	Skilled birth attendant
SCC	Source Code Control
SDAE	Stacked denoising autoencoder
SDK	Software development kit
SDSA	Semantic decision support system
SELU	Scaled exponential linear unit
SGDM	Spatial grey dependency matrix
SHS	Smart home system
SI	International system
SIFT	Scale-invariant feature transform
SiLU	Sigmoid-weighted linear unit
SMIC	Spontaneous micro-expression database
SMOTE	Synthetic minority over-sampling technique
SOM	Self-organizing map
SPIFFS	Serial peripheral interface flash file system
SPL	Software product line
SRBM	Social restricted Boltzmann machine
SS	slave system
SSA	Small Signal Analysis

SSA	Steganography Security Algorithm
STD	(ensemble) Standard Deviation of Mean Squared Errors
STT	Synchronized Traceable Time
SV	State variables
SVD	Singular value decomposition
SVM	Support vector machine
SYN	Sync flooding attacks
TC	Time commerce
TC	Total cholesterol
TDS	Time dissemination service
TERI	The Energy Research Institute
TF	Transfer function
TF-IDF	Term frequency–inverse document frequency
TG	Triglycerides
TIF	Tagged Image File
TIM	Temporal interpolation model
TKEO	Teager–Kaiser energy operator
TLN	Temporary logic network
TN	True Negative
TOR	The Onion Router
TP	True Positive
TRIAC	Triode for alternating current
TS	Thompson sampling
TSA	Timestamp Authority
TSH	Thyroid-stimulating hormone
TSS	Timestamp service
TTS	Trusted time stamping
TWh	Terawatt hour
UCB	Upper confidence bound
UCOS	Unified collaborative organizational structure
UMBC	The University of Maryland, Baltimore County
UML	Unified Modeling Language
UNFPA	United Nations Population Fund
UNICEF	United Nations International Children's Emergency Fund
URL	Uniform Resource Locator
URMP dataset	University of Rochester Music Performance dataset
USA	The United States of America
USF-HD	University of South Florida High Definition
UTC	Coordinated Universal Time
V2I	Vehicle to infrastructure
V2V	Vehicle to vehicle
VANET	Vehicular ad hoc network
VARS	Vehicle ad hoc reputation system
VGG	Visual Geometry Group

Vin	Input voltage
Vit	Vitamin
Vo-	Output voltage
VPS	Visual positioning system
VR	Virtual reality
VRET	Virtual reality exposure therapy
WAN	Wide area network
WAV	Waveform audio file
WebRTC	Web Real-Time Communication
WER	Word error rate
WHO	World Health Organization
Wi-fi	Wireless fidelity
WLAN	Wireless local area network
WMN	Wireless mesh networks
WMV	Windows Media Video
Word 2Vec	Words in vector space representation
WoS	Web of Science
WPAN	Wireless personal area network
WS	WordSim
WSN	Wireless sensor network
XP	Cross-polarization
ZRP	Zone Routing Protocol

Big Data Analytics and Business Intelligence

Experimental Comparison and Scientometric Inspection of Research for Word Embeddings

Minni Jain, Sidharth Bansal, and Yatin Gupta

Abstract Word embedding or universal embeddings are the representation of textual data into vectors of real numbers. It acts as a link between the human understanding of a text to that of the machine. It maps the high-dimensional textual data into a vector space of low dimension, which represents the complex relationships existing in the data. Thus, it boosts the performance of tasks involving natural language processing. To know about the popularity gain and the impact of this field, research literature between 2000 and 2019 is analyzed with the help of a scientometric mapping for research done in word embeddings. The paper visualizes year-wise analysis, demographic analysis, category-wise distribution, and document type-wise distribution of the publications indexed in Web of Science (WoS). The paper also comprises a comparative study of Word2Vec, FastText, global vector representation of words, and bidirectional encoder representations for transformers. Pre-trained models are used for experimental comparison. Performance is measured by calculating the deviation of the similarity score of two words given by the models from manually assigned similarity scores by experts, repeated over a list of words on various datasets. The least deviation is shown by FastText due to the usage of morphological information in the skip-gram model and n-gram architecture.

Keywords Scientometric analysis · Word embeddings · Word2Vec · BERT · GloVe · FastText · WordNet · Web of science

1 Introduction

In the contemporary world, the data is abundant and heterogeneous. There is a need to process and interpret the results from this immense data. For example, the reviews from an e-commerce Web site can be used to predict the relevant products which the

---

M. Jain · S. Bansal (✉) · Y. Gupta
Department of Computer Science and Engineering, Delhi Technological University, New Delhi, Delhi, India
e-mail: bansal.sidharth2996@gmail.com

A. Abraham et al. (eds.), *Proceedings of 3rd International Conference on Computing Informatics and Networks*, Lecture Notes in Networks and Systems 167,
https://doi.org/10.1007/978-981-15-9712-1_1

user wants to buy, understanding the semantics associated with the articles, classifying emails as promotional emails, priority emails, spam emails, social emails, etc. All of these tasks require an understanding of the natural language [1]. As the data is huge, humans find it difficult to interpret results from all the documents. Machines do not understand the words. The computer just understands the binary language or mathematical language. So, machines require special strategies for understanding and analyzing the natural language to derive conclusions [2]. One way to represent words is vectors. These vectors are called embeddings. Thus, word embeddings [3] are a mathematical way to represent natural language words and phrases so that computers can understand the natural language text.

The central inspiration of our analysis is to comprehend the research done in word embeddings, by performing a scientometric analysis and a comparison of different state of the art word embedding models. The scientometric analysis demonstrates the growing use of word embeddings, domains they are primarily used in, countries where they are widely used, etc. Then, different word embedding models are compared on various properties, and an experimental study is performed to compare and quantify their performance. Various word embedding models are proposed to date. For example, Word2Vec [4], one hot encoded vector [5], bidirectional encoder representations from transformers (BERT) [6], global vectors for word representation (GloVe) [7], FastText [8], etc. Word embeddings gained immense popularity due to its usage in machine learning applications. The comparative study enables us to choose the model for these applications.

We used the research publications indexed on the Web of Science [9] for the year 2001 to the year 2019. The paper tried to calculate the interestingness and growth in this field. We used document-type distribution, the demography-based analysis, and category-wise distribution to demonstrate the interestingness. Paper visualizes the rapid growth in research and usage of word embeddings in the last few years. It also shows the domain distribution where word embeddings are used with artificial intelligence leading by a big factor. The widespread demographic spread of these architectures was identified by interpretation of the country-wise distribution. Section 2 discusses a scientometric analysis [10] in the field of word embeddings, and Sects. 3 and 4 compare different word embedding models proposed to date based on properties by using different datasets. The study helps us to demonstrate detailed insights on different word embedding models on various datasets. Section 5 concludes the study and tells areas for future work.

2 Scientometric Analysis

This section comprises the scientometric analysis and mapping done on research papers. It comprises various tables and figures consisting of details of resultant values found in the field of word embeddings from the Web of Science portal.

Table 1. Details of dataset

Source/index	Document types	Category	Years	Total number of papers retrieved	Date of download
Web of science	The article, proceeding paper, early access, review and letter	Artificial intelligence, information systems, software engineering, interdisciplinary application, computer engineering theory methods, computer science hardware architecture, cybernetics	2001–2019	741	25.11.2019

2.1 *Details of the Dataset*

Table 1 shows the details of the research papers we used for the scientometric analysis. We analyzed a total of 741 research publications for "word embeddings" from 2001 to 2019. Out of which, 740 papers were in the English language, while a single paper was in Spanish.

2.2 *Document Type-Wise Distribution*

Figure 1 demonstrates a treemap consisting of the distribution of document types.

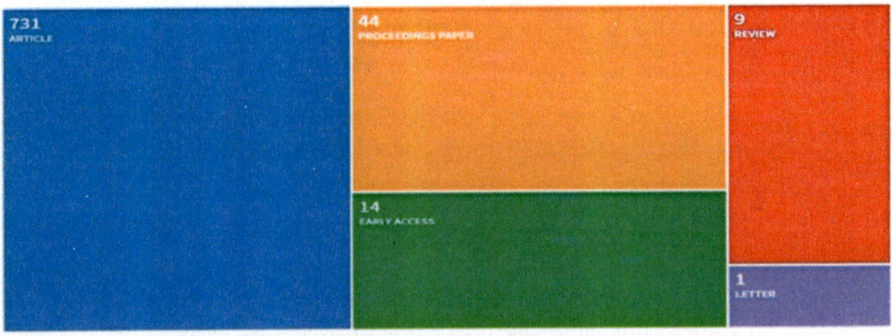

Fig. 1 Treemap demonstrating document types with their record counts

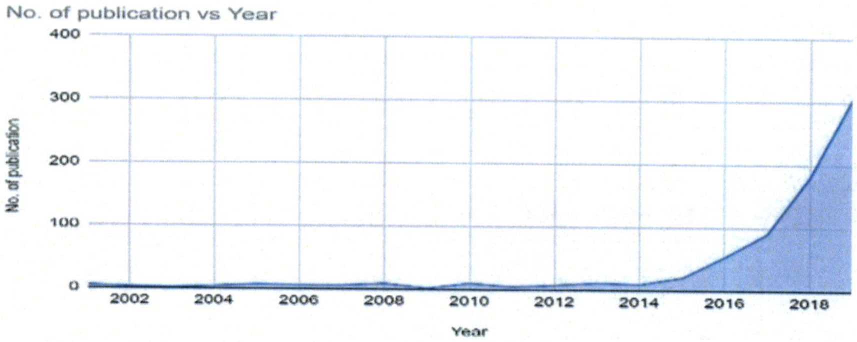

Fig. 2 Graphical representation of the number of papers published per year

Articles are the most commonly used document type. 731 articles are produced until 2019.

2.3 *Year-Wise Publications of Word Embeddings*

Figure 2 represents an increased number of publications in the field of word embeddings from 2001. The popularity of word embeddings rose to many folds in the past 20 years.

2.4 *Demographic Distribution of Word Embedding Research Publications*

We analyzed different research publications in the field of word embedding in the year 2001–2019 in Fig. 3. The top three countries contributing to this field were China, the USA, and England. China contributed 292 record counts which is 39.4% of 741 total publications.

2.5 *Category-Wise Distribution of Word Embedding Research Papers*

The most number of research publications was under "Computer Science Artificial Intelligence" comprising 21.2% of the total distribution. The number of papers was 327. Figure 4 demonstrates these statistics in the form of a pie chart.

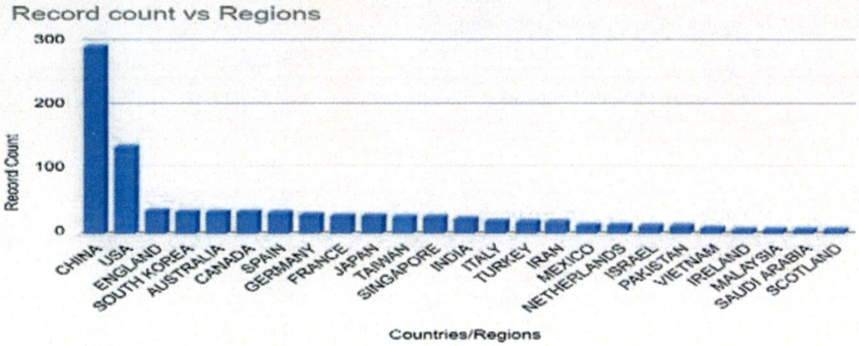

Fig. 3 Country-wise distribution of the publications

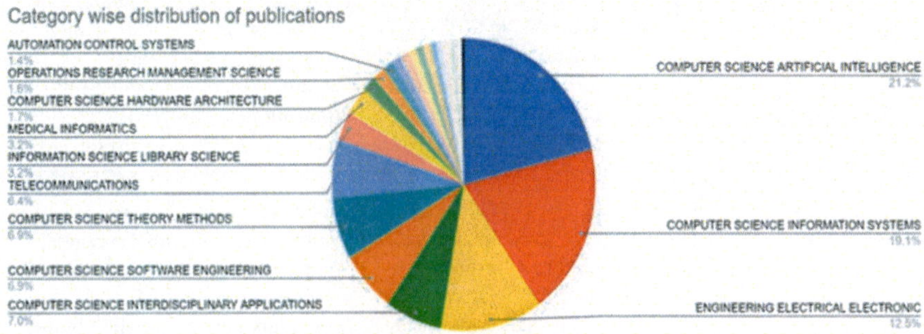

Fig. 4 Category-wise distribution of the publications

3 Comparison of Universal Embeddings Models

Different aspect-based comparison between Word2Vec, global vectors for word representation (GloVe), FastText, and BERT is done in Table 2. The table differentiates between word embedding models based on the technology used, types of the model, applications, advantages, etc.

4 Experimental Study of Word Embeddings Models

Table 3 shows the performance of different word embeddings models on various datasets. Performance is measured by calculating the deviation of the similarity score given by the model for two words from a manually assigned similarity score by experts, repeated over a list of words in different datasets. The lesser value of deviation implies the model similarity scores differ from manually assigned scores very less, and hence, the model is better. Following pre-trained universal embedding models were used. They were trained on the dataset as described below:

Table 2 Difference between word embeddings

S. No.	Property	Word2Vec	GloVe	FastText	BERT
1	Technology used	Neural networks [11]	Matrix factorization techniques [12]	N-grams [13]	Transformer [14]
2	Type of model	Predictive models	Count-based models	Predictive model	Deep learning model
3	Types	Skip-gram and continuous bag of words (CBOW)	GloVe 50D, GloVe 100D, GloVe 200D, and GloVe 300D	FastText with subword info and without subword info	None
4	Application	Question answering, named entity resolution [15], automatic summarization, sentiment analysis [16], etc.	Named entity recognition (NER) tasks, word similarity [17], and word analogy	Content tagging, content classification, sentiment analysis, spam filtering [18]	Next sentence classification [19], named entity recognition (NER), question and answer system
5	Advantages	Word2Vec CBOW tends to produce vectors that are more topically related, skip-gram pays more attention to words in the close proximity and tends to have more syntactic information as a result	Concurrent queries can be processed GloVe tends to produce vectors that are more topically related too	While learning word representations, FastText considers the internal structure of words that are useful for words that occur rarely and morphologically rich languages. Thus, it increases performance	The bidirectional context of the BERT is applied in the reconstruction process
6	Out of vocabulary words and rare words	Word2Vec is unable to represent words that are absent in the training dataset	Cannot handle it	FastText can create words that are absent in the training corpora by using its n-grams	Cannot handle it

- Word2Vec: GoogleNews Word2Vec [20] is used which was trained on 100 billion words and phrases. The length of the vector is 300 dimensions.
- GloVe: This model was trained on Wikipedia 2014 + Gigaword 5 [21]. Gigaword 5 consists of 400 K vocabulary of words and phrases, uncased, 50 dimensional, 100 dimensional, 200 dimensional, and 300-dimensional vectors, 6 B tokens. The dataset is of 822 MB.

Table 3 Deviations of different word embedding models

Dataset	GoogleNews Word2Vec (%)	GloVe 50D (%)	GloVe 100D (%)	GloVe 200D (%)	GloVe 300D (%)	FastText (%)	FastText (with subword info) (%)	BERT (%)
WS-353 [24]	53.514	27.746	29.475	36.251	43.251	27.640	26.883	34.333
WS-353-REL [25]	59.786	29.052	29.641	36.108	44.015	30.837	30.431	44.521
Mc-35 [26]	48.18	55.357	50.894	51.661	55.701	46.732	42.107	71.354
RG 65 [27]	47.633	49.190	45.950	50.846	56.016	41.670	37.548	74.118
Card-660: Cambridge rare word dataset [28]	14.443	14.351	15.877	17.022	18.221	19.255	20.113	59.553
Stanford rare word (RW) similarity dataset [29]	54.189	55.669	59.981	63.741	66.457	41.154	34.026	32.690
MEN [30]	43.769	31.501	32.423	39.075	45.795	29.006	27.668	56.114
MTURK-771 [31]	60.384	31.862	37.100	46.203	54.458	28.420	25.650	36.482

- FastText: Pre-trained on statmt.org news dataset[1] [22] and UMBC corpus. UMBC is a Web-based dataset. statmt.org dataset consists of 16B tokens.
- BERT: Pre-trained on Wikipedia and BookCorpus dataset [23]. It was trained for 1 million epochs.

The FastText model with subword information performs better than the other models due to its ability to understand directly use the morphological information. For instance, word1 = "animal" and word2 = "animals" have the same prefix and similar meanings. However, the words "man" and "management" have different meanings. The relation between the words "animal" and "animals" is the same as the relation between "reptile" and "reptiles". FastText uses this morphological information in the skip-gram model, whereas other models fail to do so. Hence, the deviation of FastText with subword information is lesser than other models.

[1] Facebook Open Source [22].

5 Conclusion

This work discussed the scientometric analysis in the field of word embeddings and a comparison between the various word embedding models. The scientometric analysis showed a tremendous increase in interest in this field over the years. Seven hundred and forty-one publications were found for "word embeddings" between the years 2000–2019 indexed on the Web of Science (WoS). Analyzing the demographic distribution of these publications, China was found the most significant contributor. Also, the topic gained fascination in the USA, England, South Korea, and many other countries. Category-wise distribution showed "Computer Science Artificial Intelligence" having the most publications. The comparative study of universal embedding models showed the different technologies used, advantages, applications, etc., provided by the different models. The experimental comparison using the similarity score generated by different models showed FastText with subword information outperforms other models due to an understanding of the morphological information and n-gram architecture. For future work, machine learning techniques trained on different word embeddings can be applied to distinct problems. Thus, a comparative study to know which word embedding outperforms in which application scenarios can be proposed. This can also be extended to various domains where natural language processing is used like artificial intelligence, virtual reality, information systems, etc. Effect of hyperparameter tuning, differences in semantic and spatial relationships among the words in each of these models can be studied.

References

1. Pauw S, Hilferty J (2016) Embodied cognitive semantics for quantification. Belgian J Linguist 30(1):251–264
2. Lai S, Liu K, He S, Zhao J (2016) How to generate a good word embedding. IEEE Intell Syst 31(6):5–14
3. Word Embeddings (2019) Available from: https://en.wikipedia.org/wiki/Word_embedding. Accessed 25 Nov 2019
4. Mikolov T, Chen K, Corrado G, Dean J (2013) Efficient estimation of word representations in vector space. arXiv:1301.3781
5. Hinton GE (1986, Aug). Learning distributed representations of concepts. In: Proceedings of the eighth annual conference of the cognitive science society, vol 1, p 12
6. Devlin J, Chang MW, Lee K, Toutanova K (2018) Bert: pre-training of deep bidirectional transformers for language understanding. arXiv:1810.04805
7. Pennington J, Socher R, Manning CD (2014, Oct) Glove: global vectors for word representation. In: Proceedings of the 2014 conference on empirical methods in natural language processing (EMNLP), pp 1532–1543
8. Joulin A, Grave E, Bojanowski P, Mikolov T (2016) Bag of tricks for efficient text classification. arXiv:1607.01759
9. Available from https://clarivate.com/webofsciencegroup/solutions/web-of-science/. Accessed 26 Nov 2019
10. Piryani R, Madhavi D, Singh VK (2017) Analytical mapping of opinion mining and sentiment analysis research from 2000–2015. Inf Process Manage 53(1):122–150

11. Beale HD, Demuth HB, Hagan MT (1996) Neural network design. Pws, Boston
12. Salle A, Idiart M, Villavicencio A (2016) Matrix factorization using window sampling and negative sampling for improved word representations. arXiv:1606.00819
13. Cavnar WB, Trenkle JM (1994, Apr) N-gram-based text categorization. In: Proceedings of SDAIR-94, 3rd annual symposium on document analysis and information retrieval, vol 161175
14. Dehghani M, Gouws S, Vinyals O, Uszkoreit J, Kaiser Ł (2018) Universal transformers. arXiv:1807.03819
15. Nadeau D, Sekine S (2007) A survey of named entity recognition and classification. Lingvisticae Investigationes 30(1):3–26
16. Maas AL, Daly RE, Pham PT, Huang D, Ng AY, Potts C (2011, June) Learning word vectors for sentiment analysis. In: Proceedings of the 49th annual meeting of the association for computational linguistics: human language technologies, vol 1. Association for Computational Linguistics, pp 142–150
17. Lin D (1998, July) An information-theoretic definition of similarity. ICML 98:296–304
18. Suchomel V (2017) Removing spam from web corpora through supervised learning using FastText
19. Hassan A, Mahmood A (2018) The convolutional recurrent deep learning model for sentence classification. IEEE Access 6:13949–13957
20. Pre-trained vectors trained on part of Google News dataset (about 100 billion words). Available from https://code.google.com/archive/p/word2vec/. Accessed 27 Nov 2019
21. Pennington J, Socher R, Manning CD (2014) GloVe: global vectors for word representation. Available from https://nlp.stanford.edu/projects/glove/. Accessed 27 Nov 2019
22. Facebook Open Source: FastText: Library for efficient text classification and representation learning. Available from https://fasttext.cc/docs/en/english-vectors.html. Accessed 28 Nov 2019
23. BookCorpus Dataset. Available from https://github.com/sgraaf/Replicate-Toronto-BookCorpus. Accessed 29 Nov 2019
24. The WordSimilarity-353 test collection. Available from https://aclweb.org/aclwiki/WordSimilarity-353_Test_Collection_(State_of_the_art). Accessed 28 Nov 2019
25. WordSim353: Similarity and relatedness. Available from https://alfonseca.org/eng/research/wordsim353.html. Accessed 28 Nov 2019
26. MC-35 Dataset. Available from https://web.eecs.umich.edu/~mihalcea/downloads.html. Accessed 29 Nov 2019
27. RG-65 test collection (state of the art). Available from https://aclweb.org/aclwiki/RG-65_Test_Collection(State_of_the_art). Accessed 29 Nov 2019
28. Pilehvar MT, Kartsaklis D, Prokhorov V, Collier N (2018) Card-660: Cambridge rare word dataset a reliable benchmark for infrequent word representation models. arXiv:1808.09308
29. Stanford rare word (RW) similarity dataset. Available from https://nlp.stanford.edu/~lmthang/morphoNLM/. Accessed 29 Nov 2019
30. The MEN test collection. Available from https://staff.fnwi.uva.nl/e.bruni/MEN. Accessed 30 Nov 2019
31. The word relatedness Mturk-771 test collection. Available from https://www2.mta.ac.il/~gideon/mturk771.html. Accessed 30 Nov 2019

R and Hadoop Integration for Big Data Analytics

Dipika Singh and Rakhi Garg

Abstract Big data is affecting every walk of life right from mobile maps, GPS, online shopping, wearable gadgets to our day-to-day activities. Data generated are in terabytes which if collected and mined properly can infer important results for decision making and planning in business, education, health care, etc. So, big data analytics are in demand. Big data analytics require a tool for data analysis and a platform for parallel computing. R is one of the most robust languages for data science, whereas Hadoop is distributed computing framework to handle big data processing. R and Hadoop integration seems to be a perfect combination for data analytics. In this paper, we have explored major big data technologies available and also compared major data analytics languages being used nowadays. We have discussed procedures for combining R and Hadoop to get best of the two in analysing big data. This paper will be beneficial for researchers who are working in the area of big data analysis.

Keyword R programming · Big data · Data analysis · RHadoop

1 Introduction

With the advancement of Internet, smart phones, IOT devices, etc., big data has become a familiar term in data analysis nowadays. It is difficult to work in society without using big data somewhere or other. It is proving to be an asset to society if handled properly. For example, GPS services that were used few years back were not as good as they are today. It is due to processing of huge amount of data that provides more accurate results.

D. Singh (✉)
Department of Computer Science, Institute of Science, BHU, Varanasi, UP, India
e-mail: dipikaa.ssingh@gmail.com

R. Garg
Department of Computer Science, MMV, BHU, Varanasi, UP, India
e-mail: rgarg@bhu.ac.in

A. Abraham et al. (eds.), *Proceedings of 3rd International Conference on Computing Informatics and Networks*, Lecture Notes in Networks and Systems 167,
https://doi.org/10.1007/978-981-15-9712-1_2

Data are being produced at astronomical rates today [1]. In the 90s, the amount of data generated in a year is now being generated in a minute because of smart phones, IOT devices, etc. The characteristics of big data were already defined with respect to 5Vs, i.e. velocity, volume, variety, value and veracity.

Moreover, big data is huge in size, so it cannot be handled by standalone system. It requires distributed computing environments like cluster, grid and cloud to handle it and massive parallel processing tools like Hadoop and Spark to work on it.

Because of complexity of handling big data, it becomes burning topic for research nowadays. When we talk about big data analytics, we need a language which is capable of data analysis, and at the same time, it should have provisions to handle big data. In this paper, we will explore R programming language for big data analytics.

In 2017, David Robinson performed a survey which stated R is emerging as one of the most rapid developing languages in recent years [2]. But, to gain efficient performance in R, data should reside in primary memory. To solve this problem, R can be combined with Hadoop. It is open source Apache framework for handling distributed processing of big data.

In this paper, we explore different distributed techniques to handle big data, in Sect. 2. We briefly describe and compare different popular data science languages, in Sect. 3. Section 4 describes procedure to combine R and Hadoop for handling big data. At last, Sect. 5 concludes the paper.

2 Distributed Computing Techniques to Handle Big Data

Because of 5 Vs characteristics of big data, it will be difficult to handle it by standalone systems and existing technology. New techniques, framework and languages are required to deal with it. Various techniques have been developed, e.g. cluster, grid and cloud computing to handle big data. Each of them has its own pros and cons. Also, different frameworks and languages are developed to enhance the performance of parallel computing to deal with big data. We will focus on each of them in brief further.

2.1 Cluster Computing

A single autonomous computer was unable to handle the demand of big data, so cluster computing was introduced. This technology is often used in LAN. Here several computers are interconnected and behave as a single computer. Computer systems used in a cluster are tightly coupled, and they are at same physical location connected with a high-speed connection. Clusters can be classified as homogeneous and heterogeneous in depending upon the similar or different types of configuration of nodes and operating systems present at nodes. Clusters are generally managed by single organization [3]. There are three types of clusters that are used: high availability (failover), load balancing and high-performance computing.

High Availability Clusters. It provides uninterrupted service to users. In case of failure of a node, the service will continue but performance will be degraded. These clusters are used for service critical applications such as database, mail and web server.

Load Balancing Clusters. It distributes incoming requests among several nodes. Each node is capable of handling similar requests. It can be combined with high availability cluster to increase reliability, availability and scalability of services.

High-Performance Computing. High-performance computing clusters have large number of computers called nodes. Maximum nodes have identical configurations. Some nodes can have physical or logical difference. The basic concept is individual tasks in parallel application should run well irrespective of node they are dispatched to.

Cluster computing has several advantages such as nodes in a cluster can easily be added and removed; it is cost-effective; and it is also fast and flexible. On the other hand, there are disadvantages too like it is difficult to manage without experience. Large clusters are difficult to debug. Working with heterogeneous clusters is more complex than homogeneous one.

2.2 *Grid Computing*

Grid computing consists of loosely coupled systems from various administrative domains to achieve a specific target. The systems in a grid can be distributed in several locations. It is used for jobs that can be divided into chunks ready to be executed on different nodes. Computers forming grid can have different operating system. Grids are owned by several organizations, with their mutual consent, e.g. CDAC [4]. It is used in simulations, predictive modelling, energy resource exploration, military research, etc.

The benefits of grid computing are resources which are used efficiently; large servers are not required as task is split into smaller subtasks. Failure rate of grid computing is low, and upgradation is easier. On the other hand, disadvantages are grid computing requires robust and fast interconnection between resources. Moreover, as different organizations are involved in grid computing, so licensing and agreements become crucial.

2.3 *Cloud Computing*

In cloud computing environment, services are provided by remote devices owned by different organizations [5]. User can have any device connected to the Internet with cloud computing software installed. It provides services ranging from email to complex data analysis programs. It provides large storage to companies without internal storage centres. Geographical bounds do not exist for clouds, wherever we are we can be served. Cloud is generally owned by one organization, and it provides services to different users. Major areas where cloud computing is used are space exploration, weather forecasting, insurance, banking, etc.

There are numerous advantages of cloud computing such as it has unlimited storage capacity, low cost and faster access. It can be accessed from any location having Internet connection. It is platform independent, and failure in local computer does not affect our data. Few disadvantages of cloud computing are that high-speed Internet connection that is required and limited features are provided by cloud computing.

Among these, clusters are best suited for research as it is easy to work with. In this paper, we will mainly focus on high-performance computing cluster (HPC) and its working, i.e. different language used to work on HPC with its pros and cons. Firstly, we discuss about different types of HPC, namely MPI clusters, Hadoop cluster and Spark cluster. The comparative analysis is done further.

MPI. It stands for Message Passing Interface. It is used for parallel and distributed computing [6]. It gives user flexibility of calling routines from different languages like C, C++, Fortran, Java and Python.

Hadoop. It is Apache framework. Its source code is open and written in Java. It facilitates processing big data in distributed environment [7].

Spark. It is extremely fast Apache cluster computing technology. It is developed in 2009 in UC Berkeley's AMPLab by Zaharia [8].

A brief comparison of MPI, Hadoop and Spark is done in Table 1.

Both Hadoop and Spark are clusters that allow us to store, process big data. But they are not sufficient for data analytics. Moreover, applications that involve pre computation on the data set bring down the advantages of Hadoop MapReduce. Also, problems that cannot be trivially partitionable or recombinable become the

Table 1 Comparison of MPI, Hadoop and spark

	MPI	Hadoop	Spark
Data read	No data locality has to send data to other node for processing	Has to read from and write to disc	In memory processing
Speed	Fast	Slower than spark	Faster
Volume of data	Less	Larger data sets can be processed	Comparatively less data sets are processed
Best features	Standardization, portability. A variety of implementations are available	Good for linear processing of huge data sets, and economical if no immediate result is required	Good for iterative processing, graph processing, machine learning

limitation of map reduce problem solving. So it is required to combine it with some programming language that has the capability of data analytics.

3 Different Data Analytics Languages

Nowadays, many languages are available for data analysis. Each language has its own ups and downs. The most popular languages for data science are as follows.

3.1 SAS

SAS stands for Statistical Analysis Software, created by SAS institute in 1960. It is used for predictive analysis, business analysis, data analysis, etc. It is platform independent [9].

3.2 SPSS

It stands for Statistical Package for Social Science, developed in 1966 by SPSS Inc. [10]. In 2009, it was acquired by IBM. It is used for complex statistical data analysis. It was primarily created for management and statistical analysis of social science data.

3.3 Java

James Gosling of Sun Microsystem developed object oriented language Java in 1991. Today, it enjoys being one of the most preferred general-purpose programming languages [11]. Java can be used for data science for variety of purpose such as data analysis, text analytics and data visualization. Although it has the capability of performing these data analytics task, it is not a preferred language for data science. The reason behind this is the existence of more robust languages like R and Python that have highly specific libraries for data analytics task.

3.4 Python

It is open source, object oriented, interpreted, high-level language. It has huge number of libraries to deal with data science task [12]. Python along with R is the most

Table 2 Merits and demerits of popular data science languages

Language	Merits	Demerits
SAS	Official support available, capable of handling large data set	High cost, slow adaptation to new technique
SPSS	Good user interface, official support available	High cost, licence requirement, slow adaptation, inefficient for big data
Java	Efficient in handling large-scale system, faster development time	Least suited to hardcore analysis, inefficient visualization
Python	Free of cost, scalability, easy to learn, vast community, general-purpose language	Weaker than R in explanatory modelling, official support and user interface not available
R	Free of cost, excellent visualization, robust community to manage library, excellent performance for explanatory and predictive modelling	Slower for big data, difficult to learn, user interface is not good

preferred language for data analytics. The most important libraries of Python for data science are Numpy, Pandas, Matplotlib, Scipy and Scikit-learn.

3.5 R

R was developed in 1992, by Ross Ihaka and Robert Gentleman at University of Auckland, New Zealand [13]. Nowadays, R is one of the most preferred languages for data science because of tremendous packages suitable for data analysis tasks. Table 2 provides merits and demerits of commonly used data analysis languages.

Among these, R is the most promising language for data science. It has large number of package dedicated for different data analysis tasks.

4 Integration of R and Hadoop

R can be integrated with other distributed frameworks to get rid of the speed and memory boundation.

4.1 R and Hadoop Combination

R is very powerful for statistical analysis, but data sets should be in primary memory for analysis. Hadoop framework allows parallel processing of massive amount of

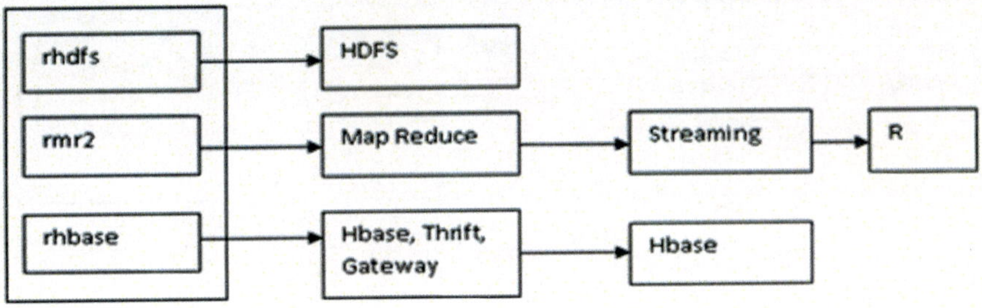

Fig. 1 Architecture of RHadoop

data but it lacks in efficient statistical analysis. Using R with Hadoop facilitates scalability of calculations in R.

Different techniques to integrate R and Hadoop are described below:

RHadoop. It is the most common, simple and cheap way of integrating R and Hadoop developed by revolution analytics [14]. It allows R users to directly use data from Hbase server and HDFS. The architecture of RHadoop is shown in Fig. 1.

It has 5 different packages for performing various operations.

rhbase. It is used for database handling by R on Hadoop database, HBase.

rhdfs. It provides connectivity of R with distributed file system of Hadoop. Major functionality of rhdfs is accessing files from HDFS from R and vice versa as shown in Fig. 2.

plyrmr. It enables various operations for manipulating big data sets in Hadoop [15].

ravro. It is used for manipulation of Avro files which are stored in row based format in local or distributed environment [15].

rmr2. It enables writing MapReduce jobs in R and provides ability to parallelize algorithms and to use big data sets without needing to sample data [16].

RHIPE. It stands for "R and Hadoop Integrated Programming Environment" [17]. Using this, Hadoop MapReduce can be created by R. Map and Reduce functions are defined in R as per requirement, and then they invoke Map and Reduce task of Hadoop using RHIPE library.

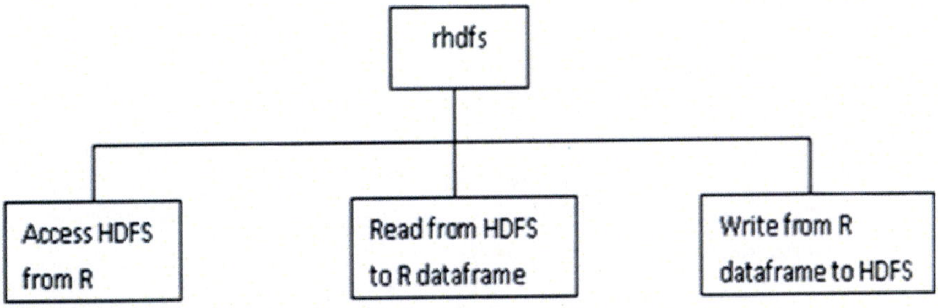

Fig. 2 rhdfs functionalities

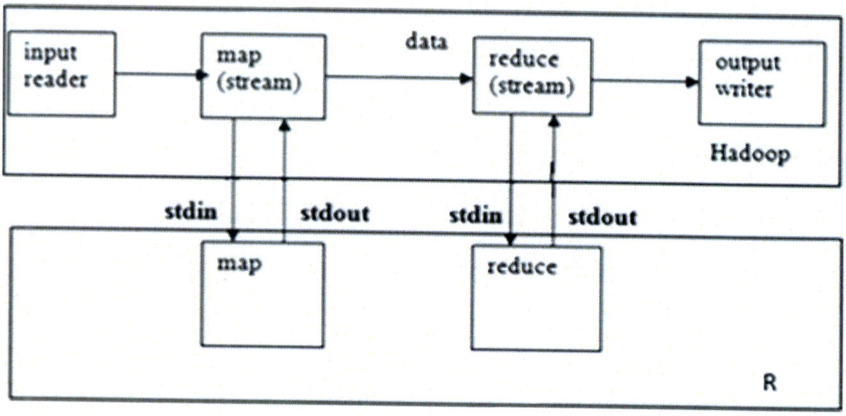

Fig. 3 Block diagram of R and Hadoop streaming

R and Hadoop Streaming. Hadoop Streaming is a utility for running map/reduce jobs on the cluster using Shell scripts, Perl, Python, R, etc. In R and Hadoop streaming, R language is used for writing map/reduce jobs. Executable reads input from STDIN and writes output to STDOUT as shown in Fig. 3 [18].

In this figure, first the mapper takes input and converts it into lines and transfer it to stdin of process and then takes output from stdout of process. Then it generates key/value pair. Same process is repeated at reducer task.

RHIVE. Apache Hive is an open-source data warehouse system built on top of Hadoop. It is basically used to eliminate the need of writing of complex map reduce jobs. It uses simple SQL queries known as HiveQL (HQL) instead of map reduce functions. Using this, Hadoop data can be processed by R which extends HiveQL to achieve this. It provides faster and parallel approach for solving the analysis problem [19].

4.2 R and Spark Combination

Apart from combining R with Hadoop, it can also be combined with Spark to handle big data. Apache Spark is a lightning fast cluster or parallel computing technology. SparkR is a combination of R and Spark [20]. Figure 4 represents the working of SparkR.

In this figure, Table 1 and Table 2, Table *n* represent huge data sets. After preprocessing of these by SparkR, processed and smaller data set are obtained represented as Table *t*1 and Table *t*2. Further analysis is done on the smaller data sets by R which is represented as Table *t*3.

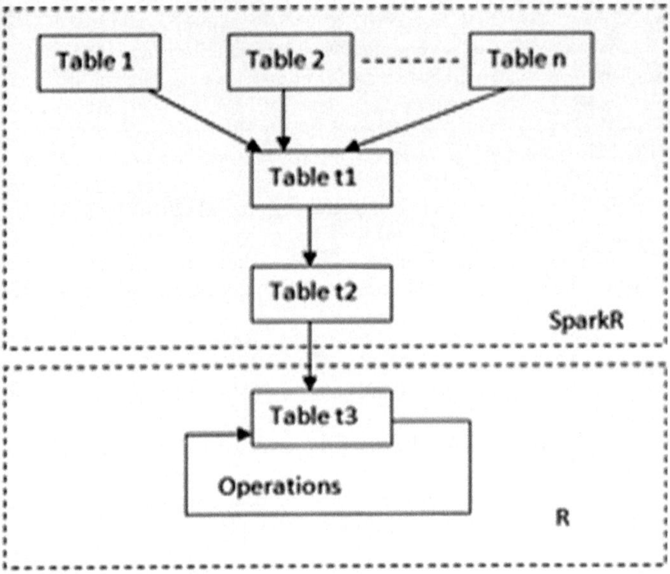

Fig. 4 Workflow of SparkR

5 Conclusion

In this paper, we have explored the combination of R and Hadoop for big data analytics. R is a robust language with large number of dedicated libraries for data analysis task, whereas Hadoop is a robust distributed framework to handle big data processing. There are multiple ways to join R with Hadoop. This combination provides robust platform for big data analytics by taking the advantage of best of the two giants, i.e. R and Hadoop. This paper will be beneficial for researchers working in the area of data science to gain an insight about different distributed platforms and data analytics languages for big data analytics and also procedure to combine R and Hadoop.

References

1. Stephens ZD, Lee SY, Faghri F, Campbell RH, Zhai C, Efron MJ, Iyer R, Schatz MC, Sinha S, Robinson GE (2015) Big data: astronomical or genomical? PLoS Biol 13(7):e1002195
2. Singh D, Garg R (2019) R, language for data analytics. In: Proceedings of international conference on sustainable computing in science, technology and management (SUSCOM), Amity University Rajasthan, Jaipur, India
3. Buyya R (1999) High performance cluster computing: Architectures and systems (volume 1). Prentice Hall Upper SaddleRiver NJ USA 1(999):29
4. Berman F, Fox G, Hey T (2003) Overview of the book: grid computing–making the global infrastructure a reality. Making Glob Infrastruct Reality 3

5. Armbrust M, Fox A, Griffith R, Joseph AD, Katz R, Konwinski A, Lee G, Patterson D, Rabkin A, Stoica I, Zaharia M (2010) A view of cloud computing. Commun ACM 53(4):50–58
6. Gropp W, Thakur R, Lusk E (1999) Using MPI-2: advanced features of the message passing interface. MIT press
7. Shvachko K, Kuang H, Radia S, Chansler R (2010) The Hadoop distributed file system. In: 2010 IEEE 26th symposium on mass storage systems and technologies (MSST). IEEE, pp 1–10
8. Zaharia M, Chowdhury M, Franklin MJ, Shenker S, Stoica I (2010) Spark: cluster computing with working sets. HotCloud 10(10–10):95
9. Spector PE (2001) SAS programming for researchers and social scientists. Sage
10. Nie NH, Bent DH, Hull CH (1975) SPSS: statistical package for the social sciences, vol 227. McGraw-Hill, New York
11. Gosling J, Joy B, Steele G, Bracha G (2000) The Java language specification. Addison-Wesley Professional
12. Van Rossum G (2007) Python programming language. In: USENIX annual technical conference, vol 41, p 36
13. Ihaka R, Gentleman R (1996) R: a language for data analysis and graphics. J Comput Graph Stat 5(3):299–314
14. Cai L, Guan X, Chi P, Chen L, Luo J (2015) Big data visualization collaborative filtering algorithm based on RHadoop. Int J Distrib Sens Netw 11(10):271253
15. Rotte AV, Patwari G, Hiremath S (2015) Big data analytics made easy with Rhadoop. Int J Res Eng Technol 4(5):9–15
16. Analytics R (2012) rmr2: R and Hadoop streaming connector. R package version 2(1)
17. Guha S, Hafen R, Rounds J, Xia J, Li J, Xi B, Cleveland WS (2012) Large complex data: divide and recombine (d&r) with rhipe. Stat 1(1):53–67
18. Prajapati V (2013) Big data analytics with R and Hadoop. Packt Publishing Ltd.
19. Cho W, Lim Y, Lee H, Varma MK, Lee M, Choi E (2014) Big data analysis with interactive visualization using R packages. In: Proceedings of the 2014 international conference on big data science and computing, pp 1–6
20. Venkataraman S, Yang Z, Liu D, Liang E, Falaki H, Meng X, Xin R, Ghodsi A, Franklin M, Stoica I, Zaharia M (2016) SparkR: scaling R programs with spark. In: Proceedings of the 2016 international conference on management of data, pp 1099–1104

Review of Text Summarization in Indian Regional Languages

Surendrabikram Thapa, Surabhi Adhikari, and Sushruti Mishra

Abstract Propelled by the advancements in the field of natural language processing, generating summaries of long texts using various NLP tools and techniques has always been a subject of great interest for scientists all over the world. Data is ubiquitous, and a large amount of data is processed every second in the digital space. For these reasons also, there is need of machine learning algorithms that can automatically shorten longer texts or documents and provide the accurate and meaningful summaries. These summaries find their applications in a wide variety of fields like medicine, market review, business analytics, etc. In other languages like English, an ample amount of study and research can be observed. However, in context of the Indian regional languages, the research in text summarization is very limited and is still in the infancy state. This paper tries to explain the research and works that have been performed in the field of text and document summarization in Indian regional languages.

Keywords Text summarization · Extractive summarization · Abstractive summarization · Indian regional languages

1 Introduction

Natural language processing (NLP) in particular deals with programming the computers to parse, process and perform analysis of huge amount of human language data that surrounds us. Today, the world is so centralized on computers, and for that

S. Thapa (✉) · S. Adhikari · S. Mishra
Department of Computer Science and Engineering, Delhi Technological University, New Delhi, India
e-mail: surenthapa5803@gmail.com

S. Adhikari
e-mail: suravi.prasiddi@gmail.com

S. Mishra
e-mail: mishrasushruti99@gmail.com

A. Abraham et al. (eds.), *Proceedings of 3rd International Conference on Computing Informatics and Networks*, Lecture Notes in Networks and Systems 167,
https://doi.org/10.1007/978-981-15-9712-1_3

reason, natural language processing is widely used these days [1]. Text summarization is a sub-domain of NLP which deals with extracting or collecting the important information from the given text or document and gives concise information regarding the text or document in the form of brief summary. Automated summarizers reduce reading time, and, in many cases, these summarizers can provide unbiased summaries than that of human beings. Text summarization has a lot of applications like customer review summary, online news article summary, a summary of the minutes of the meeting, automated research abstract, etc. Automatic text summarizers in languages such as English were in existence from 1950, but the advancement in summarization has seen a rapid pace in the last two decades. Indian regional languages have however seen good development in recent 10 years. In India, there is no one single language that is used across India as an official language. There are more than 22 official languages in India, and each of them is used for official purposes. So, our concern should not be getting too focused on a single language, but the summarization techniques in every language should also be explored. There cannot be a single system that can generalize the summarization process of all Indian regional languages. This is because all languages have their own linguistic features, and hence, each language should be dealt with independently. But works and study in the field of such text summarization in Indian regional languages are very less and are still in infancy state.

Based on the techniques used, text summarization techniques can be broadly divided into two major categories, namely extractive summarization and abstractive summarization. Extractive method of text summarization selects phrases and sentences from the source documents or text and includes such information extracted in the newly generated summary. The summary is based on key features in the text. For finding the phrases and sentences required, extractive methods make use of statistical features like sentence position, proper noun, numerical data, topic frequency, topic token frequency, normalized sentence length, cluster frequency, etc. Mostly, extractive summarization techniques make use of three tasks viz. tokenization of the text, calculating the word scores mainly TF/IDF scores, calculation of scores of the sentences based on such word scores and selection of summary comprising highest scores. In this way, the extractive summarization method, in short, extracts the most relevant information from the text document and includes them in summary. Since the phrases or sentences of generated summaries are directly extracted from the given document or text, the summaries are sometimes not meaningful and complete. However, this method is preferred because of its easy implementation. The abstractive text summarization technique, on the other hand, generates a meaningful summary, and the summary is more human-like. This method generates entirely new phrases and sentences to provide a summary of the source text. The system generates new phrases by rephrasing instead of simply extracting phrases or even using the words that are not present in the source text or document. Since the abstracted summary may contain phrases or sentences that are not in the source document or text, this is a more challenging approach. For a good abstractive summary, the built model should be given the independent ability to understand the given document, and the model will output its understanding in its own words [2]. Mostly deep learning models are

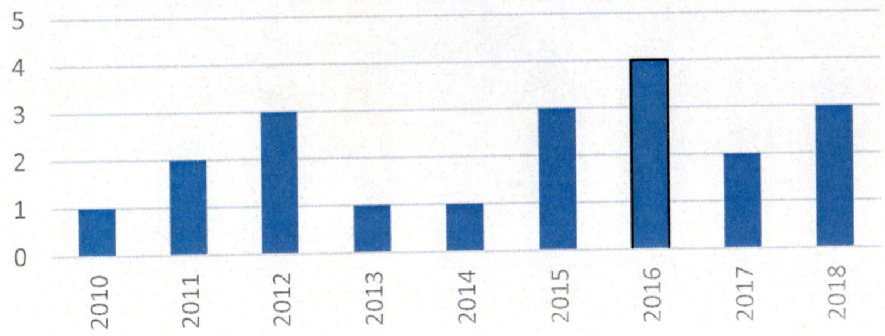

Fig. 1 Distribution of the number of papers discussed and year of publishing

used to generate an abstractive summary. It is much harder than extractive techniques, but the human-like summary outweighs its drawbacks in implementation.

In Indian regional languages, the extractive methods are very widely used. Research in abstractive methods, on the other hand, is very nominal and needs to be explored more extensively. In this paper, we present the summarization techniques that are used in various Indian regional languages in great detail. The paper describes the methods, different works carried out in various Indian regional languages and comparative analysis of different methodologies used in text summarization of Indian languages in the past 10 years, and the distribution of number of papers and year of publishing is as shown in Fig 1.

2 Related Works

With more than 122 languages and 22 designated official languages, India has a huge diversity in languages [2]. The need for text summarization in regional languages is much needed as they are used extensively in official works. A lot of effort has been made by a lot of researchers to effectively summarize the texts written in Indian regional languages. In this paper, we majorly discuss the text summarization techniques used in seven major languages used in India viz. Hindi, Bengali, Telugu, Marathi, Tamil, Urdu and Punjabi. The techniques that were used in the past 10 years for the text summarization in Indian regional languages have been discussed in this paper. Figure 2 shows the distribution of the number of papers that we have studied and the languages.

Generally, the approach followed by each of the summarizing technique can be represented by flow chart diagram as shown in Fig 3. The summarization usually starts with input of text which is followed by preprocessing steps done to it. After preprocessing of the text, there is iterative calculation based on the model of the summarizer (mostly calculations are based on word vector, word–sentence relationship and graph model). The process is followed by final step of summary generation after doing the analysis of processes done before.

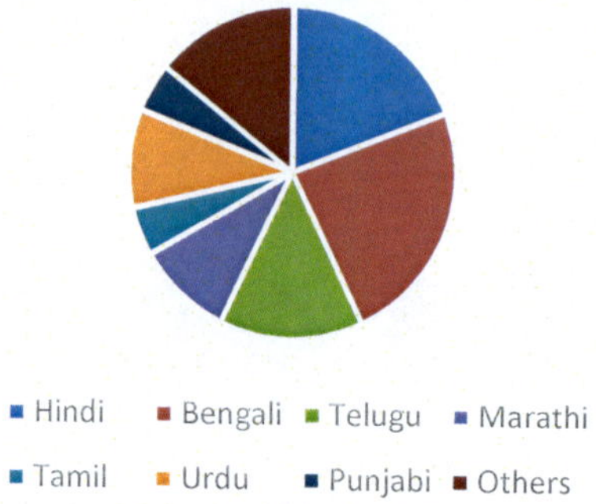

Fig. 2 Distribution of the language and the number of papers discussed

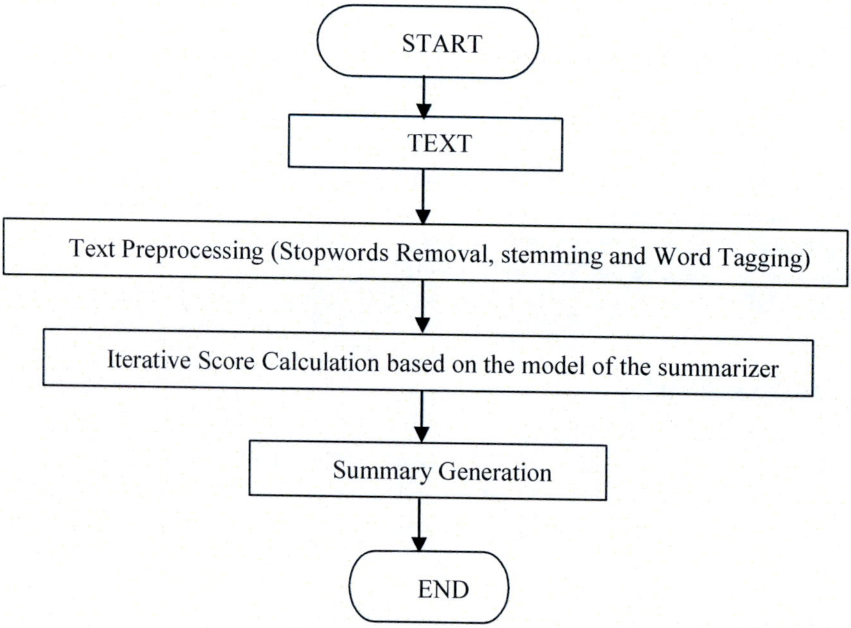

Fig. 3 Flow chart for working of text summarizer in Indian regional languages

2.1 *Hindi*

Thaokar et al. [3] discussed on how we can carry out summarization of given text or documents in Hindi language using the sentence extraction method. The technique used six statistical features along with two linguistic features. It used Hindi WordNet for tagging of the appropriate part of speech of the word to check subject–object–verb of the given sentences. The result was optimized using a genetic algorithm. Gupta et al. [4] proposed the rule-based technique for the summary generation of the

Hindi text documents based upon the linguistic rules. Five features for each sentence were discovered, and the proposed methodology was then subject to testing on thirty different documents that belonged to various domains. Input size was decreased to 60–70% with an accuracy of 96% when testing was done on given thirty given documents for generation of the summary of provided Hindi text. Kumar et al. [5] proposed the system based on an extractive approach that selected the important and meaningful sentences from the text based on some thematic approach. The system relies on the scoring based on the occurrence of root words, and the sentences with the highest scores were then included in the generated summary. Subramaniam et al. [6] in 2015 have proposed an abstractive method for the generation of summaries.

2.2 *Bengali*

The extractive method that is used in conjunction with the approach proposed by Abujar et al. [7] and a set of Bengali text analysis rules derived from the heuristics was able to give summaries nearly equivalent to human-generated. In the approach proposed by the authors [7], the importance of sentences and phrases was identified based on word scoring, sentence scoring and graph scoring to find the appropriate texts that need to be included in the summary.

Sarkar [8] discussed an effective extractive summarizer that gives a summary using sentence length feature along with TF/IDF weights (TF: term frequency and IDF: inverse document frequency). The evaluation results in the paper showed that the methodology proposed had better performance in terms of different performance measures than given three systems compared to the paper.

Akter et al. [9] proposed a method that extracts important sentences or phrases from the text document. Word score in the system was calculated by TF/IDF. After calculation of word scores, sentence scores were calculated. The scores of words that constituted the sentence were added, and value for position of sentence was also given while calculating sentence scores. Finally, the K-means clustering algorithm was used for the generation of required summary. Das et al. [10] proposed another topic-based opinion summarizer for Bengali language. Features were extracted in forms like syntactic and lexico-syntactic features. Aggregation of such topic-sentiment was then done using a clustering algorithm (K-means).

2.3 *Telugu*

Telugu is a Dravidian language and is mostly spoken by the people residing in Indian states of Telangana, Andhra Pradesh and Union Territory of Puducherry. Reddy et al. [11] proposed an extractive summarization technique that summarizes the articles in the Telugu language by using key features such as sentence's order of appearance in the given document, sentence similarity with title, word-frequency and centrality of the sentence. The sentences were then ranked by calculating scores for each sentence by taking all given features into consideration. Naidu et al. [12] have also suggested a summarization technique that summarizes text with automatic keyword extraction from the dataset used in Telugu e-newspapers. In their described technique, the researchers have used human evaluation to train the system for seeking the key phrases or keywords that have maximum probability of inclusion in summary. They were able to get great accuracy when tested with different datasets with this method. Similarly, Kallimani et al. [13] have proposed the abstractive method for summary generation of text documents in the Telugu language.

2.4 Marathi

Giri et al. [14] discussed the extractive summarization in the Marathi language by extracting relevant sentences using the application of statistical features as well as features that depended on Marathi language. Similarly, the works of Sarwadnya et al. [15] discussed the text or document summarization technique that used a graph-based model that also used the extractive approach of text summarization in the Marathi language.

2.5 Tamil

The Tamil language is another Dravidian language that is spoken predominantly in Singapore, Sri Lanka, Tamil Nadu and Puducherry. Among the very few works done in Tamil language, Priyadharshan et al. [16] have proposed the method which can automatically summarize Tamil online sports news articles using natural language processing and a generic stochastic ANN. The feature matrix was created with various linguistic features to enhance accuracy.

2.6 Urdu

Burney et al. [17] in 2012 had designed and developed an add-in for MS word which summarized text in Urdu language. The approach that used a statistical method of sentence weight algorithm was able to summarize the Urdu text to the accuracy of over 80% when one human verifier was used. Humayoun et al. [18] have discussed the effect of preprocessing setting in the accuracy of summarization of text in the Urdu language which is one of the great works done in the field of NLP in the Urdu language.

2.7 *Punjabi*

Punjabi is another widely spoken language in India. Gupta et al. [19] have described the automatic summarizer for Punjabi text for summarizing the news articles in the Punjabi language. In their proposed method, the score of the sentences was calculated by making use of a feature-weight equation, and the sentences with highest rankings were then arranged to get a summary of the news articles.

2.8 *Other Regional Languages*

NLP and the works in summarization of text are still in the premature stage in other regional languages. In the Kannada language, Geetha et al. [20] have discussed extractive text summarization using latent semantic analysis. Similarly, Kabeer et al. [21] proposed the text summarizer for Malayalam text documents in 2014. The paper implemented a statistical technique for calculating the sentence scores and made use of a semantic graph-based method for summary generation. Also, Krishnaprasad et al. [22] have described a similar extractive approach for summarization of text in the Malayalam language. The works in other languages such as Assamese, Nepali, Santali and Dogri are very limited. Works in semantic analysis have been done, but the work in text summarization has not yet been explored.

3 Discussion

From the above works, we have seen that researchers who are doing their research in text summarization of Indian regional languages extensively use extractive text summarization techniques. We have also seen that an ample amount of work has been done in the Hindi language than any other Indian regional language. Figure 4 gives the picture of distribution of the number of papers that we have studied and the methods used in those papers. Despite some inaccuracy extractive models pose, researchers have found their way to make summaries of long text more accurate by various methods. The extractive summarization process that involves the conventional process of primarily ranking sentences with scores and including the sentences with highest scores can be improved with various techniques such as improvement in preprocessing and sentence scoring methods.

Krishnaprasad et al. [22] have suggested that our summarizers can generate even great quality summaries if we can improve the sentence scoring stage. Sarkar [8] has also suggested that the system performance may be further improved by improving stemming process including more number of effective features. Similarly, the relevancy of sentences should also be measured to get better summary. Akter et al. [9] in their work suggested that the relevancy of sentences can be measured by using

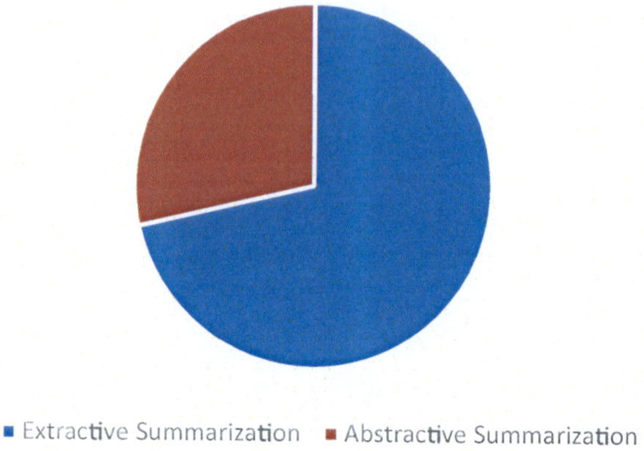

Fig. 4 Distribution of number of papers with summarization techniques used

syntactic and similarity in the future to get good accuracy. Apart from this, a lot of time and work must be done in the addition of more features to get more accurate summaries. Reddy et al. [11] who proposed their works in Telugu text summarization also suggested that with addition of more features like cue phases, the existence of some punctuation marks, day-month names, numeric, literals, etc., to the existing methodologies can boost up the quality of summary generated by the system.

The efforts have been made by authors to attain maximum accuracy with the extractive text summarization process. Sometimes, the sentences that are extremely important and are extremely important for inculcating in final generated summary are long sentences. This makes generated summaries lengthier, and we might also be missing some other relevant information that might be present in other shorter sentences when our model includes such long sentences by keeping such shorter sentences aside. Kabeer et al. [21] in their paper discuss the need for abstractive summarization processes to deal with such problems. Sunitha et al. [23] in their paper have presented experimental works that have been performed by using the abstractive summarization techniques in Indian regional language. They have also put an emphasis on promoting method of abstractive summarization in Indian regional language. With rapid advancement in deep learning and advanced researches in neural networks and various deep models, abstractive summarization can serve as a very great technique for summarizing text documents. Also, generative models can prove to be a great summarizer for text documents. Liu et al. [24] have proposed a generative adversarial network for an abstractive summary generation. Their work was concentrated on the English language but the same can also be done in Indian regional languages also. Similarly, the use of transfer learning can also facilitate the task of summarization using abstractive methods. Also, various scoring methods which provide a better method for evaluating the quality of the summary can be explored.

4 Conclusion and Future Works

The enormous amount of data that surrounds us needs summarization, and we are mostly surrounded by text that is in our regional languages. So, text summarization in regional languages should be of great priority of research. The study of above discussed paper also roughly presents that we should shift more towards abstractive summarization so that the machines can generate human-like summaries. The irrelevant information in extracted summary sometimes might lead us to confusion and that can be solved slowly with techniques that will be available with the advent of newer technologies. Also, very less works have been done in regional languages other than the abovementioned. The work should be taken with high priority especially by the native speakers of those languages. The paper has clearly explained that we lack in the field of abstractive summary and future works should be towards making summarizers for an abstractive summary generation. Similarly, further works should be done in addressing the complex morphological variations in Indian regional languages.

References

1. Collobert R, Weston J, Bottou L, Karlen M, Kavukcuoglu K, Kuksa P (2011) Natural language processing (almost) from scratch. J Mach Learn Res 12:2493–2537
2. Anthes G (2010) Automated translation of Indian languages. Commun ACM 53:24–26
3. Thaokar C, Malik L (2013) Test model for summarizing hindi text using extraction method. In: 2013 IEEE conference on information and communication technologies. IEEE, pp 1138–1143
4. Gupta M, Garg NK Text summarization of Hindi documents using rule based approach. In: 2016 international conference on micro-electronics and telecommunication engineering (ICMETE). IEEE, pp 366–370
5. Kumar KV, Yadav D (2015) An improvised extractive approach to hindi text summarization. In: Information systems design and intelligent applications, pp 291–300
6. Subramaniam M, Dalal V (2015) Test model for rich semantic graph representation for Hindi text using abstractive method. Int Res J Eng Technol 2
7. Abujar S, Hasan M, Shahin M, Hossain SA (2017) A heuristic approach of text summarization for Bengali documentation. In: 2017 8th international conference on computing, communication and networking technologies (ICCCNT). IEEE, pp 1–8
8. Sarkar K (2012) An approach to summarizing Bengali news documents. In: proceedings of the international conference on advances in computing, communications and informatics, pp 857–862
9. Akter S, Asa AS, Uddin MP, Hossain MD, Roy SK, Afjal MI (2017) An extractive text summarization technique for Bengali document (s) using K-means clustering algorithm. In: 2017 IEEE international conference on imaging, vision and pattern recognition (icIVPR). IEEE, pp 1–6
10. Das A, Bandyopadhyay S (2010) Opinion summarization in Bengali: a theme network model. In: 2010 IEEE second international conference on social computing. IEEE, pp 675–682
11. Reddy PV, Vardhan BV, Govardhan A (2011) Corpus based extractive document summarization for Indic script. In: 2011 international conference on Asian language processing. IEEE, pp 154–157
12. Naidu R, Bharti SK, Babu KS, Mohapatra RK (2018) Text summarization with automatic keyword extraction in telugu e-newspapers. Smart computing and informatics. Springer, pp 555–564

13. Kallimani JS, Srinivasa K (2011) Information extraction by an abstractive text summarization for an Indian regional language. In: 2011 7th international conference on natural language processing and knowledge engineering. IEEE, pp 319–322
14. Giri VV, Math M, Kulkarni U (2016) A survey of automatic text summarization system for different regional language in India. Bonfring Int J Softw Eng Soft Comput 6:52–57
15. Sarwadnya VV, Sonawane SS (2018) Marathi extractive text summarizer using graph based model. In: 2018 fourth international conference on computing communication control and automation (ICCUBEA). IEEE, pp 1–6
16. Priyadharshan T, Sumathipala S (2018) Text summarization for Tamil online sports news using NLP. In: 2018 3rd international conference on information technology research (ICITR). IEEE, pp 1–5
17. Burney A, Sami B, Mahmood N, Abbas Z, Rizwan K (2012) Urdu text summarizer using sentence weight algorithm for word processors. Int J Comput Appl 46:38–43
18. Humayoun M, Yu H (2016) Analyzing pre-processing settings for Urdu single-document extractive summarization. In: Proceedings of the tenth international conference on language resources and evaluation (LREC'16), pp 3686–3693
19. Gupta V, Lehal GS (2012) Automatic Punjabi text extractive summarization system. In: Proceedings of COLING 2012: demonstration papers, pp 191–198
20. Geetha JK, Deepamala N (2015) Kannada text summarization using latent semantic analysis. In: 2015 international conference on advances in computing, communications and informatics (ICACCI). IEEE, pp 1508–1512
21. Kabeer R, Idicula SM (2014) Text summarization for Malayalam documents—an experience. In: 2014 international conference on data science and engineering (ICDSE). IEEE, pp 145–150
22. Krishnaprasad P, Sooryanarayanan A, Ramanujan A Malayalam text summarization: an extractive approach. In: 2016 international conference on next generation intelligent systems (ICNGIS). IEEE, pp 1–4
23. Sunitha C, Jaya A, Ganesh A (2016) A study on abstractive summarization techniques in Indian languages. Proc Comput Sci 87:25–31
24. Liu L, Lu Y, Yang M, Qu Q, Zhu J, Li H (2018) Generative adversarial network for abstractive text summarization. In: Thirty-second AAAI conference on artificial intelligence

A Vital Approach for Smart Home System Using OWL Ontology

Ria Rawal, Kartik Goel, and Preeti Arora

Abstract Software product line (SPL) features, both variant and reusable ones, have been depicted through numerous ways. Several studies have portrayed the different approaches to model the features of software products. In our paper, we have chosen smart home as the domain to explain the features and characteristics of product line. Our paper has adopted Web ontology approach to propose the relationship between software product line conformations. The dynamic interpretation of different software product line features has been depicted using variants. Feature tree, also called the feature diagram, has been exercised as comprehensive model to model software product line variants. Thus, presenting a graphical way is to represent the feature model using Web ontology approach. The design, analyzation and verification of feature model have evolved over the years. This paper focuses on OWL-DL as a reliable language for showcasing the variant features of SPL. Our paper discusses these essential variant components of smart home domain using Web semantics.

Keywords Semantic · OWL ontology · Smart home system · Variant model · Internet of things · Software product line

Abbreviations

CCTV	Closed Circuit Television
DL	Description Logic
EM	Energy Management

R. Rawal (✉) · K. Goel (✉) · P. Arora (✉)
Bhagwan Parshuram Institute of Technology, GGSIPU, New Delhi, India
e-mail: riarawal99@gmail.com

K. Goel
e-mail: kartikgoel1999@gmail.com

P. Arora
e-mail: erpreetiarora07@gmail.com

A. Abraham et al. (eds.), *Proceedings of 3rd International Conference on Computing Informatics and Networks*, Lecture Notes in Networks and Systems 167,
https://doi.org/10.1007/978-981-15-9712-1_4

IOT Internet of Things
OWL Web Ontology Language
RDF Resource Description framework
SPL Software Product Line
SHS Smart Home System
UML Unified Modelling Language
Wi-Fi Wireless fidelity

1 Introduction

Semantic Web, also known as data Web, governs the rules of storing and manipulating data globally across the Web. It is a vision where data and information are given detailed meanings, enabling machines to automatically process, and thereafter integrate the information that is available on the Web [1]. So, we can use Semantic Web to make the machines capable of understanding data.

Internet of things (IoT) heads to connecting and controlling enormous physical devices using Internet as a medium in the process. These physical devices communicate through efficient networks to perform different actions, which are controlled using sensors and smart technology. Smart technology includes motion sensor tracing body movement, lock control preventing unauthorized access, automatic turning lights and fans on/off, etc., all this is quite advantageous but becomes a laborious task for implementation. In our paper, we have chosen smart home as the primary domain for explaining features and constraints of products. The concept of smart home revolves around the control of electronic devices, such as lighting devices, air conditioners, heaters, access control, and security devices, indirectly using remotes, machines, and mobile apps.

Different propositions for modelling variants are presented using a hierarchical diagram, called feature diagram or feature tree. Understanding the different variants in domain modelling is a strenuous task for the machine.

Feature diagrams illustrate the variant features based on relevant information gathered from the software system. Features are split into mandatory and optional categories based on different requirements and dependencies. The feature diagrams of SPL are transformed and mapped into OWL ontology, to substantiate the consistency of the diagram.

Our paper is organized into different sections, the second session summarizes literature review. Third describes the variant model. Fourth shows OWL-DL representations of axioms of feature model. Fifth and sixth sections present the conclusion and future scope, respectively.

2 Literature Review

OWL ontology [2] caters a language, which depicts classes, their properties, and the relationship between different classes. It serves the purpose of presenting formal semantics. Web ontology language [3] is divided into three branches: OWL Full, OWL-DL and OWL Lite. While OWL Full focuses on maximum possible expressiveness of language, OWL-DL focuses on description logic, and OWL Lite focuses on cardinality constraints. OWL contributes to adding more vocabulary for describing properties, cardinality, classes, and characteristics of data.

Amongst the sublanguages of OWL available today, this paper uses OWL-DL (description logic) approach. Description logic has been well known for its successful application in solving numerous complex configuration problems and to check consistency in UML diagrams.

Software product line (SPL) is developed because of reiteration of features and component designs of software product. SPL establishes a smooth transition from domain engineering to application engineering. The anchor to this approach remains the domain model, variant and reusable components, specifications, maintainability, functional requirements, and several non-functional requirements. Software product line features have grown around Web semantics [4]. Several tools have been developed to verify the features. However, no definite solution is obtained till date to satisfy the requirements and completely verify the software product line features. RDF was gradually replaced by OWL semantics [5], including all the three branches of OWL. Variants help in developing products to give maximum benefit to the manufacturers and users. The distinct variants of a product trace the development of software product line.

3 Variant Model

In our paper, we have used smart home system as our domain [6] under the software product family to explain the product line variants. Our proposed model describes the variants of the software including behaviour, characteristics, and customization of features. For depicting the variants, we have used frameworks to ascertain variability in earlier stages of product line development. The basic motive of our paper is amendment of the existing software features to ameliorate the existing features in order to provide better security, living conditions, and comfort for people staying in smart homes.

3.1 Smart Home System (SHS) Overview

We use smart home system (which is also called as home automation) as our domain to depict the variability model. Smart home systems have widely been approved for the quality and comfort they provide (Fig. 1).

The core functionality of our proposed model aims to facilitate the homeowners with ambient living conditions, proper energy management, security, and comfort by helping them to manage different devices very easily using Internet services. In our model, we include features like lock control, alarm, energy management, ambient living, Internet, and multimedia. These features are distributed into optional, mandatory, OR, and alternative. Mandatory features like lock control, alarm, Internet, and energy management are important and must be introduced while setting up a smart home. Optional features like multimedia and ambient living conditions are not very important and may or may not be present in setting up a smart home. These features and their relations form the basis for the construction of feature diagram.

Fig. 1 Types of features

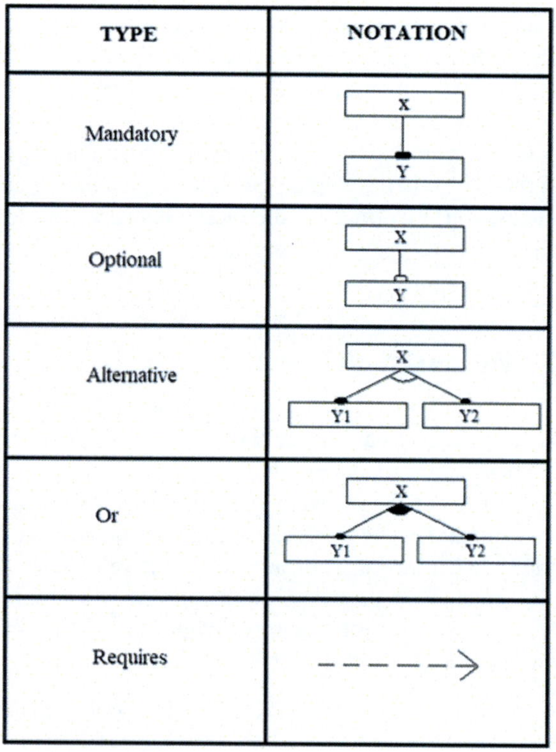

3.2 *Feature Diagram of Smart Home System (SHS)*

Feature diagram of SHS is a graphical tree-like representation depicting the organization [6] of features in the systematic form of a hierarchy. The root in the tree, symbolizes the ***concept node***, whereas all the different functionalities are represented by the other nodes. The functionalities of smart home system as depicted are as follows (Fig. 2):

Alarm

Alarms are used in residential, commercial, and many other purposes including smart home for protection against burglary (anti-theft) and can also be used to alert the users in case of any fire by detecting the presence of smoke.

Lock Control

This feature includes child features which controls the lock like authentication device which helps to authenticate a person by keypad (by manually typing a password or a key on a pad), a card reader (a card acts as a key to open the lock), or a fingerprint scanner (which allows only authenticated users with their fingerprints to enter the house).

Energy Management

A smart home can only be proved to be a boon if there is an efficient energy management which includes child features like lighting, water, and fuel.

These are the basic commodities which play a vital role in implementing a smart home. Thus, an efficient management of all these energy resources will improve the living conditions in the house.

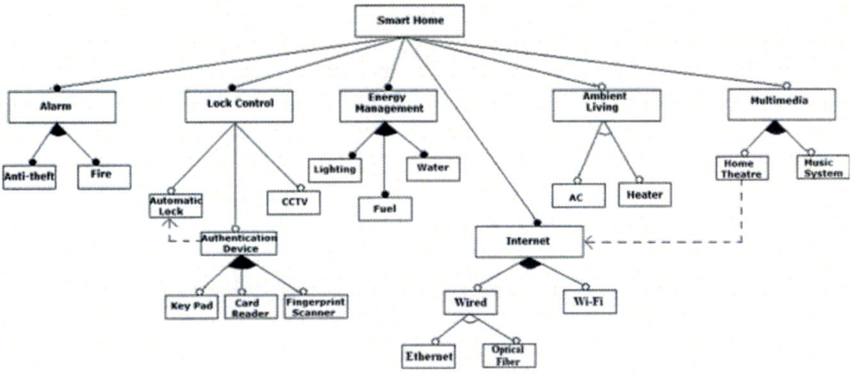

Fig. 2 Feature diagram of SHS

Multimedia

This provides entertainment to users by easily managing the music system and home theatre. A smart home is designed to enable users to access and control all their multimedia and IoT devices via a single app.

Ambient Living

SHS ensures better living conditions by having an efficient temperature control. An air conditioner or a heater can be required as per the climate of the region where smart home is made.

Internet

Smart house is called SMART because all devices can interconnect and communicate due to the use of Internet services which can be installed either through wired media (optical cable or Ethernet) or through wireless media (Wi-Fi) [7] Administrator.

4 Proposed Feature Model

Our approach aims to give a formal definition to our variability model [8] using Semantic Web-based ontological approach. Our feature diagram of SHS represents four feature relations. These relations are optional, mandatory, OR, and alternative, which have been modelled using OWL-DL (description logic). In addition to this, "***requires***" constraint is also modelled. OWL ontology for the various nodes and edges present in the feature tree is built before modelling the different feature relations.

Our paper revolves around the domain of smart home system where concept and features like multimedia, lighting facilities, etc., can be shown as classes in an ontology. In this domain, the concept node refers to the topmost class smart home represented as S. Subsequent nodes, are the feature nodes, represented through branch connections in the model.

4.1 Representation of SHS Using OWL-DL

Step 1 Identification and distribution of the nodes (as concept and feature nodes) (Fig. 3).

Step 2 Every node of the feature tree is then modelled as one of the corresponding OWL class. The OWL class is the subclass of the topmost class S. By default, it is always assumed features with variant names are distinct and mutually disjoint.

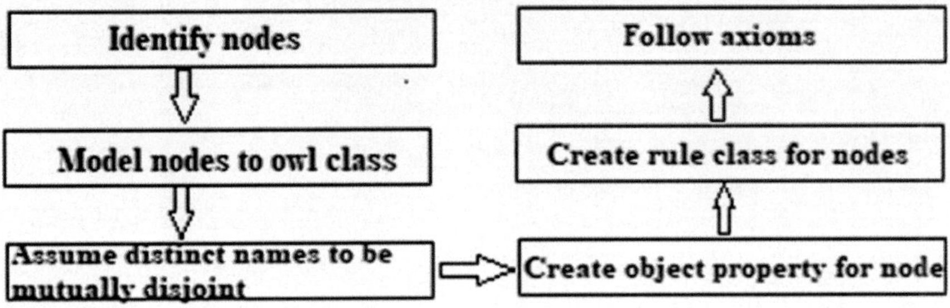

Fig. 3 Flowchart depicting steps of SHS

$$\begin{aligned}&\text{LockControl}|\text{Multimedia}|\text{Alarm}\sqsubseteq S\\&\text{LockControl}\sqcap\text{Alarm}|\text{Alarm}\sqcap\text{Multimedia}=\bot\end{aligned}\tag{1}$$

Step 3 We create an object property for every edge present in our OWL model of smart home. The concept node and feature nodes in the feature diagram are inter-related by distinct edge types as shown in the diagram.

$$S\sqsubseteq\forall\,\text{has}\,X.X\tag{2}$$

$$S\sqsubseteq\forall\,\text{has}\,Yi.Yi,\ \text{for}\,1\le i\le n\tag{3}$$

where X is parent feature and Yi child features.

Step 4 We now create a rule class for every node present in the feature tree where we impose all the restrictions.

$$X\,\text{Rule}\sqsubseteq S,\ X\,\text{Rule}\equiv\exists\,\text{has}\,X.X\tag{4}$$

$$Yi\,\text{Rule}\sqsubseteq S,\ Yi\,\text{Rule}\equiv\exists\,\text{has}\,Yi.Yi,\quad \text{for}1\le i\le n\tag{5}$$

where X is the parent feature and Yi are child features.

Now that we know the steps to follow, we are ready to start modelling our feature diagram. By following the above-mentioned steps and some axioms as stated in the table below, we convert the feature diagram into ontology (Table 1).

Table 1 Axioms for OWL-DL representation

Notation used	Explanation of the notation
S	Represents the superclass of all the OWL
$F \bigcup G$	Classes union of different classes
$F \square G$	Intersection of different classes
$F \subseteq G$	F is a subclass of G
$F \subseteq \neg G$	G and F are disjoint
$F \equiv G$	Presents equivalence of classes
$S \subseteq \forall PF$	Range depiction of different properties P in class F
$S \subseteq \exists PF$	For every instance of class that has property say P, then some/all the instances with the property are members of the class F

4.2 *Representation of SHS Using OWL-DL*

Mandatory Feature

Amongst various mandatory features, we are showing the OWL-DL representation of one of them which is the lock control feature.

$$\text{LockControl}|\text{LockControlRule} \sqsubseteq S \tag{6}$$

$$\text{hasLockControl} \sqsubseteq \text{Object Property} \tag{7}$$

$$S \sqsubseteq \text{hasLockControl.LockControl} \tag{8}$$

$$\text{LockControlRule} \equiv \exists\, \text{hasLockControl.LockControl} \tag{9}$$

$$\text{SmartHomeRule} \sqsubseteq \exists \text{hasLockControl.LockControl} \tag{10}$$

This suggests that a lock control feature is mandatory in a smart home system in order to improve the security.

Optional Feature

These are not necessarily added in the feature diagram. Their inclusion in the feature diagram is optional. For each of the optional features, $Y1, \ldots, YN$ of parent feature X, no extra statements are needed.

We are showing one of the optional features which is multimedia:

$$Yi\ \text{Rule}|\text{Multimedia} \sqsubseteq S \tag{11}$$

$$\text{has } Yi\,|\,\text{has Multimedia} \sqsubseteq \text{objectProperty} \tag{12}$$

$$Yi \text{ Rule} \equiv \exists \text{has} Yi.Yi \quad \text{for } 1 \leq i \leq n \tag{13}$$

where Y_i is an optional feature and S is smart home system (SHS).

Requires

$$X \text{ Rule} \sqsubseteq \exists \text{ has } Yi.Y \tag{14}$$

$$\text{HomeTheatreRule} \sqsubseteq S \tag{15}$$

$$\text{MultimediaRule} \equiv \exists \text{ has Internet.Internet} \tag{16}$$

Yi feature appears in configurations if *X* is present.

Using this constraint, we model the relationship present in the feature diagram in which it is shown that a home theatre requires Internet services, thus, making it easier for the users to operate home theatre.

Alternative

Alternative features denote that exactly one feature from the available set of features can be included in the model.

For example: The feature diagram of wired Web connection shows ethernet and optical fibre composing a set of alternative features.

$$X \text{ Rule} \sqsubseteq ((\exists \text{ has } Yi.Yi) \cup (\exists \text{ has } Yj.Yj)) \quad \text{for } 1 \leq i,\ j \leq n \tag{17}$$

$$X \text{ Rule} \sqsubseteq ((\exists \text{ has } Yi.Yi) \cup (\exists \text{ has } Yj.Yj)) \quad \text{for } 1 \leq i,\ j \leq n \tag{18}$$

$$\text{WiredRule} \sqsubseteq ((\exists \text{ has OpticalFibre.OpticalFibre}) \cup (\exists \text{ has Ethernet.Ethernet})) \tag{19}$$

$$\text{WiredRule} \sqsubseteq ((\exists \text{ has OpticalFibre.OpticalFibre}) \cup (\exists \text{ has Ethernet.Ethernet})) \tag{20}$$

OR

OR allows one or more features from the available set of features to be incorporated into the model. As shown in the figure of SHS, some of the OR features are energy management, alarm, multimedia, and Internet.

$$X \text{ Rule} \sqsubseteq \cup \ (\exists \text{has } Yi.Yi) \quad \text{for } 1 \leq i \leq n \tag{21}$$

Using this rule, we model OWL-DL representation for Energy Management (EM).

$$\text{EM Rule} \sqsubseteq ((\exists\, \text{has Lighting.Lighting}) \cup (\exists\, \text{has Fuel.Fuel}) \cup (\exists\, \text{hasWater.Water})) \tag{22}$$

In addition to these features, we devise two more features called as optional alternative and optional OR which aims at making the designing stages of software product development easy. It enhances the existing functionalities [9].

Optional Alternative (Optional + Alternative)

This means that exactly one optional feature can be added into the feature tree.

As shown in the feature diagram, we can see that ambient living is the alternative optional feature. It is based on both the optional and alternative constraints.

(*Optional Constraints*)

$$Y1, Y2 \sqsubseteq S \tag{23}$$

$$Y1\,\text{Rule}|Y2\text{Rule} \sqsubseteq S \tag{24}$$

$$\text{has } Y1|\text{has } Y2 \sqsubseteq \text{ObjectProperty} \tag{25}$$

$$Y1\,\text{Rule}|Y2\,\text{Rule} \equiv \exists\, \text{has } Y1.Y1/\text{ has } Y2.Y2 \tag{26}$$

Here, $Y1$ is Ambient Living, $Y2$ is Multimedia, and S is SmartHome.

(*Alternative Constraints*)

$$X\,\text{Rule} \sqsubseteq ((\exists\, \text{has } Yi.Yi) \cup (\exists\, \text{has } Yj.Yj)) \tag{27}$$

$$X\,\text{Rule} \sqsubseteq ((\exists\, \text{has } Yi, Yj) \cup (\exists\, \text{has } Yj, Yj)) \tag{28}$$

Here, Yi is Ambient Living, Yj is Multimedia, and X Rule is SmartHomeRule.

Optional OR (Optional + Or)

This means that more than one optional feature can be incorporated in the feature tree. The smart home has an authentication device in which it is optional to have keypad, card reader or a fingerprint scanner to authenticate the user the open the lock and enter the house thus ensuring great amount of safety. It will follow both optional and OR constraints.

(*Optional Constraints*)

$$Y1|Y2|Y3 \sqsubseteq S \tag{29}$$

$$Y1\,\text{Rule}|Y2\,\text{Rule}|Y3\,\text{Rule} \sqsubseteq S \tag{30}$$

$$\text{has } Y1|\text{has}Y2|\text{hasY3}\sqsubseteq\text{ObjProperty} \tag{31}$$

$$Y1\,\text{Rule}|Y2\,\text{Rule}|Y3\,\text{Rule} \equiv \exists\,\text{has}\,Y1.Y1|\text{has}\,Y2 \cdot Y2|\text{has}\,Y3,\,Y3$$

$Y1$ is KeyPad, $Y2$ is CardReader, $Y3$ is Fingerprint Scanner, and S is Smart Home as shown in the equations above.

(*OR Constraints*)

$$X\,\text{Rule}\sqsubseteq((\exists\,\text{has}\,Yi.Yi) \cup (\exists\,\text{has}\,Yj \cdot Yj) \cup (\exists\,\text{has}\,Yk,\,Yk))) \tag{32}$$

Y_i is KeyPad, Y_j is CardReader, Y_k is FingerprintScanner, and X Rule refers to AuthenticationRule as shown in (32).

5 Conclusion and Future Scope

SPL mirrors the systematic mechanism of software engineering to reuse beneficial features and solutions of software products. The outlook of the common software products directs to effective cost-cutting, prevention of redundant work, and wastage of time. This paradigm manages to combine the approaches of problem space and solution space, ranging from gathering the requirements and customer needs to presenting the product for customer use.

Our paper proposes the method to make use of Semantic Web-based approach for describing smart home system (SHS). Ontological definitions [10] used for the feature diagram have given a formal definition to our variability model. OWL-DL has been used to present our feature model in a concise and unambiguous manner. The domain of smart home was explicitly explained throughout the paper to illustrate our Web-based approach. In addition to mandatory, optional, OR, and alternative features used to enhance the characteristics of our SPL model, we have devised two more features optional OR and optional alternative. The features collectively focus on maintaining appropriate management of energy, security, comfort, and ambient living conditions for the homeowners. Through our research paper, we have focused on modelling and presenting the functional features of smart home system (SHS). Our further plans include devising a method to verify the consistency of ontology-based feature tree and to incorporate the requirements of software product line (SPL). The paper can be extended to formally verify the configuration and integration of features.

References

1. Osborne F, Motta E, Mulholland P (2013) Exploring scholarly data with rexplore. In: Alani H et al (eds) ISWC 2013, Part I, vol 8218. LNCS. Springer, Heidelberg, pp 460–477
2. Fernández M, Gómez-Pérez A, Juristo N (1997) METHONTOLOGY: from ontological art towards ontological engineering. In: Symposium on ontological engineering of AAAI, Stanford, California
3. Corcho O, Fernández M, Gómez-Pérez A (2003) Methodologies, tools and languages for building ontologies. Where is the meeting point? Data Knowl Eng 46
4. Berners-Lee T, Hendler J, Lassila O (2001) The semantic web. The Scientific American, May 2001
5. Motta E, Sabou M (2006) Next generation semantic web applications. In: Mizoguchi R, Shi Z, Giunchiglia F (eds) ASWC 2006. LNCS, vol 4185. Springer, Heidelberg
6. Cunningham H, Gaizauskas R, Humphreys K, Wilks Y (1999) Three Years of GATE. In: Proceedings of the AISB'99 workshop on reference architectures and data standards for NLP, Edinburgh, UK
7. Kietz J-U, Volz R, Maedche A (2000) Semi-automatic ontology acquisition from a corporate intranet. In International conference on grammar inference (ICGI-2000), Lecture Notes in Artificial Intelligence, LNAI
8. Motik B, Maedche A, Volz R (2002) A conceptual modeling approach for building semantics-driven enterprise applications. In: 1st International conference on ontologies, databases and application of semantics (ODBASE-2002), California, USA
9. Buitelaar P, Eigner T, Declerck T (2004) OntoSelect: a dynamic ontology library with support for ontology selection. In: McIlraith SA, Plexousakis D, van Harmelen F (eds) ISWC 2004. LNCS, vol 3298, Springer, Heidelberg
10. Bechhofer S, Volz R (2004) Patching sayntax in OWL ontologies. In: McIlraith SA, Plexousakis D, van Harmelen F (eds) ISWC 2004. LNCS, vol 3298, Springer, Heidelberg

Plagiarism Detection Using Deep Based Feature Combined with SynmDict

Ashish Varghese Muttumana, Harsh Goel, Yash Teotia, and Piyush Bhardwaj

Abstract Plagiarism in colleges is a significant issue, staying as a point for logical works for a considerable length of time. We can watch plagiarism happen in different fields like writing, scholastic, science, and music inconceivably. It very well may be likewise conceivable that one day we will get our task work in another production without legitimate reference. Plagiarism discovery systems are there, which are ordered into character-based strategy, basic based technique, characterization or group-based strategy, cross language-based methods, citation-based methods, semantic-based methods, and syntax-based methods. Different devices are accessible utilizing the above plagiarism strategies. Our tests show the viability of "deep features" in the undertaking of grouping task program entries as copy, partial-copy, and non-copy by bunching systems. Here, we have created a database containing sets of synonyms in the tabular form; it covers a variety of words containing a total of 100,000 words. This dataset helps to create an instantaneous feature for the specific dataset.

Keywords Plagiarism · SynmDict · Natural language processing · Deep learning · Deep feature · Deep deep-based feature · Machine learning

---

A. V. Muttumana (✉) · H. Goel · Y. Teotia · P. Bhardwaj
Bhagwan Parshuram Institute of Technology (affiliated to Guru Gobind Singh Indraprastha University, Delhi 110078), Rohini, Delhi 110089, India
e-mail: ashishvarghesem@bpitindia.com

H. Goel
e-mail: harshgoel@bpitindia.com

Y. Teotia
e-mail: yashteotia@gmail.com

P. Bhardwaj
e-mail: piyushbhardwaj@bpitindia.com

A. Abraham et al. (eds.), *Proceedings of 3rd International Conference on Computing Informatics and Networks*, Lecture Notes in Networks and Systems 167,
https://doi.org/10.1007/978-981-15-9712-1_5

1 Introduction

Plagiarism is viewed as scholastic dishonesty and disregard to journalistic ethics. It is liable to punishments, suspension, ejection from school or work, significant fines, and even detainment. As of late, instances of "extreme plagiarism" have been founded in the scholarly world [1]. Throughout history, it was discovered that some outstanding works were copied from past works that did not receive much fame, and the original author was deprived of credits. Thus, we require proper plagiarism detection software or systems which can detect plagiarism to the minute level hence providing the original author with the credit one deserves. In this paper, we have discussed the use of SynmDict[1] created by us in the project. We show that the state-of-the-art strategies can be effectively consolidated utilizing SynmDict via machine learning for an all the more dominant and flexible plagiarism identification apparatus. We additionally show that highlights can be developed from related fields of research and that these can help in ordering plagiarism [2]. We use SynmDict to make new information which includes these highlights that would be explicit for an informational index that exists just for that occasion. At the point when we enter two content document source and doubt, then profound element makes a component which is the mix of the considerable number of equivalent words that are in source and doubt. The size of this presented highlight is significantly less when contrasted with the total volume of SynmDict, which thus spare substantially more execution time in contrast with different calculations present. We have used the data collected from "PAN, a series of scientific events and shared tasks on digital text forensics and stylometry." The structure of paper contains five sections. The first covers the introduction, the second discusses the state-of-the-art methods used, the third elaborates our methodology, the fourth shows the results that we obtained, and the fifth section covers the conclusion and future scope.

2 State of Art

We have used the different text-based features in the method from state of the art.

2.1 Feature Engineering

Feature engineering is exceptionally vital while making machine learning models. Accuracy of predicted values of machine learning models highly depends on the feature vectors that have been chosen for the model; therefore, the real aim is to engineer such features that will help our machine learning pipeline [3].

[1]Database containing sets of synonyms in tabular form, having a total of 100,000 words.

2.2 *Jaccard Similarity*

The Jaccard similarity index checks the correlation between elements of two or more sets and gives an idea of how similar or different they are. The highest similarity is 100 percent while lowest is 0. For two sets *A* and *B*, it is calculated by an intersection *B* divided by *A* plus *B* minus *A* intersection *B*. It is used in a lot of plagiarism softwares as well as computer-vision software [4].

$$J(X, Y) = |X \cap Y| \div |XUY| = |X \cap Y| \div |X| + |Y| - |X \cap Y| \quad (1)$$

2.3 *Dependency Parser*

A dependency parser keeps a tab on the syntactic construction of a sentence, establishing a link between keywords and words which can modify these keywords. Dependency is a relationship that shows that a component, or set of components, requires other model components for their determination or implementation. The component is reliant upon the autonomous component, called the provider. At least two components right now called tuples [5].

2.4 *N-Gram*

N-gram is a set of words in sequence; it is usually collected from text or speech. It is beneficial for the detection of plagiarism as it provides a specific context which can be used for detection. There are different types of n-gram, depending on the value of n [6].

3 Methodology

3.1 *Preprocessing*

Stemming. It is a process of cutting off a part of the word either from the start or from the end to get to the root of the original word. For stemming in our research paper, we have used porter stemmer, which is very popular in stemming of words.

Tokenization. It is a process of breaking the text into smaller parts known as tokens; thus, all the.txt files in our dataset were broken into some important words known as tokens.

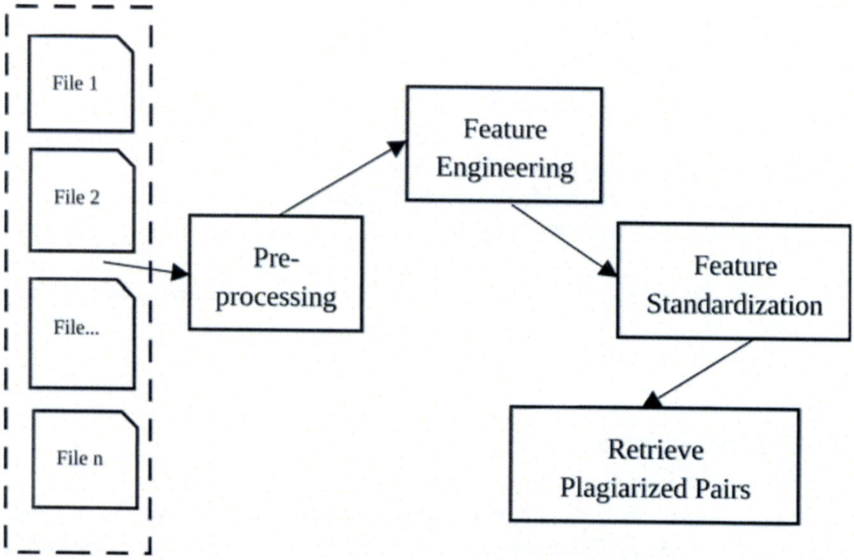

Fig. 1 Overview of the pipeline using an unsupervised learning-based approach

Stop-words. These are the words whose frequency is pretty high and does not add a sentiment value to the word; thus, all .txt files in our dataset eliminated all the stop-words, giving us the words that were of significant value.

We used regular expression and removed all the characters except for [a-zA-Z]. We converted all the words to the lowercase. Split the sentences to a set of words. We then removed all the stop-words. Stem the rest of the left words. Then, save these words in a data-frame. Repeat steps for rest of the text files (Fig. 1).

3.2 LSTM

LSTM is a neural system that is a piece of the repetitive neural system family. It is utilized in succession to arrangement expectation issues, for example, temperature-determining, hand-composing acknowledgement, and so on. We have to initially take a view at how neural systems (RNN and LSTM's) are prepared: first of all the forward propagation, then the mistake concerning the yield is estimated with which we figure angle in which this propagation proceeds. However, for different data sources the slope in deep neural systems are not steady. Along these lines, prior slopes are the result of later angles, and they will either increment or decline exponentially and in this way it cannot be stable. This is known as the exploding/vanishing tendency issue; this is the explanation why individuals use LSTM over RNN [7].

3.3 *Text-Based Features*

Difference in Length of Text (DLT). DLT captures the difference between the lengths of each text.

Similarity as Measured by Difference (SMD). SMD measures the number of lines of common terms in the original text extracted using different text.

Similarity in String Literals (SSL). SSL measures the similarity between the two sets of literals, one from each text file.

Jaccard Similarity and LCS (Longest Common Subsequence). Jaccard similarity and LCS are also used as text-based features.

3.4 *SynmDict*

It is a Dataset created which contains sets of synonyms in the tabular form; it covers almost 100,000 words.

3.5 *Deep Based Feature*

Deep Feature. A deep feature is a reliable reaction of a node or layer inside a various leveled model to info that gives a reaction that applies to the models' last yield. One feature is considered "deeper" than another relying upon how early in the decision tree or other framework, the reaction is enacted [8].

Deep based Feature. We use SynmDict to create new data features. These features would be specific for a data set that exists only for that instance. When we enter two text file sources and suspicious files, then deep feature creates a feature, which is the combination of all the synonyms that are in source and suspicious file. The size of this introduced feature table is much less as compared to the total size of SynmDict, which in turn saves much more execution time in comparison with other algorithms. This method is termed as "deep based features." The "deeper" layers can react and make their feature filters for increasingly complicated patterns in the input, for example, language blunder, evacuation of non-relevant words (such as the, a that, an, and so on), word tallying, or varieties of features processed earlier.

In Fig. 2, after the feature table, is made utilizing feature engineering and coordinating methodologies, information is sent to the RNN machine, which in turn uses LSTM strategy after classifier-based calculation giving the outcome as the group of various sorts of copy and no-copy information found in these two records and arranges as no-copy (0), partial-copy (1), copy (2).

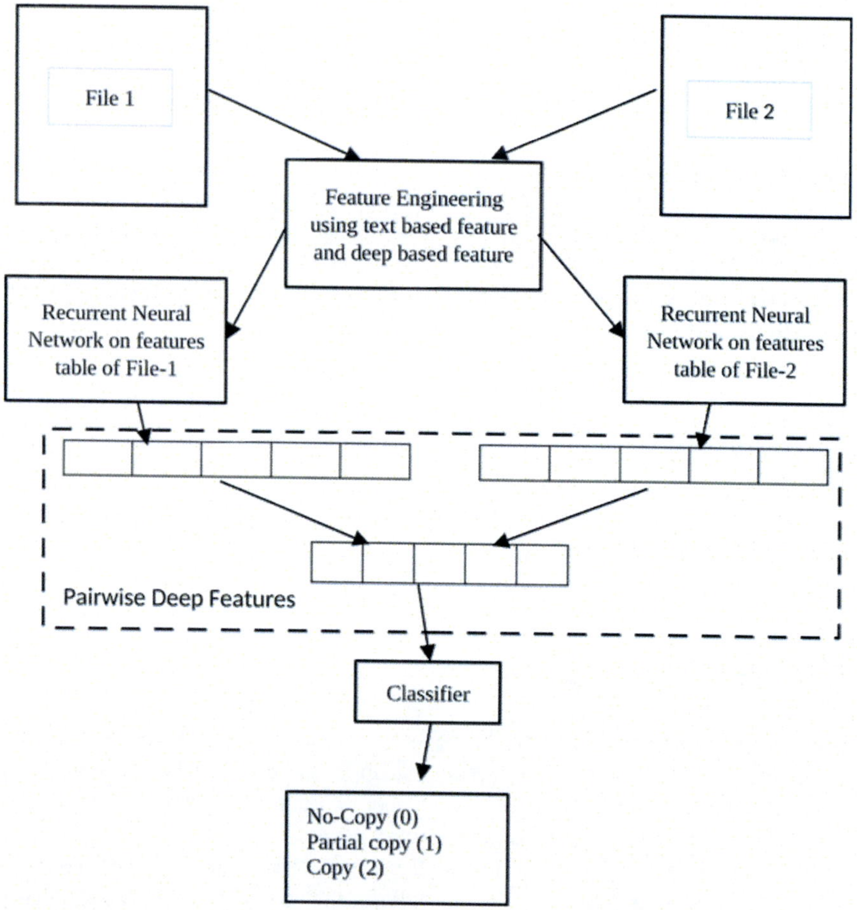

Fig. 2 RNN model approach using text and deep based features, which gives a percentage of no-copy, partial-copy, copy data from the source, and suspicious text files

4 Results

We have calculated the precision and recall value, which is further used to calculate the $F1$ score [9].

$$F1 = 2 * (\text{precision} * \text{recall})/(\text{precision} + \text{recall}) \quad (2)$$

These are the outcomes when text-based and deep based features are utilized. From the table, we can determine that when we use the text-based feature, it gives an $F1$ score of 0.42, and it gives an $F1$ score of 0.55. In any case, when we utilize both the features together, there is an exceptional change in the precision and recall value, which this way influences the $F1$ score. Here, we see that the $F1$ score gets

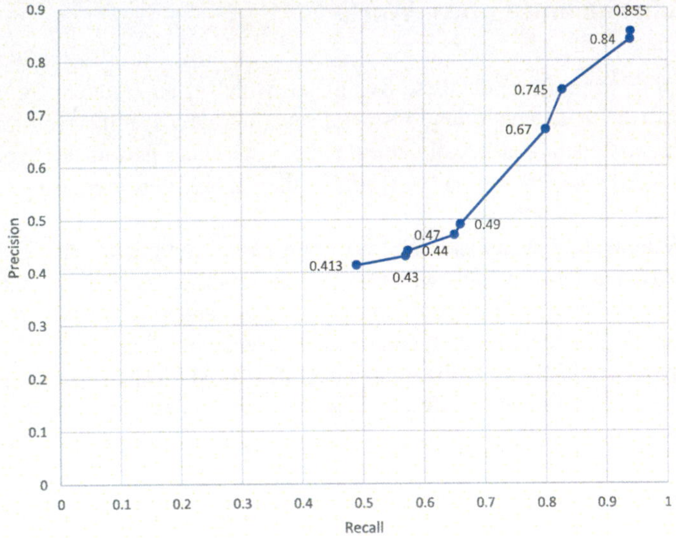

Fig. 3 Precision versus recall with each test

0.71, practically twofold than the individual strategies. In this manner, in the present state of the art, both deep feature and text feature is utilized.

4.1 *SynmDict*

Here, we have actualized content and profoundly based highlights using SynmDict in the RNN model. RNN machine uses the LSTM technique after classifier-based estimation, which gives the result as the gathering of different sorts of copy and no-copy data found in these two records and orchestrates as no-copy (0), partial-copy (1), copy (2). From the given table, we see that we achieved a pretty good improvement in the precision and recall value, which in turn provides a better $F1$ score of 0.91 in "no-copy" class.

4.2 *Graphical Representation*

Figure 3 shows different values of precision and recall values plotted in a line graph. It gives a piece of detail information about the improvements made by us during the timeline by conducting the different tests on the data provided by PAN[2] with each time improved SynmDict.

[2]PAN is a series of scientific events and shared tasks on digital text forensics and stylometry.

5 Conclusion and Future Works

In this paper, we used the database SynmDict, which is an acronym for synonym dictionary. It helps to create new instantaneous and data-specific features, which increases the precision and recall values. Only a particular part of the SynmDict is used for a dataset depending on the words in the dataset, thus increasing the speed while checking for plagiarism.

Multidimensional database. We can improve the SynmDict and increase the dimension of the table, which will increase accuracy as synonyms that are not words with the same meaning but similar meaning, and it changes with context. Multidimensional database may increase the efficiency drastically.

Tree-structured LSTM. Utilization of tree-structured LSTM, recursive recurrent neural networks can encourage us to get familiar with some portrayal to grow such detection frameworks [10].

References

1. Blum SD (2011) My word! Plagiarism and college culture. Cornell University Press
2. Kalleberg RB (2015) Towards detecting textual plagiarism using machine learning methods (Master's thesis, Universitetet i Agder; University of Agder)
3. Bengio Y, Courville A, Vincent P (2013) Representation learning: A review and new perspectives. IEEE Trans Pattern Anal Mach Intell 35(8):1798–1828
4. Niwattanakul S, Singthongchai J, Naenudorn E, Wanapu S (2013) Using of Jaccard coefficient for keywords similarity. Proc Int Multi-Conf Eng Comput Sci 1(6):380–384
5. Nivre J (2005) Dependency grammar and dependency parsing. MSI Rep 5133(1959):1–32
6. Younes N, Reips UD (2019) Guideline for improving the reliability of Google Ngram studies: Evidence from religious terms. PloS One 14(3)
7. Sherstinsky A (2020) Fundamentals of recurrent neural network (RNN) and long short-term memory (LSTM) network. Phys D Nonlinear Phenom 132306
8. Jaderberg M, Vedaldi A, Zisserman A (2014) Deep features for text spotting. In: European conference on computer vision. Springer, Cham, pp 512–528
9. Sokolova M, Japkowicz N, Szpakowicz S (2006) Beyond accuracy, Fscore and ROC: a family of discriminant measures for performance evaluation. In: Australasian joint conference on artificial intelligence. Springer, Berlin, Heidelberg, pp 1015–1021
10. Tai KS, Socher R, Manning CD (2015) Improved semantic representations from tree-structured long short-term memory networks. arXiv:1503.00075

ADS Optimization Using Reinforcement Learning

Rachna Jain, Preeti Nagrath, Sai Tiger Raina, Paras Prakash, and Anuj Thareja

Abstract Our aim is to optimize the performance of ads for company products which are being deployed at user's end to increase the revenue generated by getting more clicks and also spending less time and money in Research and Development. This is a specific application that we have chosen, but it can be applied in various fields as per the requirement of the problem. This is done by first analyzing the demands of various products by making use of various factors like product category and the time period of the year. Once it is found that the products have low sales, the ads are pushed to create interest among the users. Multiple ads which are created by the company are deployed and then optimized by analyzing the clicks that they generate over a period of time. Hit and trial way for exploring is one of the characteristic features of reinforced algorithms. For unique cases actions not only affect the present state but also the next state and the succeeding rewards. We have tried to find a way to solve the problem of multi-armed bandit (MAB) problem or N-armed bandit problem. Though several strategies have been suggested over the years, the two most prominent and commonly used are upper confidence limit (UCB) and Thompson sampling (TS). This paper explains why N-arm is preferable over A/B testing in such cases. Comparison of various approaches to solve the N-arm problem has been done. The strategies that we use for gathering of information and exploiting includes two methods, first option being arbitrary selection and the second one is that we are optimistic about uncertain machine initially and we collect the information of getting success in each round from selected machine. These actions having higher arbitrariness are favored because they provide more data advantage. We found out that Thompson sampling slightly outperforms UCB since it does a better job at manipulation.

Keywords Reinforcement learning · UCB algorithm · Thompson sampling · N-armed armed bandit problem

R. Jain · P. Nagrath · S. T. Raina (✉) · P. Prakash · A. Thareja
Department of Computer Science and Engineering, Bharati Vidyapeeth's College of Engineering, New Delhi, India
e-mail: Sai.raina@gmail.com

A. Abraham et al. (eds.), *Proceedings of 3rd International Conference on Computing Informatics and Networks*, Lecture Notes in Networks and Systems 167,
https://doi.org/10.1007/978-981-15-9712-1_6

Abbreviations

MAB	Multi-armed Bandit
ML	Machine Learning
RL	Reinforcement Learning
TS	Thompson sampling
UCB	Upper confidence bound

1 Introduction

Machine learning is a growing field that is tremendously being used nowadays to solve various problems. The machine can go in the past and learn a behavior and can perform in the same way to make predictions in the future with good precision. Human intervention is kept to the least as the machine does automatically what it has to do. In this data, it is observed or accessed from the past experience and then used to find patterns that may help us in finding a solution to the problem [1, 2].

Reinforcement learning is a class of ML algorithms other than supervised and unsupervised learning methods. It is about taking the best options to gain the maximum reward in various real-life situations [3]. These methods separate itself from labeled or supervised learning in a way that the training data has well categorized data. The model is taught with the correct key, whereas in deep learning, there is no correct answer that we know of beforehand, but the model chooses what actions to take to accomplish the task in hand. Reinforcement methods are connecting link between machine learning and deep learning procedures.

N-armed bandit problem is such a problem where we have limited set of assets which must be allotted between contending choices in such a way that make the most of their projected gain, when apiece choice's characteristics are not known completely at the time of allocation and usually becomes better as time passes by and by allocating resources to the choice [4, 5]. It is a standard problem that demonstrates the trade-off quandary between how much to explore and how much to exploit.

This paper is organized into various sections. Section 1 contains the introduction and then it proceeds to (Sect. 2) that contains the related works. The proposed work is presented in (Sect. 3), and conclusion is provided with the results in (Sect. 4).

2 Related Work

The first two papers that we read covered the basics of RL [4, 6]. Both these papers cover that how we can make the machines learn through experiences. The prime idea is to give reward and punishment based on the action the machine takes in the various scenarios. Next, we started to explore the research done in the area of the different techniques that can be used to solve our problem statement, and we had two main approaches.

The first method is MAB while the other is A/B testing. It is described in the paper below why we choose the MAB problem over A/B testing and came to this conclusion by searching online [7]. In order to select among the N-armed bandits and A/B testing, we evaluate the trade-off between exploitation and exploration. Figure 1 shows the Gaussian exploration-exploitation trade-off. With A/B testing, we have a limited period of time where we only explore and distribute the users equally among the two competitors. Once a winner is declared, only exploitation is done, where all of the customers are diverted to the winner of the competition [8, 9]. The disadvantage is that a lot of time and assets are wasted while exploring the loser trying to learn about the winning participant. In N-arm bandit testing, the tests adapt according to the present scenario and do both exploring and exploiting at the same time. They move the users slowly in the direction of the winner, instead of waiting to declare the supreme till the end of the competition. Hence, the procedure is quicker and has higher efficiency since lesser amount of resources are spent on sending users to obviously substandard variants [10].

While learning about the MAB problem, we came to know of three ways to solve our problem. The most basic of them being no exploration, and then, a slightly better method being arbitrary exploration which is not consistent and the last method

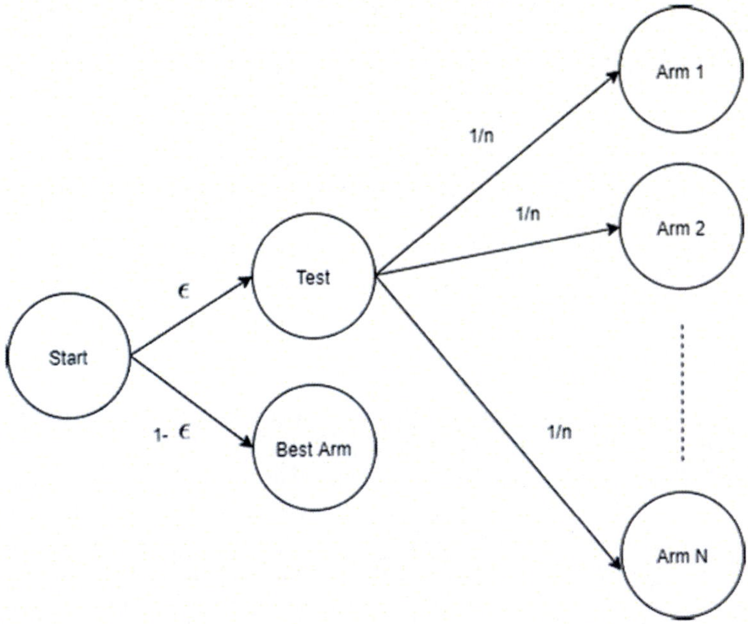

Fig. 1 Gaussian exploration-exploitation trade-off

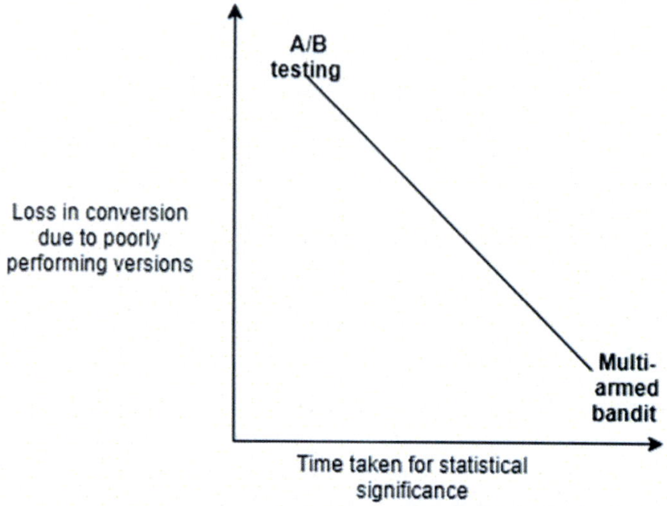

Fig. 2 Comparison between MAB and A/B testing with respect to loss in conversion

which is the best being clever exploration [7, 11]. There are many algorithms for clever exploration, but the two most popular are TS and UCB.

The next research paper explores UCB [12, 6]. The principle of this algorithm is hopefulness in the case of doubt, or in other words, it means to choose the action assuming that the bandit as nice as it is possible. There are two outcomes to the optimism. Either it was justified in this case that the bandit is acting correctly or it is not justified. The last research paper consists of the working of TS [13]. This method helped us in a positive way to increase the efficiency of finding the optimal result. In the paper, it is cited that how the method is one of the easiest to implement and also has a high rate of success. It also compares the performance with other algorithms such as UCB and Bayes approach to Gittins index over various simulated as well as real world data (Fig. 2).

Another paper published by Osband et al. [14] introduced the concept of randomized value functions in exploration. Osband [14] notes the relative deficiency in body of work concerning exploration strategies. Osband [14] introduces the idea of using actions that are greedy and aims to sample from a proxy of the posterior distribution over value functions. This sporadic behavior generally creates a positive bias and by doing so prompts exploration. This helps in the handling of Bandit problem and increases the overall efficiency of the ads prediction.

In the next research paper [15], a new method is explained to solve the k-armed dueling bandit problem which is a slight alteration of the *k*-armed bandit problem which is called relative upper confidence bound (RUCB). As such, *k*-armed dueling bandit problem is different since the feedback comes in the form of pairwise preferences. It is also a part of the framework of preference learning which focuses on learning based on relative feedback rather than on real-valued feedback. In relative feedback, only the preferred alternative of the two options are specified. Other algorithms are usually judged based on the accuracy of the returned best arm in the exploration stage but in contrast that is not the case in RUCB which does not require this input, making it more practical and useful in setting a good exploration horizon. Since it is not possible to always assign a fixed reward value to ads in our case, this approach of preference learning allows us to still choose a preferred option during the learning phase thereby solving a key limitation of our system. This also solves our dependency on real-valued data as relative data is sufficient for the learning phase in preference learning.

Barto et al. [16] published a paper that is an elaboration of the RL framework that entails the autonomous development of skill hierarchies through intrinsically motivated reinforcement learning. The paper illustrates its ability to allow an agent to learn broad aspects in a simple "playroom" environment. The conventional practice in application of RL algorithms is to formulate the problem that one wants the agent to learn how to solve and define a reward function specially made for this problem. Often ingenuity is required for tailoring reward function to suit the need of the problem. Contrary from above approach, [1] we implement RL algorithms by showcasing the lack of necessity in redesigning the internal motivation system of the agent for different environments.

3 Experimental Analysis

At the beginning of the experiment, we do not know what is the best arm out of the many arms present in front of us. So we cannot prefer any one arm over another. The UCB algorithm assumes that all arms have equal chances to return the rewards. Then, the method maintains information about confidence bound of each arm [17]. Initially, it arbitrarily chooses any of the arms. After that, two things can happen that the user will pull the arm, which gives a reward, or the user does not pull. Let us say that the arm did not have a pull or the arm was a failure. So, the observed average of the arm will go down [18]. And the confidence interval will also shrink, and the UCB will come down. If the arm is pulled, the observed average will go up, and the confidence limit would also go up. By exploiting the best arms, we are making the confidence limit to go down. As we are adding more number of samples, the chance of other arms doing well also goes up. This is the fundamental concept of UCB [12, 19].

Methods for solving bandit problem:

Based on the ways we do exploration, the several ways to solve the multi-armed bandit are as follows:

Exploration is not done: The most basic approach and not one with high returns.

Exploring arbitrarily: Sometimes good rewards, but the rewards are luck based.

Exploiting in a certain way and choosing high variance data to have the best results.

Epsilon-Greedy Method:

The epsilon-greedy method helps in picking optimal measures for most cases, but performs casual exploration to gather information infrequently [20]. For each action, cost is given according as per historical knowledge by taking the mean of the rewards similar to the target:

$$\overline{Q}_{\mathrm{t}}(b) = 1/N_{\mathrm{t}}(b)\sum_{r=1}^{t} T_{\mathrm{r}1}[b_{\mathrm{r}} = b] \tag{1}$$

where 1 shows the binary result and $N_{\mathrm{t}}(b)$ shows the number of times that action has been chosen:

$$N_{\mathrm{t}}(b) = \sum_{r=1}^{t} T_{\mathrm{r}1}[b_{\mathrm{r}} = b] \tag{2}$$

As per this method, an unexplored arm is taken by it, but otherwise $1 - \varepsilon$ the best arm is taken for which we are sure of getting a reward:

$$\hat{\mathrm{a}}t* = \arg\max a\varepsilon A\overline{Q}t(a) \tag{3}$$

UCB Algorithm Intuition:

The selection of bandits arbitrarily provides us a luxury to try out all arms for which we are uncertain about without storing this information. Also, due to the nature of this algorithm, it is likely to end up taking an action which returns punishment. Figure 3 illustrates UCB algorithm graph where limits exist for upper bound.

$\sigma(a_i)$ is the standard deviation, and $c\sigma(a_i)$ is the upper confidence bound. The constant c is an adjustable hyper-parameter [15].

For evading actions like this, one way can be reducing the value of epsilon. Other method could be to be optimistic about highly uncertain selections and hence select

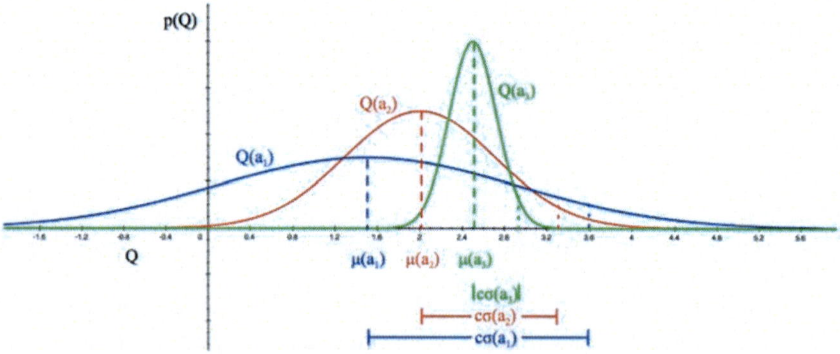

Fig. 3 Gaussian distribution of variables

actions which were not explored yet, we favor exploration of actions having strong chances to have a give us a reward or best return [10, 19].

Parameter for UCB-Hoeffding's Inequality:

We use Hoeffding's inequality when we do not want to give any knowledge in the beginning on how the curve is going to look approximately [21]. If we look at the inequality with respect to probability theory, then it gives us UCB on the probability that the sum of the independent arbitrary variables deviates from the mean value.

The theorem is a general form of Chernoff theorem and can be applied to a restricted distribution. Let X_1 …., X_t be arbitrary variables and bound to the interval [0,1].

For $u > 0$:

$$P[E[X]] > [t + u] \leq e - 2tu * u \tag{4}$$

Now, let us consider:

$R_t(a)$ Arbitrary variables,
$\underline{Q}(a)$ Real average value or mean,
$\overline{Q}_t(a)$ Sample average value or mean,
U Upper confidence limit.

$$U_t(a) = \sqrt{-\log\log p/2N_t(a)} \tag{5}$$

Algorithm

Step 1 In every round r, take two values for arms a piece into deliberation.

$P_i(r)$ Number of times that the arm was selected till round r.
$C_i(r)$ Sum of reward up till round r.

Step 2 From these two numbers, we compute:

The average reward of arm till round n

$$t_i(r) = P_i(r)/C_i(r) \tag{6}$$

The confidence limit

$$[t_i(r) - \Delta_i(r), t_i(r) + \Delta_i(r)] \tag{7}$$

At round n with

$$\Delta_i(r) = \sqrt{3\log\log(r)/2C_i(r)} \tag{8}$$

Step 3 Select the arm which has maximum upper confidence limit

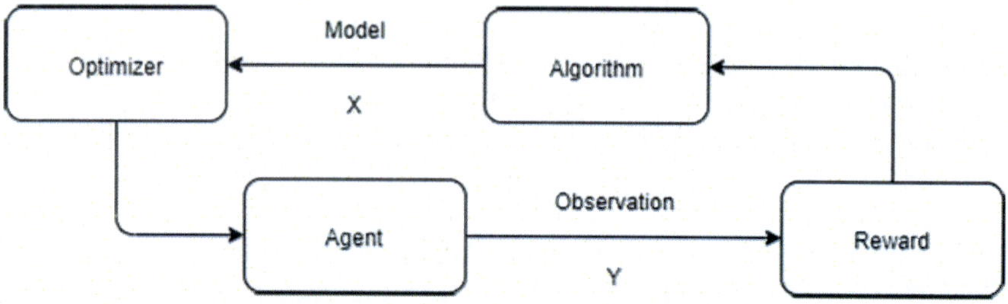

Fig. 4 Working of agent (agent gives reward and punishment based on observation)

$$t_i(r) + \Delta_i(r) \tag{9}$$

Thompson sampling:

This algorithm makes use of exploring and exploiting to increase the number of rewards obtained by performing any action. Thompson sampling also referred as posterior sampling. In the beginning, all machines are supposed to have the same chance of getting rewards. At the end of one observation, we will have a new graph showing the chances of getting a reward for every machine. The next observations are made based on these obtained graphs at the end of every round [22, 23].

After adequate rounds, every slot machine will have enough information to choose our arm, which will get us success for that round [7, 24].

Parameter for Thompson Sampling-Beta distribution:

It is one of the most widely used functions in mathematics and also has a great importance in various algorithms. It has two parameters α, γ which control the shape of the graph. We use it to describe one's vagueness for the chances of getting success from an experiment [6, 19].

The Ad optimizer model or the agent works on the principle of rewards and losses that occur during performance. Figure 4 illustrates the working of agent in learning environment.

4 Conclusion

We choose the products with the low demand and then apply reinforcement learning on the product on it. Random selection is a luck-based algorithm and can sometimes give good result. Random exploration can give results that are not up to the satisfaction level because it involves randomness. Figure 5 shows how random exploration can lead to bad results, even after a lot of runs (exploration), the system is not able to exploit, and hence, all the Ads are selected almost equally.

We used UCB and Thompson sampling which are selective reward methods. In these methods, we choose the variables which have the highest variance. The

Fig. 5 Number of times each Ad was selected versus Ad number (random selection)

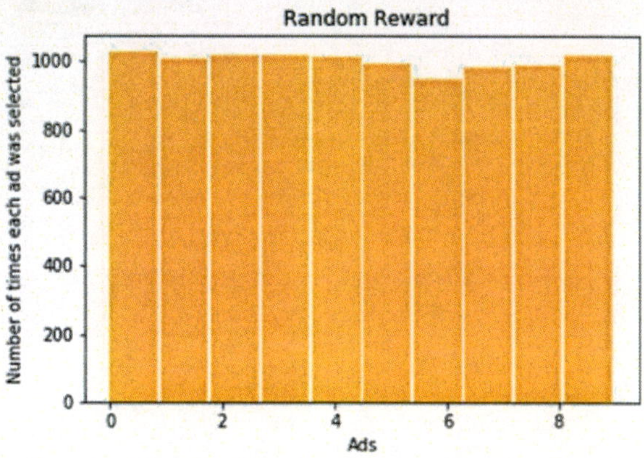

decrease in the parameter ε can help to stay away from inefficient exploration. Also being positive about the options with rare chances can also help to not explore random exploration. Usually the choice of exploration depends on which one has the most optimal value. On analyzing the graphs which have been plotted and are shown above in their respective sections, we conclude that Thompson sampling is better than UCB which can be seen from Figs. 6 and 7 that UCB has exploited the fact that Ad 4 has the highest number of people and deployed it nearly 6000 times out of possible of total of ten thousand runs while Thompson sampling has done a better job at exploiting the same Ad 8000 times (Fig. 8).

To obtain the maximum result from N-armed problem, we used three different approaches which are no exploration, arbitrary, and clever exploring method. The clever exploring method uses exploration-exploitation trade-off very well and gives a much better result than arbitrary exploration. It is also observed that Thompson

Fig. 6 Number of times each Ad was selected versus Ad number (upper confidence bound)

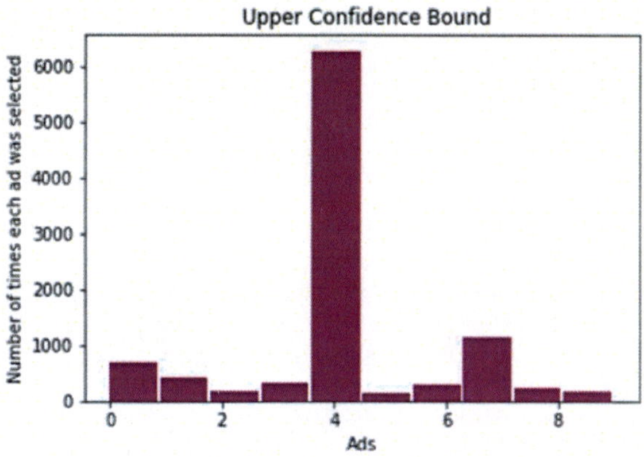

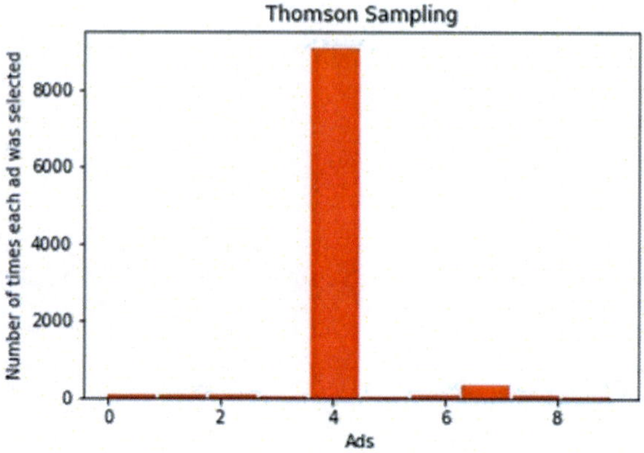

Fig. 7 Number of times each Ad was selected versus Ad number (Thomson sampling)

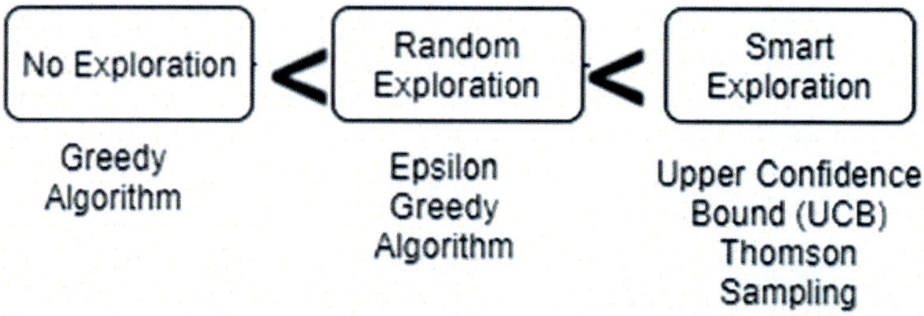

Fig. 8 Result

sampling slightly outperforms UCB as the former does better at manipulating the trade-off.

References

1. Bharat K, Henzinger M (2017) Improved algorithms for topic distillation in a hyperlinked environment. ACM SIGIR Forum 51(2):194–201
2. Barto AG (2013) Intrinsic motivation and reinforcement learning. In: Intrinsically motivated learning in natural and artificial systems. Springer, Berlin, pp 17–47
3. Kaelbling LP, Littman ML Moore AW (1996) Reinforcement learning: a survey. J Artif intell res 4:237–285
4. Whitehead S, Ballard D (1991) Learning to perceive and act by trial and error. Mach Learn 7(1):45–83
5. Rydén T (1997) On recursive estimation for hidden Markov models. Stoch Process Appl 66(1):79–96
6. Auer P (2002) Using confidence bounds for exploitation-exploration trade-offs. J Mach Learn Res 3:397–422
7. Hansen LV, Kiderlen M, Vedel Jensen EB (2011) Image-based empirical importance sampling: an efficient way of estimating intensities. Scand J Stat 38(3):393–408

8. Féraud R, Urvoy T (2013) Exploration and exploitation of scratch games. Mach Learn 92(2–3):377–401
9. Osband I, Van Roy B, Russo D, Wen Z (2017) Deep exploration via randomized value functions. arXiv:1703.07608
10. Audibert J, Munos R, Szepesvári C (2009) Exploration–exploitation tradeoff using variance estimates in multi-armed bandits. Theoret Comput Sci 410(19):1876–1902
11. Szepesvári C (2010) Algorithms for reinforcement learning. Synth Lect Artif Intell Mach Learn 4(1):1–103
12. Pelekis C, Ramon J (2017) Hoeffding's inequality for sums of dependent arbritrary variables. Mediterr J Math 14(6)
13. Littman ML (2015) Reinforcement learning improves behaviour from evaluative feedback. Nature 521(7553):445
14. Luhmann C (2013) Discounting of delayed rewards is not hyperbolic. J Exp Psychol Learn Mem Cogn 39(4):1274–1279
15. Zoghi M, Whiteson S, Munos R, de Rijke M (2013). Relative upper confidence bound for the K-armed dueling bandit problem
16. Guttman I, Scollnik D (1994) An index sampling algorithm for analysis. Commun Stat Simul Comput 23(2):323–339
17. Vorobeychik Y, Kantarcioglu M (2018) Adversarial machine learning. Synth Lect Artif Intell Mach Learn 12(3):1–169
18. Shepard R (1964) Feigenbaum E, Feldman J (eds) Computers and thought. Behav Sci 9(1):57–65 (1963)
19. Lillicrap TP, Hunt JJ, Pritzel A, Heess N, Erez T, Tassa Y, Wierstra D (2015) Continuous control with deep reinforcement learning. arXiv:1509.02971
20. Riquelme C, Tucker G, Snoek J (2018) Deep bayesian bandits showdown: an empirical comparison of bayesian deep networks for Thompson sampling. arXiv:1802.09127
21. Mei Y (2006) Sequential change-point detection when unknown parameters are present in the pre-change distribution. Ann Stat 34(1):92–122
22. Barto A (1991) Learning and incremental dynamic programming. Behav Brain Sci 14(1):94–95
23. Russo D, Van Roy B, Kazerouni A, Osband I, Wen Z (2018) A tutorial on Thompson sampling. Found Trends Mach Learn 11(1):1–96
24. Chen W, Niu Z, Zhao X, Li Y (2012) A hybrid recommendation algorithm adapted in e-learning environments. World Wide Web 17(2):271–284

Automatic Facial Expression Recognition Based on Deep Layered Representation of Convolution Neural Networks

Arun Kumar Dubey and Vanita Jain

Abstract Facial expression is one of the utmost dominant, natural and instant ways for human beings to communicate their sentiments and feelings. Automatic facial expression recognition is an exciting and challenging problem due to its variability and complexity that impacts its importance in human–computer interaction applications. This paper illustrates a novel layered extended convolution neural networks architecture named deep layered representation (DLR). In this paper, we have used Kaggle dataset FER2013 for our layered deep neural network based approach. The implementation of DLR has shown better results. Results are also analyzed and compared using five generalized activation functions: Elu, ReLu, Softplus, Sigmoid and Selu, in the last dense layer. We have also compared our Elu and ReLu-based model with GoogLeNet and VGG16 + SVM-based model on the same dataset.

Keywords CNN · Activation function · Facial expression · FER2013

1 Introduction

Ideas and emotions are easily expressible with verbal communication, but world does not adopted any common language to identify emotion of a person. However, facial expression is one of the utmost dominant, natural and instant ways for human beings to communicate their sentiments and feelings [1–3]. Facial expression is a nonverbal method for sharing expression. Common human expressions which can be easily identified by us are angry, surprise, contempt, neutral, happy, sad and disgust [4–6]. Children are also capable of recognizing these emotions despite their country, race

A. K. Dubey
USICT Guru Gobind Singh Indraprastha University, New Delhi, India
e-mail: arudubey@gmail.comm

V. Jain (✉)
Bharati Vidyapeeth's College of Engineering, New Delhi, India
e-mail: vanita.jain@bharatividyapeeth.eduu

A. Abraham et al. (eds.), *Proceedings of 3rd International Conference on Computing Informatics and Networks*, Lecture Notes in Networks and Systems 167,
https://doi.org/10.1007/978-981-15-9712-1_7

knowledge, etc. Same capability can also be developed in machine if properly being modeled [3–10].

Automatic facial expression recognition is an exciting and challenging problem due to its variability and complexity that impacts its importance in human–computer interaction applications [11–13]. Facial expression recognition (FER) has helped a lot to medical science to understand psychological behavior of patient [14]. Human eyes can easily identify emotion of a person if it exists on the face more than half second. To incorporate and mimic this human ability in human–computer-based interaction systems, facial expression recognition(FER) gravity increased in recent years [15–18].

Further, this paper is organized in five sections. Section 1 gives the introduction that covers the problem statement and its motivation. Section 2 describes the related work of facial expression recognition systems. In the Sect. 3, proposed model has been discussed. Section 4 shows the experimental results and its comparison with other models. Section 5 gives the conclusion of the paper.

2 Related Work

Facial expression is one of the favorite topics of experts in computer vision. A radical survey on existing FER has done within last decade [1, 2, 4, 19, 20]. Many extraordinary works are done on static image to detect emotion of human [5–16]. Zeng et al. [19] published a survey on audio, visual and spontaneous expressions and shown their importance and comparisons between voice and facial images. Video-based facial expression recognition (FER) is proposed by Sariyanidi et al. [21], and residue learning [22] has shown some important variants of FER.

In previous works, emotion recognition depends on the traditional two-step machine learning strategy, where in its very first step, a number of unique characteristics or features are extracted from the pictures, and in the second step, a classifier (such as SVM [23], neural network [24] or random forest) is used to identify the emotions [23–25]. Several popular handcrafted capabilities are employed for facial expression recognition [26]. CK, CK+ [23], Fer2013 were popular dataset for facial expression [27, 28]. Typical FERs focused on still face images and analyzed statistical features, while some researchers have explored the different features such as optical flow features [29] and LBP [30]. For video-based facial expression recognition, motion units in faces are identified by Walecki et al. [31]. He emphasized the temporal variation of FER. In 2019, J. Yang et al. performed action unit [32]-based facial expression recognition system which incorporates facial muscle movements that effectively reflect the changes in people's facial expressions.

Due to the inherent importance of deep learning [27, 29], this paper illustrates a novel layered extended convolution neural networks architecture named deep layered representation (DLR). Extended convolution neural networks architecture, compared to other state-of-the-art methods, has demonstrated better result after implementation.

Results are also analyzed and compared using five generalized activation functions: Elu, ReLu, Softplus, Sigmoid and Selu, in the last dense layer.

3 Proposed model

This paper illustrates a novel layered extended convolution neural networks architecture named deep layered representation (DLR). The architecture of proposed deep layered representation (DLR) model for facial expression is shown in Fig. 1. It is depicting the details of the Input, Filter, Stride (st.), Padding (pad), Output size and Parameters. For each convolutional layer, the output is given by

$$\rho_{xy}^{i} = \sum_{\alpha=0}^{k-1} \sum_{\beta=0}^{k-1} w_{\alpha\beta} q_{(x+\alpha)(y+\beta)}^{i} + \chi i \quad (1)$$

where

ρ_{xy}^{i} the output at position (x, y).
$w_{\alpha\beta}$ is the weight of the kernel and
χi is the bias on layer i.

In each 2D convolution layer, 64 3×3 filters are used. Output layer has seven labels with softmax classifier. This layered approach has used batch normalization and dropout apart from convolution and max pooling layer. Distribution of input to the layers has been changed after each batch but batch normalization standardized it and helped to reduce epochs. Dropout is used for reducing hyper-parameter and

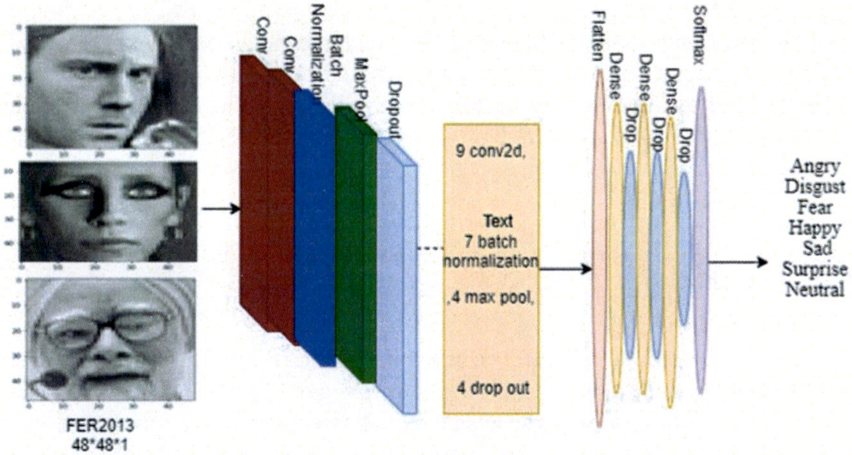

Fig. 1 Proposed deep layered representation (DLR) model for facial expression

helped the model to minimize overfitting. ReLu activation function has been used in each layer except last dense layer.

For the given sample image z, three layered process are followed. First, two normalized convolutional layers plus one fully connected (FC) layer with additional dropout layer are aligned. This generates a feature vector $\nu = \rho(z)$. In the second step, nine convolutional layers with seven batch normalization plus four fully connected layers with additional four dropout layers are aligned which converts $\rho(z)$ into a feature vector $\xi = \Psi(\rho(z) - d)$. In the last step, this combines FC layers with dropouts which finally generate deep layered features (DLF) ϕ.

$$\phi = \delta\big(\Psi(\rho(z) - d) - d'\big) \tag{2}$$

where d and d' are a difference given by dropout layer, and $d \neq d'$. δ, Ψ, ρ are responses of each convolution sets.

Last dense layer is generalized and tested with five activation functions, ReLu, Elu, Selu, Sigmoid and Softplus. Then, each model is designed as DLF + ReLu, DLF + Elu, DLF + Selu, DLF + Sigmoid and DLF + Softplus model. Additionally, zero point four dropout rate has been taken in last three dropouts—dropout 5, dropout 6 and dropout 7. It has dropped some neurons randomly. From the facial expressions, angry, disgust, fear, happy, sad, surprise and neutral emotion have been detected by this model on FER2013 dataset. Kaggle dataset is already processed and converted in to csv file. It is available on Kaggle Web site, and size of each image is 48×48 on single channel. Total parameters, trainable parameters and non-trainable parameters are 5,905,863, 5902, 151 and 3712, respectively.

4 Experiment Result

We have used google co-lab GPU platform for implementation. Model is executed on hundred epoch. Last dense layer's activation function has been kept generalized. This last fully connected layer's activation function is replaced by Elu, ReLu, Softplu, Selu and Sigmoid for hundred epoch. Python Keras framework has helped us to implement our approach.

4.1 Database

We have tested our proposed model on benchmark FER2013 dataset. There are seven emotions: angry, disgust, fear, happy, sad, surprise and neutral labeled as 0–6 in the database. The highest information available in dataset is of happy, whereas disgust is least. Table 1 presents database description as sample per emotion and sample dataset.

Table 1 Database description as sample per emotion and sample dataset

<table>
<tr><th></th><th>S. No.</th><th>Emotion</th><th>Usage</th><th>Pixels</th></tr>
<tr><td rowspan="4">Samples per emotion:
3 8989 Happy
6 6198 Neutral
4 6077 Sad
2 5121 Fear
0 4953 Angry
5 4002 Surprise
1 547 Disgust
Name: emotion, dtype: int64
Number of Labels: 7</td><td>0</td><td>0</td><td>Training</td><td>70 80 82 72 58 58 60 63 54 58 60 48 89 115 121…</td></tr>
<tr><td>1</td><td>0</td><td>Training</td><td>151 150 147 155 148 133 111 140 170 174 182 15…</td></tr>
<tr><td>2</td><td>2</td><td>Training</td><td>231 212 156 164 174 138 161 173 182 200 106 38…</td></tr>
<tr><td colspan="4">Number of pixel per emotion is: 2304
Number of examples in dataset: 35887</td></tr>
</table>

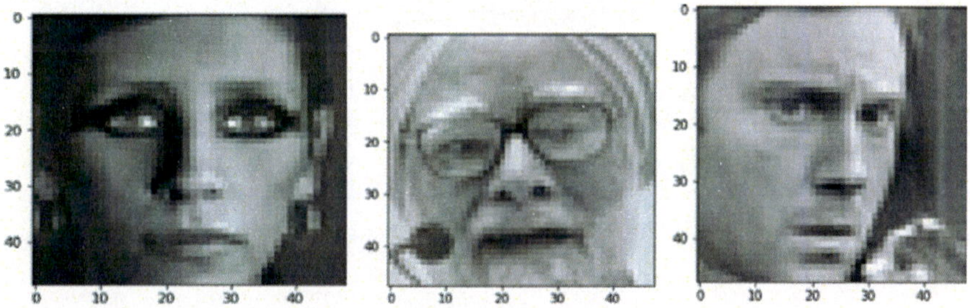

Fig. 2 Faces of dataset consist of 48 × 48 gray image

This dataset consists of set of real-world images like Man Mohan Singh, etc. These are static gray images of size 48 × 48 and processed for single channel. Sample of faces is shown in Fig. 2.

To evaluate accuracy and loss in each epoch, we have trained proposed model on above dataset as per following distribution.

We have calculated the performance of proposed model for 100 epochs using accuracy and loss graph. In each epoch, DLF + Elu model, DLF + Softplus model and DLF + ReLu model-based layered approach have performed better than DLF + Sigmoid model and DLF + Selu model. DLF + Selu model has shown 25% accuracy, and however, it is stable throughout training and validation. Figure 3 represents accuracy of each model in duration of 100 epochs.

In Fig. 4, the loss of the training and validation over 100 epochs has been shown. It is seen that Elu, ReLu and Softplus-based layered models have shown good performance for expression detection. Elu, ReLu and Softplus-based models have shown almost similar performance and outperformed the other two. All three models do not have early stopping point for overfitting.

To analyze the performance of DLR model, receiver operative characteristics (ROC) curves on different activation functions are also presented. Figure 5 shows

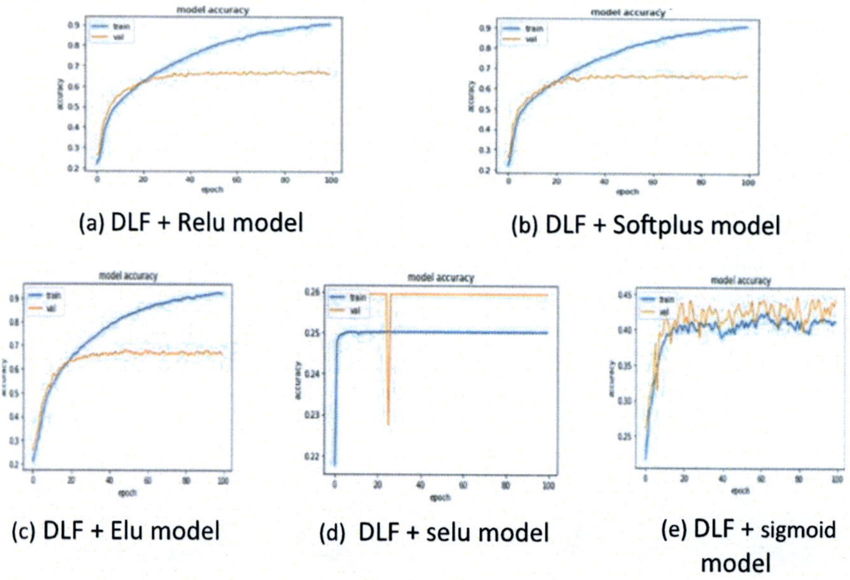

Fig. 3 Training and validation accuracy of each model over 100 epochs

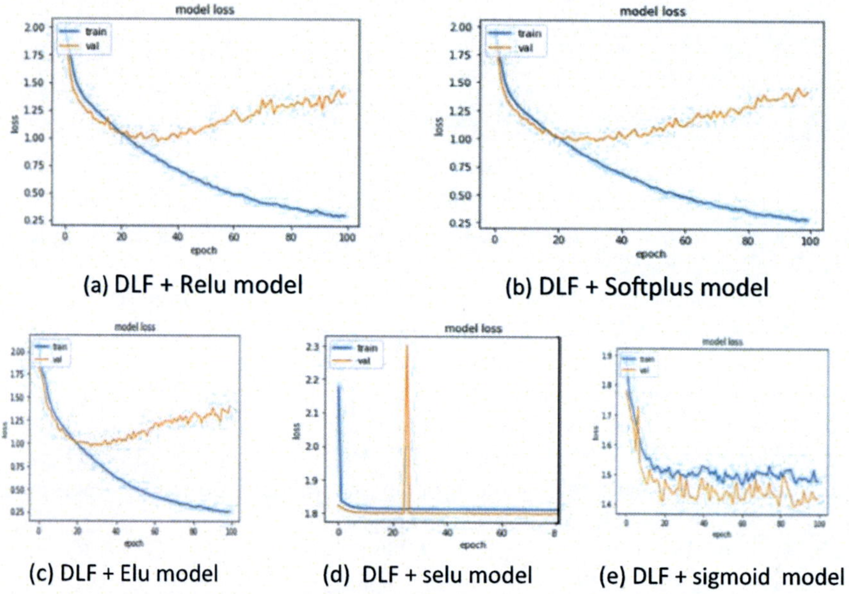

Fig. 4 Loss versus epoch graph on five activation function

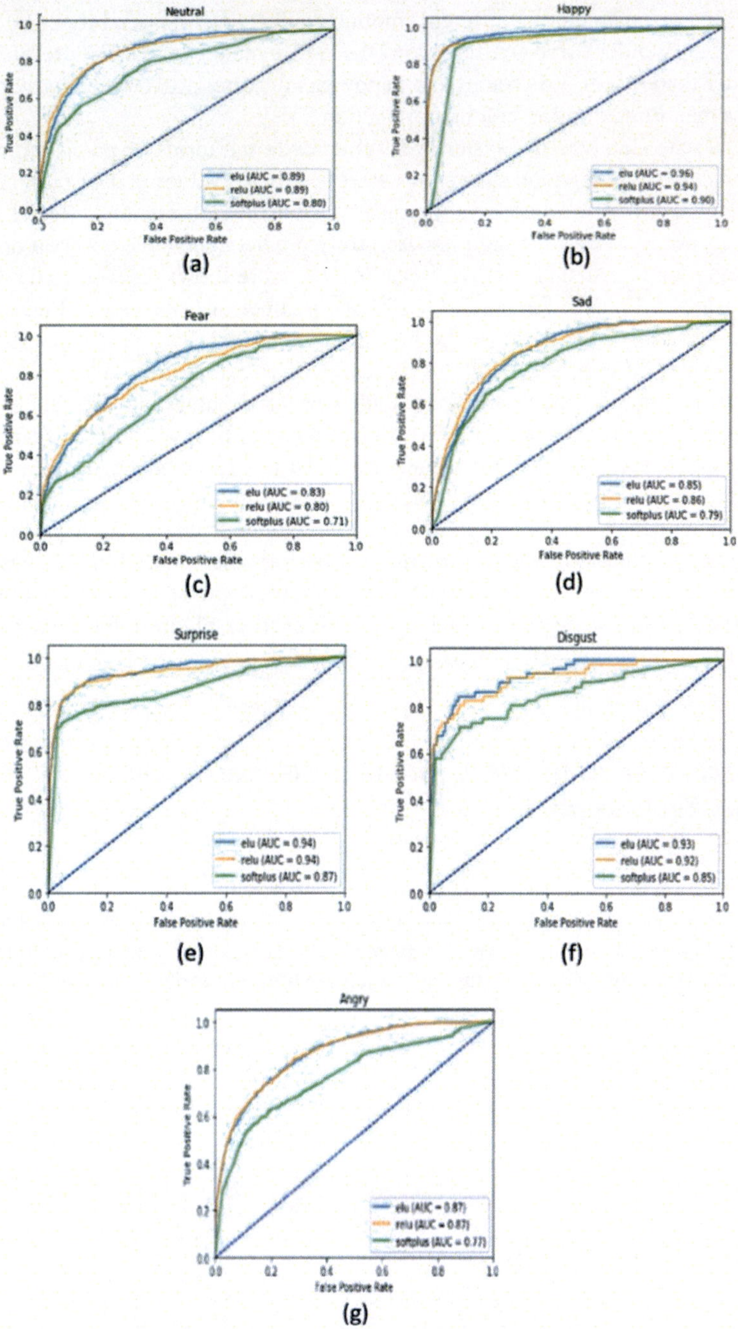

Fig. 5 ROC curve of top three activation function-based proposed approach

the ROC curve on predicting different emotion on FER2013 dataset. From Fig. 5, it is clear that area under curve (AUC) of the DLF + Elu model and DLF + ReLu model has better than others. It is found that happy facial expression prediction modeling is better than ROC of other emotion prediction.

We can also analyze the system performance in the terms of precision, recall, $F1$-score and support which shows the correct predicted values in dataset. Precision is the ratio between true observation and total positive observation. Recall is the sensitivity which shows the ratio between truly predicted positive observations and all observation in actual class, marked as Yes. $F1$-score shows the weighted average of precision and recall. Supports describe the number of occurrence of each label in tested dataset. Table 2 shows the precision, recall, $F1$-score and support-based comparison of DLF + Elu, DLF + ReLu and DLF + Softplus models.

Confusion matrix shows the matrix between true value and predicted value. Highest accurate prediction in DLF + ReLu model is happy where true label and predicted label are 735. DLF + Softplus model, DLF + Elu model and DLF + ReLu model are better in expression prediction in comparison with other two models. It is found that models-based ReLu, Elu, Softplus and Sigmoid are best for happy expression. DLF + Selu models have shown worst performance, and all the expressions are detected as happy. Accuracy and loss graph have depicted that proposed model is the best for Elu, ReLu and Softplus activation function. Figure 6 depicts confusion matrix of top three confusion matrix.

4.2 Accuracy Comparison with Other Research Work on Same Dataset

We have compared our model with other existing and latest models on fer2013 dataset. It is found that our layered approach has performed well on Elu activation function in comparison with the previous model. It has 68.11 accuracies on sixty epoch and 66.50 on a hundred epoch, which is shown in table. ReLu and Softplus models have also achieved significant accuracy result 67.99 and 67.09 on eighty-one and ninety-one epoch, respectively. Table 3 shows the comparative result with the existing model.

Result shows that our proposed layered model has achieved better results in comparison with the previous existing model on FER2013.

5 Conclusion and Future Work

Our deep layered approach has demonstrated a significant result on facial expression. Different variants are also examined based on activation function. Best, 68.11%,

Table 2 Precision, recall, $F1$-score and support-based comparison with DLF + Elu, DLF + ReLu and DLF + Softplus models

Model	Label	precision	Recall	$F1$-score	Support
DLF + Elu	0	0.569	0.556	0.562	498
	1	0.647	0.635	0.641	52
	2	0.581	0.479	0.525	545
	3	0.838	0.852	0.845	881
	4	0.529	0.549	0.539	588
	5	0.766	0.773	0.769	414
	6	0.620	0.687	0.652	611
Accuracy		–	–	0.665	3589
Macro avg.		0.650	0.647	0.648	3589
Weighted avg.		0.663	0.665	0.663	3589
DLF + softplus	0	0.604	0.488	0.540	498
	1	0.653	0.615	0.634	52
	2	0.589	0.486	0.533	545
	3	0.831	0.863	0.846	881
	4	0.517	0.590	0.551	588
	5	0.801	0.698	0.746	414
	6	0.564	0.684	0.618	611
Accuracy				0.656	3589
Macro avg.		0.651	0.632	0.638	3589
Weighted avg.		0.660	0.656	0.654	3589
DLF + Relu	0	0.593	0.536	0.563	498
	1	0.655	0.692	0.673	52
	2	0.550	0.505	0.526	545
	3	0.854	0.834	0.844	881
	4	0.521	0.580	0.549	588
	5	0.738	0.771	0.754	414
	6	0.589	0.614	0.601	611
Accuracy				0.654	3589
Macro avg.		0.643	0.647	0.644	3589
Weighted avg.		0.656	0.654	0.654	3589

accuracy achieved by DLF + Elu model on 60th epoch and showed the correct prediction of facial expression on FER2013 dataset. Elu and ReLu-based models have outperformed the others with 66.50% and 65.40% accuracy, respectively. Different quantitative and qualitative performance measures have supported our proposed

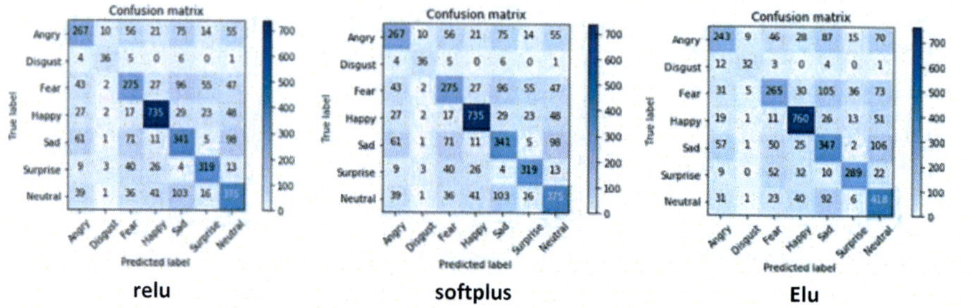

Fig. 6 Confusion matrix of proposed model on ReLu, Elu and Softplus

Table 3 Accuracy comparison with other existing model

Model/Research work	Accuracy
VGG + SVM [9]	66.31
GoogLeNet [10]	65.20
Ergen et al. [11]	57.10
Proposed layered approach on ReLu	67.99
Proposed layered approach on Elu	68.11
Proposed layered approach on Softplus	67.09

model. Our extended deep neural network approach with Elu, ReLu activation function has shown modest performance than GoogLeNet and VGG + SVM-based model.

Fusion of model is being used in the latest research era. FER accuracy can be increased with this method. This model may be fused with one or more than one model, and it can deliver some significant results. In future work, we will also try to combine two or more than two models and analyze its effect on facial expression recognition.

References

1. Pantic M, Rothkrantz L (2000a) Automatic analysis of facial expressions: the state of art. IEEE Trans Pattern Anal Mach Intell 22(12):1424–1445
2. Fasel JL (2003) Automatic facial expression analysis: a survey. Pattern Recogn 36:259–275
3. Pantic M, Rothkrantz L (2003) Toward an affect-sensitive multimodal human–computer interaction. In: Proceeding of the IEEE, vol 91, pp 1370–1390
4. Tian Y, Kanade T, Cohn J (2005) Handbook of face recognition. Springer (Chap. 11. Facial expression analysis)
5. Yacoob Y, Davis LS (1996) Recognizing human facial expression from long image sequences using optical flow. IEEE Trans Pattern Anal Mach Intell 18(6):636–642
6. Essa I, Pentland A (1997) Coding, analysis, interpretation, and recognition of facial expressions. IEEE Trans Pattern Anal Mach Intell 19(7):757–763

7. Lyons MJ, Budynek J, Akamatsu S (1999) Automatic classification of single facial images. IEEE Trans Pattern Anal Mach Intell 21(12):1357–1362
8. Donato G, Bartlett M, Hager J, Ekman P, Sejnowski T (1999) Classifying facial actions. IEEE Trans Pattern Anal Mach Intell 21(10):974–989
9. Pantic M, Rothkrantz L (2000b) Expert system for automatic analysis of facial expression. Image Vis Comput 18(11):881–905
10. Tian Y, Kanade T, Cohn J (2001) Recognizing action units for facial expression analysis. IEEE Trans Pattern Anal Mach Intell 23(2):97–115
11. Cohen I, Sebe N, Garg A, Chen L, Huang TS (2003) Facial expression recognition from video sequences: temporal and static modeling. Comput Vis Image Underst 91:160–187
12. Yin L, Loi J, Xiong W (2005) Facial expression representation and recognition based on texture augmentation and topographic masking. ACM Multimedia
13. Yeasin M, Bullot B, Sharma R (2004) From facial expression to level of interests: a spatio-temporal approach. IEEE conference on computer vision and pattern recognition (CVPR)
14. Burrows AM, Waller BM, Parr LA, Bonar CJ (2006) Muscles of facial expression in the chimpanzee (pan troglodytes): descriptive, comparative, and phylogenetic contexts. J Anat 208:153–167
15. Hoey J, Little JJ (2004) Value directed learning of gestures and facial displays. In: IEEE conference on computer vision and pattern recognition (CVPR)
16. Chang Y, Hu C, Turk M (2004) Probabilistic expression analysis on manifolds. In: IEEE conference on computer vision and pattern recognition (CVPR)
17. Hu J, Yu B, Yang Y, Feng B (2019) Towards facial de-expression and expression recognition in the wild. In: 2019 8th international conference on affective computing and intelligent interaction (ACII), Cambridge, United Kingdom, pp 157–163
18. Zhong L, Liu Q, Yang P, Huang J, Metaxas DN (2015) Learning multiscale active facial patches for expression analysis. IEEE Trans Cybern 45(8):1499–1510
19. Zeng Z, Pantic M, Roisman GI, Huang TS (2009) A survey of affect recognition methods: audio, visual, and spontaneous expressions. IEEE Trans Pattern Anal Mach Intell 31(1):39–58
20. Bengio, Courville A, Vincent P (2013) Representation learning: a review and new perspectives. IEEE Trans Pattern Anal Mach Intell 35(8):1798–1828
21. Sariyanidi, Gunes H, Cavallaro A (2017) Learning bases of activity for facial expression recognition. IEEE Trans Image Process 26(4):1965–1978
22. Yang H, Ciftci U, Yin L (2018) Facial expression recognition by de-expression residue learning. In: Proceedings of the IEEE conference on computer vision and pattern recognition, pp 2168–2177
23. Abdulrahman M, Eleyan A (2015) Facial expression recognition using Support Vector Machines. In: 2015 23nd signal processing and communications applications conference (SIU), Malatya, pp 276–279
24. Tian Y (2004) Evaluation of face resolution for expression analysis. In: CVPR workshop on face processing in video
25. Zhang Z, Lyons MJ, Schuster M, Akamatsu S (1998) Comparison between geometry-based and Gabor-wavelets-based facial expression recognition using multi-layer perceptron. In: IEEE international conference on automatic face and gesture recognition (FG)
26. Georgescu M-I, Ionescu RT, Popescu M (2018) Local learning with deep and handcrafted features for facial expression recognition. arXiv:1804.10892
27. Giannopoulos P, Perikos I, Hatzilygeroudis I (2018) Deep learning approaches for facial emotion recognition: a case study on FER-2013. In: Advances in hybridization of intelligent methods. Springer, Cham, pp 1–16
28. Tumen V, Soylemez OF, Ergen B (2017) Facial emotion recognition on a dataset using convolutional neural network. In: Proceedings of the artificial intelligence and data processing symposium, pp 1–5
29. Mase K (1991) Recognition of facial expression from optical flow. IEICE Trans Inf Syst E74-D(10):3474–3483

30. Zhao G, Pietikäinen M (2007) Dynamic texture recognition using local binary patterns with an application to facial expressions. IEEE Trans Pattern Anal Mach Intell 29(6):915–928
31. Walecki R, Rudovic O, Pavlovic V, Pantic M (2015) Variable-state latent conditional random-elds for facial expression recognition and action unit detection. In: Proceedings of 11th IEEE international conference on workshops automat. Face Gesture Recognit. (FG), May 2015, pp 1–8
32. Yang J, Zhang F, Chen B, Khan SU (2019) Facial expression recognition based on facial action unit. In: 2019 tenth international green and sustainable computing conference (IGSC), Alexandria, VA, USA, pp 1–6

Effective Background and Foreground Segmentation Using Unsupervised Frequency Domain Clustering

Shreya Kapoor, Manu S. Pillai, and Ashish Nagpal

Abstract Computer vision is the way in which the computer perceives a certain image. Background and foreground detection of an image are based on the concept of computer vision. Traditional approaches in background and foreground detection of an image imply clustering algorithms like K-means clustering, Gaussian mixture model to compute the result in the spatial domain. In spatial domain, we take into account the pixels of an image to classify them as background or foreground. In this paper, we have reviewed the process already been done in spatial domain, then we study some of the state-of-the-art background detection techniques in digital image processing and propose a novel approach to shift the domain of input images to frequency while applying the same algorithms for detection. We have used the concept in which we take into consideration the frequency rather than pixels of an image. The concept of fast Fourier transform (FFT) has been used to determine the frequency. With this solution concept, we aim at reducing the variance in input image by smoothing out the frequency domain image and experimentally demonstrate that the transition into the frequency domain outperforms the majority of techniques employed in spatial domain for background detection.

Keywords Frequency processing · Background segmentation · Fast fourier transform · Clustering

---

S. Kapoor (✉) · A. Nagpal
Dr. Akhilesh Das Gupta Institute of Technology and Management, New Delhi, India
e-mail: shreyakapoor98@gmail.com

A. Nagpal
e-mail: ashishnagpal2498@gmail.com

M. S. Pillai
Bharati Vidyapeeth's College of Engineering, New Delhi, India
e-mail: manupillai308@gmail.com

A. Abraham et al. (eds.), *Proceedings of 3rd International Conference on Computing Informatics and Networks*, Lecture Notes in Networks and Systems 167,
https://doi.org/10.1007/978-981-15-9712-1_8

1 Introduction

Background detection is considered to be an important computer vision concept. This is a mechanism where the background and foreground of an image are identified and separated. An image possesses a main object and its surrounding. When the object is extracted from the image and it is distinguished from the background part, it helps to analyse the image in a close perspective [1]. This functionality is achieved by using image segmentation. During the image segmentation process, the image is split into two sections which are background and foreground. This functionality can be achieved in many different ways but the most popular amongst them is by using clustering algorithms like K-means and Gaussian mixture model. As the name suggests, the clustering algorithm clusters or groups the similar kind of data. It is an unsupervised algorithm where the information regarding the data is not known instead the data is just segmented into different groups. Each group is termed as a class. So we basically classify the data points into groups of classes in which all the classes show distinguishing features. This approach has a varied application in many fields nowadays. It is used in the medical field for studying the tissues or any particular organ, video surveillance to detect the moving object, self-driving cars and monitoring the traffic of an area [2].

In this paper, our work is aimed at improving the accuracy of background segmentation in an image using conventional clustering algorithms. After studying deep about the topic, we have observed that the conventional clustering algorithms are applied in the spatial domain. In spatial domain, the pixels of the image are considered and the clustering algorithms are applied on the pixels to classify them as background and foreground [3]. After carefully observing, we have seen that this approach is not able to segment the background and foreground accurately. Hence, we have come up with an approach in which we specifically transform an image from spatial domain to frequency domain and perform unsupervised clustering. It has been observed that the domain shift has significantly improved the results to a great extent. In frequency domain, the segmentation of background and foreground is seen to be very precise and accurate. We experimentally show the advantage of such a domain shift increases the accuracy of conventional clustering algorithms like K-means and Gaussian mixture model in the background segmentation task. K-means and Gaussian mixture model are popular clustering algorithms which are used to cluster similar groups of data. All the data points when given as input to the clustering algorithm are divided amongst classes of similar kind. In a similar way, the background and foreground of an image are distinguished.

2 Related Work

A lot of work has already been done in the field of background or object detection. It is observed that they are done using the clustering algorithms in the spatial domain.

The most widely used clustering algorithm seem to be used is the K-means clustering algorithm which is used for detecting the background in which we find the centroid of the cluster and surround the data points around it by finding the mean of all the points and thereby finding the exact centroid [4]. Second is the Gaussian mixture model which is also widely used in which data points are obtained from Gaussian distributions [5].

There are many research works done on this topic in order to improve the existing algorithm as well as introducing new methods for background detection to acquire better results. Wagstaff et al. [6] proposed a method in which they improvised the existing K-means algorithm with the help of pair wise constraints in which they considered must link constraint and cannot link constraint to improve the accuracy. They have applied their methodology on six different data sets.

Dhanachandra et al. [7] enhanced image quality by using contrast stretching process followed by subtractive clustering algorithm and K-means clustering, and they have used this to improve the results in the medical field.

Sinha and Mareboyana [8] used a combination of simplified mean shift filter and K-means clustering for background extraction on a set of video data and showed that using mean shift will lead to 28% gain in time.

Vacavant et al. [9] introduced a method to detect the moving objects efficiently by taking into consideration the environmental condition, illumination conditions, etc.

Yedla et al. [10] have come up with an approach for finding better initial centroids in K-means without the use of additional inputs such as threshold values. They have shown an effective method of segregating data points with reduced complexity.

GaliC and LonCariC [11] showed the detection of moving object in an image which is a very important factor. They have worked on the two-dimensional feature vector. Their methodology aimed at segmentation of medical image sequences.

3 Methodology

In this paper, the existing methods of evaluation of the background are shifted in the frequency domain to enhance precision. The principal concept of this method is to improve the accuracy and enhance the existing methods. Conventional clustering algorithms are applied in the same way but on the sets of frequencies of the images instead of the pixel values. We have used K-means and Gaussian mixture model to depict the results.

K-means clustering algorithm is a very reliable approach to evaluate it. It is an unsupervised algorithm in which all the points in the cluster belong to a specific centre. The value of K is decided, and it depicts the number of clusters. This algorithm is divided into two parts. Firstly, it finds the number of centroids or the value of K and after that it assigns each data point to a centroid. Then, once all the data points are assigned, the new value of centroid is calculated again and based upon that the new distance between the centroid and the data points is computed. Equation 1 shows the evaluation of K-means [4, 12].

$$J(V) = \sum_{i=1}^{c} \sum_{j=1}^{c_i} \left(\left| \left| x_i - v_j \right| \right| \right)^2 \tag{1}$$

A Gaussian mixture model is said to be a model of probability, which indicates that all data points are obtained from a combination of a limited number of Gaussian distributions and unaccountable parameters [5, 13]. Mixture models can be viewed as generalizing K-means to incorporate knowledge on the covariance structure of data and the centres of the latent Gaussians. Gaussian distribution leverages the behaviour of Gaussian mixture model [14].

Equation 2 shows the Gaussian density function.

$$N(\mathrm{X}|\mu, \Sigma) = \frac{1}{(2\pi)^{D/2} |\Sigma|^{1/2}} \exp\left(-\frac{1}{2} (\mathrm{X} - \mu)^T \Sigma^{-1} (\mathrm{X} - \mu \right) \tag{2}$$

We have carefully compiled some images for comparison of segmentation of background and foreground of the image. K-means and Gaussian mixture model are applied on the set of images. We aim at depicting the background and the foreground differently. The background will be depicted in black colour, and the foreground will be depicted in white colour. The images are being processed through the following steps in order to get the desired results.

In spatial domain, we have converted the images from RGB to greyscale so that the mapping of these images while shifting domains is easier [15]. Then, these images are fed into clustering algorithms. Now we used K-means and Gaussian mixture model for comparison, and these clustering algorithms are trained using two cluster centres so that the resultant image gets divided into only two classes which are background and foreground. Figure 1 shows the flow chart of the algorithm that we applied in the spatial domain.

In frequency domain, the image is transformed from RGB to greyscale [15] and then the frequency of the greyscale image is evaluated by using fast Fourier transform.

Fast Fourier transform is a way in which the discrete Fourier transform of the image is computed [16]. It comes into play in order to reduce the complexity of finding the Fourier transform of the image and thereby increasing the speed. As the name says, fast Fourier transform evaluates the frequency much faster than discrete Fourier transform [17]. Equation 3 gives the Fourier transform of a discrete function.

$$F(u, v) = \sum_{x=0}^{m-1} \sum_{y=0}^{n-1} f(x, y) \mathrm{e}^{-j2\pi \left(\frac{ux}{m} + \frac{vy}{n} \right)} \tag{3}$$

Equations 4, 5 and 6 represent the phase and spectrum of the image.

$$F(u, v) = R(u, v) + jI(u, v) = |F(u, v)| \mathrm{e}^{j\varphi(u,v)} \tag{4}$$

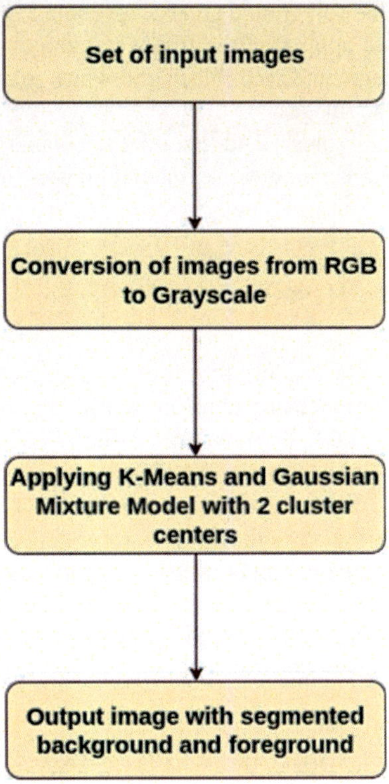

Fig. 1 Algorithm applied in spatial domain

$$|F(u,v)| = \left[R^2(u,v) + I^2(u,v)\right]^{1/2} \tag{5}$$

$$\varphi(u,v) = \arctan\left[\frac{I(u,v)}{R(u,v)}\right] \tag{6}$$

Then, by taking the frequency of the images as input, the clustering algorithms are applied in the same way like the spatial domain. The resultant is actually in frequency domain so we cannot directly mask it with the original image as we did in spatial domain because right now y (predicted class labels) is in the frequency domain. There is a need of conversion of image into spatial domain so that it can be used to mask our input image.

In an FFT image, the phase [18] of the FFT contains the structure of the image and the power spectrum [19] contains the intensity values, so we first mask the frequency using the class label images with the help of element wise multiplication which is known as Hadamard product [20]. Then, the magnitude of the resultant image is evaluated. Next step is to calculate the phase of the original image. After which this calculated phase and magnitude are combined together and the inverse Fourier transform [21] is applied in order to get the masked image which has been classified into foreground and background. Figure 2 shows the flow chart of the algorithm that we applied in the frequency domain. Output produced will be the image which is segmented into black and white colours. Figure 3 shows the frequency spectrum of the images.

Tables 1 and 2 show the comparison between the segmentation done in spatial domain and the segmentation done in the frequency domain.

4 Results

Table 1 shows the result of background detection in spatial domain and frequency domain using K-means clustering.

Table 2 shows the result of background detection in frequency domain using Gaussian mixture model.

It can be seen from the comparison of results that in spatial domain the clustering algorithms fail to accurately segment the background and foreground and the segmentation between the background and foreground is not smooth. But the transition to frequency domain resolved this issue by giving distinguishable segmentation, better and more precise results.

5 Conclusion and Future Work

We have successfully demonstrated that the transition to frequency domain in background detection significantly increases performance of traditional methods applied in spatial domain. Given the confidence level of the results obtained, there still lacks an extensive performance comparison of methods employed in background detection. Digital image processing has witnessed plethora of pre-processing and filtering methods. For our future work, we will be focusing on multiple pre-processing methods with more complex methodologies applied in frequency domain.

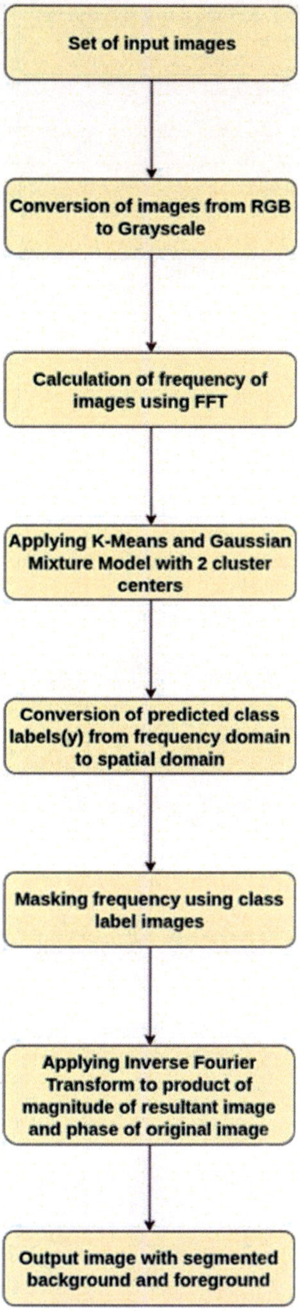

Fig. 2 Algorithm applied in frequency domain

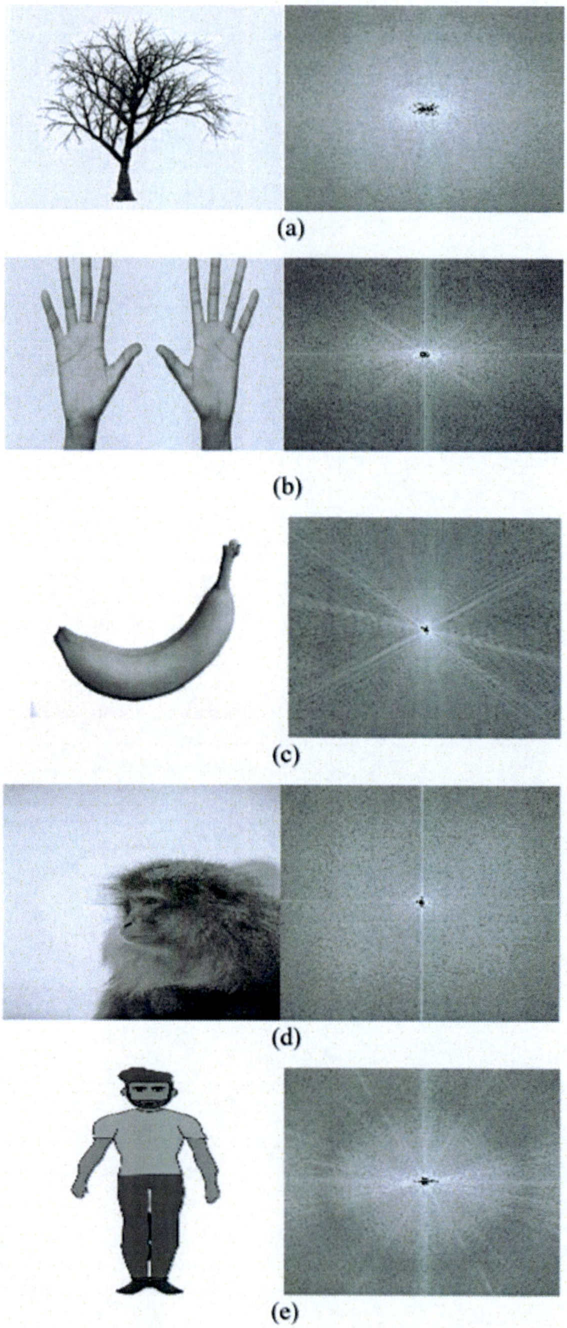

Fig. 3 Frequency spectrum of the images

Table 1 Comparison of background detection in spatial domain and frequency domain using K-means

Input image	Output image in spatial domain	Output image in frequency domain
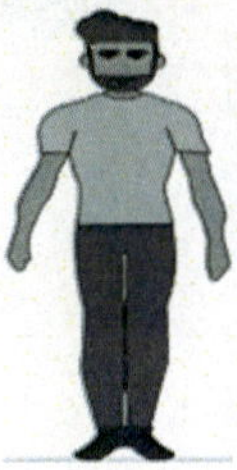	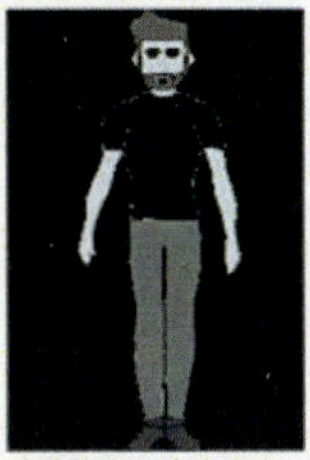	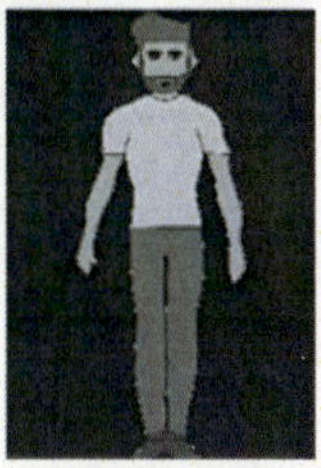
	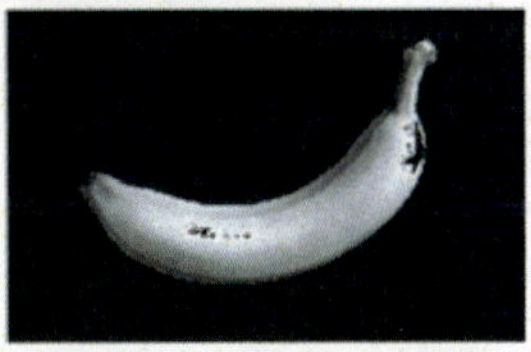	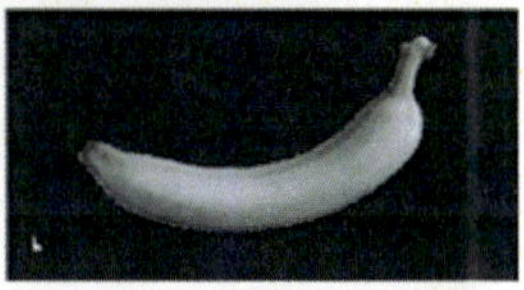

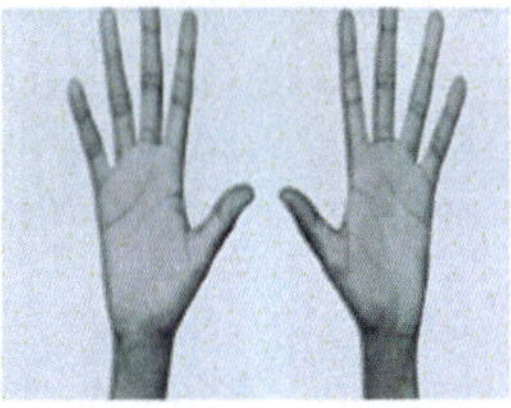	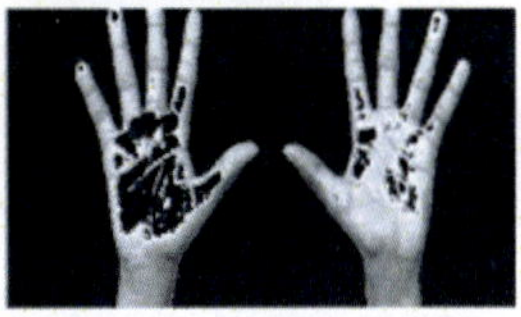	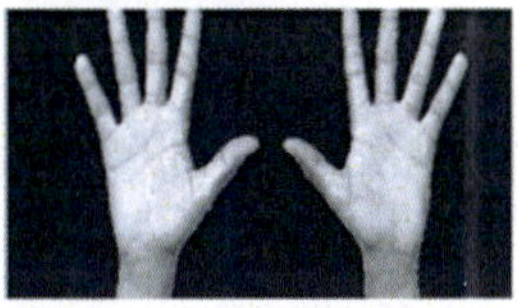

Table 2 Comparison of background detection in spatial domain and frequency domain using Gaussian mixture model

Input image	Output image in spatial domain	Output image in frequency domain
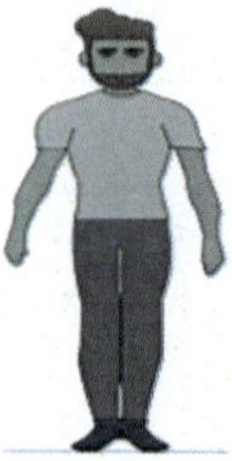	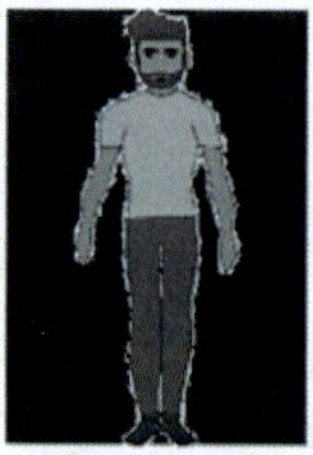	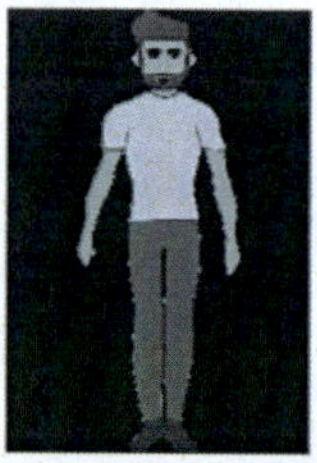
	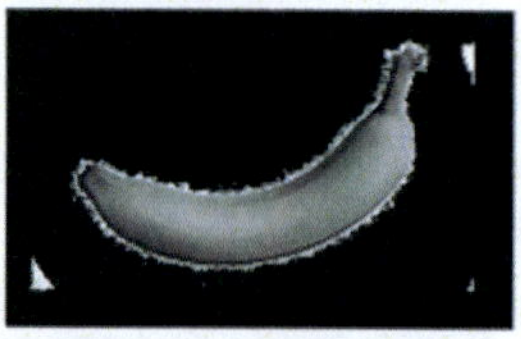	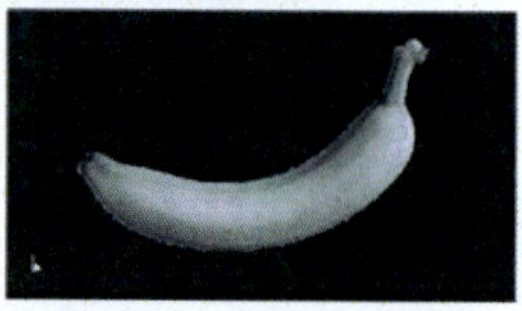

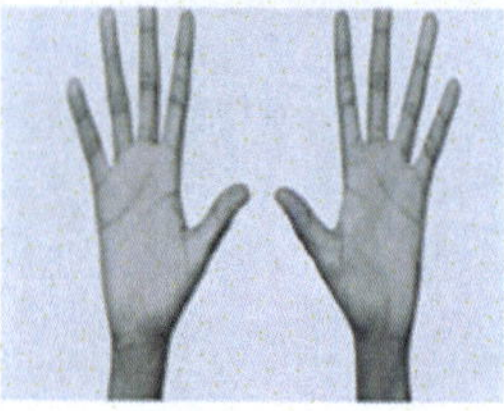	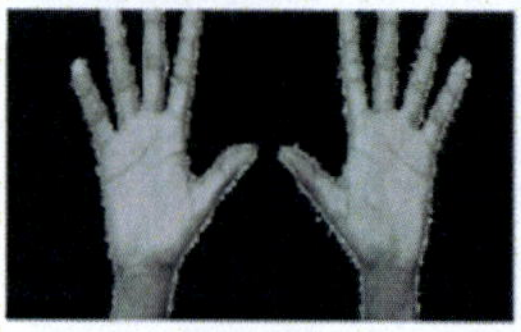	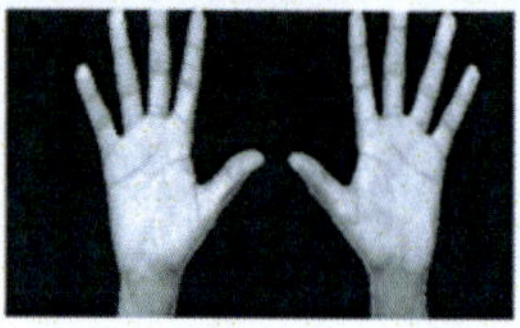

References

1. Zhu W, Liang S, Wei Y, Sun J (2014) Saliency optimization from robust background detection. In: Proceedings of the IEEE conference on computer vision and pattern recognition, pp 2814–2821
2. Chowdhury A, Chong UP (2012, Apr) Real time shadow removal with K-means clustering and RGB color model. In: Int J Multimedia Ubiquitous Eng IJMUE 7(2)
3. Elhabian SY, El-Sayed KM, Ahmed SH (2008) Moving object detection in spatial domain using background removal techniques-state-of-art. Recent Patents Comput Sci 1(1):32–54
4. Khan SS, Ahmad A (2004) Cluster center initialization algorithm for K-means clustering. Pattern Recogn Lett 25(11):1293–1302
5. Farnoush R, Zar PB (2008) Image segmentation using Gaussian mixture model
6. Wagstaff K, Cardie C, Rogers S, Schrödl S (2001, June) Constrained k-means clustering with background knowledge. ICML 1:577–584
7. Dhanachandra N, Manglem K, Chanu YJ (2015) Image segmentation using K-means clustering algorithm and subtractive clustering algorithm. Proc Comput Sci 54:764–771
8. Sinha S, Mareboyana M (2014) Video segmentation into background and foreground using simplified mean shift filter and K-means clustering. ASEE. University of Bridgeport
9. Vacavant A, Chateau T, Wilhelm A, Lequièvre L (2012, Nov) A benchmark dataset for outdoor foreground/background extraction. In: Asian conference on computer vision. Springer, Berlin, pp 291–300
10. Yedla M, Pathakota SR, Srinivasa TM (2010) Enhancing K-means clustering algorithm with improved initial center. Int J Comput Sci Inf Technol 1(2):121–125
11. GaliC S, LonCariC (2000, June) Spatio-temporal image segmentation using optical flow and clustering algorithm. In: IWISPA 2000. Proceedings of the first international workshop on image and signal processing and analysis. In conjunction with 22nd international conference on information technology interfaces. IEEE, pp 63–68
12. Zahra S, Ghazanfar MA, Khalid A, Azam MA, Naeem U, Prugel-Bennett A (2015) Novel centroid selection approaches for K means-clustering based recommender systems. Inf Sci 320:156–189
13. Permuter H, Francos J, Jermyn IH (2003, April) Gaussian mixture models of texture and colour for image database retrieval. In: 2003 IEEE international conference on acoustics, speech, and signal processing, 2003. Proceedings. ICASSP'03, vol 3. IEEE, pp III-569
14. Goodman NR (1963) Statistical analysis based on a certain multivariate complex Gaussian distribution (an introduction). Ann Math Stat 34(1):152–177
15. Padmavathi K, Thangadurai K (2016) Implementation of RGB and grayscale images in plant leaves disease detection–comparative study. Indian J Sci Technol 9(6):1–6
16. Liu QH, Nguyen N (1998) An accurate algorithm for nonuniform fast Fourier transforms (NUFFT's). IEEE Microw Guid Wave Lett 8(1):18–20
17. Weinstein S, Ebert P (1971) Data transmission by frequency-division multiplexing using the discrete Fourier transform. IEEE Trans Commun Technol 19(5):628–634
18. Vakoc BJ, Yun SH, De Boer JF, Tearney GJ, Bouma BE (2005) Phase-resolved optical frequency domain imaging. Opt Express 13(14):5483–5493
19. Akselrod S, Gordon D, Ubel FA, Shannon DC, Berger AC, Cohen RJ (1981) Power spectrum analysis of heart rate fluctuation: a quantitative probe of beat-to-beat cardiovascular control. Science 213(4504):220–222
20. Horn RA (1990, May) The hadamard product. Proc Symp Appl Math 40:87–169
21. Guan S, Marshall AG (1996) Stored waveform inverse Fourier transform (SWIFT) ion excitation in trapped-ion mass spectometry: theory and applications. Int J Mass Spectrom Ion Processes 157:5–37

Computer File Signature Analysis Through Hexadecimal Editor Software

Shshank Sourabh and Monika Chauhan

Abstract Computer file signature is a unique hexadecimal value written to a file header that acts as an identifying feature to identify a file type, as criminals are becoming more and more aware of digitalization these days; they tend to hide sensitive file information within the computer itself without destroying it using tricks to work for them. One of those tricks is to alter a file extension. Accordingly, a more adequate method of data analysis, known as file signature analysis, is needed to counter these measures, and it is done using hexadecimal editor software (HxD). A manipulated computer file can be opened in this software to get its file signature which can be further searched in online database (*File Signatures Database*) detecting the correct format (extension) of the file rendering criminals plotting to fail and by bringing the hidden information into light.

Keywords File signature · Hexadecimal editor software · Files signatures database · File toolkit · Extension · HxD

Abbreviations

FTK	File Toolkit
PDF	Portable Document Format
JPG	Joint Photographic Expert Group
PNG	Portable Network Graphics
GIF	Graphic Interchange Format
TIF	Tagged Image File
FLAC	Free Lossless Audio File
WAV	Waveform Audio File

S. Sourabh · M. Chauhan (✉)
Division of Forensic Science, School of Basic and Applied Sciences, Galgotias University, Greater Noida, UP 201310, India
e-mail: monika.chauhan@galgotiasuniversity.edu.in

A. Abraham et al. (eds.), *Proceedings of 3rd International Conference on Computing Informatics and Networks*, Lecture Notes in Networks and Systems 167,
https://doi.org/10.1007/978-981-15-9712-1_9

FLV	Flash Video
WMV	Windows Media Video
HTML	Hyper Text Markup Language

1 Introduction

Computer-based forensic study is often referred to as computer forensics, which is used to collect and analyze data as an evidence for court jurisdiction, where data is identified, extracted, documented and preserved as an evidence that includes physically copying the data (information) present in computer's hard disk by creating copy of the data, loading image to an empty or formatted disk, retrieval of data of evidential value and preparing a readable and easily understandable form of evidence for the court [1].

A computer file analysis method is used, when suspicious files are encountered on suspect's computer where file header and extension are compared to a known database of headers and extensions to verify the file in question with respect to any attempt made in order to tamper the file to hide sensitive information. Thus, in order to overcome these circumstances, file signature analysis is done.

A hexadecimal editor software works as an interface to visualize the file signature represented in the header of a file. Each computer file has their own unique file signature (which contains total of 16 bases ranging from 'A' to 'F' in alphabet and '0' to '9' in numeric value) [2], which acts as an identifiable characteristic whenever a suspicious file is encountered.

When a file's extension is changed, e.g., *from secret.docx to secret.pdf*, they do not influence any change in its hexadecimal value encoded to its header because the file was originally created in Microsoft Office Word not in PDF. So, upon opening the file a dialog box appears showing corrupt file as the extension of the file and *file signature* [3] of the file mismatches. Thus, by using hexadecimal editor software such as *HxD*, [4] file signature of the manipulated file (*secret.pdf*) can be opened deciphering the file signature of Microsoft Office Word instead of PDF and the correct extension based on the deciphered file signature by the can be obtained from online database—*file signatures database* [5]. Hence, by changing its extension back to *secret.docx,* intentionally hidden information by the suspect can be rendered.

The study proposed in the later sections of this paper emphasizes the analysis of every file present on the computer device (document, image, audio, video, etc.) through the software tool *HxD*, unlike other proposed methods that are useful in analyzing only selected files present on the computer. This software works flawlessly along with its diverse nature of analyzing different types of computer files and can effectively analyze large files without lagging.

2 Related Work

File analysis and format-specific approaches are two of the most abundant approaches that are used for the detection of suspicious files through software [6].

To carry out file analysis software such as *Forensics Toolkit (FTK)* [7] and *EnCase* [8] are used to analyze the storage media. *Forensic Toolkit* is used to extract file for determining the data type which leads to the extraction of the text. Thus, for law enforcement *EnCase* is very likely used.

To find particular application data or data type, format analysis is done using software like *Jhead* [9] which is used for JPEG image data extraction, where one can find the date and time to know when the image was captured or resolution of the image or the shutter speed. Other software is *DataLifters* [10] which is used for multitude type file extraction. The fact remains that both of these softwares are limited to analyze selected file type making these approaches non-robust.

An automated scheme for the analysis of digital files present on the computer's hard drive including image and initially tampered files was given in 2007 by Haggerty and Taylor [11], where this scheme identified the digital files of interest (multimedia) and then compares the acquired data to the file signature of the files to make sure if any tampered file is present on the host computer. The proposed scheme made it very easy to detect such files which have been tampered in an automated manner.

M. Yip in 2008 worked on computer file analysis to detect the information which was intentionally hidden by the suspect with regard to changing its file extension or deleting the file, and the process was based on detecting the signature of the files, for which he used HexEdit software [12].

3 Structure of a File

File structure is data organization that allows applications to read, write and edit data. The file structure consists of the name of the file/file name, the header/footer of the file and the content of the file [13].

Name of the file/file name—A complete title of a file and its extension is commonly known as file name, e.g., *secret.docx* is a complete filename. In some case, filename only consists of first word; *secret* and its extension *.docx* are not visible as it is attached to the property of the file making it a document file [14].

Header/footer of a file—The region where bookkeeping information is kept is called as a header of a file; it is located at the beginning of the file. It contains the information of when the file was created or updated and the size of the file. File footer is similar to file header but is often used at the offset of the file. File signature is contained in both header and footer of a file [15].

These files are opened in a software called hexadecimal editor software (HxD), from Fig. 1.; the highlighted part depicts the file signature of the word file and on the decoded text section it shows *XML* as content type of the file, proving it to be

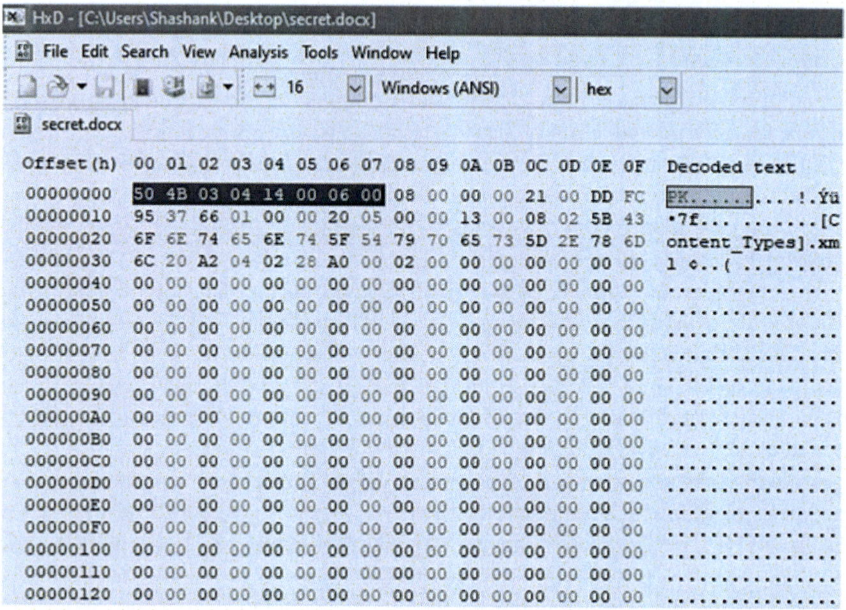

Fig. 1 File signature representation of secret.docx file

a document file. In Fig. 2, another document file is opened which shows same file signature representation. Hence, by comparing both Figs. 1 and 2, it can be clearly stated that this is the correct string of file signature associated with word file.

File content—File content may be defined as the information stored in a particular file, be it in the form of written statement, audio, image or video, respectively [16].

4 Common File Formats [17]

- *Microsoft Office Word*—'*.doc*' (Microsoft Office 97–2003), '*.docx*' (Microsoft Office 2007)
- *Microsoft Office Excel*—'*.xls*' (Microsoft Office 97–2003), '*.xlsx*' (Microsoft Office 2007)
- *Microsoft Office PowerPoint*—'*.ppt*' (Microsoft Office 97–2003), '*.pptx*' (Microsoft Office 2007)
- *Adobe pdf*—'*.pdf*'
- *Image file*—'*.jpg*', '*.jpeg*', '*.png*', '*.gif*', '*.tif*'
- *Audio file*—'*.mp3*', '*.flac*', '*.wav*', '*.m4a*'
- *Video file*—'*.mp4*', '*.avi*', '*.flv*', '*.wmv*', '*.mov*'.

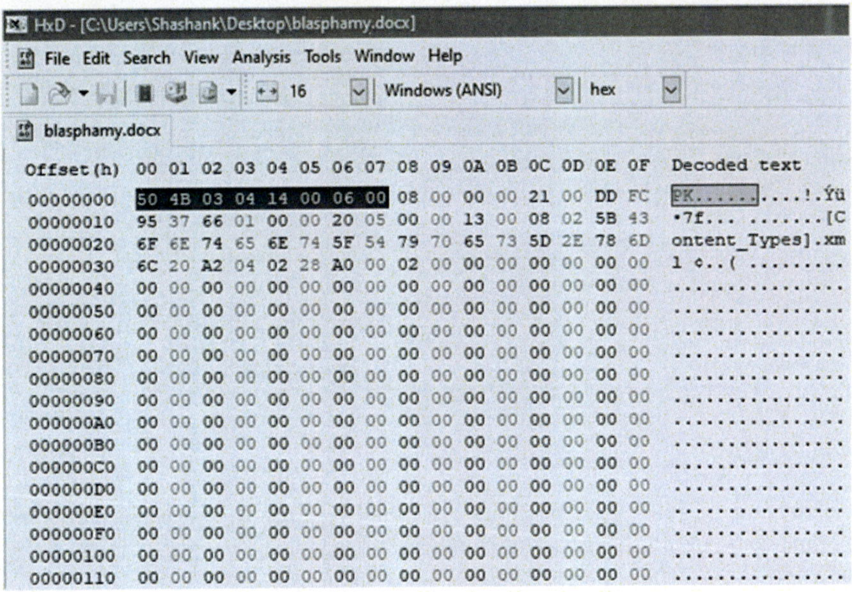

Fig. 2 File signature representation of blasphemy.docx file

4.1 File Signatures of Common File Formats [18]

- Document file
- Image file
- Audio file

Table 1 List of file signature of document file

File format	File signature (hexadecimal)
Adobe PDF	25 50 44 46
Microsoft Office 97–2003 documents	D0 CF 11 E0 A1 B1 1A E1
Microsoft Office 2007 documents	50 4B 03 04 14 00 06 00

Table 2 List of file signature of image file

File format	File signature (hexadecimal)
JPG and JPEG	FF D8 FF E0
PNG	89 50 4E 47 0D 0A 1A 0A
GIF	47 49 46 38
TIF	49 20 49

Table 3 List of file signature of audio file

File format	File signature (hexadecimal)
Mp3	49 44 33
FLAC	66 4C 61 43 00 00 00 22
WAV	52 49 46 46
M4A	00 00 00 20 66 74 79 70 4D 34 41

Table 4 List of file signature of video file

File format	File signature (hexadecimal)
Mp4	6D 6D 70 34
FLV	46 4C 56
WMV	30 26 B2 75 8E 66 CF 11
MOV	6D 6F 6F 76

- Video file

Therefore, by going through Tables 1, 2, 3 and 4, respectively, we can see the list of some common file formats along with their file signatures which are generally altered by the criminals to hide information useful to them.

5 Data Manipulation by Changing Filename and Extension

Some information is too valuable for criminals that they cannot get rid of them by simply deleting those files, thus they tend to change its filename along with its extension so that others cannot open those files and consider them to be useless and only the criminal can retrieve the original information from those files.

The most commonly used method is to change the filename and extension of the file, but it proves to be very useful in deceiving people to believe that the file is of no use, keeping this fact in mind let us see how filename and its extension could be changed:

- Right click on the file (*secret.docx*) and then click on *'rename'*.
  Figure 3 shows the process of renaming a file and changing its extension.
- Change its name from *secret.docx* to *apple.png*, a dialog box will appear asking for permission, then click on *'yes'* (grating permission).
  Figure 4 shows a dialog box asking permission to change the change and the extension of the file (*secret.docx*).
- You will see that the name of the file along with its extension has changed.
  Going through above steps, one can easily manipulate the file (*apple.png*), and when the file is opened by double clicking, the application in which it is opened shows corrupt file or a popup appears showing the same.
  Figure 5 displays the screen of Windows Photo Viewer in which the altered file (*apple.png*) is opened, which shows corrupt file due to the mismatch between the extension and file signature of the file.

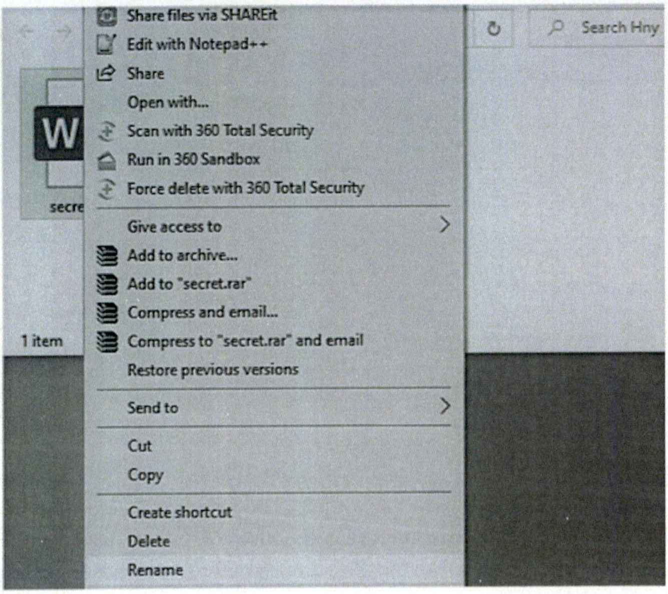

Fig. 3 Renaming file (secret.docx)

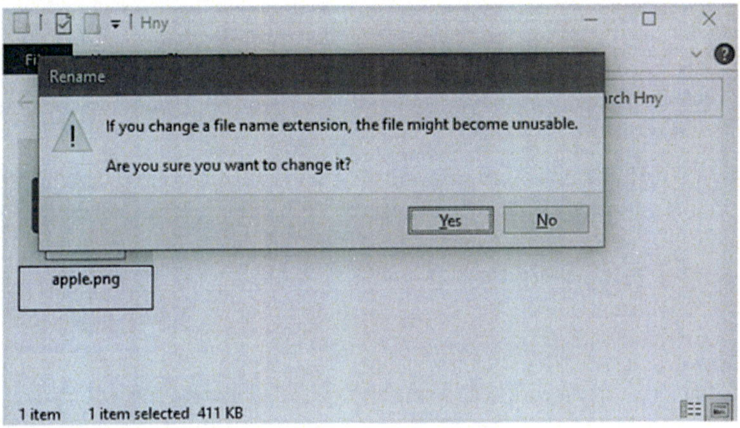

Fig. 4 Granting permission to change file name and extension

6 Methods and Methodology

As it is evident from the previous sections, a file can be easily manipulated by changing its file name and extension. Hence, by conducting file signature analysis, one can detect the extension of the tampered file, and if detected further process is carried out to decipher the intentionally hidden information.

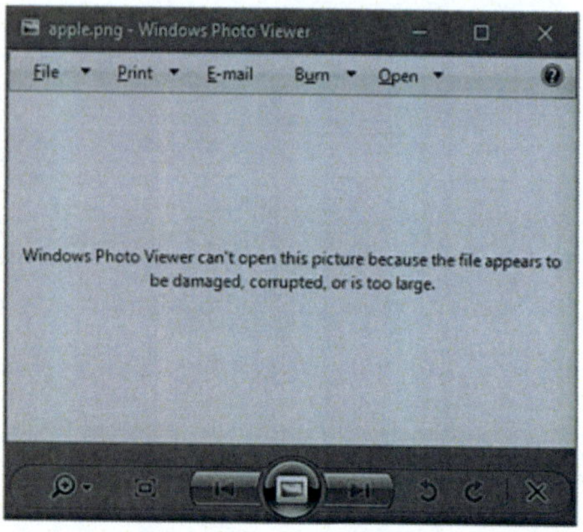

Fig. 5 File cannot be opened (displaying corrupt file)

Here, file signature analysis is done by using software—HxD. The software tool (HxD) used in analysis renders file signatures (hexadecimal value) of the original platform upon which the file was first created. Once the file signatures are obtained, they are searched on an online database (file signatures database) to find out the correct extension of the tampered file.

6.1 Detailed process of data retrieval using HxD

- Open suspected file in hexadecimal editor software (HxD) and upon opening the file its file signature is obtained.

In Fig. 6, file signature of the opened file (*apple.png*) has been deciphered, where the decoded text clearly shows that this file does not belong to an image file having extension as '*.png*'. If it belonged to an image file, then the decoded text would have shown content type as PNG. File signature of this file is highlighted on the left side which is the next searches in the online database to get the correct extension of the file.

- Note down the file signature and search it on online database called *file signature database* to get the extension of the file.

In Fig. 7, it can be seen that the file extension belongs to *Microsoft Office Document* not to the image file into which it was manipulated by the suspect.

- Rename/change the file extension from *apple.png* to *apple.docx* to decipher the hidden information.

In Fig. 8, file extension of the altered file (*apple.png*) has been changed (*apple.docx*) to decipher the information intentionally hidden by the suspect after the information gathered from online database (*file signatures database*).

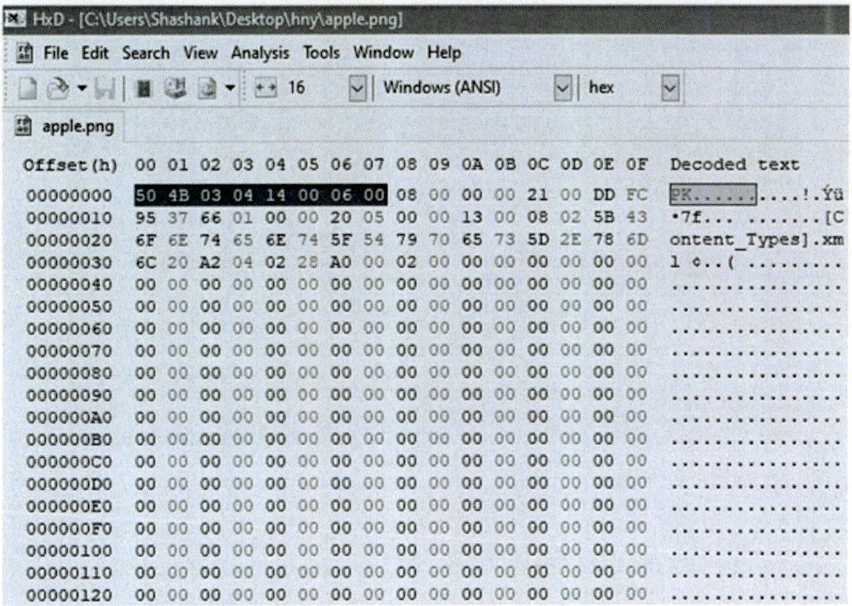

Fig. 6 File signature of apple.png

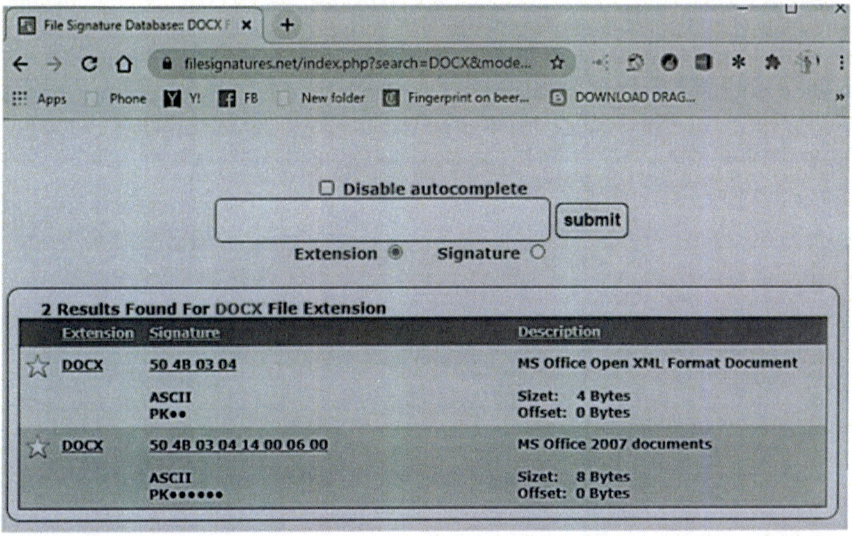

Fig. 7 File extension shown in online database (file signatures database)

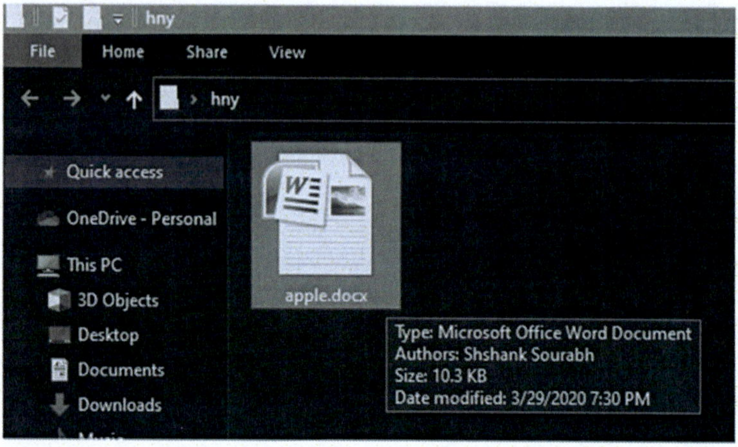

Fig. 8 File extension changed from apple.png to apple.docx

- Double click on the file to decipher the information hidden by the suspect.

In Fig. 9, the information contained in the manipulated file has been deciphered as 'Nothing is true Everything is permitted'.

Hence, it is clear that, by conducting file signature analysis intentionally manipulated files can be deciphered with ease. Similarly, by using above depicted steps, any altered file type (document, image, audio and video) can be examined.

There are several other software tools available for this analysis, one of the tool is '*HexEdit*' [11]. This tool is also used to visualize hexadecimal value of a file but in comparison with '*HxD*' it is slower and does not work efficiently with larger file.

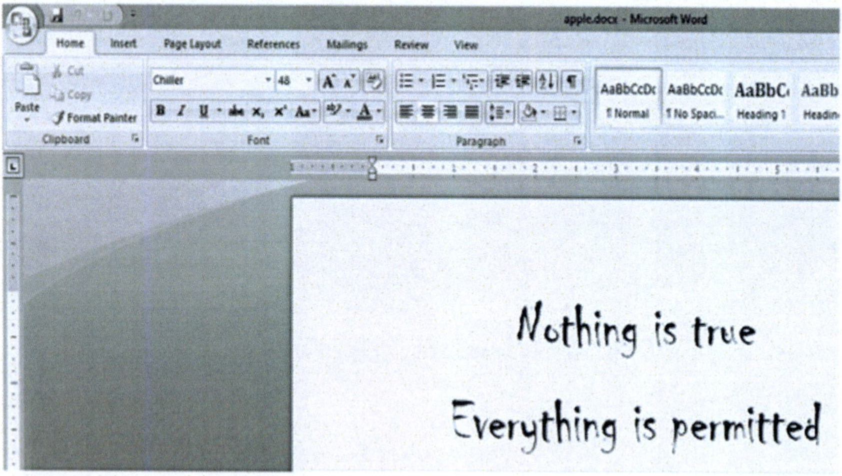

Fig. 9 Deciphered information

When working on a larger file, *HexEdit* crashes several times and it lags at multiple intervals.

7 Conclusion and Future Aspects

Computer file signature analysis involves the analysis of hexadecimal values encoded in the header and footer of a file, whether the file is document or media (audio and video). There are numerous software tools available which are able to analyze the file but they are not robust as they usually stick to identify single or selected file types only, as mentioned (Sect. 2: related work).

This paper presented HxD (file signature analysis software), an approach to identify every existing file signature at high rate of speed without crashing or lagging while working with large files with plenty of features at its disposal which can reduce the time of work unlike other available softwares.

Further work emphasis mainly more efficient handling of large data files, source code formatted output like HTML, file splitter/joiner and graphical statistics.

Acknowledgements I would like to express my heartfelt gratitude to my Head of the Department and Dean of School of Basic and Applied Sciences, Galgotias University, Greater Noida, for giving me with an auspicious opportunity to begin a project of my own. Their benevolent and supporting nature has always guided me toward the successful completion of this paper.

References

1. Carrier B (2005) File system forensic analysis. Addison-Wesley
2. Koblitz N, Menezes AJ (2007) Another look at provable security. J Crypt
3. Mohay G, Anderson A, Collie B, De Vel OM, Kemmish R (2003) Computer and intrusion forensics. Artech House, MA, USA
4. HxD Computer Forensic Software Homepage. https://www.mh_nexus.de/en/hxd. Last accessed 21 Feb 2020
5. File Signature Database Homepage. https://www.filesignatures.net. Last accessed 13 Apr 2020
6. Burr WE Cryptographic hash standards where do we go from here? In: IEEE security and privacy, pp 88–91
7. The Forensics Toolkit Homepage. https://www.accesssdata.com. Last accessed 12 Feb 2020
8. Guidance Software Encase Homepage. https://www.guidancesoftware.com. Last accessed 21 Feb 2020
9. Jhead Homepage. https://www.sentex.net/~mwandle/jhead/. Last accessed 21 Feb 2020
10. DataLifter, Computer Forensics Software Homepage. https://www.datalifter.com. Last accessed 21 Feb 2020
11. Haggerty TT (2007) FORSIGS-New approaches for security, privacy and trust in complex environments. In: Proceedings of the IFIP TC-11 22nd international information security conference, Sandton, South Africa, pp 1–12
12. Yip M (2008) Signature Analysis and computer forensics. School of Computer Science, University of Birmingham, UK, pp 1–11

13. Karresand M, Shahmehri N (2006) File type identification of data fragments by their binary structure. In: Proceedings of the 2006 IEEE workshop on information assurance, US Military Academy, West Point NY, 21–23 June 2006
14. Li X, Seberry J (2003) Forensic computing. In: Proceedings of INDOCRYPT, New Delhi, India, LNCS 2904, Springer, pp 18–35, 8–10 Dec 2003
15. Ying Z, Robertazzi TG (2014) Signature searching in a networked collection of files. In: IEEE Trans Parallel Distrib Syst 25:1339–1348
16. McClelland D, Marturana F (2014) A digital forensics triage methodology based on feature manipulation techniques. In: Proceedings of the international conference on communications workshops, Sydney, Australia, pp 676–681
17. Richard GG, Roussev V (2005) Scalpel: a frugal, high performance file carver. In: Proceedings of digital forensic research workshop, New Orleans, USA, 17–19 Aug 2005
18. GCK's File Signatures Table Homepage. https://graykessler.net/library/file_sigs.html. Last accessed 16 Apr 2020

A Systematic Literature Review of Automated Software Testing Tool

Lalji Prasad, Rashmi Yadav, and Niti Vore

Abstract Automated software testing has proven its value for software development increasingly over the past few years. Software testing is an important phase in the entire software development process. There are various automated software testing tools are available today, which are used for testing various software applications whether it is desktop-based, mobile application, or a Web-based application. Evaluating a software testing tool is rather a subjective task, depending on the evaluator's opinions rather than based on the objective approach. For this purpose, we have studied research papers, articles, journals, books, conference papers, few Web sites, etc., related with the study of software testing tools based on which we performed a survey of various automated software testing tools, i.e., Selenium, Watir, QTP, TestComplete, WinRunner, LoadRunner, SilkTest, Apache Jmeter, Wapt, Tellurium, Web Load, NeoLoad, LoadUI, Appvance, rational performance tester, SahiPro, Telerik Test Studio, Ranorex, Storm, Soap UI, TestNG, FitNese, Xebium, etc. The purpose of this research work is to summarize the existing literature and to establish an overview of the existing automated software testing tool to benefit the practice of users and for future research. We are attempting to provide detailed insight into automated software testing tools which can help the tester to choose the tool most suited to test his/her application.

Keywords Automated · Software testing · Selenium · Software development · Testing tool · Web-based application

L. Prasad (✉) · N. Vore (✉)
Sagar Institute of Research and Technology, SAGE University, Indore, India
e-mail: lalji.research@gmail.com

N. Vore
e-mail: nitiansis.mhow@gmail.com

R. Yadav
Department of Computer Science and Engineering, Acropolis Technical Campus, Indore, India
e-mail: rasneeluce@gmail.com

A. Abraham et al. (eds.), *Proceedings of 3rd International Conference on Computing Informatics and Networks*, Lecture Notes in Networks and Systems 167,
https://doi.org/10.1007/978-981-15-9712-1_10

Abbreviations

QTP QuickTest professional
IDE Integrated development environment
RC Remote control
API Application program interface

1 Introduction

The software testing process intends to distinguish all the imperfections existing in a software. Testing is no longer observed as a movement that begins simply after the coding stage, it ought to be inescapable all through the product advancement life cycle. Testing is a lot of exercises that can be arranged and directed deliberately [1]. It is the way toward accessing a framework or system parts by manual or programmed intends to confirm that it fulfills expected requirements or not. It additionally distinguishes the contrasts among expected and genuine outcomes.

Testing can be done by using automated software testing tools, which is also known as test automation tools [2] is when the tester writes scripts and uses another software to test the software. It can perform load, execution, practical and stress testing, and so on. It improves precision and sets aside time and cash in contrast with manual testing. Manual testing requires many labor-intensive tasks, running numerous program executions, and handling a great amount of information. Testing often consumes 40-50% of development efforts, and it consumes more effort for systems that require a higher level of reliability [3]. So, it will give rise to the demand for automated software testing tools. Appropriate tools can alleviate the burden of clerical, tedious operations and make them less error-prone. Sophisticated tools can support test design and test case generation, making it more effective. Direction to supervisors and analyzers on the best way to choose testing apparatuses that will be generally helpful to their association and experts is significant, as device determination extraordinarily influences testing productivity and viability.

We proceed with this paper with our writing study in which we have characterized and investigated many softwares (e.g., Selenium testing tool, Watir, QTP, TestComplete, WinRunner, Load Runner, SilkTest, Apache Jmeter, Wapt, Tellurium, Web Load, Neo Load, LoadUI, Appvance, rational performance tester, SahiPro, Telerik Test- Studio, Ranorex) dependent on different software quality attributes like understandability, modifiability, modularity, cost-effectiveness, extendibility, reliability, flexibility, reusability, browser compatibility, and OS compatibility [4–8].

Analysis of automated software testing tool gives administrators and analyzers understanding that can drive significant choices in regards to device determination for their undertaking. This data will give a superior comprehension of the apparatuses. As there are different automated testing tools which have different element yet despite the fact that they likewise have hardly any restrictions. This makes them monotonous to use for testing. There is a prerequisite of such a device that can be demonstrated its capability to do the entire testing process in less time, costs less cash, and needs less exertion and labor.

This paper will further contain Sect. 2 as a review methodology that describes the method adopted for the literature review. The literature review is described in Sect. 3. Further, the result of review work is presented in Sect. 4, and Sect. 5 presented the conclusion of all the above work.

2 Literature Review

The literature review was done to establish an understanding of the state of research in the area of automated testing tool, and the goal is to address the existing research done for automated testing tools and to present results related to these. The literature review represented in this paper is such that which identifies publications of relevance, evaluates, and selects publications for review and organizes them such that common concepts are identified and structured manuscript in different sections.

Anju Bansal [1] presented details of the flow of test information while testing an application. It also represents a different form of black box testing, white box testing, and gray box testing. Software testing can provide an independent view of the software to allow the business to appreciate and understand the risk of software implementation. This paper also provides the detailed insight into types of testing along with their advantages and disadvantages. With black box testing, one need not to have knowledge of programming language, testing can be done from user point of view, but test case is very hard to design, and some part of the backend is not tested at all. Apart from black box testing, white box testing focuses on the internal logic of code, it helps in uncovering the error from the code, but for this task, skilled testers are needed, and it is very tedious and time-consuming. While the third type, i.e., gray box testing generally combines the benefits of both types of black and white box testing.

A framework based on Selenium WebDriver and TestNG is proposed by Satish Gojare, Rahul Joshi, and Dhanashree Gaigaware [2]. As Selenium has a lack of customized report generation facility, therefore the author proposed a framework which is used to generate the customized report. The component of the proposed framework is object repository which performs the task of maintaining and repairing test cases, the input file will contain the input required by the web pages under test, and the utility section contains web page-related functions. Test suite can be formed by taking input from these three component object repository, input file, and utility file. Above all these, the test result after running the test suite can be fed to TestNG unit which will then generate a test report.

Quality of software system is defined based on quality attributes such as portability, reusability, interoperability, reliability, and performance. Software systems are checked based on their quality. Therefore, the quality characteristics are considered to be very important during the development of software. Therefore, Ritika Vern and Sanjay Kumar Dubey [4] present a study of few quality characteristics based on fuzzy logic technique. This will help the researcher, developer, and tester to decide their perspective in the area of software quality.

The quality attribute is required to capture how the functional requirement of an application is achieved. To trace the quality of the software, the quality attribute plays a very important role. Gorton, essential software architecture [6], gives a description of various quality attributes which helps to capture the quality of software, and it also sets a comparison criteria for comparing various software.

SWEBOK V3.0, [8] a guide to the software engineering body of knowledge Chap. 4 software testing, this chapter will represent the importance of software testing as a part of software development, and it also describes the basic terminologies and software quality-related attributes like maintainability, portability, usability, etc.

Nisha Gogna [9] talked about the most basic features of browser-based automation testing tool, i.e., Watir and Selenium. In this paper, the author describes mainly the details of components of Selenium such as Selenium IDE, Selenium RC, and the Selenium Grid. According to which, Selenium IDE is the integrated development environment used for preparing test cases and is used as Firefox add-ons. Firefox has many add-ons to be used with Selenium IDE which will increase its usability. The author also described briefly about Selenium RC, according to which the developer has to use programming language for writing test script to get maximum flexibility and extendibility. The Selenium Grid is another component of Selenium testing tool which scales the use of Selenium RC on multiple platforms and on multiple browsers. This paper also gives the details of the platform, browser, and languages supported by the Selenium testing tool. Also, Selenium has a deep learning curve to switch from one component to another as compared to Watir.

Based on the study of Selenium RC and Selenium WebDriver, the author Mahan Sunhare, Abhishek Tiwari [10] found that the architecture of Selenium RC is more complicated as compared to Selenium WebDriver and Selenium WebDriver has also friendlier API than Selenium RC. Software developers thought to build beautiful and attractive web pages, but they did not bother about their accessibility and presentation on various browsers. Therefore, the author describes the need of focusing on browser compatibility issue for a consistent look of web pages on all browsers.

In this paper, the author, Shaukat [11], presents the taxonomy of various automated software testing tools along with the importance of software testing methods as a part of the software development life cycle (SDLC). The author's perspective is to present with a comparative study of various automated software testing tools such that one can get a brief overview of all the tools presented in the paper. The tools which are chosen for comparative study are Selenium, Watir, Sahi Pro, HP-QTP, TestComplete, Ranorex, NeoLoad, hp LoadRunner, FitNese, SilkTest, TestNG, Apache Jmeter, Soap UI, Tellurium, WebTest, Xebium, Wapt, WebLoad Testanywhere, LoadUI, Appvance, rational performance tester, Visual Studio, and Telerik Test Studio. All these tools are studied based on operating system support, browser support, application used for testing, language support, cost-effectiveness, etc.

A study on various open source automated testing tools and a brief comparison of the same based on application support, language support, platform support, and understandability is presented in the paper by the author Bhat and Chaudhary [12]. Based on the survey, it is found that some of open source tools have some limitations while a commercial one has some good features. Therefore, the selection of the tool is completely based on the project needs and testers knowledge. According to the study of this paper, among all studied tools, Selenium has support for vast programming languages and platforms.

Chandraprabha et al. [13] talked about the systematic study of Selenium testing tool with all its components, architecture, and limitations also. As Selenium is an open source automated Web testing tools and being component-based, it is used by most of the testers. According to the study, automated testing is very advantageous as it saves time, money, and effort. It is also reliable and improves the accuracy of the testing process. Along with the details of the Selenium testing tool's component, the author also talked about the limitations of the component. According to which, the main feature of Selenium IDE is easy to record and playback, support autocomplete command, customization is allowed through plug-ins, allow to set breakpoints and debug the script, but apart from all these features, it has few limitations as it can be only used as a Firefox plug-ins, it has its own language Selenese. With Selenium RC, it is possible to run tests inside every JavaScript compatible browser using a wide range of programming language. On contrary to this feature, it is slow, struggles while running concurrent tests, and does not allow simultaneously tests across different OS and browsers. But with the help of Selenium Grid, one can run test cases across all browsers. Selenium WebDriver being a robust, open source, cost-effective and widely used tool cannot readily support new browsers.

Altaf et al. [14] specify all the pros and cons of every component of the Selenium testing tool. Selenium is a freeware open source testing tool. There are many challenges with it, and the main is that it has drawbacks, the user interface is slow for many reasons.

Mustafa et al. [15] had discussed about sets of software testing tool and classify them over the types of testing they can perform. Here, stress, load, regression, functional, unit, performance, and security testing are defined. The author had also represented the result of his/her study through the graph. One graphical representation specifies the type of application which is mostly tested, and the result shows that Web application widely tested by using software testing tool. There is another graph that shows a widely performed type of testing, and the result shows that functional testing is widely performed on applications.

3 Result of Literature Review

This section presents a brief study of most widely used testing tools based on literature review, which includes various testing tools. Individual tools such as Selenium testing tool, Watir, QTP, TestComplete, WinRunner, LoadRunner, SilkTest, Apache Jmeter, Wapt, Tellurium, Web Load, Neo Load, LoadUI, Appvance, rational performance tester, SahiPro, Telerik Test Studio, Ranorex are classified in Table 1 which represents

Table 1 Classification of automated software testing tools

Tool name	Developer (describes the developer's name)	OS support (describes about the operating system supported by the tool)	Browser support (describes about the browser supported by the tool)	Application support (describes about the application tested by tool) [29]	Testing type (describes about the type of testing performed by the tool) [30]	Language support (describes the language supported by the tool)	Open source/A commercial (specifies whether the tool is open source or not)
Selenium [9–12, 16, 17]	Developed by Jason Huggins in 2004	All platform	All major browser	Web application [28]	Integration, regression testing and functional testing	Java, C#, PHP, Ruby, Python, Perl	Open source
Watir [9, 11, 12, 18, 19]	Developed by Bret Pettichord and Paul Rogers	Cross platform (Windows, Linux)	Internet Explorer, Opera, Firefox	Web application	Regression testing and functional testing	Java, .Net, C#	Open source
Test complete [11, 16, 18–20, 33]	Developed by Smart Bear Software In. in 1999	Windows	IE, Firefox, Chrome	Desktop web and mobile applications	Regression testing and functional testing	Vbscript, Javascript, C++, Delphiscript	A commercial
QTP [11, 19, 21–23, 35]	Developed by HP in 2001	Windows Xp	IE, Firefox, Chrome	Client server application (Web)	Regression testing and functional testing	Java, Vbscript, JavaScript	A commercial
Win Runner [11]	Developed by HP	Windows, Linux	IE, Netscape	Web application	Load testing	SL Test scripting language	A commercial
Silktest [11]	Developed by Segue Software In. in 1993	Windows	IE, Firefox,	Web hybrid and Native application	Regression testing and functional testing	Java, VB, C#, VB.NET	A commercial

(continued)

Table 1 (continued)

Tool name	Developer (describes the developer's name)	OS support (describes about the operating system supported by the tool)	Browser support (describes about the browser supported by the tool)	Application support (describes about the application tested by tool) [29]	Testing type (describes about the type of testing performed by the tool) [30]	Language support (describes the language supported by the tool)	Open source/A commercial (specifies whether the tool is open source or not)
LoadRunner [11, 19, 24]	Developed by HP in 1989	Windows, Linux, MAC	Any browser	Web application	Load testing	Java, VB, C#, Vbscript, JavaScript, C	A commercial
Apache Jmeter [11, 24]	Developed by Apache software foundation in 1998	Unix, Windows	Chrome	Web application	Load testing	Java	Open source
Wapt [11, 24]	Softlogic	Windows, Unix	IE, Firefox, Chrome	Web, Mobile application	Load and stress testing	Asp.Net	Open source
Tellurium [11, 18, 19]	Developed by Jian Fang in 2007	Windows, MAC, Unix	All brower, Android, Blackberry	Web application	Load and performance testing	Java, Groovy	Open source
Web load [11, 24]	Developed by Redview Software in 1997	Windows, Linux, Unix	IE, Firefox, Chrome	Web, Mobile application	Load and performance testing	Java, JavaScript, .Net, PHP	Open source
Neo Load [11, 19, 24]	Developed by Neotys in 2005	Windows, Linux, Solaris	Chrome, Firefox	Web, Mobile application	Load and performance testing	JavaScript, .Net , PHP	A commercial

(continued)

Table 1 (continued)

Tool name	Developer (describes the developer's name)	OS support (describes about the operating system supported by the tool)	Browser support (describes about the browser supported by the tool)	Application support (describes about the application tested by tool) [29]	Testing type (describes about the type of testing performed by the tool) [30]	Language support (describes the language supported by the tool)	Open source/A commercial (specifies whether the tool is open source or not)
Load UI [11, 24]	Developed by Smart Bear Software In. in 2010	Windows, Linux, MAC	Chrome , Firefox	Web, Mobile application	Load and performance testing	Java, Javafx, Groovy	Open source
Appvance [11, 24]	Appvance Inc.	All platform	All browser web and mobile also	Web and Mobile application	Function and performance testing	PHP, Perl, Python, Groovy, Ruby, C#, Java	Open source
Rational performance tester [11, 24]	IBM	Linux, Windows	IE, Chrome, Firefox	Web and Server application	Data-driven and performance testing	Java	A commercial
Sahi Pro [11, 18, 19, 24]	Developed by Sahi in 2005	All platform	IE, Firefox	Web application	Data-driven testing	Java, PHP, Java Script, Python	A commercial
Telerik test studio [11]	Telerik	Windows	All browser	Web application and desktop application	Performance, data-driven and functional testing	C#, .Net	A commercial
Ranorex [11, 19, 20, 34]	Ranorex Gmbh	Windows	Opera, Firefox, Netscape, IE, Chrome	Desktop web and Mobile applications	Function testing	Java, Html, C#, Android	Open source

the basic details of tools such as developer' s name, year in which the tool is initially developed, OS support, browser support, application support, testing type, language support and whether the tools is open source or a commercial.

All the above-mentioned automated software testing tools are compared based on software quality attributes like understandability, modifiability, modularity, cost-effectiveness, extendibility, reliability, flexibility, browser compatibility, OS compatibility [4–8].

3.1 Software Quality Attributes

The detailed definition of all the above-mentioned software quality attributes is listed below, and their comparison based on these attributes is given in Table 2.

1. Modifiability: This defines the ability of software testing tool of being modified according to user requirement [4, 6].
2. Understandability: It is the attribute that determines how easy is the tool to be used without the need for expertise [4].
3. Extendibility: It defines the degree to that software should be extended, i.e., functionalities should be added to the existing software [7].
4. Modularity: It is the degree to which a system's component may be separated [7].
5. Compatibility: It is the usability of the same software in different environments (like different OS, browsers, etc.) [7].
6. Cost-effectiveness: It is the attribute that tells us whether the software is freely available to the user or there is purchasing cost associated with it [7].
7. Flexibility: It is the degree of ease with which a system can respond to any change [4].
8. Reliability: It is the degree to which the result of testing should be accurate or consistent. [4].

4 Conclusion

In total, from the survey and reviewed literature, we found that the tool selection depends on diverse evidence, such as development choices, evaluation objectives, execution facilities, and on so many parameters. In general, there may not be a unique tool that will satisfy particular needs, so maybe a suite of tools would be an appropriate choice. So, it gives us motivation to classify automated software testing tools and study them to explore their features and limitations. Analysis of automated software testing tool provides managers and testers insight that can drive important decisions regarding tool selection for their task. This information will provide a better understanding of the tools. As there are various automated software testing tools available which have various feature, but although they also have few limitations.

Table 2 Comparison of automated software testing tool based on software quality attributes

Tool name/year/version released in that year	Understandability	Modifiability	Flexibility	Extendibility	Modularity	Cost-effectiveness	Reliability	Browser compatibility	OS compatibility
Selenium [9–12, 16, 17, 27]	Test reports are relatively easy to understand	It is modifiable	It can withstand the changes made in it	Selenium IDE supports a wide range of extensions to enhance the capabilities of the core tool, thereby multiplying its potential	It has adopted modular approach for its development	Freely available	Selenium offers a user-friendly interface that helps to create and execute test case easily and ensure reliability in terms of result	Selenium provides support across multiple browsers, namely Internet explorer, Chrome, Firefox, Opera, safari, etc	Selenium can operate and support across multiple operating systems (OS) like Windows, Mac, Linux, unfix, etc
Selenium 2016 (version 2.49 to 3.0.0)	Improved report generation by using TestNG framework	It has capability to modify according to user requirement	It is flexible, i.e., respond well to the changes made in it	Support for wide range of extensions to improve performance but has limited support for browser extension	Modularity is supported by it, as it has various components	Freely available	It makes the testing result more reliable or consistent	Firefox is only supported till version 47.0.1 or earlier	Fully support all OS

(continued)

Table 2 (continued)

Tool name/year/version released in that year	Understandability	Modifiability	Flexibility	Extendibility	Modularity	Cost-effectiveness	Reliability	Browser compatibility	OS compatibility
Selenium 2015 (version 2.46 to 2.48)	By improving easy output generation up to some extent, it also added support for enhanced screenshot capturing	It supports modifiability	It gives good response to the changes made by user or programmer	Support for various update of technologies but with few limitations like it does not support Firefox above 47.0.0	It supported modularity	Freely available	Yes, it generates accurate result, thus supporting reliability	Supported Firefox only up to 47.0.0 do not support newer versions and all remaining browsers	Fully support all OS
Selenium 2014 (version 2.40 to 2.45)	Improve understanding by removing unnecessary dependencies	It can be modified easily	It has feature to response to change	Updating gecko driver, thus improving browser capability	It supported modularity	Freely available	It ensures reliability to generate correct output	Support Firefox only up to 33 , improving Chrome facilities and support other browsers	Yes, fully support all OS

(continued)

Table 2 (continued)

Tool name/year/version released in that year	Understandability	Modifiability	Flexibility	Extendibility	Modularity	Cost-effectiveness	Reliability	Browser compatibility	OS compatibility
Selenium 2013 (version 2.28 to 2.39)	Improved capability to take good screen shots	It supports to modify easily	It supports flexibility, i.e., quite easy to introduce any change	Improve extendibility by updating various driver supports for various browsers in the core tool	It supported modularity	Freely available	It has features that make testing more reliable	Support Firefox up to 26	Yes support Linux up to 24 versions , also added support for iphone driver
Selenium 2012 (version 2.16 to 2.27)	Improved error messages on console, support for hovering on Linux	It allows to modify few requirement according to user	Yes, it is easy to make changes in it	Test result can be extended, and it can be represented by test history	Supported modularity	Freely available	Yes, it makes testing more reliable	Support Firefox up to 17, update Opera to 0.15, and safari driver also added	It added support for hovering on Linux
Selenium 2011 (version 2.15 and before)	Improved by introducing screenshot capability	It is modifiable	Yes, it has capability to response well to changes	It support extendibility by updating various browser support	Supported modularity	Freely available	It ensures reliability to generate correct output in less time	Support only up to Firefox 11. Improved performance with Chrome and, i.e.	Yes (support for ice cream sandwich in the Android)

(continued)

Table 2 (continued)

Tool name/year/version released in that year	Understandability	Modifiability	Flexibility	Extendibility	Modularity	Cost-effectiveness	Reliability	Browser compatibility	OS compatibility
Watir [9, 11, 12, 18, 19]	It allows writing the tests which are easy to read and maintain	It allows writing the tests which are easy to modify	It is simple and flexible but limited flexibility	Support extendibility partially	Modularity is not fully supported	It is a free open source tool	Yes, it is reliable tools	Watir classic supports only Internet Explorer (limited) on Windows; supports Firefox, Internet Explorer, Opera , Chrome and also runs in headless mode (html unit)	It supports Windows, Mac, and Linux
Watir 2016 Versions (6.0.0 to 6.0.3)	Quite easy (prior knowledge of scripting required)improved error message facility	It allows easy maintenance by making modification if required	It has limited support for flexibility	It supports extendibility up to certain extent only	Modularity is not fully supported	It is a free open source tool	It is reliable in terms of result generation	Added support for Chrome as default browser	It support Windows, Mac, and Linux

(continued)

Table 2 (continued)

Tool name/year/version released in that year	Understandability	Modifiability	Flexibility	Extendibility	Modularity	Cost-effectiveness	Reliability	Browser compatibility	OS compatibility
Watir 2014 and 2015 Version (4.1.0 to 5.0.0)	Improved output facility and enhanced screenshot capability	It allows easy modification if required by user	It supports only few changes	It has limited support for addition of new features	Modularity is not fully supported	It is a free open source tool	It is reliable due to improved output generation	More flexibility with Internet Explorer and Firefox	It supports Windows, Mac, and Linux
Watir 2013 Version 3.4.0 to 4.0.1	Add capability to take screenshots	It ensures easy modification and thus removes unused processes	It responds well to few changes only	Added support for few technologies	Modularity is not fully supported	It is a free open source tool	It is reliable	Added support for Internet Explorer 10	It supports Windows, Mac, and Linux
Watir 2012 Version 3.0.0 to 3.3.0	Support for drag drop options	Modified by adding api support	It does not work well with major changes	This tool has limited support for extendibility	Modularity is not fully supported	It is a free open source tool	It makes testing process more reliable	Added support for opera	It supports Windows, Mac and Linux

(continued)

Table 2 (continued)

Tool name/year/version released in that year	Understandability	Modifiability	Flexibility	Extendibility	Modularity	Cost-effectiveness	Reliability	Browser compatibility	OS compatibility
Watir 2011 Version 1.7.0 to 2.0.4	Improved speed of execution of test cases	Remove unnecessary dependencies, thus ensuring easy modification	It has limited support for flexibility	Extendibility for fixed input values is added	Modularity is not fully supported	It is a free open source tool	Improved speed of execution of test cases making it more reliable and accurate	IE improvement and Firefox improvements	It supports Windows, Mac, and Linux
QTP/UFT [11, 19, 21–23, 35]	It is easy to use software for both people with or without programming knowledge	No, difficult to change	Difficult to change	Hard to extend	Support to modularity	Costly (not open source)	Yes it is reliable	Support multiple browser but not all (IE, Chrome, Firefox, safari)	Limited support
QTP/UFT 2016 Versions 12.54	Easy to use and work with this tool (improved drag and drop option)	Difficult to modify	Yes it supports flexibility but it is restricted as it is licensed	Functionality can be added but in limited manner	Yes it has modular approach for development	Not freely available	It is reliable as it produces the desired result	No support Chrome till version 54 and Firefox till version 49	Only support Windows system

(continued)

Table 2 (continued)

Tool name/year/version released in that year	Understandability	Modifiability	Flexibility	Extendibility	Modularity	Cost-effectiveness	Reliability	Browser compatibility	OS compatibility
QTP/UFT 2015 Versions 12.5	Improved cross-browser testing , and now result can be shared without result viewer installation on other system	It supports modifiability but only by its original developer	Flexible up to some extent	Hard to extend because it is not open source	It has modular approach for development	Not freely available	It is reliable	Cross-browser testing is introduced	Only support Windows system
QTP/UFT 2014 Versions 12 to 12.02	Easy to learn, faster installer, better documentation	Costly to modify as it is a commercial tool	Not much flexible as it allows update up to certain extent	Hard to extend	Yes it has modular approach for development	Not freely available	It is reliable due to good output generation	Yes (support for safari browser)	Only support Windows system (added support for Windows 8.1)
QTP/UFT 2013 Versions 11.51, 11.52, 11.53	Good and UFT now fully supports .xlsx format.	Modifiability do not supported by users	Yes it supports flexibility, but it is restricted as it is licensed	Hard to extend because it is a commercial tool	Yes it has modular approach for development	Not freely available	Yes it is reliable	Added support for IE 10, support for Chrome 37 and 38 (beta) and also for Firefox 32 and 33 (beta).)	Only support Windows system (added support for Windows 8)

(continued)

Table 2 (continued)

Tool name/year/version released in that year	Understandability	Modifiability	Flexibility	Extendibility	Modularity	Cost-effectiveness	Reliability	Browser compatibility	OS compatibility
QTP/UFT 2012 Versions 11.5	Good and easy to learn	Being a commercial difficult to modify	Not so flexible	Not support much extendibility	Yes it has modular approach for development	Not freely available	It is reliable	Yes support multiple browser	Only support Windows system also added support for mobile testing
QTP/UFT 2011 Versions 11.0	Good looking result and enhanced results viewer	Costly to modify as it is a commercial tool	Yes it supports flexibility, but it is restricted as it is licensed	Difficult to extend	Yes it has modular approach for development	Not freely available	Yes it is reliable	Yes (record support for Firefox is now available)	No only support Windows system
Ranorex [11, 19, 20, 34]	It is an easy to use software for both people with or without programming knowledge	Difficult to modify Ranorex	Difficult to modify Ranorex	No, hard to extend Ranorex	Modularity is not fully supported	Costly	Yes, support reliability	Support multiple browser	No, its only Window OS

(continued)

Table 2 (continued)

Tool name/year/version released in that year	Understandability	Modifiability	Flexibility	Extendibility	Modularity	Cost-effectiveness	Reliability	Browser compatibility	OS compatibility
Ranorex 2016 Versions 5.4.5 to 6.2.0	Upgrading of files, projects, and solutions has improved, creates a backup, and shows an upgrade report, view test report during test execution	Difficult to modify	It supports flexibility, but it is restricted as it is licensed	Functionality can be added but in limited manner	Modularity is not fully supported	Only trail version is free	It is reliable tool	Added support for android 7	Added support for iOS 10, added support up to Windows 10
Ranorex 2015 Versions 5.2.2 to 5.4.4	The report now signals modules with errors using an error icon	It supports modifiability but only by its original developer	Flexible up to some extent	Hard to extend because it is not open source	Modularity is not fully supported	Only trail version is free	Yes it supports reliability	Added support for Firefox 41 and higher and Android 6	Added support for Windows apps in Windows 100
Ranorex 2014 Versions 4.1.5 to 5.2.1	Improved facility for import and export functions	Costly to modify as it is a commercial tool	Not so flexible as it allows few update	Hard to extend	Modularity is not fully supported	Only trail version is free	Yes it is reliable tool	Added support for Firefox 34	Do not support all OS
Ranorex 2013 Versions 4.0.2 to 4.1.4	Image-based recording can now be enabled for mobile apps, improved report performance in Internet Explorer	Modifiability do not supported by users	It supports flexibility, but it is restricted as it is licensed	Hard to extend because it is a commercial tool	Modularity is not fully supported	Only trail version is free	Yes it is reliable tool	Added support for Firefox 26, Chrome 32, and newer versions of android also	Added object recognition support for Windows 8.1 preview apps

(continued)

Table 2 (continued)

Tool name/year/version released in that year	Understandability	Modifiability	Flexibility	Extendibility	Modularity	Cost-effectiveness	Reliability	Browser compatibility	OS compatibility
Ranorex 2012 Versions 3.2.1 to 4.0.1	Added shortcuts for many features	Being a commercial difficult to modify	Not so flexible	Not support much extendibility	Modularity is not fully supported	Only trail version is free	Yes it is reliable tool	Added support for Firefox 18 and Android also	Not fully support all OS
Ranorex 2011 Versions 2.3.8 to 3.2.0	Improved usability for cut/copy/paste , setting values, report now provides links to corresponding recording items or user code	Costly to modify as it is a commercial tool	It supports flexibility, but it is restricted as it is licensed	Difficult to extend	Modularity is not fully supported	Only trail version is free	Yes it is reliable tool	Yes added support for Google Chrome and apple, safari and Firefox 8	Only Windows but added support for 64 bit
Test complete [11, 16, 18–20, 33]	Test steps are improved	Difficult to modify	Yes it supports flexibility	Functionality can be added	Modularity is not fully supported	Only trail version is free	Yes it is reliable tool	Support for Chrome , Internet Explorer , Firefox , Opera	Support for Windows , iOS
Test complete 2016 Version 11.0 to 11.31	Test steps are improved	Difficult to modify	Yes it supports flexibility, but it is restricted as it is licensed	Functionality can be added but in limited manner	Modularity is not fully supported	Only trail version is free	Yes it is reliable tool	Support for Chrome 49, Internet Explorer 11, Firefox 45, Opera 31	Support for Windows 10, added support for iOS 9.3

(continued)

Table 2 (continued)

Tool name/year/version released in that year	Understandability	Modifiability	Flexibility	Extendibility	Modularity	Cost-effectiveness	Reliability	Browser compatibility	OS compatibility
Test complete 2015 Version 10.0-10.60	Support for new technologies which improve its working	It supports modifiability but only by its original developer	Flexible up to some extent	Hard to extend because it is not open source	Modularity is not fully supported	(only trail version is free)	Yes it supports reliability	Support for Chrome 33 and Opera 19, Firefox 26	Support for iOS testing and Windows 8.1
Test complete 2014 Versions 9.0 to 9.31	Improved gui, enhanced report generation, test summary	Costly to modify as it is a commercial tool	Not so flexible as it allows few update	Hard to extend	Modularity is not fully supported	(only trail version is free)	Yes it is reliable tool	Not support all browser fully	Support for Windows 8 and Windows server 2012 support, support for Firefox 23 and Chrome 29
Test complete 2013 Versions 8.0 to 8.70	Support for various technologies like .net, visual studio	Modifiability do not supported by users	Yes it supports flexibility, but it is restricted as it is licensed	Hard to extend because it is a commercial tool	Modularity is not fully supported	(only trail version is free)	Yes it is reliable as it produces the desired result	Not support all browser fully	Support for Windows mobile 6 series, support for Firefox 6

(continued)

Table 2 (continued)

Tool name/year/version released in that year	Understandability	Modifiability	Flexibility	Extendibility	Modularity	Cost-effectiveness	Reliability	Browser compatibility	OS compatibility
Test complete 2012 Version 7.0 to 7.52	The recording engine has been improved, quick search box for recording logs	Being a commercial difficult to modify	Not so flexible	Not support much extendibility	Modularity is not fully supported.	Only trail version is free	Yes it is reliable tool	Not support all browser fully	Now support Internet Explorer ver. 8, support for Windows 7. Now, you can run your tests under Microsoft Windows 7 support for Firefox 3.5
Test complete 2011 Version 6.0 to 6.52	Compress test results, performance has been improved	Costly to modify as it is a commercial tool	Yes it supports flexibility, but it is restricted as it is licensed	Difficult to extend	Modularity is not fully supported	Only trail version is free	Yes it is reliable as it produces the desired result	Not support all browser fully	Support for 64-bit applications

This makes them tedious to use for testing. In today's era, there is a requirement of such a tool which can be proved its capability to carry out the whole testing process in less time, costs less money, and needs less effort and man power. Also, during our study, we found few tools are widely used because of their features and performance, while some of them lag. So, in our further study, we give preference to such tools that are widely used.

References

1. Bansal A (2014) A Comparative study of software testing techniques. Int J Comput Sci Mob Comput
2. Gojare S, Joshi R, Gaigaware D (2015) Analysis and design of Selenium testing tool web driver automation testing framework. Elsevier
3. Bhadauria SS, Kothari A, Prasad L (2011) Design architecture class diagram for comprehensive testing tool. Orient J Comput Sci Technol
4. Vern R, Dubey SK (2013) A survey on evaluation of quality of software system by using fuzzy logic approach. Global J Comput Sci Technol Softw Data Eng
5. International standard for the evaluation of software quality. https://en.wikipedia.org/wiki/ISO/IEC_9126
6. Gorton (2011) Essential software architecture. Springer, Berlin Heidelberg
7. Senthilkumar R, Arunkumar T (2016) A survey on prioritization of software quality. Indian J Sci Technol 9(44)
8. SWEBOK V3.0 A guide to the software engineering body of knowledge
9. Gogna N (2014) Study of browser based automated test tool Watir and Selenium testing tool. Int J Inform Edu Technol
10. Sunhare M, Tiwari A (2016) Survey of comparative study for automated tools such as Selenium testing tool RC, Selenium testing tool web driver. Int J Sci Res Dev 4(6)
11. Shaukat K (2015) Taxonomy of automated software testing tool. Int J Comput Sci Innov
12. Bhat W, Chaudhary K (2016) A comparative study on software testing tools based on open source technology. Int J Multi Res Dev 3(1)
13. Chandraprabha, Kumar A, Saxena S (2013) Systematic study of a web testing tool: Selenium testing tool. Int J Adv Res Sci Eng 2(11)
14. Altaf I, ul Rashid F, Dar JA, Rafiq M (2015) Survey on Selenium testing tool in software testing. IEEE Computer Society
15. Mustafa KM, Al-Qutaish RE, Muhairat MI (2009) Classification of software testing tools based on the software testing methods IEEE
16. Kaur H, Gupta G (2013) Comparative study of automated testing tools: Selenium testing tool, QTP and TestComplete. Int J Eng Res Appl 3(5)
17. Singh J, Sharma M (2015) Performance evaluation and comparison of Sahi Pro and Selenium testing tool web driver. Int J Comput Appl 129(8)
18. Bhateja N (2015) A study on various software automation testing tools. Int J Adv Res Comput Sci Softw Eng 5(6)
19. Singh J, Sharma M (2015) A comprehensive review of web-based automation tools. Int J Innov Res Comput Commun Eng 3(10)
20. Dubey N, Shiwani S (2014) Studying and comparing automated testing tools: Ranorex and TestComplete. Int J Eng Comput Sci 3(5)
21. Rattan R, Shallu (2013) Performance evaluation and comparison of software testing tool. Int J Inform Comput Technol 3(7). ISSN 0974-2239
22. Gunasekaran S, Bargavi V (2015) Survey on automation testing tool for mobile applications. Int J Adv Eng Res Sci 2(11)

23. Rahate SR, Bhave U (2016) A survey on test automation. Int J Innov Res Comput Commun Eng 4(6)
24. Arora I, Bali V (2015) A brief survey on web application performance testing tools literature review. Int J Latest Trends Eng Technol 5
25. Kaur M, Kumari R (2011) Comparative study of automated testing tools: Testcomplete and QTP. Int J Comput Appl 24(1)
26. Bhadauria S, Kothari A, Prasad L (2011) A full featured component based architecture testing tool. 8(4)
27. Regina Ranstrom, Automated web software testing with Selenium testing tool
28. Bruns A, Kornstadt A, Wichmann D (2009) Web application tests with Selenium testing tool. IEEE Comput Soc
29. Angmo R, Sharma M (2014) Performance evaluation of web-based automation testing tools. IEEE Comput Soc
30. Hooda I, Chillar SR (2015) Software test process, testing types and techniques. Int J Comput Appl 3(12)
31. Dwivedi S, Gupta G (2015) Proposed framework based on performance evaluation and analysis of testing tool. Int J Innov Res Comput Commun Eng 3(4)
32. An open source platform GitHub. https://github.com
33. Tools support platform. https://support.smartbear.com
34. Ranorex. https://www.ranorex.com
35. QTP. https://www.learnqtp.com

Microexpression Analysis: A Review

Mamta Rani and Neeru Rathee

Abstract Micro expression analysis has been visually perceived as an abundance of advancement in recent years due to availability of frugal acquisition cameras and computational contrivances. Though datasets for micro expression analysis are available but even this advancement has still not reached to that caliber where we can built an automatic micro expression system just like automatic macro expression recognition system. Researchers have put their best efforts to develop the system for automatic facial expression analysis to recognize basic emotion which include jubilant, sad, irate, penitence, fear and surprise. The micro expression analysis task is quite challenging and fascinating due to advacement in automaticity in many fields of life. To address these challenges in a systematic manner, authors have endeavored to present a detailed analysis of the work done till date in micro expression field. The detailed description includes preprocessing and feature extraction. The advantages and disadvantages listed in this paper provide an impetus toward the future work and avail in culling an area of research in this field.

Keywords Microexpression Analysis · Face expression analysis · Review · Feature extraction

1 Introduction

Micro expression is fleeting countenance that occurs within 1/25th of a second. Micro expression is uncontrolled expression that occurs without the cognizance of an individual and showcases the true emotions even if the person is endeavoring to conceal an emotion. The pioneering work in micro expression apperception was

M. Rani (✉) · N. Rathee
Maharaja Surajmal Institute of Technology (affiliated to Guru Gobind Singh IndraPrastha University), C-4, Janakpuri, New Delhi, India
e-mail: mamtatholia@msit.in

N. Rathee
e-mail: neeru1rathee@gmail.com

A. Abraham et al. (eds.), *Proceedings of 3rd International Conference on Computing Informatics and Networks*, Lecture Notes in Networks and Systems 167,
https://doi.org/10.1007/978-981-15-9712-1_11

done by [11] in their study of psychotherapeutic interviews. According to the study, the emotions appear as a result of repression and could not be apperceived in authentic time [11]; however, Ekman and Friesen [3, 7] later showed that with congruous training anyone could learn to visually perceive micros as they occurred. A micro expression training implement (METT) was developed to amend the micro expression recognition abilities [7]. Ekman and Friesen first reported about finding micro expressions while examining the motion picture of a mentally disturbed person. In the video, patient seemed jubilant all the time but while reviewing the recording in slow kineticism a momentary distress lasting only two frames was spotted, which was later attested by the patient that she was endeavoring to obnubilate her plan to commit suicide [8]. The study of micro expression is very paramount as it gives the true account of the authentic emotion of a person no matter what he/she is verbalizing because nobody can suppress micro expression at their willpower. The area of facial expression recognition is quite popular, and researchers have quoted high accuracies for the same on databases acquired in studio environment as well as 'in the wild' [30]. Despite of this automatic detection of micro expression has a long way to go. Because firstly spotting and then apperceiving a particular micro expression is very arduous due to following reasons. [10] (i) Micro expression is too short and subtle to be apperceived by unclad eyes. Micro expression is ephemeral so very good resolution high speed of cameras is required for micro expression spotting. (ii) Micro expression may occur with other more protruding expression like blinking of eyes and kineticism of head and thus may be arduous to recognize accurately. (iii) Availability of few spontaneous datasets for micro expression analysis because it is very arduous to induce micro expression in the lab controlled environment. There are several steps involved from spotting to recognizing a micro expression. The typical processing of facial expression includes preprocessing, feature extraction, post processing and classification. Each of these steps is done by a variety of algorithm which further has number of performance metrics such as accuracy, recognition rate, F1 score and f measure to get the good results. So, a number of permutation and combination exist, and it is very difficult to pursue the same, to make it clear, a systematic review of preprocessing and feature extraction done in micro expression recognition is presented here.

2 Preprocessing

This is the first step toward recognizing a micro expression from the given video or image. Two main steps involved in preprocessing are face detection and micro-expression spotting. Viola and Jones [33] were first to detect facial images utilizing machine learning with cascade classifier. But the model [33] was up to the mark only for pre subsisting and was not prosperous in the wild imagery [47]. Next paramount step in facial detection is landmark localization and converting these landmark points into a modal face. In this way, the most important model used was ASM [14, 15, 34, 39, 40, 43]. ASM lacks the fine face alignment so to amend in this direction [41] used 'coarse alignment and face cropping' which involves LWM transformation [26] The accuracy was further amended utilizing DRMF which is a benchmark locator that can explicitly locate all the points in landmark localization automatically [1, 19, 20, 23]. Accuracy of landmark detection was enhanced further by [44] utilizing tree predicated face detector in DRMF that additionally fits well in the wild imagery. It is rather a challenge to descry subtle details not only for bare ocular perceiver but also for micro expression recognition tasks. These remote variations in micro expression can be amplified for better recognition [38]. EVM is one such tool purposed by [38]. It involves computing the full Laplacian pyramid [2] that disintegrates the video frame sequences into discrete frequency bands which are passed from any of the band pass filters such as Ideal, Butterworth, Second-order IIR, chebyshev to excerpt the frequency bands of interest that are amplified at different spatial level by some amplification. Conclusively, the amplified signal is appended with the pristine to eventually get the motion magnified video [17, 36] to discern different subtle forms of kineticism. Though EVM enhances forms of kineticism and colors to further ease up feature extraction task, but even then there is dearth of magnification work still to be done till date. [10]. Another paramount task assigned in preprocessing is up sampling and down sampling of the image frames and by exploring the co-cognation between them. One such method proposed by researchers is TIM [23]. Earlier it was utilized in lip reading [46], but later it was well explored and implemented for subtle emotion recognition [14, 27]. TIM abstracts the superfluous image frames that does not have any emotions and interpolate lesser number of frames, but this technique does not boost the prosperity rate of recognition [10]. Rather [41] used linear interpolation that locates and crop out the bounding box abstracting kineticism pattern with errors. Additionally optical continuance computation is utilized to quantify pixel level movement. To obtain the fine scale alignment, expedite algorithm predicated on 1D histogram is utilized. However, [16] asserted that while removing redundant frames TIM additionally incline to remove intermittent information that may be stored in these frames and proposed dynamic mode decomposition (DMDSP) to surmount the downside of TIM.

3 Feature Extraction

Feature extraction is the next step to preprocessing. Feature extraction is rudimentally a process that converts raw detected images into set of data containing features that are explicit for any particular countenance, thereby reducing the size of the data by extracting only non-redundant and consequential information. Though there subsists a number of feature extraction methods which are quite prominent in the area of countenance analysis, but in the micro expression analysis, the methods that have been popularly accepted by researchers include local binary pattern-three orthogonal planes (LBP-TOP), 3D histograms of oriented gradients (3D HOG), histogram of oriented optical flow (HOOF) and deep features. The scope of deep features is illimitable but the main quandary associated with deep features is that it requires a prodigiously and sizably voluminous dataset for training purport which lacks in the micro expression field.

3.1 HOG

The histogram of oriented gradients (HOG) is a feature descriptor technique utilized in image processing, which includes the number of gradient orientation in an image. [28] pioneered in the field by giving 3D gradient descriptor. The first frame of the video recording was separated into 12 regions of interest (ROI). Histograms of these ROIs were then calculated predicated on the orientation and gradient analysis of every frame for detecting facial motion. They utilized a posed dataset that was acquired utilizing very high speed video (200fps), and only front view of the images was considered. [29] explored HOG with k mean classifier to take into account the timing characteristics of the spontaneous frames that can be utilized for differentiating between posed and spontaneous datasets. Further extension to HOG was done by [32] proposing HOGTOP. They additionally introduced a variant of HOG as histogram of image orientation into three orthogonal planes (HIGO-TOP) that neglects the magnitude and includes the replication of histogram. [5] used HOG along with other feature descriptor to have a comparative analysis.

3.2 LBP

LBP introduced by [25] is prominent technique for texture relegation in the field of computer vision. An extension of this technique was proposed by [45] by elongating LBP to three orthogonal planes (LBP-TOP) for micro expression descriptor. The innate advantage of LBP-TOP is its robustness to illumination variations and image transformation, as it compares the intensities of central pixel with circumventing pixels to get a consummate histogram out of three planes XY (spatial) XT & YT (temporal). This is the reason it has been well explored by researchers for micro expression analysis [19, 24, 32, 36]. The fascinating work was presented by [6] by exploring LBP-TOP on the ocular perceiver zone rather than the consummate face, utilizing the pertinent information and reducing the redundant details. But the recognition accuracy was not amended for all micro expressions because some expressions such as repression and surprise may be more prominent on the facial muscles other than eyes. Additionally, different classifiers gave varied recognition accuracies for different micro expressions. One classifier, for example, random forest which gives better precision for blissful expression in eyes region gives worst precision results for other micro expression. Thus, this method of extracting features from eyes region was suited for jubilant, disgust and sadness and failed entirely for the repression expression.

He et al. [13] proposed LBP-mean orthogonal plane (LBP-MOP) by linking together LBP feature from three mean images derived from pooling the stacks from three orthogonal planes [48]. Other variants of HOG feature descriptor are LBP-TOP [45], LBP with six intersection points(LBP-SIP) [37], spatio temporal local binary pattern with integral projection (STLBP-IP) [14] . Among the all above, LBP-SIP has least computational involutions.

3.3 HOOF

Chaudhry et al. [4] first instigated the utilization of HOOF in representing human activities. HOOF uses distributions of optical flow represented by normalized histograms that are non-Euclidean and independent of direction and scale of moving person, and this is the reason that it is utilized as prominent feature descriptor in micro expression recognition task [5, 20–22, 42]. However, there is a major drawback of HOOF. In general, histogram uses Euclidean distance measure unlike general relegation task where each dimension is considered thoroughly different, histogram considers elements of neighboring dimension proximately cognate. So, when histogram as feature vector is utilized during relegation even a remote variation in histogram due to transmute in intensity will engender an immensely colossal difference because of Euclidean distance measure. This quandary can be solved in two ways. First, by utilizing a relegation algorithm predicated on other distance measures other than Euclidean distance such as earth mover's distance (EMD) [31], diffusion distance [18] and Bhattacharya distance, etc. Second is to transmute histogram by assigning virtually homogeneous values to every dimension of histogram to get homogeneous vicissitudes in the resulting histogram. Predicated on the second way [12] proposed an elongated variant of HOOF called fuzzy histogram of optical flow orientation (FHOOF) which utilizes angular histograms of motion vector considering start and terminus of each histogram connected circularly while computing fuzzy histogram. They also proposed an extension of the FHOOF called fuzzy histogram of optical flow orientation (FHOFO) which utilizes only motion direction ignoring the marginal motion magnitudes that arises due to transmute in intensity of micro expression making it robust for intensity variations.

3.4 Deep Features

Deep learning approach in feature extraction is gaining popularity day by day, it is inhibited by the fact that it requires an immensely colossal database for training purport which is not so available currently. Peng et al. [26] were first to explore deep feature for micro expression recognition by proposing neural network called dual temporal scale convolutional neural network (DTSCNN) in cooperating only four convolution layers eliminating the desideratum for a more astronomically immense dataset. Wang et al. [35] used only 560 micro expression clips for feature extraction utilizing CNN. Gan and Liong [9] applied CNN on the six diverse features extracted from the Apex frame namely vertical I& horizontal optical flow, magnitude and orientation, and distinction between apex and onset frame. However, the best recognition efficiency of 80% was achieved with vertical and horizontal optical flow features.

4 Conclusion

As of now the automatic facial expression recognition on macro level, i.e., detecting the basic emotions has been achieved prosperously, it is a long way to go for automatic micro expression recognition task. The research in the area of automatic micro expression recognition is hindered by many factors such as non-availability of spontaneous database. Also, it is very consequential to authenticate the results of micro expression recognition in databases from a well-trained psychologist and the subject itself, because some of the expressions are habitual and do not suggest any emotion. We have endeavored to present a consummate and exhaustive review about the work done till date in micro expression recognition field. Detailed information on the sundry steps involved during preprocessing and feature extraction in micro expression apperception has been discussed here. As different set of preprocessing techniques, feature extraction, relegation and post-processing give different results in terms of apperception precision, F measure, f1 score and true positive rate. For future work, each of these steps can be expounded citing different examples from the various research works with the possible outcome and challenges.

References

1. Asthana A, Zafeiriou S, Cheng S, Pantic M (2013) Robust discriminative response map fitting with constrained local models. In: Proceedings of the IEEE conference on computer vision and pattern recognition, pp 3444–3451
2. Burt P, Adelson E (1983) The laplacian pyramid as a compact image code. IEEE Trans Commun 31(4):532–540
3. Chao WL, Ding JJ, Liu JZ (2015) Facial expression recognition based on improved local binary pattern and class-regularized locality preserving projection. Sig Process 117:1–10. http://www.sciencedirect.com/science/article/pii/S0165168415001425
4. Chaudhry R, Ravichandran A, Hager G, Vidal R (2009) Histograms of oriented optical flow and binet-cauchy kernels on nonlinear dynamical systems for the recognition of human actions. In: 2009 IEEE conference on computer vision and pattern recognition, pp 1932–1939. IEEE
5. Davison A, Merghani W, Yap M (2018) Objective classes for micro-facial expression recognition. J Imaging 4(10):119
6. Duan X, Dai Q, Wang X, Wang Y, Hua Z (2016) Recognizing spontaneous micro-expression from eye region. Neurocomputing 217:27–36
7. Ekman P (2002) Microexpression training tool (mett). University of California, San Francisco
8. Ekman P, Friesen WV (1969) Nonverbal leakage and clues to deception. Psychiatry 32(1):88–106

9. Gan Y, Liong ST (2018) Bi-directional vectors from apex in cnn for micro-expression recognition. In: 2018 IEEE 3rd international conference on image, vision and computing (ICIVC), pp 168–172. IEEE
10. Goh KM, Ng CH, Lim LL, Sheikh U (2018) Micro-expression recognition: an updated review of current trends, challenges and solutions. Vis Comput, pp 1–24
11. Haggard EA, Isaacs KS (1966) Micromomentary facial expressions as indicators of ego mechanisms in psychotherapy. In: Methods of research in psychotherapy, pp 154–165. Springer
12. Happy S, Routray A (2017) Fuzzy histogram of optical flow orientations for micro-expression recognition. IEEE Trans Affect Comput
13. He J, Hu JF, Lu X, Zheng WS (2017) Multi-task mid-level feature learning for micro-expression recognition. Pattern Recogn 66:44–52
14. Huang X, Wang SJ, Zhao G, Piteikainen M (2015) Facial micro-expression recognition using spatiotemporal local binary pattern with integral projection. In: Proceedings of the IEEE international conference on computer vision workshops, pp 1–9
15. Jain DK, Zhang Z, Huang K (2018) Random walk-based feature learning for micro-expression recognition. Pattern Recogn Lett
16. Le Ngo AC, See J, Phan RCW (2017) Sparsity in dynamics of spontaneous subtle emotions: analysis and application. IEEE Trans Affect Comput 8(3):396–411
17. Li X, Hong X, Moilanen A, Huang X, Pfister T, Zhao G, Pietikäinen M (2017) Towards reading hidden emotions: a comparative study of spontaneous micro-expression spotting and recognition methods. IEEE Trans Affect Comput 9(4):563–577
18. Ling H, Okada K (2006) Diffusion distance for histogram comparison. In: 2006 IEEE computer society conference on computer vision and pattern recognition (CVPR'06), vol 1, pp 246–253. IEEE
19. Liong ST, See J, Phan RCW, Wong K, Tan SW (2018) Hybrid facial regions extraction for micro-expression recognition system. J Sig Process Syst 90(4):601–617
20. Liong ST, See J, Wong K, Phan RCW (2018) Less is more: Micro-expression recognition from video using apex frame. Sig Process Image Commun 62:82–92
21. Liong ST, Wong K (2017) Micro-expression recognition using apex frame with phase information. In: Asia-pacific signal and information processing association annual summit and conference (APSIPA ASC), pp 534–537. IEEE
22. Liu YJ, Zhang JK, Yan WJ, Wang SJ, Zhao G, Fu X (2015) A main directional mean optical flow feature for spontaneous micro-expression recognition. IEEE Trans Affect Comput 7:1–1
23. Liu YJ, Zhang JK, Yan WJ, Wang SJ, Zhao G, Fu X (2016) A main directional mean optical flow feature for spontaneous micro-expression recognition. IEEE Trans Affect Comput 7(4):299–310
24. Oh YH, Le Ngo AC, Phari RCW, See J, Ling HC (2016) Intrinsic two-dimensional local structures for micro-expression recognition. In: 2016 IEEE international conference on acoustics, speech and signal processing (ICASSP), pp 1851–1855. IEEE
25. Ojala T, Pietikäinen M, Mäenpää T (2002) Multiresolution gray-scale and rotation invariant texture classification with local binary patterns. IEEE Trans Pattern Anal Mach Intell 24(7):971–987
26. Peng M, Wang C, Chen T, Liu G, Fu X (2017) Dual temporal scale convolutional neural network for micro-expression recognition. Front Psychol 8:1745
27. Pfister T, Li X, Zhao G, Pietikäinen M (2011) Recognising spontaneous facial micro-expressions. In: 2011 international conference on computer vision, pp 1449–1456. IEEE
28. Polikovsky S, Kameda Y, Ohta Y (2009) Facial micro-expressions recognition using high speed camera and 3d-gradient descriptor. IET Research
29. Polikovsky S, Kameda Y, Ohta Y (2013) Facial micro-expression detection in hi-speed video based on facial action coding system (facs). IEICE Trans Inf Syst 96(1):81–92
30. Yan WJ, Wu Q, Liu YJ, Wang SJ, Fu X (2013) Casme database: a dataset of spontaneous micro-expressions collected from neutralized faces. In: 2013 10th IEEE international conference and workshops on automatic face and gesture recognition (FG), pp 1–7

31. Rubner Y, Tomasi C, Guibas LJ (2000) The earth mover's distance as a metric for image retrieval. Int J Comput Vis 40(2):99–121
32. Tran TK, Hong X, Zhao G (2017) Sliding window based micro-expression spotting: a benchmark. In: international conference on advanced concepts for intelligent vision systems, pp 542–553. Springer
33. Viola P, Jones M et al (2001) Rapid object detection using a boosted cascade of simple features. CVPR 1(1):511–518
34. Wang L, Zhang D, Wang Y, Chen C, Han X, M'hamed A (2016) Sparse mobile crowdsensing: challenges and opportunities. IEEE Commun Mag 54(7):161–167
35. Wang SJ, Li BJ, Liu YJ, Yan WJ, Ou X, Huang X, Xu F, Fu X (2018) Micro-expression recognition with small sample size by transferring long-term convolutional neural network. Neurocomputing
36. Wang Y, See J, Oh YH, Phan RCW, Rahulamathavan Y, Ling HC, Tan SW, Li X (2017) Effective recognition of facial micro-expressions with video motion magnification. Multimedia Tools Appl 76(20):21665–21690
37. Wang Y, See J, Phan R, Oh YH (2015) Lbp with six intersection points: reducing redundant information in lbp-top for micro-expression recognition
38. Wu HY, Rubinstein M, Shih E, Guttag J, Durand F, Freeman W (2012) Eulerian video magnification for revealing subtle changes in the world
39. Xia Z, Feng X, Peng J, Peng X, Zhao G (2016) Spontaneous micro-expression spotting via geometric deformation modeling. Comput Vis Image Underst 147:87–94
40. Xiaohua H, Wang SJ, Liu X, Zhao G, Feng X, Pietikainen M (2017) Discriminative spatiotemporal local binary pattern with revisited integral projection for spontaneous facial micro-expression recognition. IEEE Trans Affect Comput
41. Xu F, Zhang J, Wang JZ (2017) Microexpression identification and categorization using a facial dynamics map. IEEE Trans Affect Comput 8(2):254–267
42. Yan WJ, Chen YH (2018) Measuring dynamic micro-expressions via feature extraction methods. J Comput Sci 25:318–326
43. Yan WJ, Li X, Wang SJ, Zhao G, Liu YJ, Chen YH, Fu X (2014) Casme ii: An improved spontaneous micro-expression database and the baseline evaluation. PloS one 9(1):e86041
44. Zhang J, Shan S, Kan M, Chen X (2014) Coarse-to-fine auto-encoder networks (cfan) for real-time face alignment. In: European conference on computer vision, pp 1–16. Springer
45. Zhao G, Pietikainen M (2007) Dynamic texture recognition using local binary patterns with an application to facial expressions. IEEE Trans Pattern Anal Mach Intell 6:915–928
46. Zhou Z, Zhao G, Pietikäinen M (2011) Towards a practical lipreading system. In: CVPR 2011, pp 137–144. IEEE
47. Zhu X, Lei Z, Yan J, Yi D, Li SZ (2015) High-fidelity pose and expression normalization for face recognition in the wild. In: Proceedings of the IEEE conference on computer vision and pattern recognition, pp 787–796
48. Zong Y, Huang X, Zheng W, Cui Z, Zhao G (2018) Learning from hierarchical spatiotemporal descriptors for micro-expression recognition. IEEE Trans Multimedia

Network and Information Security

Cryptographic Analysis of DES and RSA Algorithm Using the AVISPA Tool and WSN

Shailendra Singh Gaur, Megha Gupta, and Gautam Gupta

Abstract With the world-changing toward IoT, wireless sensors networks (WSN) are becoming more important than ever. WSN is a complex distributed system accustomed to several technological fields especially in the area of security. Wireless sensor network deals with several research issues concerning security and energy efficiency. WSN uses cryptography as a method to ensure network data security and authentication. This paper reconsiders the algorithms of cryptography and improves them by eliminating any flaws in it, using the AVISPA tool. In this paper, symmetric and asymmetric key cryptographic algorithms like DES, RSA and the improved RSA were compared using the AVISPA tool. The objective is to enhance the existing algorithms in order to remove the flaws and to provide a more secure algorithm. To augment the security, we used two sessions as compared to one in the case of existing algorithms. To conclude, the improved algorithm is more secured as compared to existing algorithms.

Keywords Wireless sensors networks (WSN) · Clustering · Low-Energy adaptive clustering hierarchy (Leach) protocol · Cryptography · DES (Data encryption Standard) · (Rivest Shamir Adleman) RSA · Automated validation of internet security(AVISPA) tool

Abbreviations

AVISPA Automated validation of internet security

S. S. Gaur (✉) · M. Gupta
Bhagwan Parshuram Institute of Technology, New Delhi, Delhi, India
e-mail: shailendrasinghgaur@bpitindia.com

M. Gupta
e-mail: meghagupta307@gmail.com

G. Gupta
Bharati Vidyapeeth's College of Engineering, Delhi, India
e-mail: gautamgupta1811@gmail.com

A. Abraham et al. (eds.), *Proceedings of 3rd International Conference on Computing Informatics and Networks*, Lecture Notes in Networks and Systems 167,
https://doi.org/10.1007/978-981-15-9712-1_12

CH	Cluster head
DES	Data encryption standard
LEACH	Low-energy adaptive clustering hierarchy protocol
RSA	Rivest Shamir Adleman
WEP	Wired equivalent privacy
WPA	Wi-Fi protected access

1 Introduction

Network security incorporates the practices and policies used for restricting and monitoring unauthorized alteration, misuse, access and disowning of network-accessible resources and computer networks. Computer networks [1] can be both public and private, which find uses in day-to-day chores performing, governing transactions, communications amid businesses, government agencies and individuals. It protects the network, as well as secures and oversees the operations that are performed. Assigning a unique name and password to a network resource is the most communal and straightforward way of protecting it [2]. Attacks on networks can be executed from various malicious sources. These attacks can be of two categories: passive, i.e., when a network intruder captures the data traveling through the network and active, i.e., when an intruder initiates commands to dislocate the network's usual operation or to conduct examination and side movement to explore and gain access to the assets available through the network. Security is now an essential requirement since global computing is becoming integrally insecure. As our data flows from one point to another over the Internet, it passes through several other points over this route, which gives other users an opportunity to seize or alter it. There is a lot of information on the Internet uploaded/accessed by different institutions, organizations, agencies, governments, etc. This may be anything from rudimentary information to delicate information. Hence, it needs to be secured from malicious users over the Internet.

2 Related Work

2.1 Clustering Outline

Clustering refers to the process of distributing the data points into different categories such that the data points of the given category are similar to the other points of data in the same group. The main objective is to treat the trait and segregate them into such a category to form clusters.

2.2 Clustering Types

There are two types of clustering:

Hard Clustering: This clustering technique either involves the data point that belongs to a cluster or it does not involve at all.

Soft Clustering: In this clustering, the probability of using data point for cluster formation is assigned, instead of forming a single cluster.

2.3 LEACH Algorithm

LEACH algorithm is a basic clustering algorithm that is generally used in WSNs. In this algorithm, using random selection, a cluster head is selected. LEACH protocol provides many advantages such as data aggregation and energy efficiency. This protocol arbitrarily selects the cluster head nodes and accordingly provides the most optimal cluster number. The next step is to select a high energy cluster head using the algorithm of randomized rotation of cluster nodes and the disseminated energy considering the networks that are formed with the help of sensor nodes. The setup phase and the steady phase are the two main phases that form the LEACH protocol [3].

Setup Phase. Amongst the cluster nodes, cluster head (CHs) is selected. Randomly 0 and 1 is selected by each sensor node. The condition for cluster head to be selected is if it has values 0 and 1 that is less than the threshold frequency. Depending on signal strength, each node joins a cluster.

Steady phase. The steady phase provides the concept that the data from cluster head to the base station/sink is imparted and is obtained and accumulated from sensor nodes.

2.4 Low-Energy Adaptive Clustering Hierarchy—LEACH

LEACH is a technique, depending upon TDMA-based MAC protocol that is unified with cluster formation and a routing protocol that is used in wireless sensor networks. The objective of LEACH algorithm is to create and maintain clusters energy in an effort to enhance the existence of sensor nodes [4].

Protocol explanation. LEACH is a protocol where the nodes in a cluster transmit the facts to the cluster head, and these accumulated and compressed data is subsequently forwarded to the base station. The base station is sometimes called sink also. On the basis of the algorithm, cluster head is selected. Each node has to reach to the base station or cluster head on the basis of the intensity of radio used. It is also required to save power and prevent the wastage of energy. After achieving the required percentage of *P*, there will be a probability of 1/*P* to make that node, a cluster head again.

There is one more condition for those nodes who do not become a cluster head that is the node select the nearest cluster head and join it. After this, a scheduling mechanism is followed for each node for transmitting data.

The main application of wireless sensor networks is to form the position of sensor nodes for making the surveillance of a particular area. Using a secured network is the elementary requirement of the application. A challenging issue is to provide security to sensor networks keeping it energy efficient. Several security threats can malfunction the operation of these networks. Designing WSN consists of forming layer. Sensor networks need to satisfy some of the requirements for facilitating better and secure communication [5]. Availability, confidentiality, integrity and authentication are universal security requirements of WSNs. These requirements protect against those who attacked the transmitted information in a given network.

2.5 Cryptography

It is the practice of hiding information. It is a technique deployed to keep information secret and safe. Modern cryptography is an amalgam of computer science, mathematics and electrical engineering. Cryptography [6] is employed in computer passwords, ATM (Bank) cards and while shopping over the Internet. In cryptography, when a message is sent, it is ciphered or encrypted before being sent. The method of changing this plain text is called "encryption" or more precisely a "cipher". The changed text is referred to as ciphertext. This renders the message difficult to read. The message must be changed back before it could be read.

Creating written or generated codes that allow the information to be kept undisclosed is comprised of cryptography. The data or plain text is converted into a format that cannot be read by an unauthorized user. Thus, the data can be conveyed without anyone decoding back into a readable format. This technique is used by information technology at various levels. Without the key to decrypt the information, it cannot be decrypted into readable form. During transmission and storage, integrity of information is maintained. Non-repudiation is also aided by cryptography. Thus, neither the creator nor the receiver of the information can claim they did not create or receive the information or data.

2.6 AVIPSA

The objective of automated validation of Internet Security protocols and applications (AVISPA) is the development of push-button, business-strength era for large-scale net security-sensitive applications and protocols.

2.7 Objectives

To create a vivid description of language for security goals, validating protocols, threat and risk models of business complexity. To enhance the state-of-the-art in programmed deduction strategies to step as much as this complexity. To build a device grounded on these techniques with the intention to allow the industry and standardized organizations to mechanically substantiate or stumble on faults in their products. The tool is tuned for demonstrating its concept for various protocols used in industry, education, etc. To migrate this technology into the industry. The main objective of this project is to assimilate this technology into a robust automated tool, changed on real-world, significant problems and migrate it to standardized forms, whose engineers are in dreadful need in today's era.

According to the provided problem statement, the cryptographic algorithms are implemented using the AVISPA tool. For the Internet security, AVISPA tool is used. For this, we use a formal language for classifying the protocols and their security problems and combine them with a wide range of cryptographic strategies. Experimental results show implementation of AVIPSA tool on an Internet security protocols is suitable for this kind of evaluation [7].

3 Wireless Sensor Network

Nowadays, the WSN [8] has been engaged in widespread areas like tracking, controlling and monitoring. Wireless security provides protection from unauthorized access or destruction to computers, while the user is accessing wireless communication. The most common security used in our communication is the wired equivalent privacy (WEP) and the Wi-Fi protected access (WPA). WEP provides fragile security and standards to crack the password with the help of software tools which are widely available over the Internet and even with elementary laptop computer [9].

Replaced by WPA or Wi-Fi protected access, the WEP technique is an outdated IEEE 802.11 standard since 1999. WPA offered a rapid alternative to enhance the level of Internet security over WEP [10]. Currently, a standard known as WPA2 is being used. WPA2 uses an encrypting technique that encrypts the information using a 256-bit key, hence improving the security over WEP because of having a longer key length.

The network comprises the huge number of sensor nodes with a prescribed capacity along with transmitters that form the distributed system. Each sensor node possesses the capabilities to calculate, detect and communicate. These nodes are randomly circulated in the environment to be observed and can identify each other and cover the entire area. The WSN applications are not only limited to military applications, but also in terms of Internet security. The traditional mechanisms of WSN used in security have constrained sources of the sensor nodes. Hence, the security mechanisms of WSNs should be designed bearing in mind the limited resources and malicious sensors [3]. Therefore, the security requirements of the wireless sensor include integrity, confidentiality and authentication [11].

4 Simulation Results

4.1 Rivest Shamir Adleman (RSA)

This algorithm is the primary set of rules used in asymmetric cryptography algorithm. Asymmetric denotes that it works on two distinctive keys, i.e., public key that is given to everyone and private key that is kept private.

4.2 Asymmetric Cryptography Has a Few Instances

The public key is sent by the client (say browser) to the server, and request is generated for some data. Using the client's public key, server encrypts the data and then sends this encrypted data. With the help of private keys, data received by the client is decrypted. In asymmetric, the data can be decrypted by no one else except the browser despite any other parties having the public key of the browser.

4.3 Data Encryption Standard (DES)

The National Institute of Standards and Technology is known as "NIST" [12] published the block cipher symmetric key that is the technique known as data encryption standard [13]. Data encryption standard is an application of a Feistel Cipher. A 16 round Feistel structure is used by this algorithm. This technique uses a block size of 64 bits [14]. This encryption technique has a 56 bits key length, despite it has 64 bits of key length, as 8 of these 64 bits are check bits only [15].

In Figs. 1 and 2, an improved version of RSA is simulated, and the outcomes are compared with the existing algorithms, i.e., DES and RSA.

The replicated results are shown in Fig. 2, Alice sends the message request to Bob, and the acknowledgment is received by Bob as a response. The single session was used in DES and RSA, whereas, in the improved algorithm, two sessions were used.

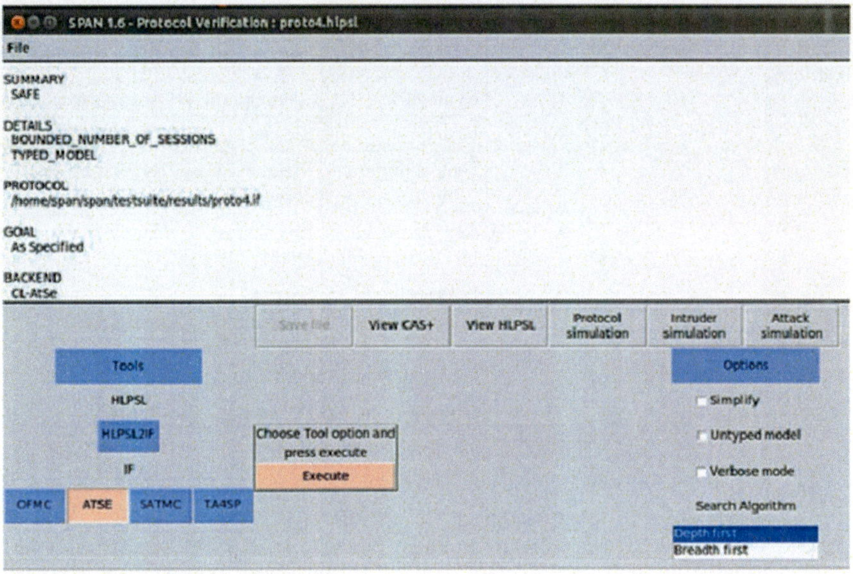

Fig. 1 Summary of improved RSA

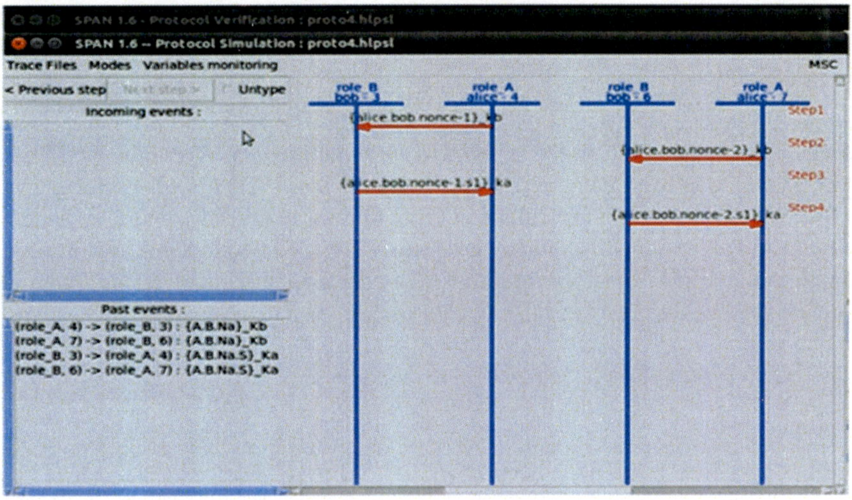

Fig. 2 Simulation of improved RSA

Table 1 Comparison of DES, RSA and the improved algorithm

S. No	DES	RSA	Proposed algorithm
1.	Symmetric	Asymmetric	Asymmetric
2.	Faster	Slower	Faster than RSA
3.	Data encryption	Symmetric key encryption	Symmetric key encryption
4.	Key size is 64 bit	Key size is 2048	Varies
5.	Less secure	Secured, as compared to DES	More secured than RSA
6.	Uses only one session	Uses only one session	Uses two sessions

5 Results

Symmetric and asymmetric key cryptographic algorithms were compared with regard to authentication, integrity and non-repudiation of a message using one or two sessions in the AVISPA tool. The result of which is as follows (Table 1) [16].

6 Conclusion

In this paper, symmetric and asymmetric key cryptographic algorithms like DES, RSA and the improved RSA were compared using the AVISPA tool. The objective is to enhance the existing algorithms in order to remove the flaws and to provide a more secure algorithm. To augment the security, we used two sessions as compared to one in the case of existing algorithms. To conclude, the improved algorithm is more secured as compared to existing algorithms.

References

1. Pranggono Abdullahi Araboand Bernardi (2013) Mobile malware and smart device security: trends, challenges and solutions. IEEE, The Oxford Internet Institute (OII), UK
2. Gaw S, Felten EW (2006) Password management strategies for online accounts. In: Proceedings of the second symposium on Usable privacy and security, pp 44–55
3. Jiang Z, Fu C, Wei W, Wei A (2013) An energy balanced algorithm of LEACH protocol in WSN. Int J Comput Sci Issues 354–359
4. Kishan Rao K, Brahma Reddy B (2013) A mmodified clustering for LEACH algorithm in WSN. IJACSA 4(5). VBIT, India and Vaagdevi College of Engineering, Warangal, India

5. Lin H, Zhou Y, Zhang Q (2010) CL-PKC based secure multicast communication for P2P network. In: 2010 second international conference on network security, wireless communications and trusted computing. IEEE, pp 154–157
6. Dey H, Datta R (2012) Monitoring threshold cryptography based wireless sensor networks with projective plane. IEEE, Indian Institute of Technology, Kharagpur, India
7. Au MH, Chen J, Liu JK, Mu Y, Wong DS, Yang G (2007) Malicious KGC attack in certificate less cryptography. In: Proceedings of ACM symposium on information, computer and communications security. ACM Press
8. Karthikeyan B, Veluman M, Kumar R, Inabathini SR (2015) Analysis of data aggregation in wireless sensor network. IEEE, School of Electronics Engineering, VIT University, Vellore, India
9. Sheng Z, Leung VCM, Mahapatra C, Zhu C (2015) Recent advances in industrial wireless sensor networks toward efficient management in IoT. IEEE, vol 3, pp 622-637
10. Ahmed MH, Alam SW, Qureshi N, Baig I (2011) Security for WSN based on elliptic curve cryptography. IEEE, pp 75–79
11. Zhu H, Liao Q (2013) An energy balanced clustering algorithm based on LEACH protocol. In: Proceeding of the 2nd international conference on system engineering and modeling vol 13, pp 0072–0077
12. Riyami SA, Paterson K (2003) Certificateless public key cryptography. In: Proceedings of Asiacrypt-03, pp 452–473
13. Paar C, Eisenbarth T, Uhsadel L, Kumar S (2007) A survey of lightweight cryptography implementations. In: 2007 IEEE test of computers, pp 0–11
14. Sumanthi P, Roseline RA (2011) Energy efficient routing protocol and algorithms for wireless sensor networks—a survey. Global J Comput Sci Technol 11
15. Kumar V, Singh J, Kumar R (2015) An RSA based certificateless signature scheme for wireless sensor networks. In: ICGCIoT-2015, IEEE, pp 443–447
16. Kaur S, Bharadwaj P, Mankotia S (2017) Study of multi-level cryptography algorithm: multi-prime RSA and DES. Int J Comp Netw Inf Sec 11(9):22

Device Level Authentication Protocol for Wireless Body Area Networks

Ashish Joshi and Amar Kumar Mohapatra

Abstract The number of wearable electronic devices in the modern world is growing steadily. The amalgamation of these devices worn by a single person to form a network is known as body area networks. Traditionally, for ensuring the security of this type of network authentication protocols are used. Many protocols has been proposed for authenticating the mobile station with various service providers. The problem arises in Intra-BAN authentication. Specifically, when these devices area changed frequently. This paper presents authentication protocol specifically designed for operation in conditions of unstable connection with a certification center (center certification) also keeping in mind frequent change in devices. Similar scenarios may occur in the case of increased load in the existing network and in places with low cellular coverage or when operating in hard-to-reach (remote) areas. We have reviewed the closest analogues of the protocol, its relative advantages are analysed functioning scenario under consideration. A detailed description of the protocol is given. Our Protocol proved to ensure integrity and confidentiality about the user of the device.

Keywords Authentication protocol · Body area networks · Wireless connection · Insecure channel · AVISPA

1 Introduction

Throughout the development of cellular communications and electronic commerce, ensuring security in data transmission is an important task [11]. Currently, authentication of users, devices of the Internet of things, events and systems is carried out

A. Joshi (✉)
University School of Information Communication and Technology,
GGSIPU, Dwarka, Delhi, India
e-mail: ashishium@gmail.com

A. K. Mohapatra
Indira Gandhi Delhi Technical University for Women, Kashmere Gate, Delhi, India
e-mail: mohapatra.amar@gmail.com

A. Abraham et al. (eds.), *Proceedings of 3rd International Conference on Computing Informatics and Networks*, Lecture Notes in Networks and Systems 167,
https://doi.org/10.1007/978-981-15-9712-1_13

using authentication protocols [15]. A distinctive feature of authentication protocols used in modern IP networks is the need for a certification authority (CA), which is also responsible for routing messages between remote nodes [17]. The main advantage of such systems is the absolute trust of the CA, which takes responsibility for the generation, distribution, updating and revocation of keys/certificates for all users of the system [2]. At the same time, the development of the wearable electronics market dictates new trends in the world of wireless technologies and information security [3]. However, relatively small devices, such as wearable electronics and smart home objects, still have low computing power and are limited in control capabilities [18]. The market of wearable electronics has just begun to develop, so the target audience of users has not yet reached a large scale [1]. The public is not ready to purchase new expensive products of wearable electronics. In this regard, to increase the consumer base, short-term rental of such devices can serve as a profitable scenario. On the other hand, according to the standards of modern manufacturers, a user change can only be carried out if the device is completely cleaned to the factory state [14]. This scenario—a temporary transfer of device control—can occur both in places with excellent cellular coverage (cities) and in remote areas [7, 9] (mountain resorts, sea cruises, etc.). In the second case, the need for a permanent connection with the CA can be a serious problem, which entails the difficulty of unequivocally confirming ownership of the device, updating transfer rules, and also extending the lease terms. Thus, the objective of this study is to provide the possibility of temporary transfer of control of devices in an unstable connection with the CA in order to provide guarantees of data integrity and confidentiality. However, a complete rejection of interaction with the CA and the transition to fully distributed solutions is also impossible. For example, renting devices in the context of such business models as B2B or B2C implies the need for formal confirmation of the transfer and return of wearable electronics not only at the lower level of interaction, but also on a centralized node (server). The main goal of the study is to develop a family of information security protocols that eliminates the above problem.

2 Review of Related Literature

Unfortunately, modern scientific literature does not consider the problem of device delegation in the context of an unstable connection with a CA. At least the following is an overview of works that are closest in terms of problems. In [13], a protocol was presented that allows one to reduce the amount of computation for devices with a significantly low level of computing resources. In this protocol, an aggregation node (AU) is responsible for issuing keys for child devices. In this way, energy costs are reduced on devices with a less capacious battery. The main advantage of the proposed solution is the transfer of the time-consuming procedure of handshaking the key generation to the AU. On the other hand, the disadvantage is the need for a stable connection with CA. Work in the presence of such a connection requires a minimum of 7 and a maximum of 15 handshakes in the association phase. In [16], an algorithm based on elliptic curves was presented, which allows authentication of devices using wireless LAN communication technologies. The main disadvantages of this technique are: the need for a permanent secure channel between the authenticated device and the CA, as well as the inability to verify the history of mutual authentications. The protocol proposed in this article was developed taking into account the problems described above. Its main task is to provide authentication for the subsequent delegation of electronic devices in both cases—in the presence and absence of a secure channel to the CA. Due to the ability of the system to work even in the event of a connection failure with the CA during the establishment of communication between the device owner and the potential tenant, the protocol eliminates the disadvantages of the known technical solutions. The system is aimed at ensuring data integrity and counteracting a number of attacks on the transmitted device from both the owner and the tenant of the device.

3 Proposed Authentication Protocol

The proposed protocol provides operation in scenarios with a potentially unstable connection from the AU to the CA, as well as in the case of the possibility of making direct connections between the AU and the delegated device and the AU of different users. The problem is solved by combining the solutions considered in systems with public key infrastructure, which are responsible for the initial generation of keys and certificates, as well as solutions that allow direct connection in the absence of communication with the public key infrastructure. The main options for cryptographic solutions for potential use in this protocol are proposed below. To obtain the result of the data hash function, GOST R 34.11-2012, SHA-2, SHA-3, BLAKE2 [5] algorithms can be used, for the CU it is possible to use the classic public key cryptosystem protocols [8, 12]. Protection against man-in-the-middle attacks (an intermediary attack) can be implemented using the classic Diffie-Hellman protocol [6] or the Elliptic curve Diffie-Hellman (ECDH) [6]. The authentication protocol for the delegation of rights to use electronic devices consists of certain operational steps:

1. Initial association of a new device.
2. Transfer the device for a specific time interval.
3. Return for the use by the device owner.
4. Dissociation of the device.

3.1 *Initial Association of a New Device*

The initial data is a unique user identifier ID (for example, user@mail.com), which is used for interaction between protocol participants (CAs, system users, and the delegated device). The participants of the protocol at this stage are the previously unused delegated electronic device (DED) and the aggregating node (for example, the user's smartphone). Further, the CA issues certificates for the user AU and the new DED, assuring the integrity of the DED firmware and its unambiguous belonging to the AU.

The protocol steps of this stage are shown in Fig. 1. Hereinafter, the owner of the DED is the user Alice; the device's tenant is Bob's user. A secure channel defines a connection that is resistant to the "man in the middle" attack. Each user has a unique IID_A type alice@mail.com, their own secret and public keys S_K and PK_A. The CA stores the public keys of all users and the corresponding certificates $\text{cert}_A = \text{sign}_{\text{CA}}(\text{PK}_A)$. Each DED has a pre-installed factory software and a factory key for resetting to the initial settings. It assumes the presence of a secure timer and secure storage, which are considered trusted. The following functions are assigned to the protected part of the device:

1. **Safe timer/clock**—providing a trusted time value. Without this property, it is impossible to track the main parameter of the delegation—time.
2. **Secure storage**—necessary to store confidential information, keys, etc.
3. **Trusted device functionality**—functions directly related to information security, such as, for example, hash functions, pseudo-random sequence generators, etc.

Also, during the operation of the protocol, the S_A key associated with the Alice user is entered. This key allows symmetric encryption of the data transmitted and stored on the DED, and also serves as an additional protection against leakage of confidential Alice user information stored on a separate DED. Further protocol steps are indicated by the corresponding numbers in Fig. 1.

1. The Alice user generates the S_A secret key for the new DED w_i and transmits it over the secure channel.
2. DED w_i transfers the result of the execution of the hash function from its own software to the CA using a secure channel $\text{hash}(S_{\text{wi}})$
3. User Alice transfers his public key PK_A and ID_A to the CA.
4. The CA generates a CA certificate for Alice. The certificate has the form $\text{cert}_{\text{CA}} = \text{sign}_{\text{CA}}(w_i, \text{ID}_A, \text{hash}(S_{\text{wi}}))$. This step is performed to ensure the integrity of system data.
5. The CA, using a secure channel, transfers the cert_{CA} to the Alice user.
6. The Alice user signs the certificate received from the CA with his secret key $\text{cert}_A = \text{sign}_A(\text{cert}_{\text{CA}})$.
7. The Alice user submits the certificate to the DED.

3.2 *Transferring the Device for Some Time to Use by Another User*

The functioning of the system can occur in two modes: in the presence of a connection to the CA and in the absence of a connection. In the first case, the CA is fully

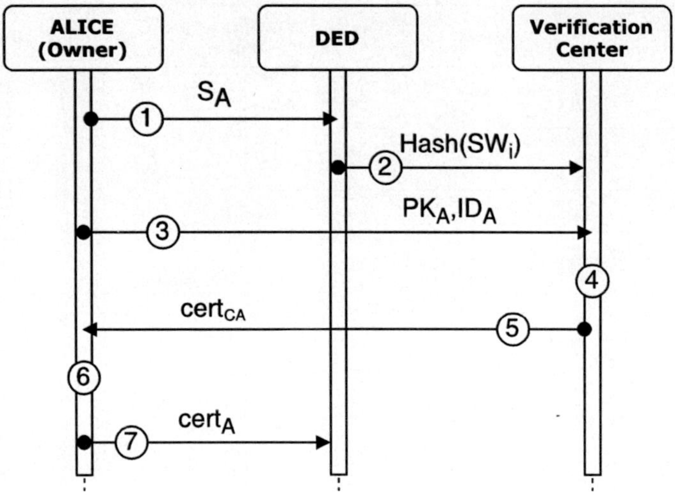

Fig. 1 Initial association of a DED

responsible for authentication. In the absence of a stable connection with the CA, the delegation protocol requires a direct connection between the user agents and the delegated DED. The protocol steps performed at this stage are shown in Figs. 2 and 3 for 2A and 2B, respectively. This scenario describes the lease of an electronic device, indicating the conditions and time of use of the DED.

Protocol 2A—presence of a stable connection for both the users

1. User Alice sets the maximum delegation time t_d for DED w_i using a service message signed by SK_A, as $m[D]_A = \mathrm{sign}_A(w_i, t_d, \mathrm{ID}_A, \mathrm{ID}_B, \{\text{other delegation conditions}\})$.
2. User Alice transfers $m[D]_A$ to the CA using a secure channel.
3. The CA verifies the authenticity of the message from Alice using PK_A. If the test fails, the protocol stops working.
4. The CA signs the delegation message $m[D]_{\mathrm{CA}} = \mathrm{sign}_{\mathrm{CA}}(m[D]_A)$.
5. The CA transmits $m[D]_{\mathrm{CA}}$ and cert_A for Bob.
6. User Alice sends the service message$[m[C(S_A)]_A$ to the DED, deleting the private key S_A from the DED.
7. If the user Bob does not trust Alice, The DED is reset to the factory settings. The reset occurs with the $m[D]_{\mathrm{CA}}$ and $\mathrm{cert}_{\mathrm{CA}}$ certificate stored in a secure store. These values were obtained at step 6 of the initialization protocol in order to preserve data integrity and confirm ownership in the absence of connection to the CA. User AU Bob compares the result of the hash function from the current DED software with that stored in $\mathrm{cert}_{\mathrm{CA}}$. If they do not match, the algorithm stops execution. Thus, the Bob user loses the ability to use the DED device, since it is assumed that the software could be compromised by the owner. It is

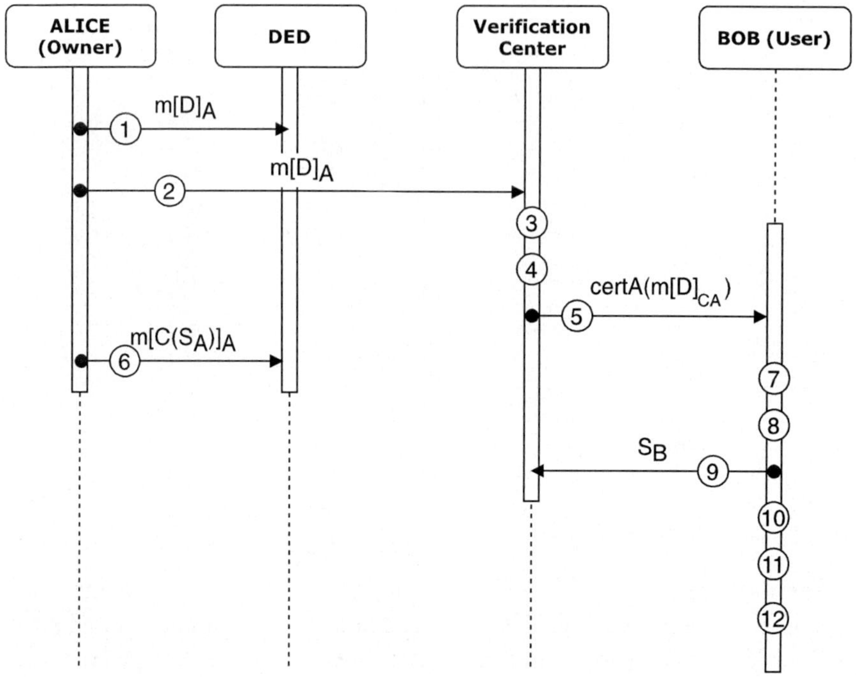

Fig. 2 Transfer of a DED in the presence of a stable connection with the CA

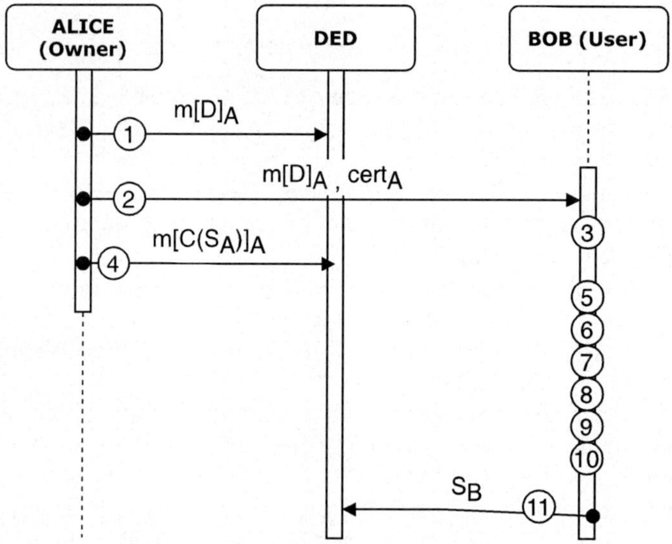

Fig. 3 Transfer of a DED in the absence of a stable connection with the CA

important to note that the protected timer and storage remain unchanged even when reset to factory settings.

8. If the Bob user trusts Alice, the DED software component remains unchanged, and the temporary user is able to use the device owner's software.
9. User Bob generates an S_B secret key for direct interaction with the DED.
10. User Bob sends the S_B to the DED via a secure channel.
11. To ensure data integrity, Bob calculates a new value of $\text{sign}_B(w_i, S_{\text{wi}})$.
12. In case of expiration of the t_d delegation timer on the DED side, the device parameters are reset to the factory settings while maintaining the contents of the protected storage. The timer can be remotely updated if there is a simultaneous connection with the user's CA and the owner when using the service message $m[D]_A = \text{sign}_A(w_i, t_d, \text{ID}_A, \text{ID}_B, \{\text{other delegation conditions}\})$.

Protocol 2B—the absence of a stable connection for at least one of the users

1. User Alice sets the maximum delegation time t_d to the DED wi using the service message signed by SK_A as $m[D]_A = \text{sign}_A(w_i, t_d, \text{ID}_A, \text{ID}_B, \{\text{other delegation conditions}\})$.
2. User Alice passes cert_A to user Bob via a secure channel.
3. Bob authenticates cert_A using cert_{CA}. If the test fails, the protocol stops working.
4. User Alice sends the service message $[m[C(S_A)]_A$ to the DED, deleting the private key S_A from the DED.
5. If the user Bob does not trust Alice, The DED is reset to the factory settings. The reset occurs with the $m[D]_{\text{CA}}$ and cert_{CA} certificate stored in a secure storage. These values were obtained at step 6 of the initialization protocol in order to maintain data integrity and confirm ownership in the absence of connection to the CA. User AU Bob compares the result of the hash function from the current DED software with that stored in certCA. If they do not match, the algorithm stops execution. Thus, the Bob user loses the opportunity to use the DED device, since it is assumed that the software could be compromised by the owner. It is important to note that the protected timer and storage remain unchanged even when reset to factory settings.
6. If the Bob user trusts Alice, the DED software component remains unchanged, and the temporary user has the opportunity to use the device owner's software.
7. User Bob generates an S_B secret key for direct interaction with the DED.
8. The user Bob transfers the S_B to the DED via a secure channel.
9. To ensure data integrity, Bob calculates the new value of $\text{sign}_B(w_i, S_{\text{wi}})$.
10. In the case of the expiration of the t_d delegation time on the DED side, the device parameters are reset to the factory settings while maintaining the contents of the protected storage. The timer can be updated using the service message $m[D]_A = \text{sign}_A(w_i, t_d, \text{ID}_A, \text{ID}_B, \{\text{other delegation conditions}\})$ if there is a direct connection between users.
11. The user Bob sends the S_B to the DED via the secure channel.
12. To ensure data integrity, Bob calculates the new value of $\text{sign}_B(w_i, S_{\text{wi}})$.

3.3 Returning the Device to the Owner After Temporary Use of the DED by Another User

The functioning of the system can occur in two modes: in the presence of a connection with the CA and without such a connection. The corresponding steps are shown in Fig. 4 (protocol 3A) and Fig. 5 (protocol 3B). In the first case, similarly to delegation, CA is fully responsible for authentication. In the absence of a stable connection with the CA, the device return algorithm requires direct connections between user accounts and the delegated DED.

Protocol 3A—the presence of a stable connection for both users

1. User Bob generates a service message signed by SK_B as $m[R]_B = \mathrm{sign}_A(\mathrm{wi}, R)$.
2. User Bob transmits $m[R]_B$ to the CA using a secure channel.
3. The CA verifies the authenticity of the message from Bob using PK_B. If the test fails, the protocol stops working.
4. The CA signs a return message $m[R]_{\mathrm{CA}} = \mathrm{sign}_{\mathrm{CA}}(m[R]_B)$.
5. The CA transmits the $m[R]_{\mathrm{CA}}$ to Alice.
6. User Bob sends the service message $m[C(S_B)]_B$ to the DED, deleting the S_B secret key from the DED.
7. If the Alice user does not trust Bob, The DED is reset to the factory settings and the data is stored in a secure storage. Alice's user account compares the result of the hash function from the current DED software with that stored in $\mathrm{cert}_{\mathrm{CA}}$. If they do not match, the protocol stops working. Thus, the Alice user loses the ability to use the DED device, since it is assumed that the software could be

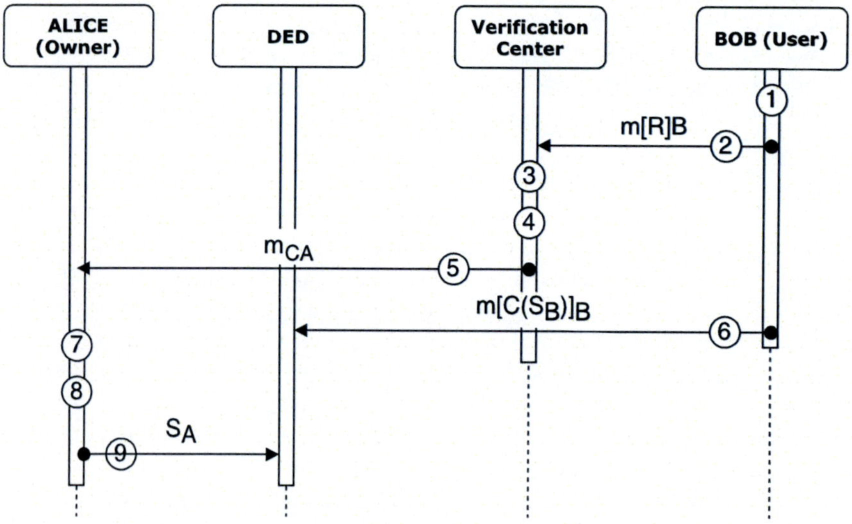

Fig. 4 DED return phase in the presence of a stable connection with the CA

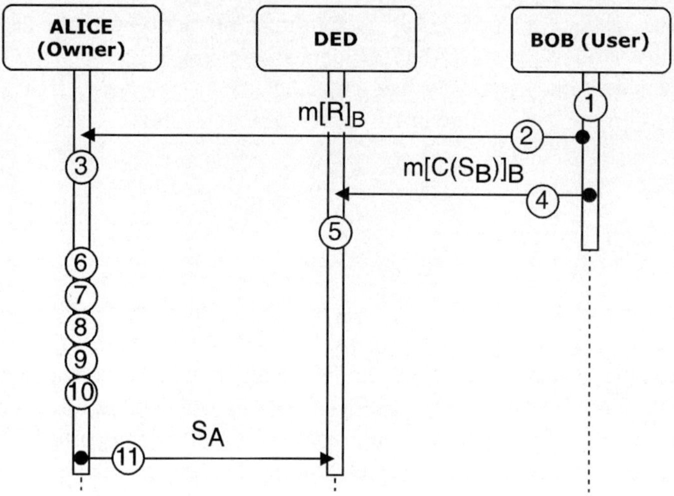

Fig. 5 DED return phase in the absence of a stable connection with the CA

compromised by the owner. It is important to note that the protected timer and storage remain unchanged even when reset to factory settings.

8. If the Alice user trusts Bob, the DED software component remains unchanged, and the device owner has the opportunity to use the software installed during the delegation process.
9. The Alice user generates a secret key S_A for direct interaction with the DED.
10. User Alice transfers S_A to the DED using a secure channel.
11. To ensure data integrity, the Alice user calculates a new value $\text{sign}_A(w_i, S_{\text{wi}})$. Only the cert_A and cert_{CA} are in the secure DED repository.

Protocol 3B—the absence of a stable connection for at least one of the users

1. User Bob generates a service message signed by SK_B as $m[R]_B = \text{sign}_B(wi, R)$.
2. User Bob transmits $m[R]_B$ to user Alice over a secure channel.
3. Alice authenticates $m[R]_B$ using cert_{CA}. If the verification fails, the protocol stops its operation.
4. User Bob sends the service message $m[C(S_B)]_B$ to the DED, deleting the S_B secret key from the DED.
5. If the Alice user does not trust Bob, The DED is reset to the factory settings and the data is stored in a secure storage. Alice's user account compares the hash function result from the current DED software with that stored in cert_{CA}. If they do not match, the protocol stops working. Thus, the Alice user loses the ability to use the DED device, since it is assumed that the software could be compromised by the owner. It is important to note that the protected timer and storage remain unchanged even when reset to factory settings.
6. If the Alice user trusts Bob, the DED software component remains unchanged, and the owner has the opportunity to use the device's temporary user software.

7. User Alice generates a private key S_A for direct interaction with the DED.
8. User Alice sends S_A to the DED through a secure channel.
9. To ensure data integrity, the Alice user calculates a new value $\text{sign}_A(w_i, S_{\text{wi}})$. Only the cert_A and cert_{CA} are in the secure DED repository.

3.4 Device Dissociation

In this case, there are two possible options for the functioning of the protocol: manual and automatic dissociation (resetting the device to factory settings). In manual dissociation, the owner and the DED communicate in a direct channel. Owner AU sends a control message that resets the DED to the state of factory settings. Automatic dissociation is possible according to a predefined timer, which performs a similar procedure for the expiration of the delegation time of the device in case of non-return.

Protocol 4A—Manual Dissociation of DED

1. Owner Alice generates a service message signed by SK_A as $m[F]_A = \text{sign}_A(w_i, F)$.
2. User Alice sends $m[R]_B$ to the DED through a secure channel.
3. The DED is reset to factory settings with the clearing of the secure storage.
4. The DED device can be restored only when using the factory key (for example, PIN) and if there is a connection with the CA.

Protocol 4B—Automatic Dissociation of DED
This protocol is executed when the device was not returned according to the conditions of delegation, lost or stolen. All personal data of the owner and user must be destroyed in order to prevent potential malicious use.

1. During the initial association of the device, the owner of Alice optionally generates a service message signed by SK_A, as $m[E]_A = \text{sign}_A(w_i, t_e)$, where t_e is the value of the countdown timer for automatic dissociation. DED refers to a timer during operation in order to detect potential time-out.
2. In the case of time-out, the DED is reset to the factory settings, clearing the protected storage.
3. The DED device can be restored only by using the factory key (for example, PIN) and with a connection to the CA.

4 Security Analysis Using AVISPA

Automated Formal verification is one of the effective techniques for security analysis and protocol design assurance. One of the laws. Most of the formal verification techniques use formal language to model network security protocols. They generally employ analytical methods and assumptions to demonstrate the security of the

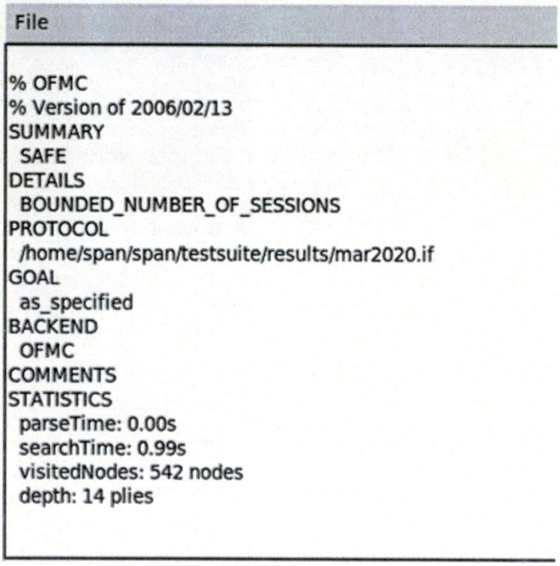

Fig. 6 Output of OFMC

designed protocol [4]. AVISPA is a type of validation automation tool which is used to verify various types of security-related network protocols [10]. The AVISPA toolset uses High level protocol specification language (HLPSL) to describe the implementation of the entire protocol. HLPSL is a rich, modular, role-based formal language. The proposed protocol was modeled using HLPSL and validated over AVISPA backends. We use OFMC and CL-ATSE to validate our protocol. Both the backends validated our protocol and considered it safe for bounded number of sessions. Figures 6 and 7 show the result of OFMC and CL-ATSE respectively.

Although our protocol was found to be safe by AVISPA but after a detailed review of the proposed protocol, it is important to note potential attacks on its operation. One such attack is phishing. In this case, the attacker Eve may appear to be Bob's trusted user with ID_B during the delegation process between Alice and Bob. If Alice cannot verify the authenticity of the source of the request, the attack is considered successful. Primary attack area is located in the field of wearable electronics, because devices of this type often do not have an additional visual channel (for example, a display) to further confirm the delegation procedure. In the modern world, there is no solution that fully protects against phishing attacks, however, multifactor authentication techniques can reduce the likelihood of an attack success.

Another important attack on the protocol is the classic intermediary attack. In our case, the attacker Eve requests $m[D]_A$ for Bob (ID_B) from Alice, pretending to be Alice (ID_A), and at the same time passes $m[D]_A$ for user Bob. As a result, Eve does not receive the ability to use the device, however, it can monitor the delegation process.

One of the most interesting attacks is an attack using an infected device. Consider the following example. After an attacker intercepts the correct $m[D]_A$ about device

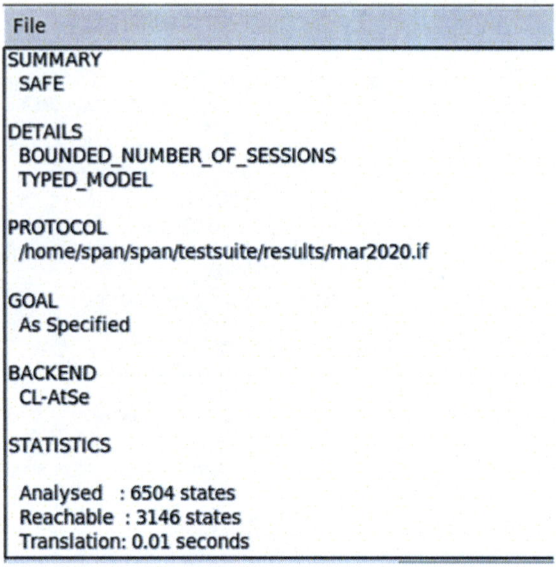

Fig. 7 Output of ATSE

delegation w_k, Eve creates an infected device that appears to be w_k and always passes the correct hash(SW_k) value. This device, for example, allows you to monitor Bob user activity. At the same time, digital signature protocols are particularly vulnerable to confidentiality, as they tend to involve unambiguous confirmation from a trusted entity. In other words, the user cannot later deny the fact of delegation. Protocols based on zero-disclosure evidence rely on this feature to enhance security, but cause additional confidentiality complications.

5 Conclusion and Future Work

An authentication protocol for rapid provisioning of electronic devices is presented. This protocol is developed for use in conditions of unstable communication with a certification authority (CA). Algorithms that implement the stages of the functioning of the proposed protocol can be implemented both in the form of software for a universal computer of any architecture, and in the form of hardware for a specialized computer of any architecture. The protocol can be used in places with no infrastructure, because for the implementation of delegation does not require the constant presence of a connection with the CA. We have not implemented the protocol over any test-bench to find out the association and disassociation time. Researchers may implement and verify the efficiency of our protocol using a suitable hardware/software.

References

1. Abrardo A, Fodor G, Tola B (2015) Network coding schemes for device-to-device communications based relaying for cellular coverage extension. In: 2015 IEEE 16th international workshop on signal processing advances in wireless communications (SPAWC). IEEE, pp 670–674
2. Andreev S, Galinina O, Pyattaev A, Gerasimenko M, Tirronen T, Torsner J, Sachs J, Dohler M, Koucheryavy Y (2015) Understanding the IoT connectivity landscape: a contemporary m2m radio technology roadmap. IEEE Commun Mag 53(9):32–40
3. Arias O, Wurm J, Hoang K, Jin Y (2015) Privacy and security in internet of things and wearable devices. IEEE Trans Multi-scale Comput Syst 1(2):99–109
4. Armando A, Basin D, Boichut Y, Chevalier Y, Compagna L, Cuéllar J, Drielsma PH, Héam PC, Kouchnarenko O, Mantovani J, et al (2005) The AVISPA tool for the automated validation of internet security protocols and applications. In: International conference on computer aided verification. Springer, pp 281–285
5. Aumasson JP, Neves S, Wilcox-OHearn Z, Winnerlein C (2013) BLAKE2: simpler, smaller, fast as MD5. In: International conference on applied cryptography and network security. Springer, pp 119–135
6. ElGamal T (1985) A public key cryptosystem and a signature scheme based on discrete logarithms. IEEE Trans Inf Theory 31(4):469–472
7. He D, Zeadally S (2014) An analysis of RFID authentication schemes for internet of things in healthcare environment using elliptic curve cryptography. IEEE Internet Things J 2(1):72–83
8. Hu C, Li H, Huo Y, Xiang T, Liao X (2016) Secure and efficient data communication protocol for wireless body area networks. IEEE Trans Multi-scale Comput Syst 2(2):94–107
9. Hummen R, Ziegeldorf JH, Shafagh H, Raza S, Wehrle K (2013) Towards viable certificate-based authentication for the internet of things. In: Proceedings of the 2nd ACM workshop on Hot topics on wireless network security and privacy, pp 37–42
10. Joshi A, Mohapatra AK (2019) Security analysis of wireless authentication protocols. Int J Sens Wirel Commun Control 9(2):247–252
11. Kim HJ, Chang HS, Suh JJ, Shon Ts (2016) A study on device security in IoT convergence. In: 2016 international conference on industrial engineering, management science and application (ICIMSA). IEEE, pp 1–4
12. Li X, Ibrahim MH, Kumari S, Sangaiah AK, Gupta V, Choo KKR (2017) Anonymous mutual authentication and key agreement scheme for wearable sensors in wireless body area networks. Comput Netw 129:429–443
13. Li X, Peng J, Kumari S, Wu F, Karuppiah M, Choo KKR (2017) An enhanced 1-round authentication protocol for wireless body area networks with user anonymity. Comput Electr Eng 61:238–249
14. Ometov A, Zhidanov K, Bezzateev S, Florea R, Andreev S, Koucheryavy Y (2015) Securing network-assisted direct communication: the case of unreliable cellular connectivity. In: 2015 IEEE Trustcom/BigDataSE/ISPA, vol 1. IEEE, pp 826–833
15. Saadeh M, Sleit A, Qatawneh M, Almobaideen W (2016) Authentication techniques for the internet of things: a survey. In: 2016 cybersecurity and cyberforensics conference (CCC). IEEE, pp 28–34
16. Shen J, Chang S, Shen J, Liu Q, Sun X (2018) A lightweight multi-layer authentication protocol for wireless body area networks. Future Gen Comput Syst 78:956–963
17. Sicari S, Rizzardi A, Grieco LA, Coen-Porisini A (2015) Security, privacy and trust in internet of things: the road ahead. Comput Netw 76:146–164
18. Wei J (2014) How wearables intersect with the cloud and the internet of things: considerations for the developers of wearables. IEEE Consum Electron Mag 3(3):53–56

MASSS—Multi-agent-Based Steganography Security System for VANET

Vinay Gautam

Abstract The contemporary development in vehicular ad hoc network (VANET) has delivered an emergent field for entrepreneurs and academician. VANET supports inter and intracommunication between vehicles and infrastructure. Just because of wireless medium, the secure and efficient data/messages transmission from hop to hop in VANET is a foremost challenges because it is exposed to many attacks. Subsequently, the attacks deceived the operation of network, and therefore, safety is mandatory for effective deployment of such technology. Therefore, this paper proposes a steganography-based multi-agent approach to transmit data or message securely from terminal to terminal. The agents are mounted on terminal to transmit/receive messages. The agent at one terminal performs encryption/decryption of messages with appropriate encryption scheme of steganography and effectively transmits the messages/data to other terminal. This scheme improves the proficiency of VANET communication.

Keywords VANET · Security · Steganography · Multi-agent · Intelligent transport system

Abbreviations

GPS	Global Positioning System
ITS	Intelligent Transportation System
IBS	ID-Based Signature
LSB	Least-significant bit
MSU	Message Steganography Unit
MPU	Message Processing Unit

V. Gautam (✉)
Chitkara University Institute of Engineering and Technology, Chitkara University, Rajpura, Punjab, India
e-mail: vinay.gautam@chitkara.edu.in

A. Abraham et al. (eds.), *Proceedings of 3rd International Conference on Computing Informatics and Networks*, Lecture Notes in Networks and Systems 167,
https://doi.org/10.1007/978-981-15-9712-1_14

OBU	On Board Unit
RGB	Red, Green and Blue
RSUs	Roadside Units
SSA	Steganography Security Algorithm
VANET	Vehicular Adhoc Network
VARS	Vehicle Adhoc Reputation System
V2V	Vehicle to Vehicle
V2I	Vehicle to Infrastucture

1 Introduction

Wireless communication is an important part of any universal network such as vehicular networks (VANET). The VANET uses the concept of continuously moving and communicating multi-hops or nodes. These nodes such as vehicle in VANET can move around and communicate without any restrictions of direction and speed.

The VANET includes different type of communication such as vehicle-to-vehicle (V2V), vehicle-to-roadside (V2R) or vehicle-to-infrastructure (V2I) communication as discussed in [1]. The communication in VANET is supported by different infrastructure units such as onboard unit (OBU) and roadside units (RSUs). The OBUs empower small capacity wireless ad hoc network between communicating nodes. Every node in VANET includes an entity to identify correct position using Global Positioning System (GPS). The roadside unit (RSU) is installed athwart the path for communication organization. The counts of RSU depend upon the communication protocol. It manages message passing among moving node in order to evade hazardous condition such as accidents, jams, speed control and unseen obstacles as described in [2]. Inter and intracommunication of moving hops in VANET makes it is an intelligent transportation system (ITS).

In intelligent transportation systems (ITS), each node broadcasts the message into network and which are received by other hops in the network and used the same as per their requirement such as guarantee harmless and jam-free traffic as discussed in [3]. A intelligent transport support system is proposed in [4] uses different types of communication such as inter-vehicle, vehicle-to-roadside and routing-based communications for effectively and efficiently management of hops in the network, and all these activities require precise and up-to-date neighboring statistics. Any intelligent transport system supports three different types of communication as discussed below.

1.1 Inter-vehicle Communication

This arrangement supports multi-hop/broadcast communication between nodes in an intelligent transport system. The two message transmission methods such as Naïve

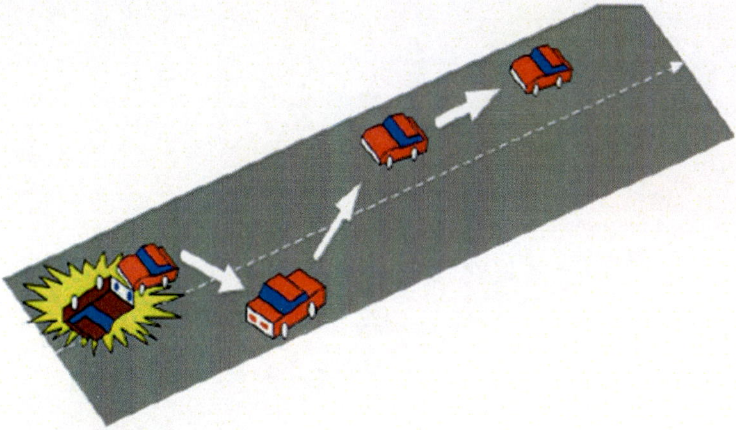

Fig. 1 I-V communication

and intelligent broadcasting are used in [4] for inter-vehicle communications as shown in Fig. 1.

1.2 Vehicle-to-Roadside Communication

This kind of arrangement uses single node broadcasting with a good extent of bandwidth link between communicating parties. In this arrangement, the roadside unit is an important hardware unit which maintains extreme load for right communication and also tackles the speed of vehicle as elaborated in Fig. 2.

1.3 Routing-Based Communication

The arrangement shown in Fig. 3 elaborates the use of routing protocol. Here, the moving node A is transmitting message to node C with the help of intermediated node B.

This section covers three different ways of transmitting a message from hop to hop, but security is a major issue in all the arrangements. The next section covers different security schemes which have been incorporated for secure communication in VANET but sill security an issue in VANET. Therefore, this paper proposes a scheme which uses steganography technique to securely transmit the message from node to node with the help of multi-agents. The agent is software component which is used to steganography technique to encrypt/decrypt the message transmit in and out of network.

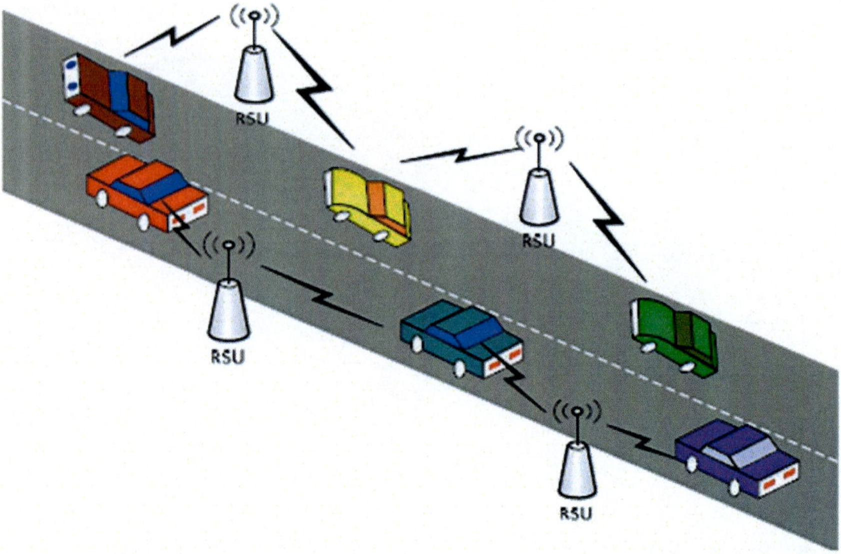

Fig. 2 V-R communication

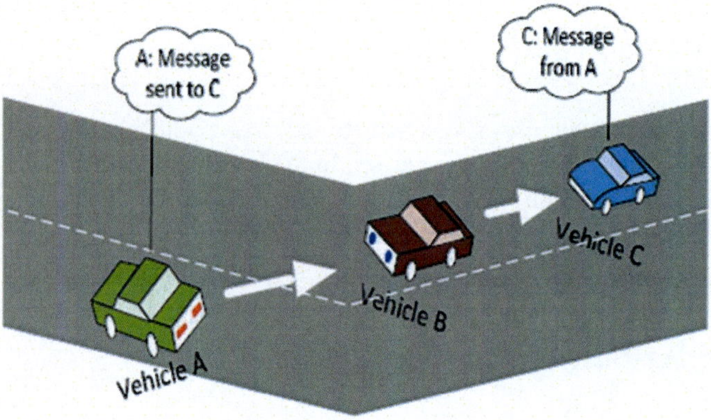

Fig. 3 Routing-based communication

This paper proposes a multi-agent-based steganography approach to secure transmission of message in VANET. The paper uses steganography algorithm to hide the message within the image. This approach uses multiple software agents to implement steganography algorithm within the VANET terminals. The layout of complete paper is as follows: The Sect. 2 explains different schemes used to secure VANET. The working of complete scheme is described in Sect. 3. The Sect. 4 describes the result and analysis. The Sect. 5 is conclusive section.

2 Related Work

Lots of advancement have been used in technology which are used in current VANET, but still, it is facing lots of security challenges in the form of attaches. Therefore, this section covers different attacks which are classified based on different ways as given below.

The network outbreaks are categorized into three dimensions such as "insider versus outsider", "malicious versus rational" and "active versus passive" as described in [5]. The attacks on message are categorized as follows: "Bogus Information", "Cheating with Positioning Information", "ID disclosure", "Denial of Service" and "Masquerade". In [6], the author has described a model to categories attacks, its risk level and its application, control, monitoring and social. Different categories of network attaches are described in [4, 7] that are based on five pillars of network security such as availability, authentication/identification, confidentiality, privacy and non-repudiation as given in Fig. 4.

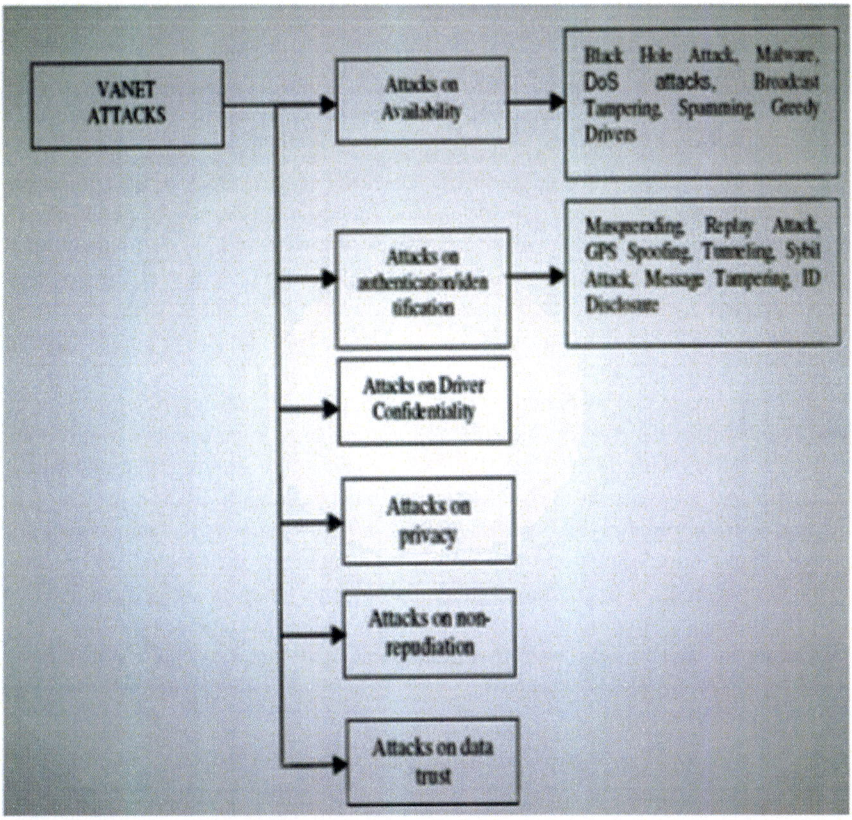

Fig. 4 Attacks in vehicular network

The current section has discussed different classification of attack which is detected over VANET. The next section will cover different solutions of security in VANET. The different security solutions are proposed in [8–25, 30] for VANET as given below: In [8], system uses four different functions to compute confidence and trust threshold for safe transmission in VANET. The computed value is improved whenever a vehicle identifies an event within its range. A vehicle ad hoc reputation system (VARS) is described in [9] which uses opinion piggybacking method for secure communication in VANET. This allows forwarding node to affix an opinion about the message and transmit further. The node opinion is based on some metrics such as trust and reputation related to geo-situation. A trust-based security employed in [10, 11] to transmit secure messages and protect application in VANET. The ID-based online/offline signature (IBOOS) technique is explained in [12] for validation of roadside units (RSU) and vehicles ID-based signature (IBS). The MLGS a new confidentiality methodology is presented in [13] to provide anonymous authentication. This methodology assumes that all nodes in network are honest and used inception contrivance as an initial countermeasure. A trust management system (RaBTM) is presented in [14] which computes trust values which is used to check whether an incoming message is trustworthy or not. An identity-based cryptosystem is described in [15] which uses pseudonym for secure communication in VANET. A secure scheme such as advanced encryption standard (AES) is used in [16] to secure VANET communication. A pairing efficient decentralized revocation (EDR) protocol is proposed in [17] for secure message passing in VANET. A conditional privacy and non-repudiation are provided using certificate-based cryptography (CBC) presented in [18] which uses hash function for secure data transmission over network. An algorithm for perceiving LSB steganography is proposed in [21]. A mathematical structure based on complex-theoretic view is proposed in [22] to seek the limits of steganography. Hash value-based scheme described in [23]. The key-based scheme is proposed in [24], where the message is encoded through a strategic and can be decoded with several other keys. A machine learning method is described in [25] to show images as suspicious or non-suspicious. An entropy-based technique is proposed in [26] for perceiving the appropriate parts of the image where data can be implanted with tiniest alteration. A new privacy preserving and authentication scheme is proposed in [31] to authenticate access of information in vehicular network. Fog computing is associated with VANET for secure data transmission [32]. A new scheme edge computing is used to share data over the network to transmit information securely [33]. Different methodology and schemes are proposed above to solve different problems of VANET, but each tackles individual problems. This paper proposes a new multi-agent-based steganography scheme which uses multiple agents for encryption and description to secure message transmission in VANET as described in next section.

3 Multi-Agent-Based Steganography Security System (MASSS) for Secure Communication

The MASSS working depends on the agents used to encrypt and decrypt the data at each terminal of the vehicular network. The agents use steganography algorithm to encrypt/decrypt the message. The detailed architecture of secure data communication is given in Fig. 5:

3.1 MASSS Architecture

As brought out above, a VANET infrastructure has two different units such as road-side unit (RST) and moving vehicles. Each terminal in the VANET is installed with multiple agents or software agents. The agents are used to encrypt/decrypt and transmit/receive the message. The complete architecture is presented in Fig. 5. The complete working of agent-based system is explained in next section. Each agent in a terminal has two different units:

a. Message Steganography Unit (MSU)
b. Message Processing Unit (MPU).

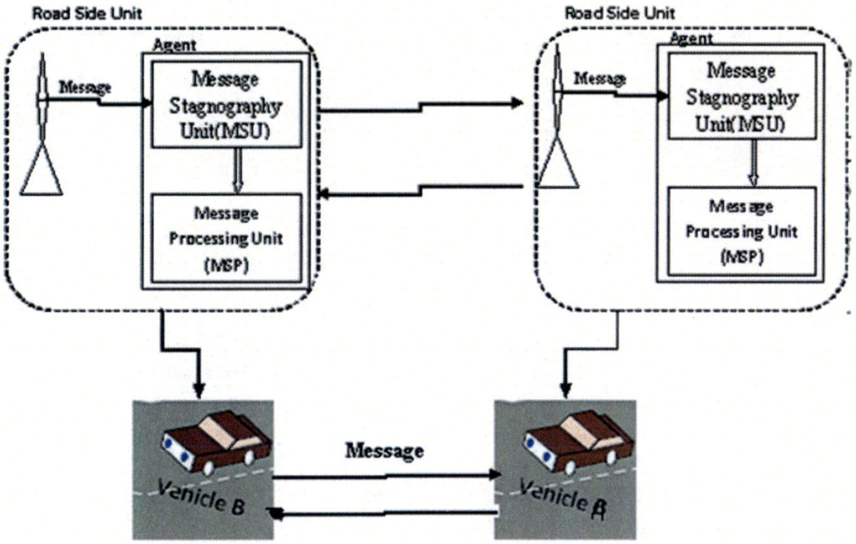

Fig. 5 MASSS architecture

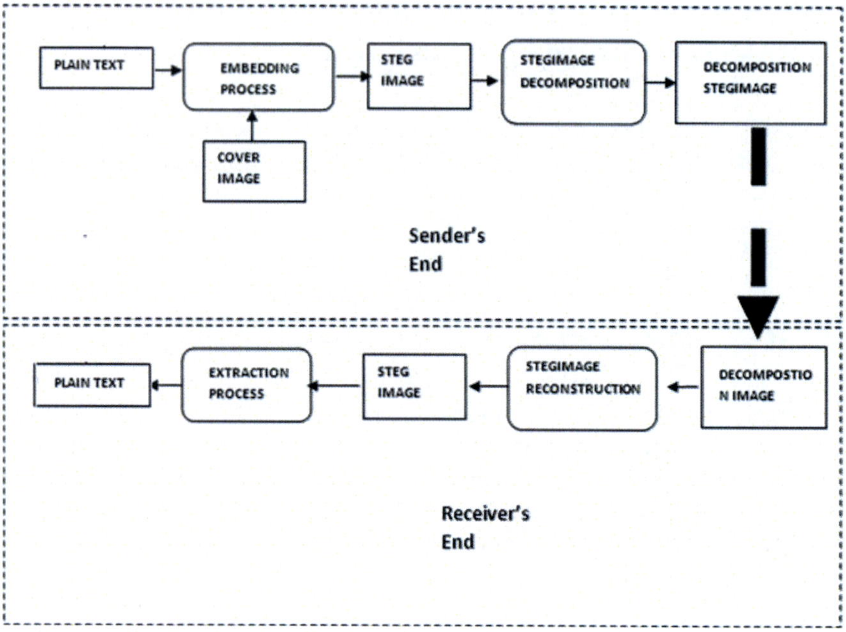

Fig. 6 Message steganography using steganography security algorithm

3.1.1 Message Steganography Unit

The MSU implements steganography security algorithm on original messages as shown in Fig. 6.

Steganography Security Algorithm (SSA)

The steganography security algorithm uses a mixture of least significant bits (LSB) and color sequence to spread message. If steganography scheme uses one LSB bit, then it has limit to encrypt secret data and include only color of RGB in the replacement. If steganography scheme uses two or more bits, only one RGB color is used for encryptions. Although SSA can utilize entire RGB of LSB, but it restricts to one bit for encoding more evidence. Seemingly, the suggested scheme is capable to encrypt one or more LSB bits and straight to four bit every RGB color pixel that depend upon the magnitude of the data as shown in Fig. 7.

- If the information is 15% of the specified image, then only LSB is used to encrypt data.
- If the information is partial to the cover, at that moment, the maximum LSBs are used to encode with secret data.

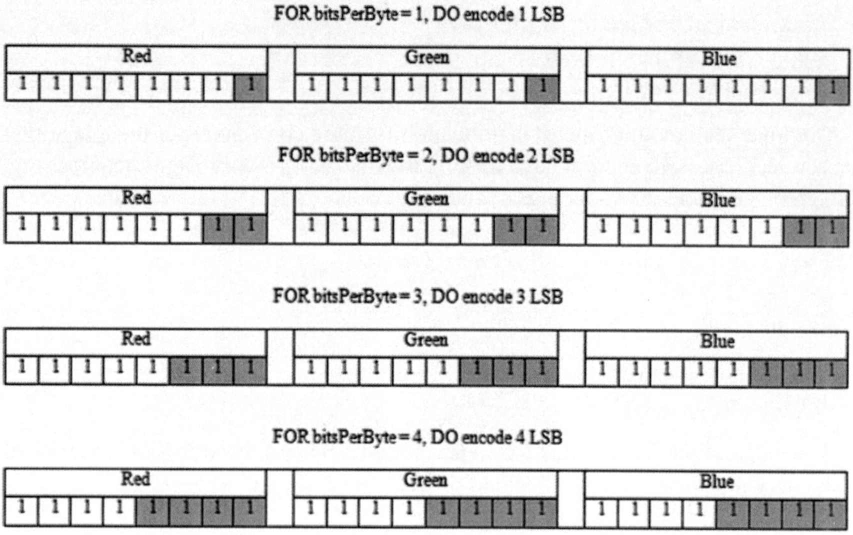

Fig. 7 Algorithm scheme

The proposed algorithm utilizes the entire LSBs to encrypt secret data of each pixel consecutively converting into binary string. Therefore, the facts of several nature can be entrenched in images as transmitted data without escalating the source. The replacement efficiently averts the thoughtfulness of the social visual system since identifying distinction transformation in an image. This algorithm can be enhanced to use multiple images to message transmission.

The complete algorithm is laid down in two parts:

Scrambling steps

The encoding process follows the steps listed below:

1. Compute the magnitude of the data and image.
2. Identify RGB bits to encrypt the data.
3. Reads the data and transforms the quantity the secret data from integer to binary string.
4. Divide the pixel into RGB.
5. Get the entire least significant bit or bits desirable for encryption.
6. Compute the groups of the bitmap of cover image to keep the facts. The n amount of pixels will be castoff to store data.

$$p = m - s \tag{1}$$

where

p Maximum size of the file.
m Represents pixel after six pixel.
s The last bit of pixel.

Compute the amount of least significant bits in the color pixels of the image that are actually castoff to encrypt data. Case, if the image is a color image encompassing red, green and blue pixels, is Figs. 5 and 6.

$$d = 8 - c \tag{2}$$

Denote

d Unused bits afterward encrypting.
c Bit castoff to encrypt data; $c = 1$–4.

The calculation acquires the total encrypted LSBs and the worth of a new pixel with encrypted data:

Entire encrypted least significant bits:

$$(x1 - y1) + (x2 - y2) + (x3 - y3) \tag{3}$$

The computed pixel value with encrypted data = old byte + entire encrypted least significant bits.

where

$y1, y2, y3$ Unused bit of RGB pixel.
$x1, x2, x3$ Numbers of RGB pixel.

Deciphering steps

After scrambling the transmitted data into the image, the data is taken to be interpreted for reclamation. The following are steps for translating:

1. Compute the magnitude of the steganography image.
2. Reads the bytes of the steganography image. Convert bits of the steganography image to binary string.
3. Decide encrypting technique for the amount of LSBs to be used
4. Compute and recognize the amount of pixels in the steganography image and also the entrenched data correspondingly using 1–3.
5. Use the latest three byte values to find the offset of the string.

•

$$\text{Ist offset} = 16 - \text{encrypted LSB} \tag{4}$$

•

$$\text{IInd offset} = 24 - \text{encrypted LSB} \tag{5}$$

•

$$\text{IIIrd offset} = 32 - \text{encrypted LSB} \tag{6}$$

6. Compute and associate the outstanding least significant bits at each pixel in case in directive to rescue rear the real data.

$$\text{Decrypted value} = (4) + (5) + (6) \tag{7}$$

From (7), the decrypted sequence is facts recovered.

3.2 Message Processing Unit (MPU)

This is the main unit of an agent which is used to transmit/receive messages and process it on the node. This unit is a part of design to progression message after steganography algorithm. The unit in the moving node acts as translator to display message on panel.

4 Experimental Results and Analysis

The proposed algorithm has been implemented using NETBEAN environment with Java to provide developers of container. The experimental test has been carried out on a 2.0 GHz i3 notebook with 8 GB RAM running on Windows OS. The VANET mobility or node communication has been implemented or simulated on simulator. The analysis of security of message transmission is given below. The results are shown in Figs. 8 and 9. Here, Fig. 8 shows the encryption of message which sender wants to send, and at receiver end decryption, the same is done. Now, in the proposed technique, the original message is hid behind an image, and at receiver end, the message is recovered. The proposed algorithm is very efficient and quick to enrapt and decrypt the message. The accuracy of the message after decryption is very good as compared to other method.

5 Conclusions and Future Work

A multi-agent-based steganography approach is proposed to securely transmit a message in VANET. This paper uses color scheme-based steganography algorithm to hide the message within image. The algorithm used here used LSB bits of image

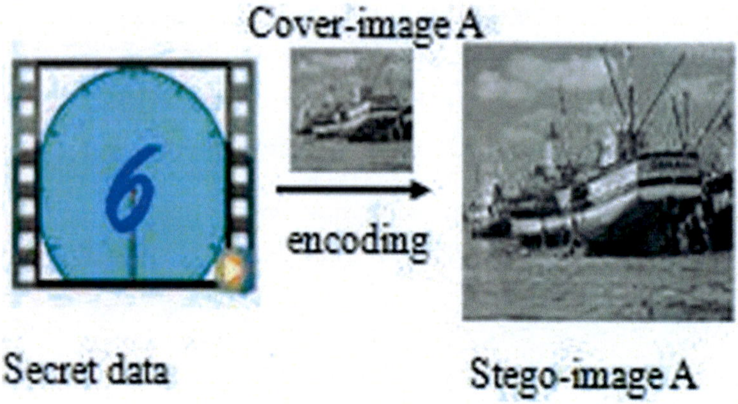

Fig. 8 Encryption

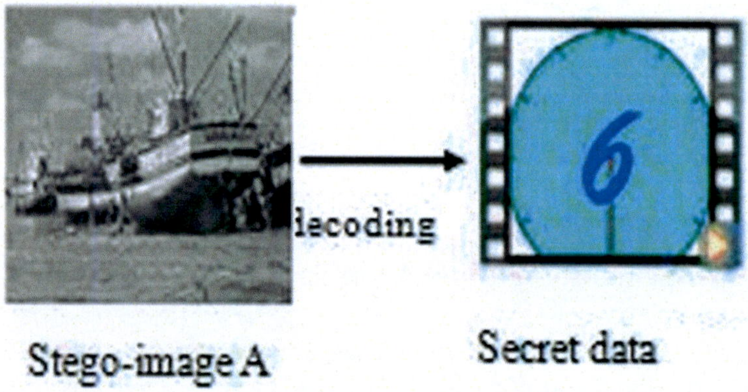

Fig. 9 Decryption

to store the ample message transmitted through network. The complete system uses different software agents to implement the steganography algorithm on original message. Each agent installed on terminal has two parts such as message steganography unit (MSU) and message processing unit (MPU) to encode/decode and transmit/receive message. The MSU uses steganography algorithm to embed message within color image based on LSB bits. The other unit MPU is used to transmit the message on proper route. The contemporary Least Significant Bits are arranged after consuming and integrating stego one LSB from four Stego LSB and one LSB from stego color cycle. As a result, the proposed scheme is able to encode up to four least significant bits in the each of the RGB pixels according to the contents of the data without visually degrading the image. Future work will focus on diversifying the various types of cover medium and to increase the capacity of payload using other alternative methods.

References

1. Kawashima H (1990) Japanese perspective of driver information systems. Transportation 17(3):263–284
2. Harsch C, Andreas F, Panos P (2007) Secure position-based routing for VANETs. In: 66th IEEE vehicular technology conference. IEEE, pp 26–30
3. Sun JCZ, Yuguang F (2007) An id-based framework achieving privacy and non-repudiation in vehicular ad hoc networks. In: IEEE military communications conference, MILCOM. IEEE, pp 1–7
4. Zeadally S, Hunt R, Chen Y-S, Irwin A, Hassan A (2010) Vehicular ad hoc networks (VANETs): status, results, and challenges. J Telecommun Syst
5. Raya M, Hubaux J (2005) The security of vehicular ad hoc networks. In: Proceedings of the 3rd ACM workshop on security of ad hoc and sensor networks, pp 1–11
6. Sumra IA et al (2011) Classes of attacks in VANET. In: Electronics, communications and photonics conference, SIECPC. IEEE, pp 1–5
7. Fuentes et al (2010) Overview of security issues in vehicular ad-hoc networks
8. Nai-Wei L, Hsiao-Chien T (2009) A reputation system for traffic safety event on vehicular ad hoc networks. EURASIP J Wireless Commun Netw
9. Dotzer F et al (2005) VARS: a vehicle ad-hoc network reputation system. In: Symposium on a world of wireless mobile and multimedia networks, pp 454–456
10. Raya M et al (2008) On data-centric trust establishment in ephemeral ad hoc networks. In: INFOCOM—conference on computer communications. IEEE, pp 1238–1246
11. Chen TM (2005) Dempster Shafer theory for intrusion detection in ad hoc networks. IEEE Internet Comput 9(6):35–41
12. Lu H, Li J, Guizani M (2012) A novel ID-based authentication framework with adaptive privacy preservation for VANETs. In: Computing, communications and applications conference (ComAp). IEEE, pp 345–350
13. Wu Q, Domingo-Ferrer J, Gonzalez NU (2010) Balanced trustworthiness, safety, and privacy in vehicle-to-vehicle communications. IEEE Trans Veh Technol 59(2):559–573
14. Wei Y, Chen Y (2012) An efficient trust management system for balancing the safety and location privacy in VANETs. TrustCom:393–400
15. Sun J et al (2010) An identity-based security system for user privacy in vehicular ad hoc networks. IEEE Trans Parallel Distrib Syst 21(9):1227–1239
16. Wang NW, Huang YM, Chen WM (2008) A novel secure communication scheme in vehicular ad hoc networks. J Comput Commun 31(12):2827–2837
17. Albert W, Shen X (2009) EDR-efficient decentralized revocation protocol for vehicular ad hoc networks. IEEE Trans Veh Technol 58(9):5214–5224
18. Wang X, Liu T, Xiao G (2012) Certificate-based anonymous authentication protocol for vehicular ad-hoc network. IETE Tech Rev 29
19. S-Tools. https://digitalforensics.champlain.edu/download/s-tools4.zip
20. Chandramouli R, Memon N (2001) Analysis of LSB based image steganography techniques. In: Proceedings of ICIP. Thessaloniki, Greece
21. Dumitrescu S, Xiaolin W, Wang Z (2003) Detection of LSB steganography via sample pair analysis, vol 2578. LNCS, Springer-Verlag, New York, pp 355–372
22. Ahn LV, Hopper NJ (2004) Public-key steganography. In: Lecture notes in computer science, vol 3027/2004 of advances in cryptology—EUROCRYPT. Springer-Verlag Heidelberg, pp 323–341
23. Pang, H.H., K.L. Tan and X. Zhou. : Stenographic schemes for file system and b-tree, IEEE Trans. on Knowledge and Data Engineering, 16: 701–713, (2004).
24. Dobsicek M (2004) Extended Stegano-graphic system. In: 8th international student conference on electrical engineering. FEE CTU
25. Mittal U, Phamdo N (2002) Hybrid digital analog joint source-channel codes for broadcasting and robust communications. IEEE Trans Info Theory 48:1082–1102

26. Pavan S, Gangadharpalli S, Sridhar V (2005) Multivariate entropy detector based hybrid image registration algorithm. In: IEEE international conference on acoustics, speech and signal processing, pp 18-30
27. Yee P-L, Kiah M-L-M (2010) Shoulder surfing resistance using Penup event and neighboring connectivity manipulation. Malaysian J Comput Sci 23(2):121–140
28. Por L-Y (2012) Frequency of occurrence analysis attack and its countermeasure. Int Arab J Inf Technol 10(1,2):4143–4147. Advance online publication, available at https://www.ccis2k.org/iajit/PDF/. Last Visited (2012)
29. Por L-Y, Wong K-S, Chee K-O (2012) UniSpaCh: a text based data hiding method using unicode space characters. J Syst Softw 85(5):1075–1082
30. Abassi R (2019) VANET security and forensics: challenges and opportunities, Wires Forensic Science. Willy
31. Wang B, Wang Y, Chen R (2019) A practical authentication framework for VANETs. In: Hindawi security and communication networks
32. Erskine SK, Elleithy KM (2019) Secure intelligent vehicular network using fog computing. J MDPI
33. Pan J et al (2019) Secure data sharing scheme for VANETs based on edge computing. EURASIP J Wireless Commun Netw

Conceptualization Model for Cyber Secure National Time Dissemination System

Amutha Arunachalam, K. Seetharaman, and Ashish Agarwal

Abstract "Time" and "timestamp" lead to a term "Time Commerce" in cyberspace. Disseminating accurate time in cyberspace and distribution of timestamp and event identification initiate the need to have Synchronized Indian Standard Time (IST) to guarantee Government of India's Digital India Program. The paper discusses "time" with its unit "second" as one of the quantities defined in Indian Legal Metrology Act 2009. Timestamp defines at least the time and correct date of action as well as identity of the device or person which receives or sends in the provisions of the Indian Information Technology Act 2000. CSIR-National Physical Laboratory (NPL) as National Measurement Institute (NMI) of India develops "Primary Time Scale" with international traceability to Coordinated Universal Time (UTC). The paper conceptualizes the design model for a collaborative national organizational structure within regulatory framework to establish National Time Dissemination System.

Keywords Timestamp · Cyber security · Time synchronization · Time dissemination · Indian standard time

Abbreviations

BIPM	Bureau International des Poids et Mesures
CSIR	Council of Scientific and Industrial Research
CIPM	Certificate in Investment Performance Measurement
CCA	Controller of Certifying Authorities
CA	Certifying Authorities
DST	Day light Savings Time

A. Arunachalam (✉) · A. Agarwal
Time and Frequency Metrology,, National Physical Laboratory, New Delhi, India
e-mail: amutha@csir.res.in

K. Seetharaman
Department of Computer and Information Sciences, Annamalai University, Chidambaram, India

A. Abraham et al. (eds.), *Proceedings of 3rd International Conference on Computing Informatics and Networks*, Lecture Notes in Networks and Systems 167,
https://doi.org/10.1007/978-981-15-9712-1_15

GPS	Global Positioning System
GMT	Greenwich Mean Time
IST	Indian Standard Time
LM	Legal Metrology
MRA	Mutual Recognition Arrangement
NTCA	National Time Commerce Authority
NMI	National Measurement Institute
NPL	National Physical Laboratory
NZMT	New Zealand unit of time
NABL	National Accreditation Board of Laboratories
NTA	National Time Authority
NTA	National Time Authority
RRSL	Regional Reference Standard Laboratories
RCAI	Root Certifying Authority of India
SI	International System
STT	Synchronised Traceable Time
TSA	Timestamp Authority
	Timestamp Authority Time dissemination service
TSS	Timestamp service
TC	Time Commerce
TDS	Time dissemination service
TTS	Trusted Timestamping
UCOS	Unified Collaborative Organisational Structure
UTC	Coordinated Universal Time

1 Introduction

The government [1] aims to facilitate the Indian Standard Time (IST) service to the entire nation for the benefit of common man and various strategic sectors. For example, "time" with different levels of accuracies is used in our society in various applications ranging from cybersecurity, communication, transportation, weather forecast, operations of electric power grids, accurate positioning of satellites, detecting the location of enemy missiles or targets, satellite-based navigation, disaster detection and management and so on. A common man also uses the precise timing system without knowing it; for example, the cell phones and all broadcasting services rely on it. Accurate timestamping is inherently important for synchronized functioning of the IT security device such as gateway routers, network switches and servers along with the communication channels. Few of the areas which need immediate attention are cybercrimes, weather forecast and disaster management. In case of inaccurate timestamping, any cybercrime becomes difficult to detect and remains untraceable in most of the cases. Thus, it is essential to synchronize all time-reliant devices used in digital communication to Indian Standard Time. This is extremely important for socioeconomic as well as national security. In this paper, we identify challenges in operational timing infrastructure and coordinated management structure for smooth dissemination of Indian Standard Time (IST) across the country.

2 Indian Legislation for "Time"

2.1 *Indian Legal Metrology Act 2009—Regulating "Time"*

Consumer Affairs Department's **SI unit** is as part of "Ministry of Consumer Affairs, Food & Public Distribution" [2]. This has been well-grooved as per a different department in Gregorian calendar month 1997 because this needed to possess a different department for relinquishing a positive stimulus to a movement of rising consumer in the nation. This department was commended that are evident in Weights and Measures Standard's Implementation—The Legal Metrology Act, 2009, Training in Legal Metrology, Internal Trade, Consumer Cooperatives, National Test House, Implementation of Consumer Protection Act, 2019 and others.

3 Defining Indian Legal Metrology and Its Roles: "Legal Metrology"

"Legal Metrology" that measured the units of measurement and weighment, techniques of measurement and weighment as well as instruments used for measuring and weighing with relation to the required legal and technical terms which are mandatory that objects for ensuring public guarantee in terms of accuracy and security of measurements as well as weighment [2].

Measurement accuracy and precision have a very important part in our daily life. The economical and clear legal system of scientific disciplines raises high confidence in terms of shoppers, industry, as well as trades which leads to a harmonious setting to conduct a business using the techniques of (a) contributing to country's economy which increases in revenues under different sectors, (b) having a crucial part is decreasing the losses of revenues in railways, petroleum, industries, mines and coal; and (c) reducing wastage and loss in the infrastructure sector.

The social control of legal scientific discipline laws is completed by the state governments by the controller of legal scientific discipline and alternative legal scientific discipline officers according to the act. The 2009 Legal Scientific Discipline Act is enforced to impact from first Gregorian calendar month; In 2011, the preamble of the act is given as: "An act to determine and enforce standards of weights and measures, regulate trade and commerce in weights, measures and alternative commodities that are sold-out or distributed by weights, live or variety and for matters connected therewith or incidental thereto."

The main feature of this act as defined below is each weight unit or must be as per the weights and measures system supported the International System of Units. The base unit under this Legal Metrology Act (i) meter defines the length, (ii) kilogram defines mass, (iii) second is for time, (iv) ampere defines electric current, (v) Kelvin is for physics temperature, (vi) candela defines aglow intensity, and (vii) mole defines the substance quantity.

Legal Metrology Act, 2009 has been added with rules for the effective implementation: (a) The Legal science (Government Approved check Center) Rules, 2013, (b) The Legal science (Government Approved check Center) Rules,2013, (c) The Indian Institute of Legal Metrology Rules, 2011, (d) The Legal Metrology (National Standards) Rules, 2011, (e) The Legal Metrology (General) Rules, 2011, (f) The Legal scientific discipline (Numeration) Rules, 2010, (g) The Legal Metrology (Approval of Models) Rules, 2011, and (h) The Legal Metrology (Packaged Commodities) Rules, 2011.

Regional Reference Standard Laboratories (RRSLs) were formed for setting up the requirement of Legal Metrology for consumers, industries, and state governments, of the country. These measure all of the five RRSLs, located in Guwahati, Faridabad, Bhubaneswar, Bangalore, and Ahmedabad. The RRSLs also offer to form a link among the states weights and measure laboratories and the National Physical Laboratory for making sure the exact measurements and weights in transaction and trade. More of these RRSLs square measures are situated in Varanasi and Nagpur. All these RRSLs of Guwahati, Faridabad, Bhubaneswar, Bangalore, and Ahmedabad are under the accreditation of National Accreditation Board of Laboratories (NABL), where all departments of Legal Metrology Division and subordinate offices are certified already under ISO 9001. RRSLs upgradation, the Bangalore branch will make on part having the best International Laboratories. The national test houses and RRSLs serve as Reference Standards having traceability to National Standards.

These laboratories are accountable to verify the measure for secondary authority standards, weights and measures models testing, sophisticated weighing Calibration, instruments measuring, as well as setting up some consumer awareness program. Indian Institute of Legal scientific discipline, Ranchi [3], is that the National Center for impartation skilled coaching to the Legal scientific discipline provides coaching to Indian and foreign participants of the neighboring/developing countries. They offer all of the facilities of organizing seminars and training. The organization also offers a basic course of training in Legal Metrology field which imparts knowledge pertaining to an Act of Legal Metrology and Rules along with its implementation in the field. Having understood the National Legal Metrology structure, it is essential further to connect its traceability with the International Organization.

3.1 *National Metrology Institute (CSIR-NPL) within the Board of Legal Metrology Act 2009 is the Generator of Indian Standard Time (IST)*

The responsibilities of CSIR-National Physical Laboratory as a signatory to Certificate in Investment Performance Measurement (**CIPM**) under Mutual Recognition Arrangement (MRA) of BIPM which maintains, develop, as well as disseminate the national standards of measurements required for the national requirement, and also makes sure the international recognized traces of calibration and measurements for instruments used for measuring and offers a base for activities like the conformance testing, trade metrology services, and the calibration services in every sector. The outcome of this is that a higher awareness is required for the discussion, improvement, and comparing the abilities of the countries for establishing, improving, and maintaining the infrastructural practices, along with recognition and compatibility internationally in such regions.

CSIR-National Physical Laboratory (CSIR-NPL) [4] is National Metrology Institute (NMI) of India which also has the mandate for unit's establishment, maintenance along with dissemination of physical dimensions depended on the International Systems (SI units) beneath the Legal Metrology Act 2009.

One of the unit "time" measured in "second" is disseminated to the nation as the *Indian Standard Time* (IST). CSIR-NPL, internationally known as UTC (NPLI) and the *Timekeeper of India,* is the "Primary Reference Clock" that can be traced to the Universal Time Coordination (UTC) as stated by BIPM situated in Sevres, France. As stated by Dennis [5], the development of UTC remains till today with the existing assumption in terms of its characterization, specific realization and applications. The IST (i.e., UTC-NPLI plus 5.30 h) is generated by a bank of cesium clocks, and hydrogen maser has current uncertainty of 20 nanoseconds. The traceability of the "Indian Standard Time" is globally recognized as UTC (NPLI) which contributes to universal coordinated time (UTC).

4 "Timestamp" in Indian Legislation within Indian Information Technology Act, 2000

The 2000 Information Technology Act states regulations which mandated that "Controller of Certifying Authorities (CCA) [6] shall provide Timestamping Service for its subscribers." Establishment of timestamps on any time reference requires a timestamp. This reference can be contracts, certificates, transactions, or documents. Operation of a digital signature could be integrated to the method of timestamping to link it mathematically with the time reference derived from the selected supply of national time for coming up with a stamp [6]. The timestamps then undergo verification for establishing the attestation time needed for the references. The services of timestamping are operated under Certifying Authorities (CAs). Likewise, this service of timestamping can also be managed under personnel who is trustworthy, to be operated under highly secure surroundings which were subjected to compliance as well as audit.

4.1 Timestamp

This indicates at least the exact time and date on which the action took place along with the identification of the device or the person who received or has sent the time (IT Act). Timestamp defines a signature created digitally for the service of timestamping which details the integrity submitted to a subscriber on exact time and date.

4.2 Timestamping

This service helps in asserting a timestamp using a digital signature denoting a time at which the reference is received from the subscriber. The subscriber then submits the request which is referred to the service of timestamping and acquires a response which is signed digitally with a timestamp. The source of time to implement this is having traceability of NPLI—UTC.

5 Encounters and Cybersecurity Difficulties of Unit "TIME" AND "Timestamp" in Digital Transactions

The Indian Legal Act by making a law to mandatorily accept the Indian Standard Time (IST—UTC (NPLI) as official Legal Time shall facilitate the consumers/industries/organizations and all the "Time Consumers" legal endowment to support the digital economy for end-user protection.

1. The legal provisions handled for **"time"** by Metrology Act and for **timestamp** by Information Technology Act are held by two different ministries; it requires greater amount of understanding and coordination for the purpose of the lawfulness of electronic commerce and protecting intellectual property rights for the benefit of consumers.
2. The complexity of the **"time requirements"** extends to various digital information in the form of electronic storage, electronic transactions, and real-time applications. Such requirements to assure integrity in the electronic form by the use of digital signatures and timestamps serve as electronic evidence in courts. Such evidences can be provided with a help of Nationally Synchronized Legal Time through networks.

3. For "time" as a source, understanding the dissemination chain and calculating the uncertainty is an essential requirement for demonstrating traceability. Each link along with associated uncertainty evaluation requires documentation, and the total uncertainty of timestamp or time output must be determined and recorded.
4. Spoofing: This can be illegitimate act in the mask of a right person. Few prior researchers [7–9] and real-world attacks in GPS spoofing have specified the GPS signal vulnerability and probability of spoofing the receivers in GPS.
5. Alteration: the illegitimate person trying to alter the digital document.
6. Repudiation: denial of contradiction of the fact of having done something in the electronic execution process [10].
7. Nowadays, the eruption in digital economy realizes the creation half of trade transactions electronically creating a new revolution. Technologically, the actual problem is essentially in mastering the protection along with the date and time traceability from the source of energy of theirs to the end user. This is vital to cure cyber-attacks which inhibit with the time message.
8. Secrecy in digital channel, authenticity of interchanging documents, and verification of document are crucial to security of transaction.

We understand "time" as the serious substructure of information society. The danger management with appropriate cybersecurity methods in digital transactions needs to be determined through time and synchronizing clocks of multiple computers working in collaboration.

6 Conceptualization Model for Cyber Secure National Time Dissemination System

The authors propose the new term "Time Commerce" for a collaborative effort that needs to be tailored under **"National Time Commerce Authority"** to remove the operational limitations and legislative augmentations to suit the growing needs in digital economy [1]. National Timing infrastructure requires to be built within international standards that have safety, health hazard free, environment, and consumer friendly.

The immediate concerns for enabling the digital economy is to have governance by National Level *Collaborative Organizational Structure* as authors designed the model proposed in Fig. 1 for national timing requirements with Legislation, Technical Regulation, Standardization, Accreditation, Training, Testing and Certification and Infrastructure Creation to cater the entire population. In this paper, we conceptualize plan to design, operate and maintain the accurate National Level Timing Infrastructure involving national metrology institute, regional metrology laboratories and designated Institutes to industrialize and disseminate time with high level of security and traceability across the country. This shall help to fulfill the ultimate goal of serving the digital economy with complete confidence.

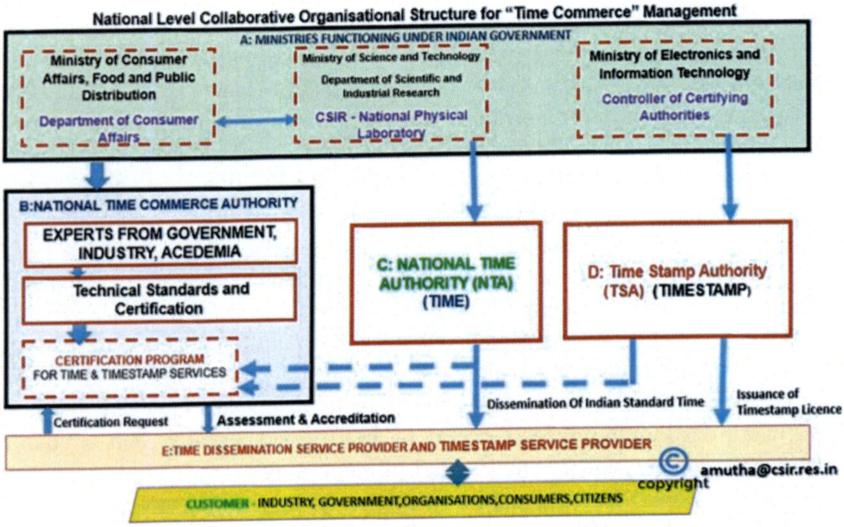

Fig. 1 Conceptualization model for cyber secure national time dissemination system. *Note* In the figure, Representation of **a** Ministries functioning under Indian Government, **b** National Time Commerce Authority, **c** National Time Authority (NTA), **d** Timestamp Authority (TSA), **e** Time dissemination service (TDS) and Timestamp service (TSS)

We propose this model for establishment of a National Time Commerce Authority that enables to build a National Level Timing Infrastructure supported by legislation, standardization, metrology, accreditation and conformity assessment. The authority will help the competiveness in *Indian Time Industry* to be competitive and at par with International Market. Here, we identify and elaborate features and areas of responsibility for building the Timing Infrastructure at national level. It is evident that variety of areas required at national level is functioning under different department necessities and greater harmonization with fixed responsibilities for a qualitative outcome in compliance to the international and regional agreements.

A. Ministries functioning under Indian Government

Are within the structure as shown in Fig. 1a and have specific business allocation roles under different departments in Indian Government and their legislation responsibilities. In context to this, Table 1 specifics the roles and legislation of the ministries that are currently engaged in the "Time Commerce" activities. And these organizations form three pillars of the Indian Legislation Authorities to disseminate the Indian Standard Time with all administrative, operational, and financial provisions at national level.

B. Describing the National Time Commerce Authority

Digital trade that uses **"time"** and **"timestamp"** information will improve the safety and reliability of the electronic information such as various electronic data, documents, and information, and we term it as **"Time Commerce"** in this paper. To

Table 1 Roles of ministries in Indian Legislation that are currently involved in the "Time Commerce" activities

Ministry and Department under Indian Government	Business allocation	Legislation responsibilities
Ministry of Consumer Affairs, Food and Public Distribution [11] Department of Consumer Affairs [2]	Underneath the Ministry of Consumer Affairs, the Department of Consumer Affairs is actually among the two departments, i.e., Food and Public Distribution. In June 1997, this was established as a distinct department as it had been regarded as required to escalate a distinct department to make a nascent fillip to the customer action in the nation	Execution of Consumer Protection Act, 1986 Execution of Bureau of Indian Standards Act, 2016 Execution of Weights and Measures Standards—The Legal Metrology Act, 2009
Ministry of Science and Technology (9) **Department of Scientific and Industrial Research** [12] Council of Scientific and Industrial Research [13] CSIR—National Physical Laboratory- [4]	DSIR features a mandate to hold out the tasks associated with indigenous know-how promotion, advancement, transfer, and utilization CSIR is actually the National R&D business providing scientific manufacturing exploration for India's economic welfare as well as economic growth. It is a country-wide community of 80 area and forty laboratory facilities covering applied and fundamental R&D in most aspects of science as well as technologies barring atomic investigation, nurturing and establishing S&T human resource for the nation through additional mural assistance and encouraging scientific skill through awards, fellowships, etc. National Physical Laboratory is actually the National Metrology Institute of India along with a Premier Research Laboratory under the area of Physical Sciences	National Physical Laboratory has the duty of sighted the products of actual physical dimensions depending on the System International (SI devices) underneath the secondary legislations of weights and Measures Act 1956 (rereleased in 1988 **underneath the 1976 Act and 2009)**

(continued)

Table 1 (continued)

Ministry and Department under Indian Government	Business allocation	Legislation responsibilities
Ministry of Electronics and Information Technology [14] Controller of Certifying Authorities [6]	In order to market e-Governance for empowering people, advertising the sustainable and inclusive progress of the Electronics, IT and ITeS industries, improving India's function in Internet Governance, following a multipronged strategy which has improvement of human resources, promoting Innovation and R&D, improving effectiveness through digital solutions and making sure a safe cyber space To trust development in Electronic Transactions	**Information Technology Act 2000** Notification under IT(Amendment) Act, 2008, IT (Amendment) Act 2008 IT Act 2000 Rules for the Information Technology Act 2000 Report of the Expert Committee on Amendments to IT Act 2000 Under Section 18(b) of the IT Act, the Controller of Certifying Authorities (CCA) has proven the Root Certifying Authority (RCAI) of India to digitally sign the public keys of Certifying Authorities (CA) of the nation

build a safe and secure digital society, it is necessary to establish **"National Timing Commerce Authority"** to deliver trustworthy time shown in Fig. 1b.

Functions Identified for the Functioning of National Time Commerce Authority

1. Build, operate, and maintain National Time Infrastructure.
2. Creation of National Time Service Models
3. Identifying Technology Trends for "Time Industry."
4. National Guidelines for Time (operations, administration, training, audit, certifications, etc.)
5. National level Identification of all issues relating to "Time Commerce"

Time dissemination and timestamp services are required for "accurate time and reliable time" in the system of electronic commerce, electronic application, etc. As CPSs interrelate with their environs, synchronized necessities and significance are of essential nature, since there are time limits to be conversant with structure reactions [15]. The authority shall help to develop significant public infrastructure for cutting-edge information and communication networked society. We foresee digital threats in the form of "fraud" and the "Information system failures" that use the characteristics of electronica data with more CPSs interconnecting to the surrounding network, and the threat to impairment that caused deliberately is greater than ever. Consequently, as stated by Schneider [16] there is a demand to synchronize safety and reliability engineering such that the difficult risk of harm as a result of either faulty or spiteful intention is sufficiently addressed. Hence, we propose a system that shall "certify the originality of electronic data," "precisely hold transaction time," and "deliver to the third party." To achieve this, we need to setup suitable administrative and technology operational organizational structure having experts from government, industry, and academia. The NTCA shall have compliance to international standards and world trade agreements. Under this authority, operational and auditing skills for the timing services are to be built with testing and training capabilities in the form of accreditation program specific to the time dissemination and timestamp services. The program shall need to be nationally and internationally recognized by accredited organizations. Figure 1b identifies the proposed National Time Commerce Authority with the given workflow.

C. Role of National Time Authority (NTA) shown in Fig. 1c

The term NTA is drawn from the International Standard ISO/IEC 18014, where the role of NTA is defined as "to generate, maintain, and distribute national standard time." NTA further distributes Time to Time Authority (TAs) which is a trusted Third party for providing time for further dissemination.

D. Role of Timestamp Authority (TSA) shown in Fig. 1d

The term Timestamp Authority is drawn from the International Standard ISO/IEC 18014, and TSA is defined "to produce and issue timestamp token for data submitted from the customer." Further, TSA is also a performer for verification of timestamp token. The TSA is also a confidential third party as designated by the Controller of Certifying Authority.

E. Role of Service Providers for Time and Timestamp shown in Fig. 1e

Time dissemination service (TDS) is a distribution of reliable time information as a service by designated institutes / service operators to the National Critical Infrastructures and other organizations that require highly accurate and reliable time and to citizen-centric services.

Timestamp service (TSS) proves the time when transactions and procedures on the Internet/leased, etc., were performed and the date and time when electronic transaction/ electronic documents existed. TSS also specify whether the particular transaction/object document has been tampered with, but also prove that it has certainly existed at a definite point in the past and the same time can verify that it has not been changed.

7 Classification of Official Methods for Time and Timestamp Certification

In Figure 2 below, we identify the measurable parameters of "time" and "timestamp." This detailing shall help the ***National Time Commerce Authority*** for the ***certification*** process of the ***time and timestamp service providers***. Specifically, focusing on the identified terms with integrity will establish quality of services at national level.

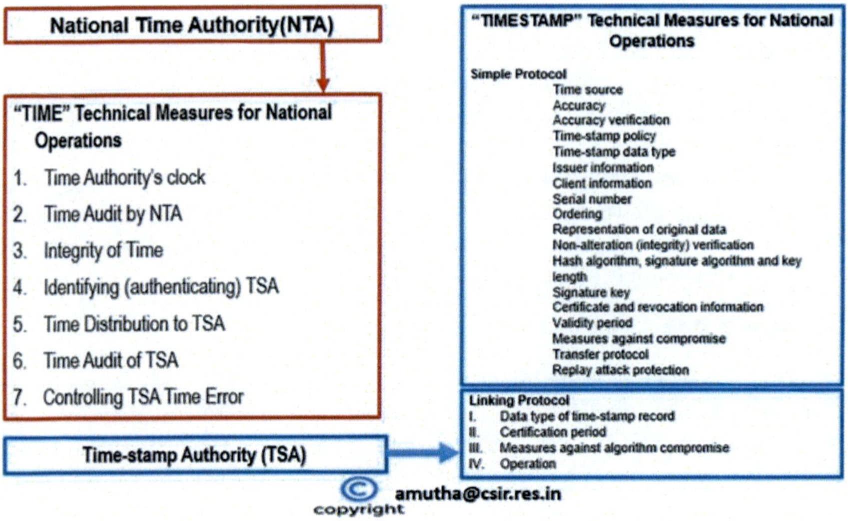

Fig. 2 Cataloguing technical measures for time and timestamp certification

Accreditation system under the National Time Commerce Authority will help to maintain fairness with the experts from government, industry, and academia for certification with the facility of the associated designated ministries.

The authors are working on this for more refinement. This is the primary study for the Ph.D. doctoral work of the first author. We intend to present in this conference the conceptualized model for establishment of National Time Commerce Authority which can be taken for consideration within the current legislation. The inter-ministerial coordination requirements are being addressed in setting up the national authority. The services for time and timestamp with trusted certification program will assure the trust to customers. In the future work, we propose to build the time infrastructure with Cyber Secure Architecture for National Distribution of Time as per international standards for quality and cybersecurity.

Acknowledgements We acknowledge our thanks for support by DG CSIR, Council of Scientific and Industrial Research, New Delhi, Director CSIR-NPL, National Physical Laboratory and colleagues of the CSIR-NPL IST Division to this work.

References

1. https://meity.gov.in/writereaddata/files/india_trillion-dollar_digital_opportunity.pdf
2. https://consumeraffairs.nic.in/organisation-and-units/division/legal-metrology/overview (oct 2019)
3. https://www.nth.gov.in/nth/index.htm (Oct 2019)
4. https://www.nplindia.in/indian-standard-time-metrology-division-d60 (Oct 2019)

5. Dennis DM (2019) The development of coordinated universal time. In: Proceedings of the 2019 international technical meeting of the Institute of Navigation, Reston, Virginia, pp 8–52
6. https://www.cca.gov.in/timestamping.html (Oct 2019)
7. Wang K, Chen S, Pan A (2015) Time and position spoofing with open source projects, vol 148. Black Hat Europe
8. Tippenhauer NO, Popper C, Rasmussen KB, Capkun S (2011) On the requirements for successful GPS spoofing attacks. In: Proceedings of the 18th ACM conference on Computer and communications security, pp 75–86
9. Zhuang's Z, Gong S, Dimitrovski AD, Li H (2013) Time synchronization attack in smart grid: impact and analysis. IEEE Trans Smart Grid 4(1):87–98
10. Qin J, Li M, Shi L, Yu X (2017) Optimal denial-of-service attack scheduling with energy constraint over packet-dropping networks. IEEE Trans Autom Control
11. https://fcamin.nic.in/home.aspx (Oct 2019)
12. https://www.dsir.gov.in/ (Oct 2019)
13. https://www.csir.res.in/ (Oct)
14. https://www.meity.gov.in/ (Oct 2019)
15. Baheti R, Gill H (2011) Cyber-physical systems. Impact Control Technol 12:161–166
16. Schneider D, Armengaud E, Schoitsch E (2014) Towards trust assurance and certification in cyber-physical systems. Springer, Berlin

Machine Learning and Soft Computing

Se.Re.Ne.: Stress Detection Using EEG and ECG

Deepali Virmani, Akshat Minocha, Lakshay Goyal, Megha Malhotra, and Megha Gupta

Abstract According to the World Health Organization, stress is a considerable problem that affects both the mental well-being and physical health of people, so stress detection becomes an important task. Various stress detection methods based on the human brain and human behavior exist, but none of them uses both brain signals and heart signals together to detect stress. In this paper, a novel approach to detect stress using EEG and ECG signals is proposed. The proposed Stress Recognition by Neuroanalysis (Se.Re.Ne.) method is validated for k-nearest neighbors (KNN) and decision tree (DT) using the correlation method. Results evaluated using Se.Re.Ne. with KNN detect stress with a precision of 0.87, recall of 0.71, and $f1$-score of 0.78 with total accuracy of 68%, whereas Se.Re.Ne. with DT detects stress with a precision of 0.85, recall of 0.84, and $f1$-score of 0.84 with a total accuracy of 75%.

Keywords Electroencephalography (EEG) · Electrocardiography (ECG) · Stress detection · Decision tree (DT) · K-nearest neighbors (KNN)

D. Virmani · A. Minocha (✉) · L. Goyal · M. Malhotra · M. Gupta
Department of Computer Science and Engineering, Bhagwan Parshuram Institute of Technology, New Delhi, Delhi, India
e-mail: akshatminocha@gmail.com

D. Virmani
e-mail: deepalivirmani@gmail.com

L. Goyal
e-mail: lakshay761@gmail.com

M. Malhotra
e-mail: megha.malhotra153@gmail.com

M. Gupta
e-mail: meghagupta307@gmail.com

A. Abraham et al. (eds.), *Proceedings of 3rd International Conference on Computing Informatics and Networks*, Lecture Notes in Networks and Systems 167,
https://doi.org/10.1007/978-981-15-9712-1_16

Abbreviations

AUC	Area Under the Curve
CNN	Convolutional Neural Network
DT	Decision Tree
EEG	Electroencephalography
ECG	Electrocardiography
KNN	K Nearest Neighbours
MFI	Multidimensional Feature Image
MIST	Montreal Imaging Stress Task
ML	Machine Learning
NN	Neural Network
PSD	Power Spectral Density
PAD	Pleasure, Arousal and Dominance
PFC	Prefrontal Cortex
ROC	Receiver Operating Characteristic
SVM	Support Vector Machines
TKEO	Teager–Kaiser energy operator

1 Introduction

In day-to-day life, everyone feels stressed either because of high workload or relationship dispute. These high brain activities for a prolonged period of time can lead to chronic stress which can further cause mental disorders like depression or anxiety. Stress can be detected through facial expressions, behavioral patterns, and varying voices, but people can deliberately hide these features which can result in false analysis. Gradually, researchers have shifted their focus on technologies like electroencephalogram (EEG) and electrocardiogram (ECG) as signals come directly through the human brain and heart, respectively.

This paper focuses on EEG and ECG signals which are recorded noninvasively to detect stress. EEG is a common test used for the recording of brain activities. It uses small metal disks known as electrodes that are placed on the scalp, and when brain cells send messages, they pick electrical signals. In this paper, out of the 64 electrode locations, 14 locations are used. The 10–20 system, which is an internationally recognized system to describe the electrode placement on the scalp, is referred in this paper.

The '10' and '20' refer to the actual distance between adjacent electrode locations which is either 10 or 20%, and for this, the brain is divided into four regions—frontal cortex, parietal cortex, temporal cortex, and occipital cortex.

ECG is used to measure the electrical activity of the heart by recording the heart rhythm. ECG waves, as in Fig. 1 consist of P-wave that detects alteration of atria, Q-R-S complex which provides information of alteration of ventricles, and T-wave

Fig. 1 ECG wave pattern

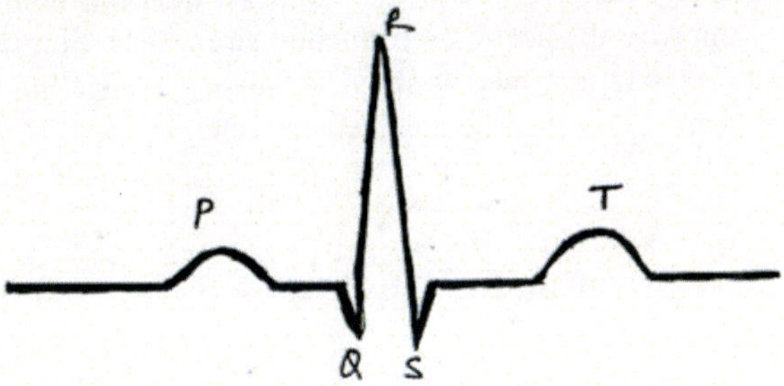

which shows ventricular repolarization. When a person is in a stressful state, his heart rate changes, causing changes in the ECG pattern [1], and these changes are used to detect stress.

According to the PAD emotional state model proposed by Mehrabian and Russell [2], three numerical dimensions, Pleasure, Arousal, and Dominance, can be used to represent all emotions such as pleasant (joy and happiness) and unpleasant (fear and anger) using Pleasure, energetic(boredom and rage) using Arousal, dominant (anger) and submissive (fear) using Dominance.

2 Literature Review

Lahane and Thirugnanam [3], in their journal, demonstrated how emotion detection can be carried out with the help of the DEAP dataset using the Teager–Kaiser Energy Operator(TKEO) approach with the k-nearest neighbor (KNN), neural network(NN), and other classifiers. This study explains how using TKEO improves feature extraction and proves a better way of emotion detection as compared to other conventional methods. Li et al. [4], in their study, recognized human emotions by EEG signals in two steps; first, it integrates the spatial attributes, frequency attributes, and time attributes of the EEG signals which map these into 2-D images. Then, it builds a series of EEG-MFI sequences to show emotion variation with EEG inputs. Secondly, it deals with the formed images by constructing a hybrid deep NN along with CNN. Kalas [5] in their research paper demonstrates decision tree, SVM classifier, and linear regression model on the patterns obtained by eye blinking via EEG using the data collected by activities like vehicle driving and heavy equipment operation where conciseness plays an important role. Patel et al. [6] use neuroimaging to differentiate between depressed and non-depressed patients and propose ways for treatment. It analyzes the past methodologies and suggests ways for future research. Subhani et al. [7] in their journal use a very famous experimental paradigm Montreal Imaging Stress Task (MIST) that consists of computerized arithmetic challenges. It has three levels—rest, control, and experimental, through which they induce stress among individuals which were then detected by applying ML models—support vector machine (SVM), logistic regression and Naive Bayesian classification on EEG data. Al-Shargie et al.

[8] study the effect of mental stress on the prefrontal cortex subregion of the brain to improve the accuracy of mental stress detection using EEG. The result of their studies was that mental stress is subregion specific, and for better accuracy, right ventrolateral PFC is a suitable candidate.

3 Motivation for the Proposed Approach

There has been a lot of research on stress and emotion detection using EEG [9], and separate studies have been done on stress detection using ECG. This paper aims at improving the accuracy of stress detection by making a hybrid model that predicts stress on the basis of both EEG and ECG. Furthermore, this paper refers the Russel's Circumplex model of affect [2] which provides a promising way to classify stress.

4 Proposed Stress Detection Method (Se.Re.Ne.)

In this paper, an ensemble method to detect stress using EEG and ECG signals (Se.Re.Ne.) has been proposed. Se.Re.Ne. works in four phases—data gathering, data analysis, stress classification, and statistical analysis. Figure 2 represents the block diagram of this work.

4.1 Data Gathering

Dataset used in Se.Re.Ne. is DREAMER [10] dataset which is made from EEG and ECG signals recorded during audio and visual stimuli used to entice specific emotions. Signals from 23 individuals were documented with their self-evaluation scores in the category of Valence, Arousal and Dominance [11] for each of the 18 clips shown. For EEG signals, theta, alpha, and beta power spectral density (PSD) for each electrode [12] was taken. For preprocessing, the dataset was stored in a '.mat' format which had to be converted to a suitable format of '.csv' for evaluation, and the emotional features from the dataset, namely Valence and Arousal, were used to classify stress using the Russell's Circumplex model of affect which uses a 2D plane to classify various emotions on the basis of Valence and Arousal. Stress is described at the top-left quadrant of Russell's plane with Valence represented at X-axis and Arousal represented at Y-axis. So, values of Valence and Arousal are taken as 2 and 4, respectively (on a scale of 5), and stress classification is done according to the conditions shown in Fig. 3.

After applying the proposed stress detection method, the data were divided into two classes—stressed (7979 data points or values) and not stressed (32,017 data points or values).

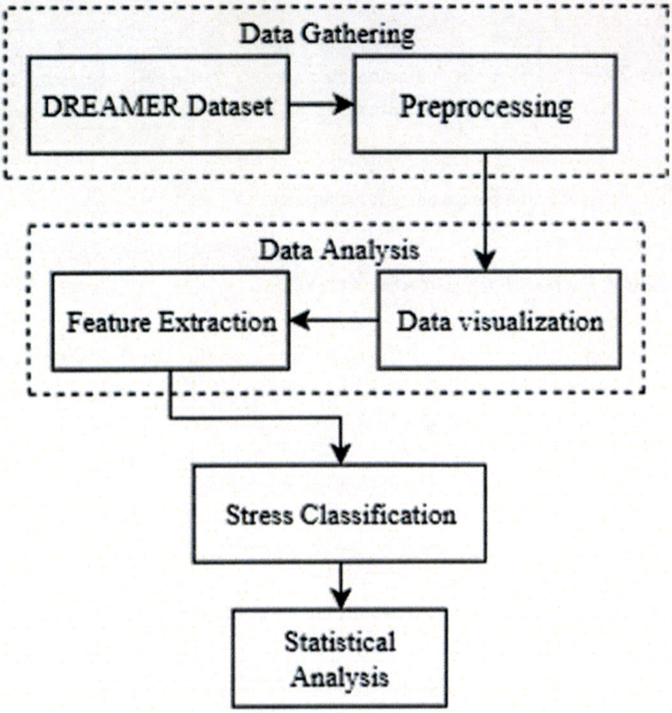

Fig. 2 Workflow of Se.Re.Ne.

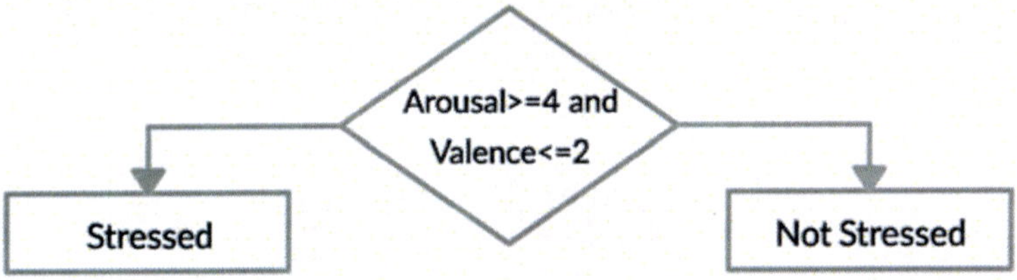

Fig. 3 Criteria for stress classification

Due to this uneven distribution, upsampling of data was required to make the dataset even for both classes. Figure 4 shows the data points for target attribute before and after the sampling.

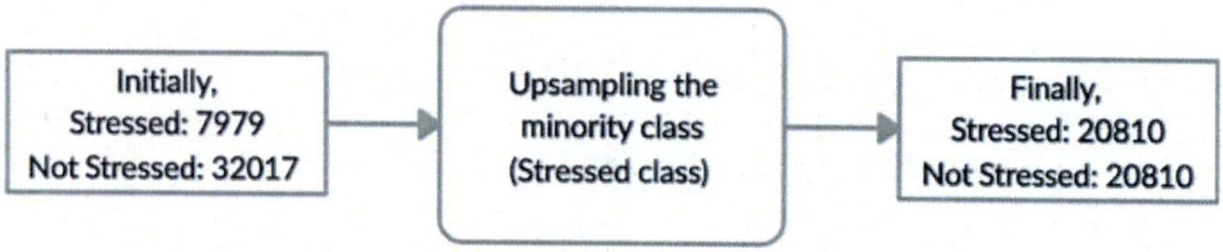

Fig. 4 Data point distribution before and after sampling

4.2 Feature Extraction for Se.Re.Ne.

Feature extraction is done by using the correlation method in this paper, and results of which are represented in Table 1. It was observed that variation in age did not correlate with change in the stress class. Also, out of two ECG channels and 14 channels of EEG signals which were considered for this paper positions of which are shown in Fig. 5, EEG_7, EEG_10, and ECG_0 have a negative correlation with stress showing that these attributes are inversely related to stress. The negative correlation of Valence with stress is in alignment with our approach of lesser the value of Valence, more stressed the person will be.

The following 14 EEG channels are considered in this paper.

Table 1 Correlation with stress

Attributes	Correlation with stress
Valence	−0.569315
Arousal	0.492630
Dominance	0.419894
EEG_0 (**AF3**)	0.014019
EEG_1 (**F7**)	0.003840
EEG_2 (**F3**)	0.009375
EEG_3 (**FC5**)	0.008402
EEG_4 (**T7**)	0.001896
EEG_5 (**P7**)	0.046115
EEG_6 (**O1**)	0.013466
EEG_7 (**O2**)	−0.009979
EEG_8 (**P8**)	0.009243
EEG_9 (**T8**)	0.001597
EEG_10 (**FC6**)	−0.012111
EEG_11 (**F4**)	0.017682
EEG_12 (**F8**)	0.006601
EEG_13 (**AF4**)	0.007597
ECG_0	−0.000899
ECG_1	0.000890

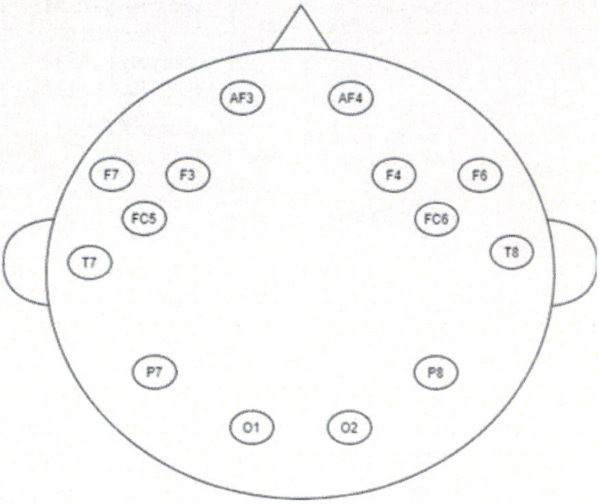

Fig. 5 EEG electrode positions selected for Se.Re.Ne.

4.3 Stress Classification

Many algorithms for mood analysis have been proposed, but to date, to the best of our knowledge, none of the algorithms has been used to detect stress. In this paper, we use two methods to detect stress.

The first method uses KNN [13], a lazy and non-parametric algorithm that is required for nonlinear EEG data. The value of k was taken as '7' for stress classification by hit and trial.

The second method uses DT, a white box type statistical algorithm used for classification. It is a non-parametric algorithm and does not rely upon the probability distribution theory. Attribute selection is done on the basis of measures such as the Gini index.

$$\text{Gini}(G) = 1 - \sum_{i=1}^{m} Pi^2 \tag{1}$$

where *pi* is the probability that a tuple in *G* belongs to Class Ci and is estimated by |*Ci*, *G*|/|*G*|.

For each attribute, the possible binary split is considered on the basis of the maximum Gini index which will be there after splitting. A part of the decision tree of height 2 is shown in Fig. 6.

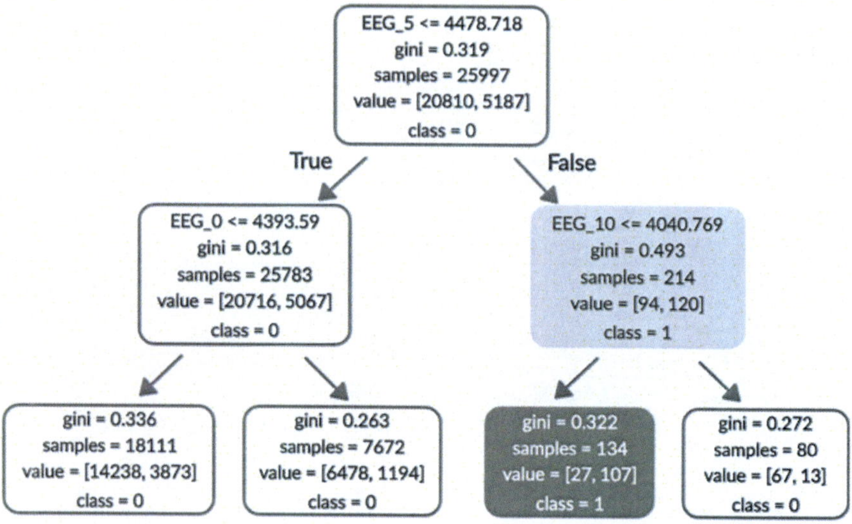

Fig. 6 Decision tree of depth 2

Table 2 Classification results

Algorithm	Output	Precision	Recall	$F1$-score	Accuracy
KNN	Stressed	0.33	0.57	0.41	0.68
	Not stressed	0.87	0.71	0.78	
DT	Stressed	0.37	0.38	0.38	**0.75**
	Not stressed	0.85	0.84	0.84	

4.4 Results Using the Proposed Method

Statistical analysis of the proposed stress classification method (Se.Re.Ne.) is conducted to get the results as in Table 2.

The ROC curve shown in Fig. 7 also helps to identify better algorithms [14] on the basis of the area under the curve (AUC). The area under the curve tests separation of data into two class labels. The more the area, preferable the result and better the algorithm. So, this plot helps us to identify DT as a better algorithm as compared to KNN which is also proved by the statistical results in Table 2.

5 Conclusion

In this paper, Se.Re.Ne., an ensemble method to detect stress using EEG and ECG signals, has been proposed. The proposed method works in four phases of data

Fig. 7 ROC curve

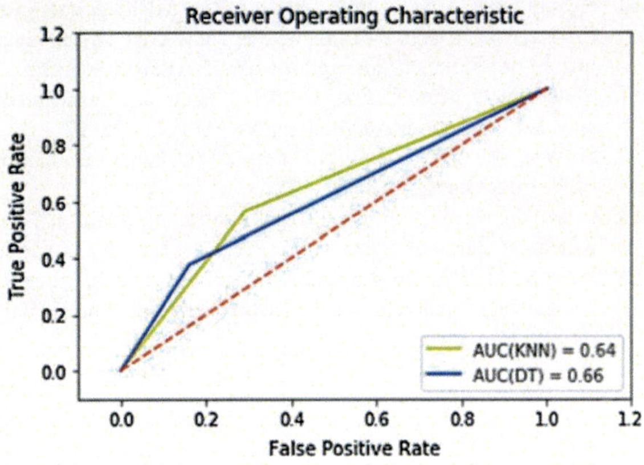

gathering, data analysis stress classification, and finally statistical analysis. Se.Re.Ne. records both audio and video signals to entice emotions along with self-evaluation scores. The correlation method is used to observe the variation in signals. Se.Re.Ne. stress classification is done using two methods KNN and decision trees. Results obtained through statistical analysis show that Se.Re.Ne. using KNN detects stress with a precision of 0.87, recall of 0.71, $f1$-score 0.78, and total accuracy of 68%, whereas Se.Re.Ne. using decision tree detects stress with a precision of 0.85, recall of 0.84, $f1$-score 0.84, and total accuracy of 75%. Hence, evaluation proves that extreme stress conditions Se.Re.Ne. detects using EEG and ECG signals with decision tree better than KNN, whereas normal stress condition Se.Re.Ne. detects using EEG and ECG signals with KNN better than decision tree.

References

1. Goel S, Kaur G, Tomar P (2017) A novel technique for stress recognition using ECG signal pattern. Current Pediatr Res 21(4)
2. Russell JA (1980) A circumplex model of affect. J Pers Soc Psychol 39(6):1161
3. Lahane PU, Thirugnanam M (2019) Human emotion detection and stress analysis using EEG signal
4. Li Y, Huang J, Zhou H, Zhong N (2017) Human emotion recognition with electroencephalographic multidimensional features by hybrid deep neural networks. Appl Sci 7(10):1060
5. Kalas MS, Momin BF (2018) Modelling EEG dataset for stress state recognition using decision tree approach, pp 82–88
6. Patel MJ, Khalaf A, Aizenstein HJ (2016) Studying depression using imaging and machine learning methods. NeuroImage Clin 10:115–123
7. Subhani AR, Mumtaz W, Saad MNBM, Kamel N, Malik AS (2017) Machine learning framework for the detection of mental stress at multiple levels. IEEE Access 5:13545–13556
8. Al-Shargie F, Tang TB, Kiguchi M (2017) Assessment of mental stress effects on prefrontal cortical activities using canonical correlation analysis: an fNIRS-EEG study. Biomed Opt Express 8(5):2583–2598
9. Liu Y, Sourina O (2013, Oct) EEG databases for emotion recognition. In: 2013 international conference on cyberworlds. IEEE, pp 302–309

10. Katsigiannis S, Ramzan N (2017) DREAMER: A database for emotion recognition through EEG and ECG signals from wireless low-cost off-the-shelf devices. IEEE J Biomed Health Info In press. https://doi.org/10.1109/JBHI.2017.2688239
11. Reuderink B, Mühl C, Poel M (2013) Valence, arousal and dominance in the EEG during game play. Int J Autonom Adapt Commun Syst 6(1):45–62
12. So WK, Wong SW, Mak JN, Chan RH (2017) An evaluation of mental workload with frontal EEG. PloS One 12(4)
13. Li M, Xu H, Liu X, Lu S (2018) Emotion recognition from multichannel EEG signals using K-nearest neighbor classification. Technol Health Care 26(S1):509–519
14. Faraz S, Ali SSA, Adil SH (2018, Dec) Machine learning and stress assessment: a review. In: 2018 3rd international conference on emerging trends in engineering, sciences and technology (ICEEST). IEEE, pp 1–4

Enactment of tf-idf and word2vec on Text Categorization

Monika Arora, Vrinda Mittal, and Priyanka Aggarwal

Abstract Text categorization has a variety of applications, such as sentiment analysis of user's tweet, categorizing blog posts into different categories, etc. The real-time data available for categorization is usually unstructured. An efficient algorithm for preprocessing the data can help to achieve better accuracy. Term frequency–inverse document frequency (tf-idf) and word2vec word embedding techniques are used widely before applying the text classification model. In order to show the enactment of these techniques on text categorization, we are comparing the accuracies of different multi-class text categorization algorithms such as Support Vector Machine (SVM), Logistic Regression and K-Nearest Neighbor (KNN) on these techniques. TagMyNews dataset is used to train the model. The results indicate that word2vec is efficient word embedding technique as it possesses higher accuracies for all the classification methods (KNN: 79.38%, SVM: 93.59%, Logistic Regression: 87.46%) as compared to tf-idf (KNN: 73.37%, SVM: 84%, Logistic Regression: 73.98%).

Keywords Text categorization · Word embedding techniques · Feature extraction · tf-idf · word2vec

1 Introduction

From the beginning of newspapers, everything changed, from the ink to type of paper used, but the categorization of news into different genres carried over by generations. Newspaper articles covers a wide scope of interests from legislative issues to sports to education, etc. But in this digital age, data available for categorization is unstructured.

M. Arora (✉) · V. Mittal · P. Aggarwal
Information Technology, BPIT, New Delhi, India
e-mail: monikaarora@bpitindia.com

V. Mittal
e-mail: vrindayash@gmail.com

P. Aggarwal
e-mail: priyankaagg1209@gmail.com

A. Abraham et al. (eds.), *Proceedings of 3rd International Conference on Computing Informatics and Networks*, Lecture Notes in Networks and Systems 167,
https://doi.org/10.1007/978-981-15-9712-1_17

Fig. 1 News genre classification model

In order to convert it into a structured format, some word embedding and feature extraction techniques can be used before applying the text classification model.

Word embedding techniques have a significant impact on the accuracy of the model [1]. The process of cleaning the data, removing irrelevant text from data, and feature selection affects the accuracy of vector space created and it depends on the word embedding technique used to extract these vector spaces.

In this project, we have used the TagMyNews dataset for training our model. The data is pre-processed by applying tf-idf and word2vec as word embedding and feature extraction techniques following which some multi-class categorization techniques such as SVM, KNN, and logistic regression were applied to classify the news headline into different genres. Since word embedding techniques can have an impact on the accuracy of algorithms, the accuracies obtained for all the algorithms are compared and analyzed on tf-idf and word2vec.

In order to validate the results for text categorization, we handpicked around 500 news snippets (in image format). The text was extracted from each image using pytesseract. The text obtained is tested on the model (as shown in Fig. 1) and results are matched with the already existing y-labels, which indicate the accuracy and efficiency of the model.

2 Related Work

A number of classification methods has been approached by researchers. There are a variety of algorithms available for text classification, but finding the efficient one is a tedious task. The text classification techniques that gained the interest of researchers are K Nearest Neighbor (KNN), Support Vector Machine (SVM), Logistic Regression, Naive Bayes, neural networks etc. These classification algorithms are often compared by the researchers in order to find the efficient ones.

Lilima Pradhan et al. compared the classification accuracies of various classification algorithms such as SVM, Naive Bayes, Decision Trees, KNN and Rocchio Classifier [2]. According to their result, SVM proves to be the strongest classification algorithm, followed by Naive Bayes and other algorithms.

Seyyed Mohammad Hossein performed news classification on popular datasets, BBC News and five groups from 20NewsGroup [3]. They used tf-idf as the extraction technique and classified the news using SVM classification algorithm. They obtained the classification accuracies of 97.84% and 94.93% on BBC News and 20NewsGroup respectively and concluded that tf-idf along with SVM propose a good text classification algorithm.

A recent research in this field shows that, classification algorithms behave differently due to many reasons. As stated in [4], the classification of news articles using naive bayes and SVM differs due to different feature extraction techniques used. Choosing the feature extraction techniques is also a tedious task, as it highly depends on the size of dataset, different features are extracted from different dataset, and hence the behavior of classification algorithm differs. In [5], authors performed sentiment analysis on Turkish twitter messages. They compared the performance of various feature extraction techniques such as BOW weighted by tf-idf and word2vec. The impact of these techniques was tested by applying various text classification algorithms such as Linear Regression, SVM, Decision Tree, Random forest etc, and achieved the highest accuracy by Random forest along with Word2vec model, i.e. 66.40%. This states that word2vec can be efficient as compared to tf-idf feature extraction technique.

In [6], the researchers used TagMyNews dataset to train their model. They classified the news articles into different categories such as sports, entertainment, etc. using SVM, Naive bayes and Softmax regression as the classification algorithms. Due to some false negative values present in the health category, Naive Bayes was inefficient as the multi class text classifier algorithm. SVM proves to be an efficient algorithm for classification as it achieved better classification accuracy as compared to both the methods. So, it can be inferred that there might be impact of different text pre-processing and feature extraction methods an on text classification methods. In this work, we are comparing the text classification accuracies of SVM, KNN and Logistic Regression on two word-embedding techniques—term frequency–inverse document frequency (tf-idf) and word2vec.

3 Dataset and Pre-processing

3.1 Dataset

TagMyNews Dataset was used to train the model. The corpus includes 32,602 training examples, where each training example includes a title, a description, a news article

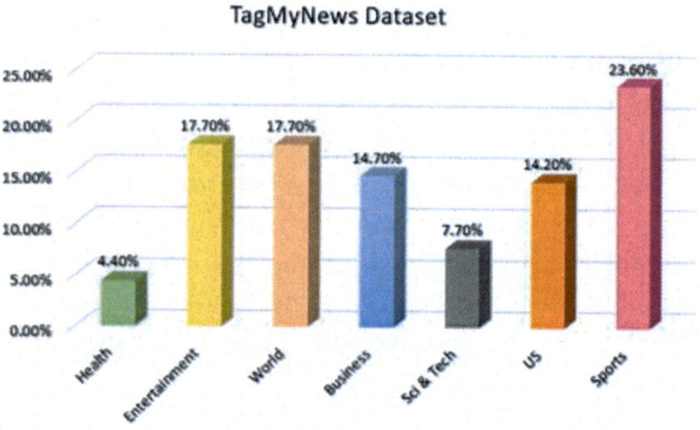

Fig. 2 TagMyNews dataset

link, an ID, the date of the publication, news article source and the subject category to which it belongs [7]. The dataset consists of six subject categories as listed in Fig. 2.

Since the corpus consists of a single file, each news article's title, description and category is segmented into seperate text file.

For validating the robustness of the model, the categories of the news snippets (image format) were used. For this purpose, a seperate database of 500 news headlines was created.

3.2 *Text Pre-processing*

Data present in the text files mentioned earlier is in a format that machines cannot understand. So, in order to make data operatable by the machine it has to be converted into a format on which machine learning models could be implemented. Text preprosessing techniques are used to serve the purpose. In this project three such techniques are used and their accuracy on the given dataset are analysed.

3.3 *Word Embedding and Feature Extraction Techniques*

There are many word embedding techniques used to represent words from the document into feature vectors. Techniques like bag of words where each word present in the document is separately converted into vectors are not able to establish semantic relationship between them [9]. Hence, in this project we tend to compare the effect of better techniques like tf-idf and word2vec on performance of a classifier.

3.3.1 tf-idf

Term Frequency inverse document frequency approach vectorizes a document using a weighting factor. It considers the frequency of a word in a document as well in the complete dataset. Inverse frequency term is used to adjust weight of terms that occurs often but does not provide much information. This reduces the weights of the words which are not significant for the category of the document [4]. Term frequency inverse document frequency method is able to weight common words by its IDF part:

$$\mathrm{idf(t)} = \lg \frac{1 + n^x}{1 + \mathrm{d}f(x, t)} + 1 \tag{1}$$

In Eq. (1), t is the term in the given document x, while n and df(x, t) expresses the whole document count and the number of documents containing t respectively. Thus, if a word is frequent in most of the documents, the denominator and numerator gets close to each other and IDF score approaches zero. Thus words which is not discriminative enough get close to zero.

3.3.2 Word2Vec

Before word2vec, traditional feature extraction technique such as countvectorizer considered the words as numbers. In contrast, word2vec considers distributional semantics, i.e, it understands the meaning of the word by looking to the context of the neighboring words, i.e, the word's vector will be similar to the vector of the context words of the text.

$$\frac{\partial J\theta}{\partial V_c} = -u_0 + \sum_{x=1}^{v} P(x|c) * u_x \tag{2}$$

The first part of Eq. (2) is the current representation of the context word. The second part is an expectation of what the context word should look like. The subtraction of the actual and expected representations gives the direction in which we should move and change the weight vector V_c so that we can maximize the likelihood.

4 Methodology

Initially, the news headlines snapshots were collected to form a dataset. Each image from the dataset is converted into text using pytessaract. The textual data is preprocessed and features were extracted using two feature extraction techniques, tf-idf and word2vec. In order to differentiate between the two techniques, and find out the efficient one, the features vectors obtained from both the techniques are

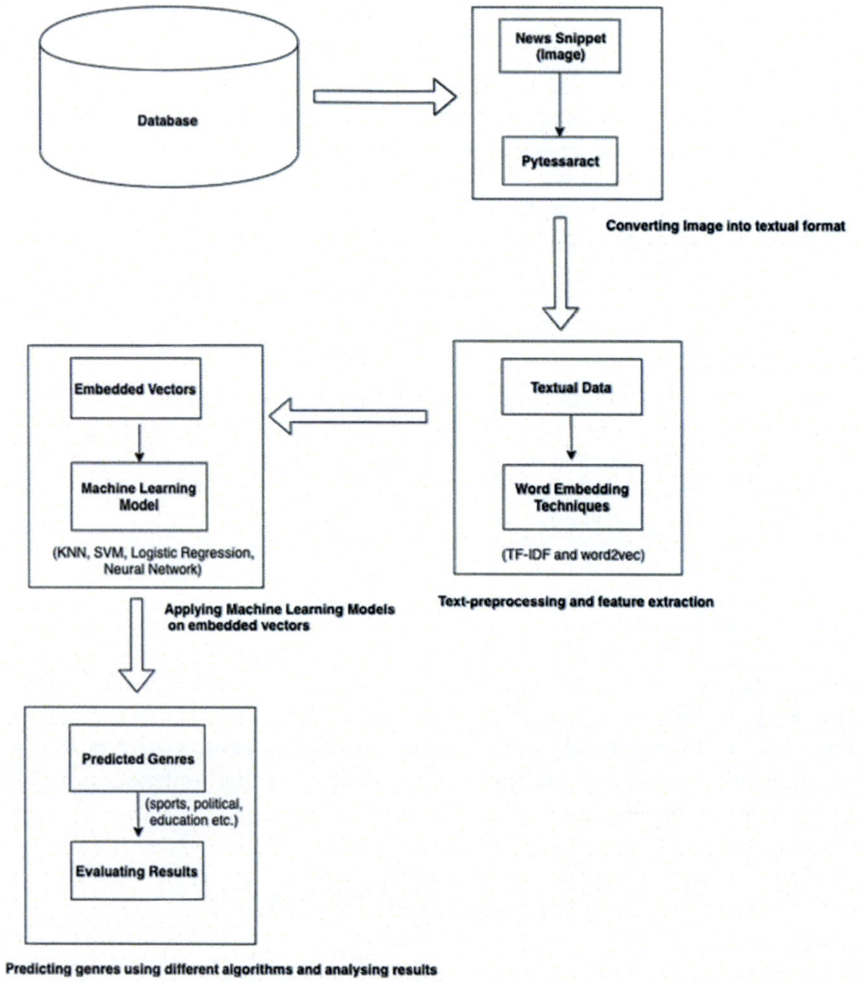

Fig. 3 Data model

trained on different multi-class classification techniques such as KNN, SVM, Logistic Regression. The classification accuracies of the different algorithms were analyzed.

Figure 3 given illustrates the data model of our project. It shows the stages followed by the data once input to the classification model.

5 Results

On studying the impact of word embedding techniques on different text classification algorithms, we observed that SVM prove to be better classification model as compared to logistic regression and KNN and word2vec turns out to be better word embedding technique than tf-idf. Table 1 illustrates the accuracies of SVM, logistic regression and KNN on application of tf-idf and word2vec word embedding techniques. word2vec along with SVM shows higher accuracy of 93.59.

As per the table, SVM and logistic Regression are proved to be better classification method as compared to k-nearest neighbours. Also, word2vec proved to better word embedding technique as compared to count vectorizer and TF-IDF transformer.

The following results can also be demonstrated using a double bar chart (Fig. 4).

We plotted the heatmaps using seaborn library in python. A heat map is a graph used to represent numerical data. Each value in the matrix represents the frequency of data as compared to overall size of the dataset. Higher frequencies are shown with darker colors as compared to lower frequency values.

Figure 5 shows that the results predicted using the SVM classifier. Correct predictions are shown along the diagonal. Intensity of the color depicts the number of testing examples belonging to respective categories. Since, the dataset contains most examples of sport category and least of sci-tech, correct predictions for sports is shown with yellow color (lighter color) and sci-tech with royal blue color (darker color). Similar heatmaps are drawn for logistic regression and KNN classifiers.

On observing the two heatmaps in Figs. 5 and 6. As heatmap for SVM (word2vec) shows lighter color for health category as compared to corresponding value in

Table 1 Comparsion of accuracies of different classification techniques

Text classification	Word embedding techniques		
Techniques	KNN	SVM	Logistic regression
tf-idf	73.37	84	73.98
word2vec	79.38	93.59	87.46

Fig. 4 Double bar chart

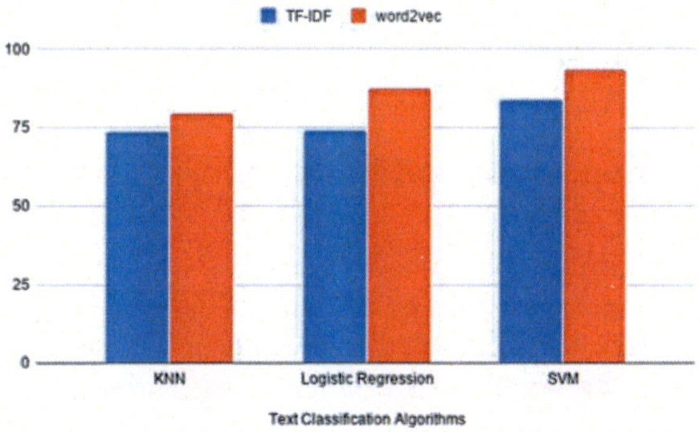

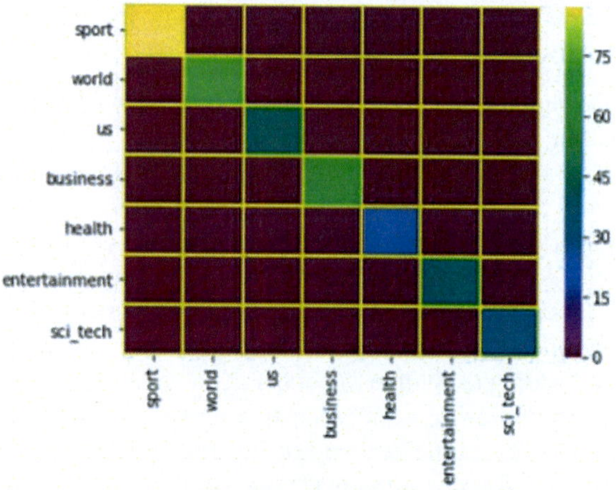

Fig. 5 Heatmap for SVM (tf-idf)

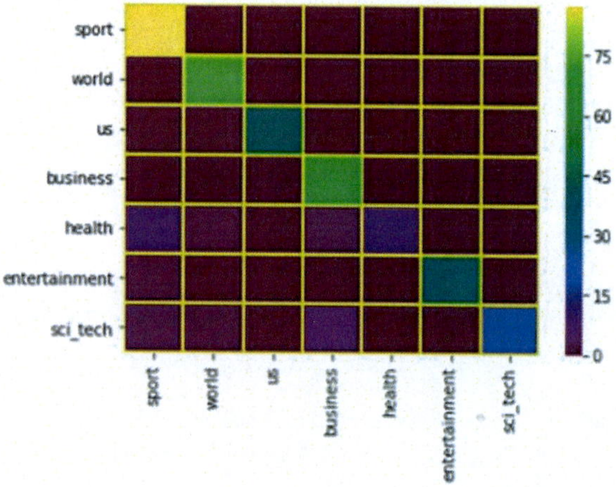

Fig. 6 Heatmap for logistic eegression(tf-idf)

heatmap of Logistic Regression (word2vec). Hence, it can be concluded that SVM made more correct predictions for health category as compared to logistic regression.

Figure 7 shows that correct predictions made in case of KNN were less as compared to other two algorithms.

6 Conclusion and Future Scope

We extracted the text from the news snippets given as input to the model, preprocessed and converted them into feature vectors, which were further trained by different classification algorithms. On analysing the results, (refer Table 1), we concluded that word2vec is an efficient word embedding technique as compared to tf-idf

Fig. 7 Heatmap for KNN (tf-idf)

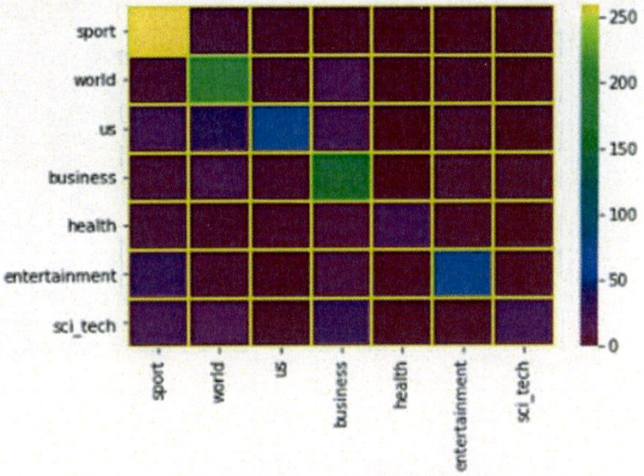

Fig. 8 Heatmap for SVM (word2vec)

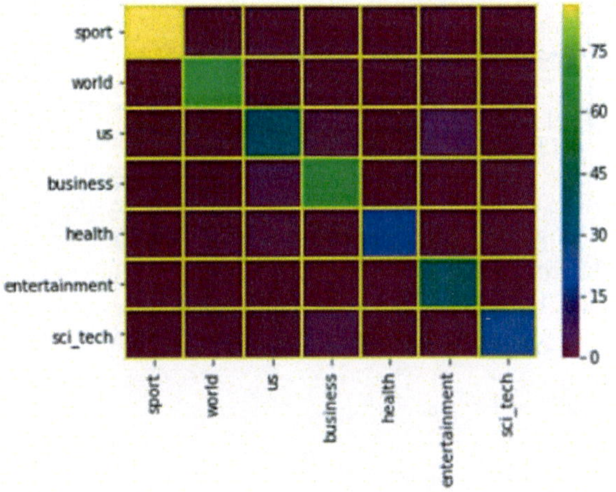

Fig. 9 Heatmap for logistic regression (word2vec)

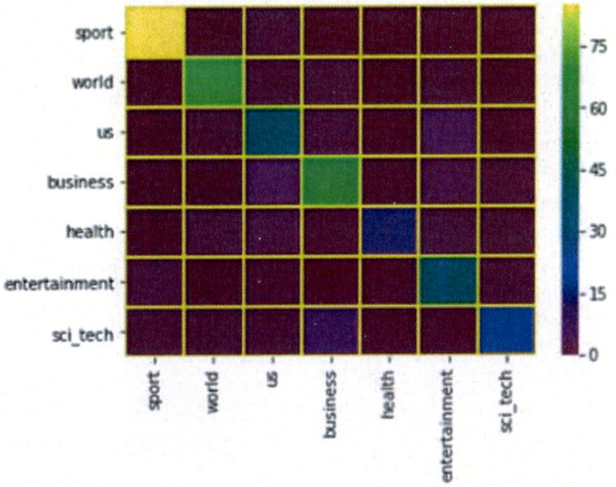

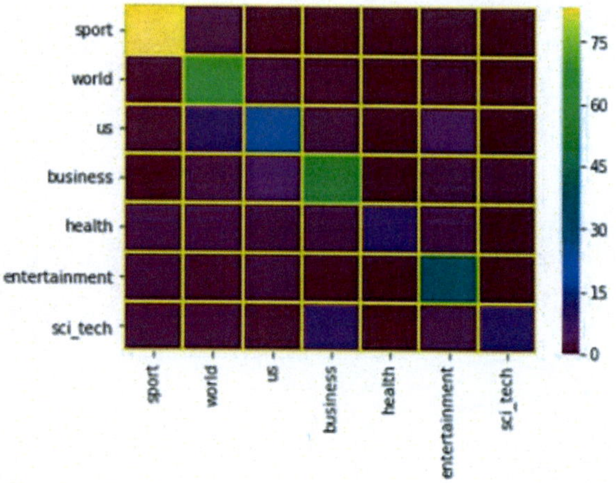

Fig. 10 Heatmap for KNN (word2vec)

as word2vec can understand the semantics of the words and hence shows better classification rate. Moreover, SVM along with word2vec possess the highest accuracy as compared to logistic regression and KNN, i.e. 93.59 Results may vary on different dataset (specially larger), as feature extracted are different on every dataset. Different word embedding techniques can also be tried to ensure that there is an impact of word embedding techniques before applying the classification model. Other text classification techniques such as CNN or LSTM can be applied (Figs. 8, 9 and 10).

References

1. Blum AL, Langley P (1997) Selection of relevant features and examples in machine learning. Artif Intell 97(1–2):245–271
2. Pradhan L et al (2017) Comparison of text classifiers on news articles. Int Res J Eng Technol 4(3):2513–2517
3. Dadgar SM, Araghi MS, Farahani MM (2016) A novel text mining approach based on TFIDF and support vector machine for news classification. In: IEEE international conference on engineering and technology (ICETECH), IEEE
4. Suleymanov U, Rustamov S (2018) Automated news categorization using machine learning methods. In: IOP conference series: materials science and engineering, vol 459, no 1. IOP Publishing
5. Effrosynidis D, Symeonidis S, Arampatzis A (2017) A comparison of pre-processing techniques for twitter sentiment analysis. In: International conference on theory and practice of digital libraries. Springer, Cham
6. Romero F, Koochak Z (2015) Assessing and implementing automated news classification
7. News Category Dataset. https://www.kaggle.com/rmisra/news-category-dataset. Last accessed 21 Dec 2019
8. Virmani D, Taneja S (2019) A text preprocessing approach for efficacious information retrieval. In: Smart innovations in communication and computational sciences. Springer, Singapore, pp 13–22

9. Grootendorst M, Vanschoren J (2019) Beyond bag-of-concepts: vectors of locally aggregated concepts. In: Joint European conference on machine learning and knowledge discovery in databases. Springer, Cham
10. A math-first explanation of Word2Vec. https://medium.com/analytics-vidhya/maths-behind-word2vec-explained-38d74f32726b. Last accessed 15 Dec 2019

Senti_ALSTM: Sentiment Analysis of Movie Reviews Using Attention-Based-LSTM

Charu Gupta, Geetansh Chawla, Karan Rawlley, Kritarth Bisht, and Mahak Sharma

Abstract Association with the customers is one of the principal elements for businesses. It is essential for these firms to recognize exactly what clients' opinions about the latest as well as established products and services. One such business is film industry. Movies are a dominant source of entertainment. Yield of a movie relies on a great extent to the reviews on social media, blogs, forums, IMDB, and movie viewing platforms. In this paper, to achieve revolutionary performance in sentiment analysis, attention-based long short-term memory, *'Senti_ALSTM,'* model is proposed which improves the neural networks by assigning memory to it and further reducing the vanishing gradient problem of recurrent neural network. *Senti_ALSTM* experiment with attention in long short-term memory model which takes into account the information reflected by the input sequence, thus remembering the context, and establishes the relation between input and output sequence. Further, the success of the benchmark data confirms the efficiency of the proposed model over the other models.

Keywords Attention · Sentiment analysis · Natural language processing · Deep learning · LSTM

C. Gupta · G. Chawla (✉) · K. Rawlley · K. Bisht · M. Sharma
Department of Computer Science and Engineering, Bhagwan Parshuram Institute of Technology, New Delhi, Delhi, India
e-mail: geetansh.chawla@gmail.com

C. Gupta
e-mail: charugupta@bpitindia.com

K. Rawlley
e-mail: karan8798rawlley@gmail.com

K. Bisht
e-mail: jollybsht9@gmail.com

M. Sharma
e-mail: sharmamahak018@gmail.com

A. Abraham et al. (eds.), *Proceedings of 3rd International Conference on Computing Informatics and Networks*, Lecture Notes in Networks and Systems 167,
https://doi.org/10.1007/978-981-15-9712-1_18

Abbreviations

AI	Artificial Intelligence
CNN	Convolutional Neural Network
GloVe	Global Vectors
IMDB	Internet Movie Database
LSTM	Long Short Term Memory
NBSVM	Naive Bayes—Support Vector Machine
RNN	Recurrent Neural Network
SVM	Support vector machine
WEKA	Waikato Environment for Knowledge Analysis

1 Introduction

Sentiment analysis refers to using natural language processing to identify and extract subjective information from documents. It is intended for the determination of an author's attitude toward a topic or the complete polarity of a source material [1–4]. Sentiment analysis is the action of detecting a piece of writing (in this study, it refers to textual reviews) which is bound to be positive, negative, or neutral. Earlier machine learning approach and lexicon-based approach were used for determination of the sentiments in a document. Machine learning approach includes dictionary-based approach, corpus-based Approach, decision tree classifier, linear classifier, and lexicon-based approach [5–7].

These approaches have limitations with respect to named entity recognition issue, anaphora resolution (pronoun refers to which of the noun occurred), sarcasm, irony and word ambiguity. To tackle the mentioned problems, deep learning can be used for sentiment analysis. Deep learning is a function of AI that copies the operation of the human brain while handling data and forming patterns to make certain decisions. Deep learning is a subset of AI, having networks that can learn from unstructured data with no supervision.

In this paper, an attention-based long short-term memory (LSTM) model, *Senti_ALSTM,* is employed to examine the sentiments in the movie reviews. The aim of this research is to implement deep learning approach in order to analyze the polarity of the movie reviews and compare the performance of various pre-defined machine learning models. LSTM associates both long-term and short-term memory to leverage the artificial neural networks and eliminates the gradient vanishing suffered in RNN. Attention layer further reduces the information loss by allowing the translator to observe the information provided by the original sentence, thus remembering the context.

2 Literature Survey

After the notion of sentiment analysis was raised by Nasukawa [8] in the early 2000s, a lot of rigorous researches have been put through by a large number of researchers. In [9–11], the work compared numerous ways for the use of semantic data to enhance the sentiment analysis performance. The conventional approaches were not considering the semantic associations among sentences or document contents. In [12], the method gave deep knowledge of the sentiment analysis with drug reviews containing numerated data. They processed sentences containing numerical terms for the classification on the basis of opinionated and non-opinionated sentences and further indication of polarity depicted by the use of fuzzy set theory. A large number of doctors from various hospitals, clinics, and medical centers were interviewed in order to develop the knowledge base.

Studies have shown NBSVM model [13], by integrating the Naïve Bayes and support vector machine having a decent performance in a variety dataset. In [14], the developed method employed a number of semantic attributes and support vector machine classifier to perform sentiment analysis. Zeng An exemplary advancement in sentiment analysis by employing convolutional neural network (CNN) to classify relations was studied in [15, 16]. Studies in the literature have recognized the part of tweets in political ballot. The method remarked that the previous works done for the mentioned purpose needed to check whether the tweets were predictive or reactive. It was discovered that the tweets were more reactive than predictive [17, 18]. Wang and Zhang [19] proposed bidirectional recurrent neural network which acquire sequences and relations from the unprocessed data. Studies have remarked that a review can be positive and negative; at the same time, she used WEKA and discovered Naïve Bayes performed the best out several other models [20].

3 Proposed Methodology (Senti_ALSTM)

In this section, details related to dataset and technical approaches used in the proposed model of *Senti_ALSTM* are explained. The steps of the proposed work are depicted in Fig. 1.

3.1 Materials and Methods

This subsection provides the details of deep neural networks used in designing the *Senti_ALSTM*. A comprehensive discussion on recurrent neural network, LSTM, and the proposed attention mechanism, is presented in this section.

Recurrent Neural Network (RNN). The conventional machine learning models show decent accuracies but suffer with the loss of sequence of words in a given

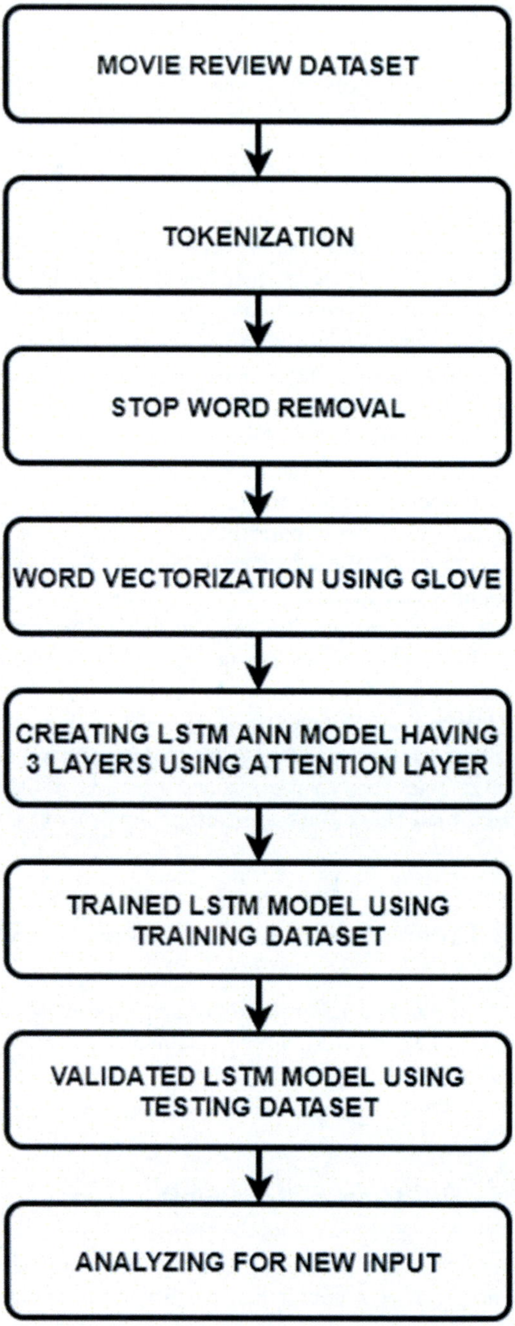

Fig. 1 Flowchart of the proposed model (Senti_ALSTM)

sentence, so they are unable to capture the fine meaning from the reviews. RNN, on the other hand, enables the user to have a hold of the fine semantics. Though recurrent neural networks show increased classification accuracies, they are unable to depict certain relations. RNN endures with the problem of vanishing gradient; that is the gradients of the loss function approaches '0' with the increase in the number of neural network layers. This makes the model difficult to train. RNNs have a defining role for sequential data modeling.

Long Short-Term Memory (LSTM). The RNNs are ineffective to learn from long-term dependencies because of the exploding problems or gradient vanishing. The LSTM networks are proposed and developed based on the RNNs to address those weaknesses. The basic structure of LSTM contains three gates and one cell memory state. The gates are namely input, output, and forget gates. In conventional LSTM, forget gate and peephole connections were absent, and the unit biases were excluded. LSTM can process single data as well as sequences. The specific schematic is illustrated in Fig. 2.

Attention. Attention removes the encoder–decoder structure by securing the median outputs from the LSTM encoder at every step of the input sequence along with instructing the model to pay attention to the mentioned inputs and connect them to the output sequence. Attention is a vector, which are normally the outputs of dense layer which uses softmax function. Prior to attention layer, translation depends on reading and compressing entire sentences to a fixed-length vector. This will lead to information loss due to compression of a sentence of heap of words into a sentence of few words. Attention resolves the mentioned problem to an extent. This enables machine translator to view the information of the actual sentence followed by generation of the genuine word in accordance to present word it is working on and the context. It also allows translator to focus on local or global features. Attention is only an interface formed by parameters and fine mathematics. On its addition to any model, the overall accuracy increases.

Softmax Layer. Softmax layer is generally the last layer used in a neural network. The function of mentioned layer is to limit the output of the function into the range of 0–1. This function is normally employed to evaluate losses which can occur when

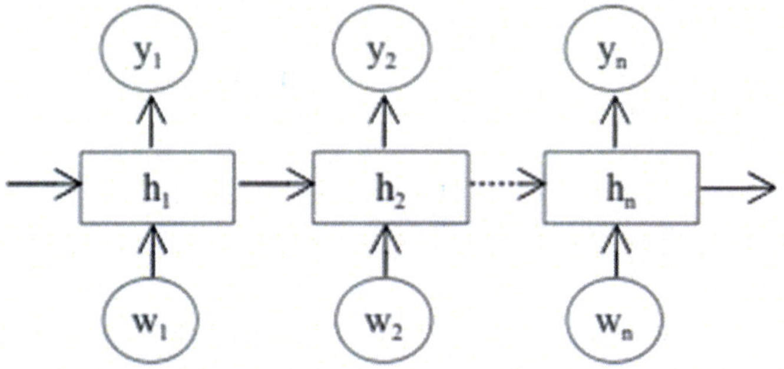

Fig. 2 Architecture of LSTM network

a dataset is being trained. Softmax regression finds its use in discriminative models like cross-entropy and noise contrastive estimation. The mentioned are only two techniques which seek to optimize the present training set to improve the probability of predicting the right word or sentence.

3.2 Experimental Settings

In this section, the experimental settings with data pre-processing details of *Senti_ALSTM* are presented with a gentle discussion of model generation.

Dataset. The dataset is a collection of 50,000 bipolar movie reviews. There are additional data, as well. This dataset is extracted from large dataset of film reviews used by Stanford University's computer science department. The aforementioned dataset is obtained from IMDB and consists of 50,000 reviews marked with positive (1) and negative (0) polarities.

Data Pre-processing. The used dataset consisted of various anomalies like stop words, punctuations, connectors, special characters, etc., which were not required for the further phases. In data pre-processing stage, the string is the input and this stage processes it by removing the anomalies mentioned, and the processed string is returned at last. First, the tags are modified by replacing the text between the opening and closing tags by replacing the characters by spaces followed by the removal of all the characters except capital and small letters resulting in long meaningless strings. The single characters are then replaced by space. At last, extra spaces are removed from the string. Further, the word encoding is the initial layer of the neural network model used; for this, tokenizer class is used to create a word to index dictionary. In the word to index dictionary, every word present in the collection is used as a key and a correlating distinctive index is used as the value for the key. GloVe is used for creation of embedding layer for word to vector conversion. GloVe model is algorithm to get word vector representations by mapping words into vector space and distance between vectors that indicate degree of similarity [21].

The conventional long short-term memory is unable to distinguish between the essential knowledge and the irrelevant one for sentiment categorization. In this task, use of attention mechanism is proposed to resolve the mentioned problem in LSTM as it can secure the main part of sentence by allowing the translator to remember the information provided by input sequence and pay special attention to the context. The mechanism of attention builds a vector of attention weight and a hidden weighted one. Attention mechanism allows the model to secure the most applicable part of a sentence when it considers different features.

4 Results and Comparisons

For the classification of movie reviews into positive and negative, a number of models like vanilla neural network, CNN, LSTM, etc., were tested. SVM was found unable to integrate the feature sets, and therefore, no satisfactory results were achieved. In comparison with the rest of the models, *Senti_ALSTM* model out-performed the other models applied with an accuracy of nearly 88%. Also, vanilla neural network was found to have least accuracy of nearly 75% out of all the models. The order of performance according to the classification accuracy was found to be *Senti_ALSTM* > Bidirectional LSTM > LSTM > CNN > Vanilla Neural Network.

Table 1 shows the validation accuracy achieved by applying different algorithms and neural networks over the above-mentioned pre-processed dataset. The training setting is shown in Fig. 3. After getting these results, it can be noted that attention layer increases the performance of general LSTM network. Combining attention with LSTM gives better results than bidirectional LSTM.

Table 1 Comparison of the proposed methodology with the existing work

S. No.	Method/technique	Accuracy
1	Vanilla neural network [22]	74.7
2	Convolutional neural network (CNN) [15]	82.3
3	Long short-term memory networks (LSTM) [23]	85.04
4	Bidirectional LSTM [24]	86.56
5	**Proposed work: Senti_ALSTM**	**87.43**

```
model.fit(X_train, y_train, epochs=6, batch_size=64, validation_data = (X_test,y_test))

WARNING:tensorflow:From C:\Users\rawll\.conda\envs\tensorflow\lib\site-packages\keras\backend\tensorflow_backend.py:422: The na
me tf.global_variables is deprecated. Please use tf.compat.v1.global_variables instead.

Train on 40000 samples, validate on 10000 samples
Epoch 1/6
40000/40000 [==============================] - 62s 2ms/step - loss: 0.4353 - accuracy: 0.7923 - val_loss: 0.3727 - val_accurac
y: 0.8274
Epoch 2/6
40000/40000 [==============================] - 62s 2ms/step - loss: 0.3416 - accuracy: 0.8489 - val_loss: 0.3374 - val_accurac
y: 0.8489
Epoch 3/6
40000/40000 [==============================] - 63s 2ms/step - loss: 0.3088 - accuracy: 0.8672 - val_loss: 0.3175 - val_accurac
y: 0.8611
Epoch 4/6
40000/40000 [==============================] - 65s 2ms/step - loss: 0.2796 - accuracy: 0.8823 - val_loss: 0.3053 - val_accurac
y: 0.8651
Epoch 5/6
40000/40000 [==============================] - 64s 2ms/step - loss: 0.2502 - accuracy: 0.8978 - val_loss: 0.2937 - val_accurac
y: 0.8743
```

Fig. 3 Training of the proposed model

5 Conclusions and Future Scope

In this paper, from the experimental results, it can be concluded that attention can be used for enhancing implementation of recurrent neural networks for retrieving underlying context from movie reviews for sentiment analysis. With this combined approach of RNN, the accuracy of the model can be increased substantially. The proposed methodology consists of GloVe 300 dimensions word embedding which is observed to be way better than the generic one hot encoding. It consumes less space for storing word vectors and consumes less time for creating the embedding matrix. Further, it is observed that there is a very small difference between the training accuracy and test accuracy which means that the proposed model does not over-fit. However, the limitation of *Senti_ALSTM* is that the result obtained is bipolar (positive and negative only).

The future of sentiment analysis is going to understand the importance of social media interactions and the depiction of customer behind the screen. Further, implementation of attention-based bidirectional LSTM can be used for enhancing results.

References

1. Sentiment analysis. In: En.wikipedia.org. https://en.wikipedia.org/wiki/Sentiment_analysis. Accessed 2 Aug 2019
2. Gupta C, Jain A, Joshi N (2019) A novel approach to feature hierarchy in aspect based sentiment analysis using OWA operator. In: Proceedings of 2nd international conference on communication, computing and networking. Springer, pp 661–667
3. Gupta C, Jain A, Joshi N (2019) DE-ForABSA: a novel approach to forecast automobiles sales using aspect based sentiment analysis and differential evolution. Int J Inf Retrieval Res (IJIRR) 9:33–49
4. Jain A, Nandi B, Gupta C, Tayal D (2019) A hybrid framework based on PSO and neutrosophic set for document level sentiment analysis. In: International conference on information technology and applied mathematics. Springer, pp 372–379
5. Baziotis C, Pelekis N, Doulkeridis C (2017) DataStories at SemEval-2017 Task 4: deep LSTM with attention for message-level and topic-based sentiment analysis. In: Proceedings of the 11th international workshop on semantic evaluation (SemEval-2017). Association for Computational Linguistics, pp 747–754
6. Rudolph S, Giesbrecht E (2010) Compositional matrix-space models of language. In: Proceedings of the 48th annual meeting of the association for computational linguistics, pp 907–916
7. Zhou P, Shi W, Tian J, Qi Z, Li B, Hao H, Xu B (2016) Attention-based bidirectional long short-term memory networks for relation classification. In: Proceedings of the 54th annual meeting of the association for computational linguistics, vol 2: short papers. Association for Computational Linguistics, pp 207–212
8. Nasukawa T, Yi J (2003) Sentiment analysis: Capturing favorability using natural language processing. In: Proceedings of the 2nd international conference on Knowledge capture, pp 70–77
9. Märkle-Huß J, Feuerriegel S, Prendinger H (2013) Improving sentiment analysis with document-level semantic relationships from rhetoric discourse structures. In: Proceedings of the 50th Hawaii international conference on system sciences, pp 1142–1151

10. Socher R, Huval B, Manning C, Ng A (2012) Semantic compositionality through recursive matrix-vector spaces. In: Proceedings of the 2012 joint conference on empirical methods in natural language processing and computational natural language learning. Association for Computational Linguistics, pp 1201–1211
11. Socher R, Perelygin A, Wu J, Chuang J, Manning C, Ng A, Potts C (2013) Recursive deep models for semantic compositionality over a sentiment Treebank. In: Proceedings of the 2013 conference on empirical methods in natural language processing. Association for Computational Linguistics, pp 1631–1642
12. Yazdavar A, Ebrahimi M, Salim N (2017) arXiv:1701.00798
13. Wong S, Manning C (2012) Baselines and bigrams: Simple, good sentiment and topic classification. In: Proceedings of the 50th annual meeting of the association for computational linguistics, vol 2: short papers, pp 90–94
14. Rink B, Harabagiu S (2010) Utd Classifying semantic relations by combining lexical and semantic resources. In: Proceedings of the 5th international workshop on semantic evaluation, pp 256–259
15. Zeng D, Liu K, Lai S, Zhou G, Zhao J (2014) Relation classification via convolutional deep neural network. In: Proceedings of COLING 2014, the 25th international conference on computational linguistics: technical papers, pp 2335–2344
16. Denil M, Bazzani L, Loris H, de Freitas N (2012) Learning where to attend with deep architectures for image tracking. Neural Comput 24:2151–2184
17. Murthy D (2015) Twitter and elections: are tweets, predictive, reactive, or a form of buzz? Inf Commun Soc 18:816–831
18. Jiang L, Yu M, Zhou M, Liu X, Zhao T (2011) Target-dependent twitter sentiment classification. In: Proceedings of the 49th annual meeting of the association for computational linguistics: human language technologies, pp 151–160
19. Zhang D, Wang D (2015) Relation classification via recurrent neural network. arXiv:1508.01006
20. Shaziya H, Kavitha G, Zaheer R (2015) Text categorization of movie reviews for sentiment analysis. Int J Innov Res Sci Eng Technol 4:11255–11262
21. GloVe (machine learning). In: En.wikipedia.org. https://en.wikipedia.org/wiki/GloVe_(machine_learning). Accessed 3 Aug 2019
22. Goyal A, Parulekar A (2015) Sentiment analysis for movie reviews. [ebook] Available at https://cseweb.ucsd.edu/classes/wi15/cse255-a/reports/fa15/003.pdf. Accessed 29 Aug 2019
23. Hochreiter S, Schmidhuber J (1997) Long short-term memory. Neural Comput 9:1735–1780
24. Zhang S, Zheng D, Hu X, Yang M (2015) Bidirectional long short-term memory networks for relation classification. In: Proceedings of the 29th Pacific Asia conference on language, information and computation, pp 73–78

DDYDAS: Driver Drowsiness, Yawn Detection and Alert System

Bhawna Suri, Moksh Verma, Kanika Thapliyal, Akshay Manchanda, and Abhishek Saini

Abstract Lack of attention during a driving session can be fatal. These resulting fatalities not only affect us as an individual but they also affect our society and our family members. It is the responsibility of the driver to ensure that all the passengers including the driver are safe. Responsible driving ensures the safety of oneself and the vehicles driven by others on road. Dozing off while driving has been observed to be one of the primary causes of road accidents. With the availability of the latest hardware and software technologies, we aim to propose a potential solution to drastically reduce the number of accidents that occur due to an individual's drowsiness. Facial features like eyes are monitored to detect closing of eyes for a period that is substantially more than the time taken for an eye blink, and similarly, yawning patterns of a person are analysed as well to conclude the driver's drowsiness.

Keywords Drowsiness detection · Driver sleepiness detector · Eye detection · Yawn detection

B. Suri · M. Verma · K. Thapliyal · A. Manchanda · A. Saini (✉)
Department of Computer Science, Bhagwan Parshuram Institute of Technology, New Delhi, Delhi, India
e-mail: sainicooldude@gmail.com

B. Suri
e-mail: bhawnasuri12@gmail.com

M. Verma
e-mail: mokshverma98@gmail.com

K. Thapliyal
e-mail: kanikathapliyal17@gmail.com

A. Manchanda
e-mail: a.manchanda97@gmail.com

A. Abraham et al. (eds.), *Proceedings of 3rd International Conference on Computing Informatics and Networks*, Lecture Notes in Networks and Systems 167,
https://doi.org/10.1007/978-981-15-9712-1_19

Abbreviations

USA	United States of America
DDYDAS	Driver Drowsiness, Yawn Detection and Alert System
EEG	Electroencephalogram
PERCLOS	Percentage of Eyelid Closure
AI	Artificial Intelligence
NIR	Near Infrared
DCT	Discrete Cosine Transformation
HOG	Histogram of Oriented Gradient
SVM	Support Vector Machine
ROI	Region of Interest
EAR	Eye Aspect Ratio
MAR	Mouth Aspect Ratio
OpenCV	Open Source Computer Vision Library
FPS	Frames Per Second

1 Introduction

Extended sessions of driving result in fatigue and, consequently, affect a person's response time. Road safety while driving has been a serious concern. Many drivers lose their lives or cause fatal accidents on the road during rash driving that causes many lives to be lost. Car accidents associated with driver fatigue are more likely serious and can lead to serious injuries and even death. According to a recent study, one in five road accidents is due to driver drowsiness and its resulting ignorance on monotonic roads [1]. Another significant problem that challenges the safe driving experience on roads is driver drowsiness and inattention [2].

To minimize the frequency of road mishap, it is important to analyse drivers and their driving behaviour to warn them about their potential state of distraction. According to the report [3], by monitoring and predicting a driver's behaviour, the frequency of car crashes is expected to decline by around 10–20%. The National Sleep Foundation of the USA stated that more than 50% of adult drivers exhibited slumberous behaviour and 28% of these drivers had admitted to have dozed off at least once while driving [4]. It has been observed that driving performance declined with increased drowsiness and resulting in accidents. Exhaustion resulting from sleep deprivation or sleep disorders has been observed to be a prominent factor in accidents [5]. Reports [6, 7] show that the mental stress and impatience of the driver may put their life at risk.

To detect and avoid such incidents from happening, various methods have been proposed based upon image processing, EEG-based techniques, artificial neural network-based techniques, vehicular-based measures and subjective measures.

Image processing is further divided into eye blink method, yawning-based technique, PERCLOS-based and template matching technique [8].

Moreover, the Viola–Jones algorithm can be employed to identify many facial regions such as eyes and mouth, which are sometimes falsely detected [9].

This paper aims to propose a prototype DDYDAS—Driver Drowsiness, Yawn Detection and Alert System. A system that will observe the driver's physical behaviour in real time with a rather practical accuracy and send out an alert when required. The indication of a driver's exhaustion can be successfully detected in advance, by constantly observing the facial movements, to avoid any potential casualty; which can be done using an array of images containing the face of the driver.

2 Literature Review

Driver fatigue and drowsiness while driving is a field of interest for many researchers in recent years. These researchers have focused on drowsiness using a wide range of techniques and procedures. Techniques ranging from a computer vision-based machine learning model to a dedicated hardware implementation.

Our safety is always our priority in doing anything. Road accidents are increasing in number day-by-day due to a driver's lack of vigilance which is becoming a serious problem. These accidents happen due to various reasons from which drowsiness, fatigue and distraction of the driver are casual factors behind a road accident. We propose an economically beneficial solution by observing the somnolence level of the driver and sending out an alert to the driver about their condition. Various techniques are being used by the researchers to detect drowsiness such as image processing, EEG, AI, driver's response, vehicle movements.

In the work by Hua Gao et al. [10], a NIR camera was used to create an array of images of the vehicle operator and compensate for the poor lighting effects. A three-dimensional cylindrical head model (CHM) was applied, and a holistic affine warping method was used to normalize the images of the driver using the eye coordinates. After fixing the eye centres, local DCT feature representation was extracted or the local descriptors were extracted directly via facial landmarks.

According to [11], the high-end models of the Volvo cars introduced a system called driver alert control in 2008 which used various sensors and cameras to monitor the driver drowsiness and movement of the car within the road lane. A similar system (ATTENTION ASSIST) [DaimerAG 09] was marketed/commercialized by Mercedes Benz in class-E vehicles.

Irtija et al. [12] describes how facial features of the driver subject are detected and drowsiness is identified by processing images of the face. Facial landmark points are first detected that are used to locate eyes and mouth on the image captured of the driver. It is done as the changes that are detected by the system in the eye and the mouth. The facial landmarks detection model consisted of histogram of oriented gradients (HOG), a linear SVM that is included in the Dlib library. Then, the landmark points on the boundary of eyes and mouth are used to calculate the distance between the respective points. These distances calculated are then passed to an SVM classifier, compared to a threshold value to detect whether the face detected in the image has drowsiness or not. It concludes that further modification of the method can allow to have face detected from live video streams to develop a more accurate system.

A method proposed in [5] is built in four stages:

1. Localization of facial region
2. Localizing the region around the eyeballs
3. Tracking the above region in each frame
4. Recognition of localization failure.

Since the face is symmetric in most of the cases, the symmetry value is described as.

$$\mathrm{SV}(a) = \sum\sum[\mathrm{abs}\, I((a, b - c) - (a, b + c))] \tag{1}$$

This is computed for $A \in [k, \text{size-}k]$, i.e. for every pixel-column in the greyscale version of the image. The centre of the face is determined by the 'a' corresponding to the lowest value of $\mathrm{SV}(a)$. Then for the location of eyes, raster scan algorithm is used which are tracked by darkest pixel in the predicted region. Horizontal histogram across the pupil is monitored for detecting the state of the driver's eyes.

In [4, 9], an algorithm called the Viola–Jones algorithm was used to detect the current state of the eyes of the target user.

Major steps performed in the Viola–Jones algorithm are:

1. Face detection
2. Face extraction
3. Identification of the region of interest (ROI)
4. Extraction of the eyes
5. Determination of the current state of eyes.

This algorithm was used to train a model using various classifiers based to get maximum accuracy which came out to be at around 80% for just the eye open/closed detection.

Faces in an image are detected using the Viola–Jones object detection algorithm owing to its rapid execution time. The key steps of the algorithm are:

1. Integral image computation
2. Feature analysis and detection
3. AdaBoosting: to reject superfluous features and
4. Classifying the detected characteristics using a series of classifiers.

Viola–Jones algorithm is fast and robust to use in the current case, but since the maximum accuracy maxes out at around 80%, the number of false alarms is not something to turn a blind eye to.

One interesting observation that can be made is that most of the existing projects assume working conditions of ample illumination of the driver's face. While the implementations are not flawed in any major form, adjusting the current project for minimal illumination of the target's face would essentially make the project even more practical and applicable in day-to-day life.

3 Methodology

Many drowsiness detection methods have appeared in the literature review, and the best method is based on the visual features. To detect the drowsiness, DDYDAS tracks the eyes and mouth of the driver in real time. The framework of the proposed system is described in Fig. 1.

The system works in two phases—pre-processing phase and the continuous loop capturing and focusing the driver while driving till the end of his journey. In the pre-processing part, a phone camera mounted onto the dashboard, or anywhere where the driver's face is visible clearly. The system first begins by asking the user to calibrate the camera dock accordingly for the system to have successful face detection. The systems then confirm successful face detection at the normal driving position of the driver subject. Next, as the driving session begins, the driver's face is then continuously monitored by recording the video via the phone's camera and then feeding to the system frame by frame to process these captured frames for facial analysis of the driver. The interior is not always well lit. Since camera hardware available on various smartphones might not perform well under excessive background lighting,

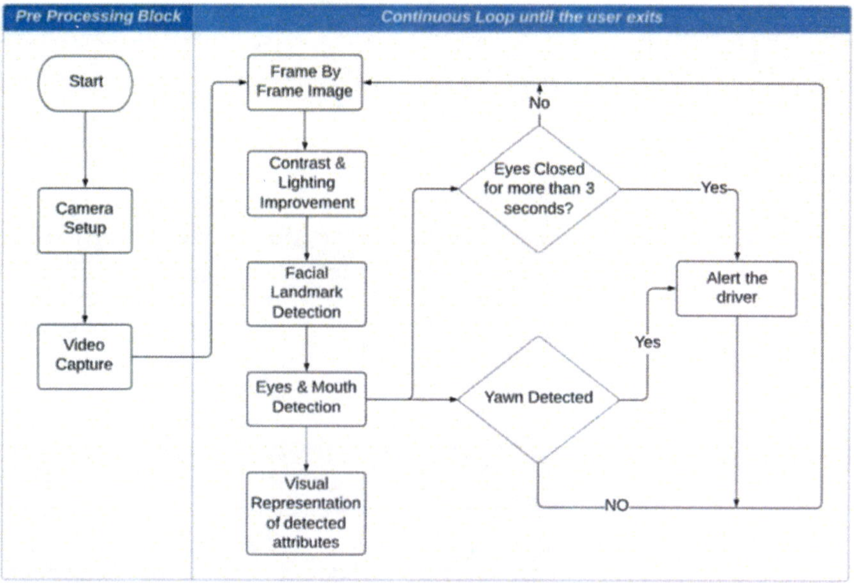

Fig. 1 Proposed framework

the driver's face may not be completely visible. So, an artificial contrast and lighting improvement technique is employed in the project to rectify this problem so that the operator's face can be easily detected and the video frames captured can be processed.

To detect the target face, DDYDAS uses a cross-platform software library (Dlib) to detect facial landmarks as shown in Fig. 2. The localised landmarks for mouth and eyes are shown in Fig 3 and Fig 4 respectively. As each frame goes through contrast improvement, the relative positions of facial landmarks in their localized regions are used to detect the mouth and eyes of the driver and are used for calculation of eye aspect ratio (EAR) [13] and mouth aspect ratio (MAR).

$$\text{Eye Aspect Ratio} = (|\,P2 - P6\,| + |P3 - P5|)/(2 * |P1 - P4|) \tag{2}$$

$$\text{Mouth Aspect Ratio} = (|P8 - P14| + |P10 - P12|)/(2 * |P7 - P11|) \tag{3}$$

Fig. 2 Dlib's facial landmarks

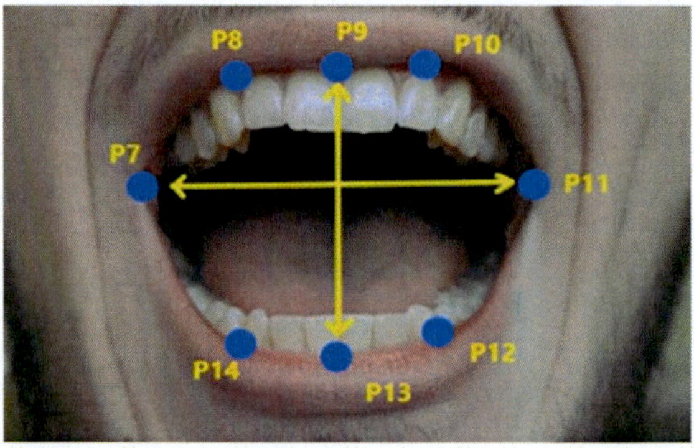

Fig. 3 Mouth landmarks

Fig. 4 Eye landmarks

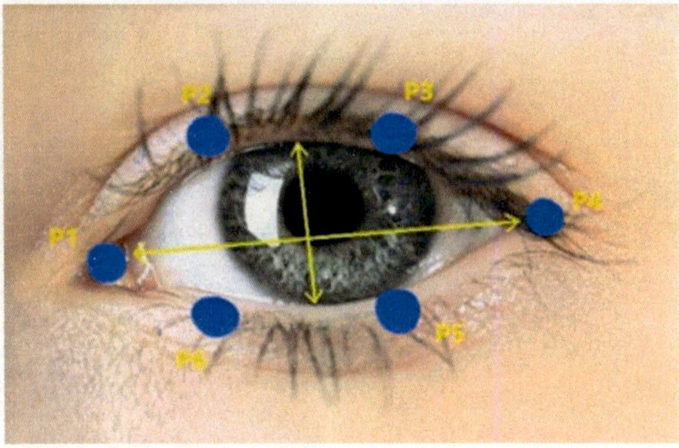

A threshold average value is then set which allows us to determine the state of the driver's eyes and mouth for more than a given time frame. The threshold value was calculated by averaging the EAR and MAR for a handful of people from different backgrounds. Two variables are set: *eyeThreshold* and *mouthThreshold*. These variables are essential to keep track of whether the target person blinked or yawned. Therefore, we set the value of *eyeThreshold* as 0.32 and *mouthThreshold* as 0.28. Setting values more or less than the mentioned values caused false alerts or the system to not detect at all. In case of eye-blinking, whenever the eyes are found to be closed for more than 3 s, the user is alerted through an alarm-like sound to make sure the user stays awake. Yawn count employs monitoring the distance between the lower part of the upper lip and the upper part of the lower lip. Whenever the distance between them increases more than the threshold value, the yawn counter is incremented.

4 Experimental Results

On implementing the above methodology, the results obtained along with their description are given as follows:

A simulation is shown in Fig. 5 of a successful yawn detection by the system. As the driver passes a minimum set threshold value of MAR, which corresponds to variable *mouthThreshold*, then the system alerts the driver for a potential drowsiness incident. Figure 6 demonstrates a successful eye blink detection by the system. Just like in the case of yawn detection, the system takes appropriate measures. Figure 7 is an example of simultaneous yawning and eye-blinking. Here all the three possible cases are handled, and the comparison with the existing system is shown in the following section.

Fig. 5 Yawn detection

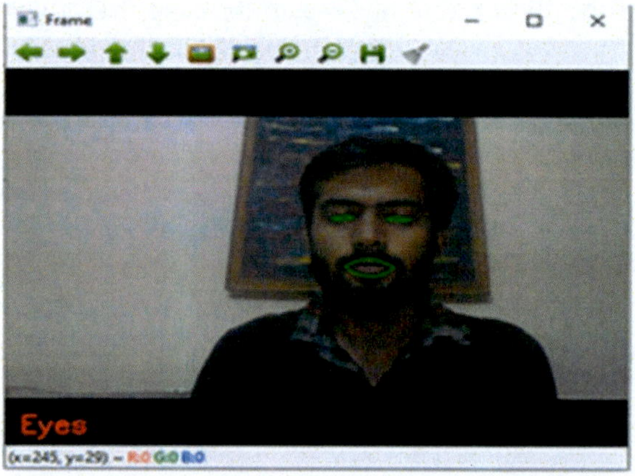

Fig. 6 Eye blinking

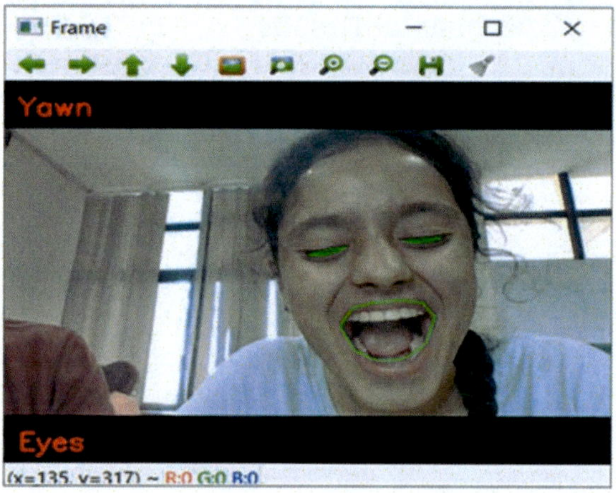

Fig. 7 Yawn detection and eye blinking

5 Comparison and Analysis

To compare our implementation with one of the most prominently used facial detection algorithms, we compare our proposed system DDYDAS, that uses Dlib HoG, with Viola–Jones algorithm for facial detection as mentioned in the literature review. We used the Haar Cascading implementation of OpenCV for comparison.

For the comparison, we have applied the two algorithms, Dlib HoG and OpenCV Haar (Viola--Jones) on the same video clips, and the following results were obtained.

From the Figs. 8a, b and 9a, b, it can be seen that frames per second (FPS) for Dlib is very high as compared to OpenCV Haar (Viola–Jones). It can also be observed that there are many false detections in the case of OpenCV as in Figs. 8a and 9a, which are not seen in Figs. 8b and 9b. Thus, we can say that Dlib is more accurate in terms of face detection. Dlib is a lightweight model which works well for frontal and slightly non-frontal faces, and it works under occlusion, which does not work in the case of OpenCV (Viola–Jones) (Fig. 10).

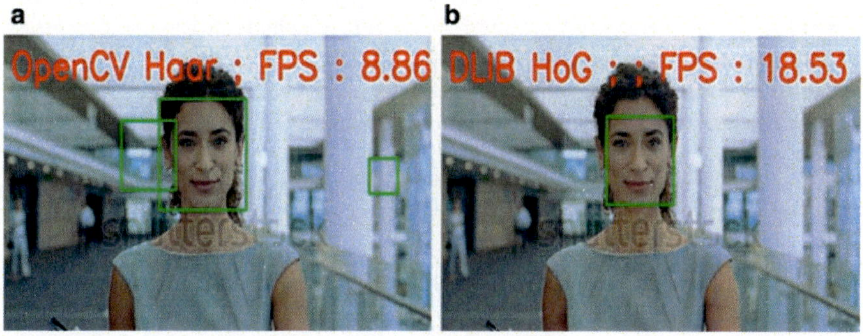

Fig. 8 **a** OpenCV haar, **b** Dlib HoG implementation example 1

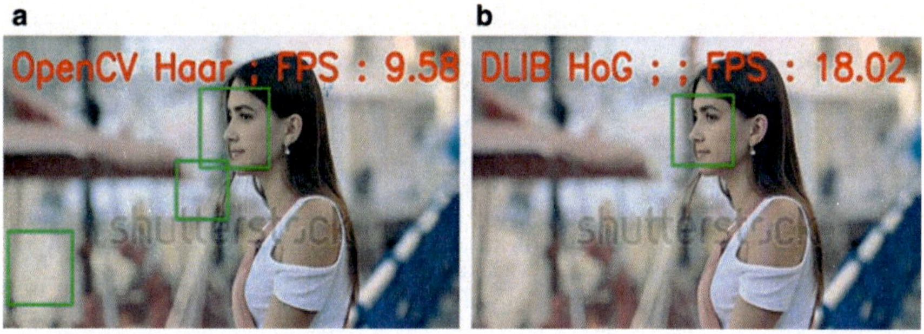

Fig. 9 **a** OpenCV haar, **b** Dlib HoG implementation example 2

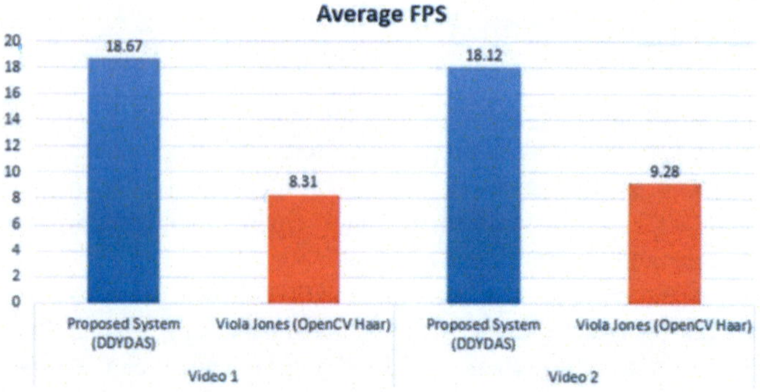

Fig. 10 Comparison between Dlib and Viola–Jones

6 Conclusion and Results

The proposed system successfully detects and monitors the driver's drowsiness. The system uses a phone camera to capture video frames, which are passed to contrast improvement process. These frames are then used by Dlib software library for face detection. At the end, the facial features are then used to calculate the opening of eyes and mouth. These two calculations, when compared to threshold values, that is, 0.32 for eyes and 0.28 for mouth to decide whether the face detected is in a drowsy state or not. This approach assumes that the driver's face is properly illuminated and any excessive lighting is minimized to the best possible levels by the system. The system successfully alerts the driver with a loud sound and thus prevents a potential mishap.

The efficiency and practicality of the application can be improved by employing a high-definition webcam or NIR (near-infrared) camera for night time. However, some falsely detected instances of yawning and eye-blinking might be there due to the rapid head movement. For better portability, a dedicated Raspberry Pi hardware can be used.

References

1. Dagli R (2016) Driver fatigue and road safety -implication in an Indian context. Int J Adv Eng Technol 9:502–506
2. Liang Y (2009) Detecting driver distraction. Ph.D. thesis, University of Iowa
3. Bayly M, Fildes B, Regan M, Young K (2007) Review of crash effectiveness of intelligent transport system. Traffic Accident Causation in Europe (TRACE)
4. Poursadeghiyan M, Mazloumi A, Nasl Saraji G, Baneshi MM, Khammar A, Ebrahimi MH (2018) Using image processing in the proposed drowsiness detection system design. Iran J Public Health 47(9):1371–1378
5. Singh HP, Mehul J, Yadav PB (2014) Drowsy driver warning system using image processing. Electronics and Communication, GEC, Bharuch, Gujarat, pp 78–83

6. Nass C, Jonsson I-M, Harris H, Reaves B, Endo J, Brave S, Takayama L (2005) Improving automotive safety by pairing driver emotion and car voice emotion. Int Conf HCI
7. Lisetti CL, Nasoz F (2005) Affective intelligent car interfaces with emotion recognition. 11th Int Conf HCI
8. Choudhary P, Sharma R, Singh G, Das S (2016) A survey paper on drowsiness detection and alarm system for drivers. Department of Computer Engineering, AISSMS COE, Pune, Maharashtra, India, pp 1433–1437
9. Khan MT, Anwar H, Ullah F, Rehman AU, Ullah R, Iqbal A, Lee B-H, Kwak KS (2019) Smart real-time video surveillance platform for drowsiness detection based on eyelid closure. Department of Information and Communication Engineering, Inha University, Republic of Korea, pp 2036818–2036827
10. Gao H, Yuce A, Thiran J-P (2014) Detecting emotional stress from facial expressions for driving safety. signal processing laboratory (LTS5). In: IEEE international conference on image processing (ICIP), Paris, pp 5961–5965
11. Kusuma KBM (2014) A real time driver drowsiness detection system. In: International conference on information and communication technologies (ICICT), pp 71–21
12. Irtija N, Sami M, Ahad MDAR (2018) Fatigue detection using facial landmarks. submission no. C000041, ISASE-MAICS-2018, pp 1–6
13. Soukupová T, Cech J (2016) Real-time eye blink detection using facial landmarks. Center for Machine Perception, Department of Cybernetics Faculty of Electrical Engineering, Czech Technical University in Prague, pp 1–8

A Utility System for Farmers to Build Decision Support System on Agrometeorological Data Using Machine Learning Algorithms

B. J. Sowmya, Aniket Singh, Zaifa Khan, B. Sathvik, S. Seema, and K. G. Srinivasa

Abstract Agriculture is the foundation of the Indian economy; however, the business is in need of more help than some others due to the rapidly changing environment and physical conditions. This evolution is happening both by the growth of humans and some explorations in agriculture field. The advancement in the area of data analytics and Internet of things also facilitates the growth the domain of data analytics (DA), and Internet of things (IoT) is very much facilitating this growth in the field of agriculture which is directly contributing to the gross domestic product (GDP). The continuous changes in the weather conditions and improper irrigation techniques are the main challenges in the field of agriculture. To solve the issues, farmers should use the benefits of modern tools and techniques. Here we have made an attempt to solve the problems in the field of agriculture by using the machine learning techniques. In our work, an attempt is made to analyse the influences of the different agrometeorological data using machine learning algorithm. Rainfall, humidity and temperature are some of the variables that determine the type of crop. So, the task of prediction of crop type given the factors using support vector machines

B. J. Sowmya · A. Singh (✉) · Z. Khan · B. Sathvik · S. Seema
Department of Computer Science and Engineering, Ramaiah Institute of Technology, Bangalore, India
e-mail: singhaniket406@gmail.com

B. J. Sowmya
e-mail: sowmyabj@msrit.edu

Z. Khan
e-mail: khanzaifa37@gmail.com

B. Sathvik
e-mail: sathvikbk123@gmail.com

S. Seema
e-mail: seemas@msrit.edu

K. G. Srinivasa
Department of Information Management and Coordination, NITTTR, Chandigarh 160019, India
e-mail: kgsrinivasa@gmail.com

A. Abraham et al. (eds.), *Proceedings of 3rd International Conference on Computing Informatics and Networks*, Lecture Notes in Networks and Systems 167,
https://doi.org/10.1007/978-981-15-9712-1_20

(SVM) is implemented, and the accuracy of the models is computed. Some of the critical parameters of SVM are tuned and applied grid search to improve the accuracy.

Abbreviations

DA	Data Analytics
IoT	Internet of Things
GDP	Gross Domestic Product
SVM	Support Vector Machine
ANN	Artificial Neural Network
KNN	K-Nearest Neighbours
GKVK	Gandhi Krishi Vigyan Kendra
RBF	Radial Basis Function
CSV	Comma Separated Values

1 Introduction

Due to recent changes in climate due to global warming and the increasing emission rate, farmers are facing more loss than ever because of planting the wrong crop for a particular season and physical conditions. With the boom of big data, a lot of meteorological and agricultural data has been made available in recent years via several sensors and IoT devices along with historical data. Hence herein, we have used the power of machine learning in order to predict the most suitable crop to be grown for the given weather conditions, using historical data. Such predictions can be used by the farmers, optimize their growth planning increase yields and minimize losses irrespective of the weather or physical conditions. Data analytics have become increasingly efficient and can be well-utilized to solve a vast multitude of problems ranging from science to daily needs, finance and marketing. Hence, these techniques can solve the issue of sub-optimal crop yield due to wrong decision-making. Forecasting the most suitable crop based on the present conditions including factors like rainfall, humidity which are meteorological contributors can vastly boost the growth of crops and ultimately the economy.

After doing a well balanced study of the different machine learning algorithms and techniques, we chose SVM (support vector machine) as the most suitable algorithm as it can efficiently classify data irrespective of the separating boundary (i.e. if the boundary is linear or nonlinear). The performance can then be boosted further by fine-tuning the hyperparameters and therefore vastly the problem at hand can be solved. A model's hyperparameter is a value that is independent of the data being applied to the model. Hence, accuracy boost by finding the best hyperparameters will give objective and non-biased results. Hence, we aim to use grid search and find the optimal hyperparameters for the SVM model. Randomized search is another method in which random values of hyperparameters are tested and the best scored one among them is selected. This was also a path taken, which improved accuracy of the model.

2 Literature Survey

A Support vector machine (SVM) works by finding the largest separating boundary, also called as hyperplane [1]. Most of the data for the real-world problems are not linearly separable; hence, we need nonlinear hyperplanes. The problem is regarding the training efficiency of the support vector machine, meaning that the training time for the SVM is too long. Thus, to reduce this, many approaches are used. One way is to obtain a low order nuclear matrix by applying greedy algorithms. The dataset could also be sampled to reduce the amount of training data, thus increasing training time. These methods might work well for linearly separable data; however for nonlinearly separable data, they do not work so well. Thus, radial bias method was used to reduce the training dataset, which lead to satisfactory results for the SVM [2]. Support vector machines were introduced as a solution to pattern analysis, mapping data to a higher dimensional space, and using a separating hyperplane for separation, involving quadratic equations. To improve generalized results while incurring low costs, least square SVM was used where in the Mercer's condition was applied and even typical two-spiral problems were solved with SVM and RBF Kernel, showed promising results [3]. Support vector machines can provide great results for classification of linearly or nonlinearly separable data even text and can be labelled as a powerful classifier. However, different prediction/classification paradigms have varying levels of complexity and a commonly taken approach to improve the performance of such classifiers is to use an ensemble with other algorithms. It is well proven for boosting weak classifiers like decision tree classifiers and neural networks, but an attempt to obtain results on SVM, a powerful classifier was made.

Even though new technologies have increased our ability to collect data from different fields, climate and agrometeorological data still present a challenge [4]. Hence, there was a need for a thorough study of techniques new and old in areas of data collection with special consideration of remote sensing techniques. Database management of US Department of Agriculture was the chosen framework where data access is open to both national and international communities. Data available include US data as well as from countries with co-operative relations with US. Data and information present focus on the research community requirements for crop or weather forecasters as well as facilitating the farmers in their production [5]. Several data mining techniques have been applied in the field of agriculture for various types of predictions. Machine learning algorithms have been applied to predict the probable yield of crops given various parameters like precipitation and location. In cases where labelled data is given, classification algorithms such as neural networks, SVMs, k-nearest neighbours were used.

Neural networks and SVMs being early learners need some amount of training time. On the other hand, k-nearest neighbours, being lazy learners need no training time and can directly be used for classification. If there was an absence of labelled data, clustering methods such as k-means clustering was used.

A survey was made of the different data mining techniques to find ways through which machine learning was used to improve the agricultural field [6]. It was found that learning algorithms were used for a variety of different applications in meteorological as well as agricultural data. K-means method was used for the prediction of pollution in the atmosphere. KNN method was used to predict daily precipitation. The same algorithm was also used to classify plants and soils. Apples running on conveyer belts were analysed through a k-means clustering technique to find if there is water core, to discard off the bad apples. Through artificial neural networks (ANN), wine fermentation through taste sensors could be monitored. SVMs were trained with sensors which could smell milk, so that milk that had gone bad could be discarded. SVMs were also used to detect the presence of nitrogen stress and weed in corn. Thus, machine learning algorithms have been extensively used in the field of agriculture.

A survey was conducted to find the methods in which artificial intelligence was used to find diseases in crops [7]. A support vector machine was used to predict the presence of parasites inside strawberries and also to detect the presence of a fungus "Microbotyum silybum". It was also used in the detection of Bakanae disease, Fusarium fujikuroi, in rice seedlings. These methods could be used to filter out the bad crops. An ANN was used to detect yellow rust infected wheat canopies (with high accuracy). A general convolutional neural network was used for detection and diagnosis of plant diseases, which used images of leaves of healthy and diseased plants, with an accuracy of 99.53%. Thus, these methods could inform the farmer of such diseases and appropriate action could be taken by him.

Since data mining techniques have enabled researchers to leverage useful information from the vast amounts of data available, we can use them to solve critical bio-socio system associated issues like agricultural yields and estimation, and produce forecast [8, 9]. Different algorithms including K-means (used during non-availability of training data), K-nearest neighbours (KNN, used when training data is available), ANN (simulates the working of human brain, trying to learn incrementally over huge number of inputs and outputs) and SVM (which is a powerful classifier based on the separating hyperplane) were tried and tested, to suggest exploration of vast assets available in both the fields of data and analytics.

3 Design

The system is being developed as a model that uses machine learning, extracts most prominent features in pre-processing stage and trains different models for obtaining certain analysis. The type of machine learning is supervised, which is used for classification.

The modules that are identified for this work are as follows:

- Acquisition and Extraction of Dataset
  In this module, the dataset is acquired from a reputed source. For this work, dataset is collected from various sources such as from, University of Agricultural Sciences—Gandhi Krishi Vignana Kendra (GKVK) and other online repositories. The dataset thus collected all are in required format.
- Data Pre-processing
  Here, pre-processing of the dataset is carried out on the data collected. Usually, the pre-processing involves cleaning of the dataset, i.e. finding out the missing values and performing appropriate operation necessary, conversion of data formats to required format and sub-setting of datasets to smaller size so only the required data is used rather than all unwanted variables that could impact the computation.
- Train and Test
  In this module, the pre-processed data is split into training and testing data. As usual, a 70–30 split is carried out, wherein 70% of the data corresponds to training and the remaining to testing set. The testing set is used for cross-validation and thereby helping in computing the performance.
- Predictions
  Here the models created based on the training set is evaluated on the testing set, and the results thus predicted is cross-validated with the testing set, and the performance of the model is computed.
- Analysis
  There are different implementation modules, which do not go through the train and test, prediction phases, in turn would fall under the analysis category where a visual representation for them is generated based on the pre-processed data.

3.1 Support Vector Machine (SVM)

Support vector machine (SVM) is suitable for both regression and classification problems, but is widely used in regard to classification. The expected result from SVM is to locate within N-dimension space, where the n features are represented, a hyperplane leading to precision classification of the data. Hyperplanes are decision boundaries, and their dimension is a direct outcome of the data and number of features it contains. Support vectors, which are a major part of SVM, represent points in the feature space located in proximity of the hyperplane and strongly affect the location and direction of the plane. Elimination of these points leads to alteration in the position of the hyperplane. The SVM with multiclass aspires to designate labels to several instances with the help of support vector machines, where the labels building a decision support system for selected agriculture parameters using data analytic techniques are selected from a set which is finite and has several elements. SVMs are generally two-class classifiers, meaning they can deduce the data to be belonging to one class or another. A classic SVM explores to locate such a margin which isolates all positively classified and negatively classified samples with respect to a particular class. Notwithstanding, such an approach may lead to models yielding inaccurate and incorrect results, if any examples under consideration in the dataset are misclassified or outlier data is present. There are also some situations, wherein a nonlinear region can segregate the groups more effectively. As far as the data that has been used in this work is nonlinear. So, the concept of nonlinear SVM is better suited in this case. SVM handles the grouping of nonlinear regions by making use of a kernel function (related to nonlinear) to project the input data in a distinct space, within which a hyperplane (linear) if employed to perform the segregation will be unsuccessful. Thereby, the kernel function allows, to form a linear function in the high-dimensional space representing the features from a nonlinear separating function, even though capabilities of this system is governed by a constraint variable that becomes independent of the space dimensionality commonly known as kernel trick. The linear classifier of SVM is expressed as a dot product between vectors of data point as shown in equation. For a function to be called kernel function, it must have a positive definite gram matrix, should be continuous and also should be symmetric. The most common Kernel function is the Gaussian radial basis function (RBF) which is equivalent to mapping the data into Hilbert space, an infinite dimensional, which is shown in the equation:

$$K\left(\vec{x}_i, x_j\right) = \vec{x}_i^{T} . \vec{x}_j$$

$$k(xi, xj) = \exp(-\gamma \|xi - xj\|2), \text{ where } \gamma > 0$$

where $k(xi, xj)$ is the kernel function, xi and xj are the data vectors and γ defines how far the influence of a single training example reaches. A radial basis function may contain building a decision support system for selected agriculture parameters using data analytic techniques features that take circles (hyperspheres) as hyperplanes, but the decision boundaries turn much more complex if there are interactions of multiple such features.

3.2 Enhanced SVM

The parameters cost (C) and gamma (γ) are the criteria for an RBF that can be tuned to enhance the existing SVM model to attain more accuracy. Here, the cost C manages the trade-off between the accurate classification of training points and a smooth boundary. Learning algorithms are mainly focussed on understanding or learning from input data and are not concerned with the means of learning although different means have variety of impact. Due to the bane of dimensionality, the training data pertaining to multi-dimension (usually large number of dimensions) can often be interpreted clearly by over fitting the model. Therefore, generally it is preferable to explicitly concede some training points to be misclassified in order to have an "overall better" position of the separating hyperplane. A high-cost value will lead the model to choose greater number of supporting vectors and increasing possibility of getting variance increasing highly while the bias dips, leading to the problem of overfitting. Conversely, a lower cost value causes the model to choose decreased number of feature points as supporting vectors and thereby obtaining a dip in variance and spike in bias leading to underfit model. So, the objective is to find the balance between "not too strict" and "not too loose" value of the cost. Cross-validation and resampling, along with grid search and randomized search, are some of the efficient ways to procure the best value for cost. As said earlier, the gamma (γ) hyperparameter decides the extent to which any singular sample from training data reaches, where a dipped value indicates being further apart and spiked value indicates closeness. The model selects the radius up to which samples hold influence which become the supporting vectors and the inverse of the same can be understood as gamma. If gamma exhibits a high value, then the decision boundary of the model will depend on data points close to the decision boundary and the closer data points carry more weights than far away points, due to which the decision boundary becomes wigglier. If gamma exhibits low value, then far away data samples carry more weights than the closer data points and thus the decision boundary looks more like a straight line.

4 Implementation and Results

As shown in Fig. 1, the decision support system on agrometeorological data is built and implemented in following modules:

Collection and Pre-processing of Data: The data which is gathered from different sources such as University of Agricultural Science—GKVK and other online repositories are converted into required format of comma separated values (CSVs). The data are then loaded and converted into feature matrix X, and dependent variable y. It consists of 1500 rows with 3 features each, namely temperature, rainfall and humidity and one dependent variable, namely Name of Crop. Now, the matrix X is checked for missing values and the missing values are filled by imputation.

For getting the notion of shape and variation of data, several graphs are plotted as shown in figures. The box plot gives an idea about the outliers. Temperature and humidity have very less variance of 1.169 and 9.961 and hence almost no outlier. Whereas rainfall has a high variance; hence, it has outliers. These outliers were removed by dropping the corresponding rows. After the completion of this step, the data are ready to be feed into machine learning algorithm. The dataset is split in the ratio of 70:30 for training and validation (Figs. 2, 3 and 4).

Model Construction and parameter tuning: In model construction, SVM is applied as machine learning algorithm, to fit and predict the values on given dataset. The algorithm decides a hyperplane which results in classification of a given feature vector into a crop.

Hyperparameter tuning refers to the process of choosing the values of parameter of a machine learning algorithm in such a way that it provides the most suitable and best-fit model. For this process, we have used two techniques: grid search and

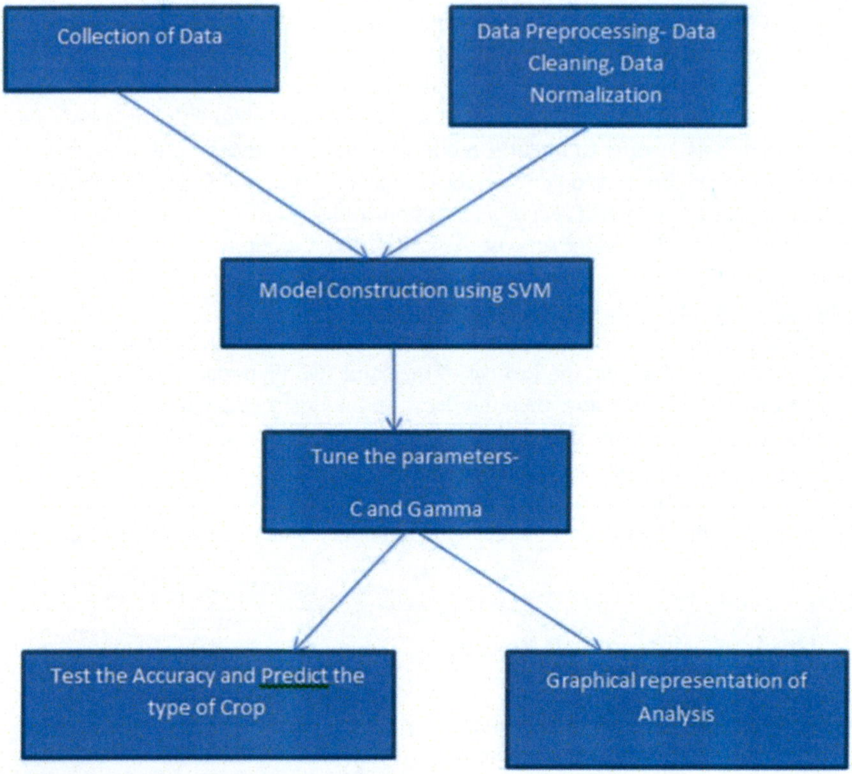

Fig. 1 Design of proposed work

Fig. 2 Box plot of feature matrix

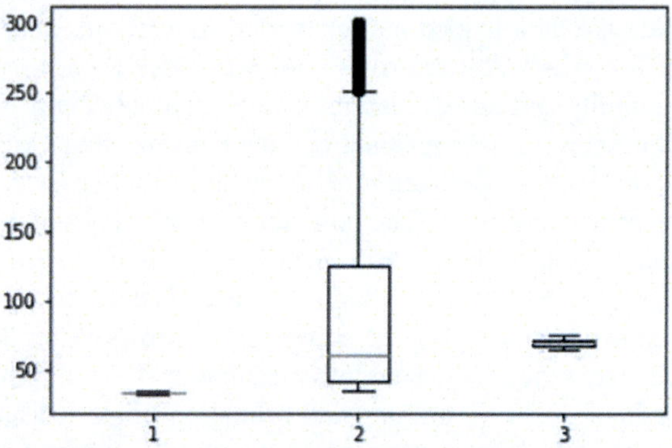

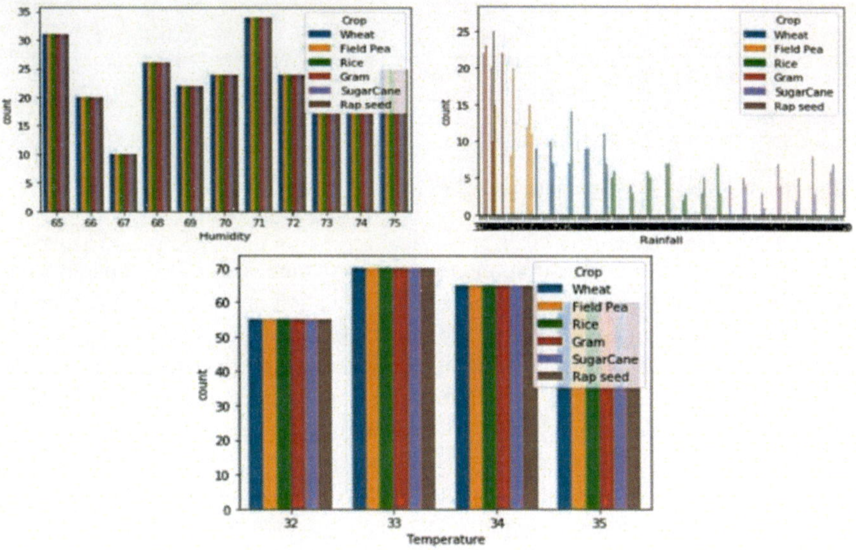

Fig. 3 Count of each type of crop for each attribute in all ranges

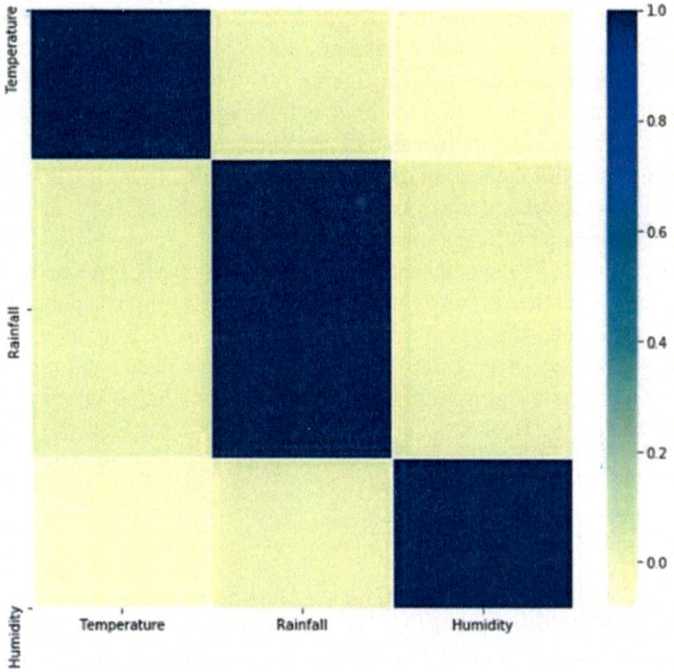

Fig. 4 Correlation matrix for input matrix

gamma / C	0.1	1.0	1.5	2.0
0.5	(0.5,0.1)	(0.5,1.0)	(0.5,1.5)	(0.5,2.0)
1.0	(1.0,0.1)	(1.0,1.0)	(1.0,1.5)	(1.0,2.0)
1.5	(1.5,0.1)	(1.5,1.0)	(1.5,1.5)	(1.5,2.0)
2.0	(2.0,0.1)	(2.0,1.0)	(2.0,1.5)	(2.0,2.0)

Fig. 5 Matrix for grid search and randomized search

randomized search. In grid search, the model is provided with a list of parameters and all the combinations are fitted for each combination and accuracy is tested. It consumes significant amount of time if number of parameters is high. Whereas, in randomized search all the combination of parameters is not tried out instead a fixed number of combination of parameters is tried out from the specified distribution. This process of trying random combination is repeated for a pre-defined number of times. Since randomized search does not try all the combination, it is less time-consuming as well as more effective than grid search in general. The parameters tuned here are kernel coefficient of kernel, i.e. gamma (γ), regularization parameter, C. A 4×4 matrix is used for tuning parameters which is formed by the combination four values of gamma and four values of C (Fig. 5).

After getting the best set of parameters for the combination of gamma and C from grid search and randomized search, two SVM models are trained. The trained models are now tested with K-fold cross-validation. In K-fold cross-validation, the number of times a model is trained is "K", by taking randomly and selecting a portion of given dataset as training set and remaining as training set. In this way, whole of the dataset is used for training the model at least once. After this step, the models are ready and verified with the testing data which was divided in the beginning.

Results: Now, graphical analysis of various outputs of both the models is performed to compare both the models (Fig. 6).

It is clear from the above graph that randomized search performs better than grid search. It provides higher accuracy than grid search. Maximum accuracy in case of grid search is 76.54%, whereas in case of randomized search it is 77.71% (Fig. 7).

The time taken by grid search is generally greater than randomized search. Minimum time taken by grid search is 0.05106, whereas minimum time taken by randomized search is 0.04986.

5 Conclusion

We found that SVM with hyperparameter tuning increased our accuracy when we performed k-fold cross-validation. The types of crop without any hyperparameter tuning can be predicted with an accuracy of 74.67%, whereas after hyperparameter tuning and optimization, it is increased to 76.54% and then to 77.71%. Hence, decision support can be achieved via the analysis of agrometeorological data. Since the data we considered had three primary features rainfall, humidity and temperature,

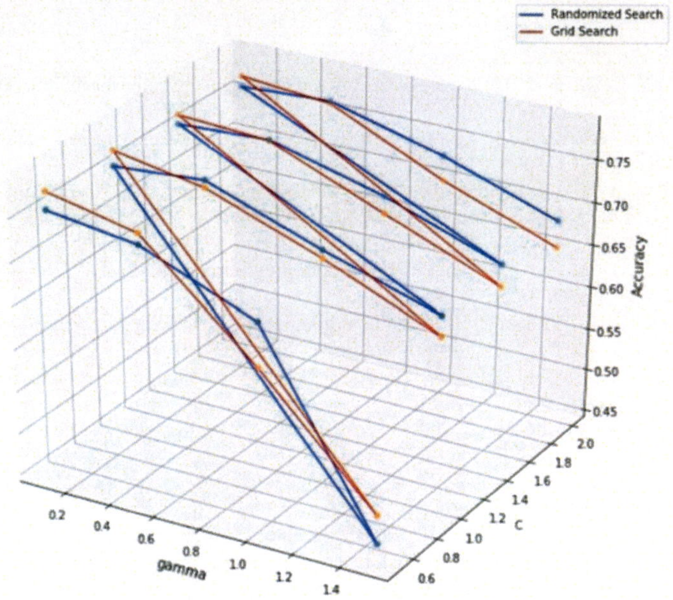

Fig. 6 Accuracy plot in three-dimension

Fig. 7 Plot of estimated time for grid search and randomized search

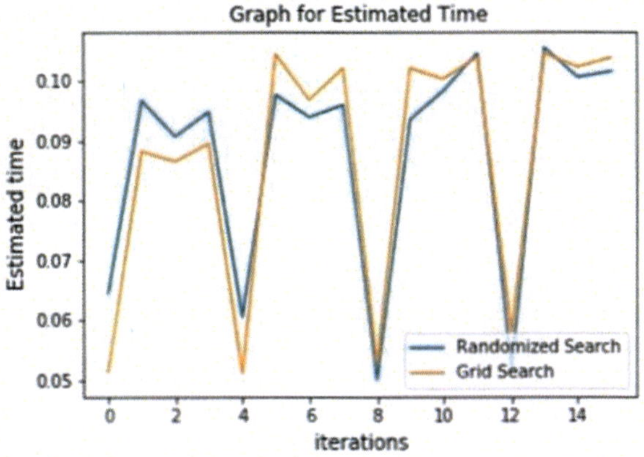

the current limitation is that we have not used other aspects of data associated with agricultural activities.

Hence as a future implementation, we can incorporate data from different facets including spatial, social, economic and other factors, perform a similar routine and obtain the most suitable choice for increased yields. Further even though such analysis powers are available at hand very few of their benefits hardly trickle down to the users, i.e. farmers, and we wish to help the community by working and boosting research enough such that the outcome is positively forwarded.

References

1. Durgesh KS, Lekha B (2010) Data classification using support vector machine. J Theor Appl Inf Technol 12(1):1–7
2. Suykens JAK, Vandewalle J (1999) Least squares support vector machine classifiers. Neural Process Lett 9(3):293–300
3. Dong Y-S, Han K-S (2005) Boosting SVM classifiers by ensemble. In: Special interest tracks and posters of the 14th international conference on World Wide Web, pp 1072–1073
4. Doraiswamy PC, Pasteris PA, Jones KC, Motha RP, Nejedlik P (2000) Techniques for methods of collection, database management and distribution of agrometeorological data. Agric Forest Meteorol 103(1–2):83–97
5. Ramesh D, Vishnu Vardhan B (2013) Data mining techniques and applications to agricultural yield data. Int J Adv Res Comput Commun Eng 2(9):3477–3480
6. Mucherino A, Papajorgji P, Pardalos PM (2009) A survey of data mining techniques applied to agriculture. Oper Res 9(2):121–140
7. Liakos KG, Busato P, Moshou D, Pearson S, Bochtis D (2018) Machine learning in agriculture: a review. Sensors 18(8):2674
8. Medar RA, Rajpurohit VS (2014) A survey on data mining techniques for crop yield prediction. Int J Adv Res Comput Sci Manage Stud 2(9):59–64
9. Shi L, Duan Q, Ma X, Weng M (2011) The research of support vector machine in agricultural data classification. In: International conference on computer and computing technologies in agriculture. Springer, Berlin, Heidelberg, pp 265–269

A Machine Learning-Based Approach to Detect Credit Card Frauds

Shweta Taneja and Sarthak Chandna

Abstract Nowadays, the amount of online transactions has increased exponentially due to the increasing usage of Internet, thereby causing steep increase in frauds. The purpose of this paper is to propose a technology led model that can help detect credit card fraud cases. The credit card dataset is highly imbalance in nature; therefore, we have applied balancing technique on the dataset. The proposed approach is implemented and tested using four classifiers: random forest, Gaussian Naïve Bayes, logistic regression and K-nearest neighbour. We have taken a standard credit card dataset of a European bank. The performance of these classifiers is measured using precision, recall, $F1$ score, ROC curve. Results have shown that random forest has performed best with $F1$ score of 0.92.

Keywords Credit card frauds · Machine learning · Classification algorithm · Balancing technique · Data pre-processing

Abbreviations

KNN	K-Nearest-Neighbors
ROC	Receiver Operating Curves
SVM	Support Vector Machine
SMOTE	Synthetic Minority OverSampling Technique

S. Taneja
Department of Computer Science and Engineering, Bhagwan Parshuram Institute of Technology, GGSIPU, Dwarka, India
e-mail: shwetataneja@bpitindia.com

S. Chandna (✉)
Department of Computer Science and Engineering, Maharaja Agrasen Institute of Technology, GGSIPU, Dwarka, India
e-mail: chandna.sarthak1998@gmail.com

A. Abraham et al. (eds.), *Proceedings of 3rd International Conference on Computing Informatics and Networks*, Lecture Notes in Networks and Systems 167,
https://doi.org/10.1007/978-981-15-9712-1_21

1 Introduction

Due to digitization and growing usage of Internet, the number of online transactions has increased exponentially. Accordingly, the cases of frauds have also increased proportionally. Therefore, a major issue of concern is detection of these frauds. Our focus is on credit card frauds. Machine learning has got the solution to our problem. There are descriptive and predictive models in machine learning. In predictive modelling, we can make a prediction about the unknown or future data. In credit cards, we can identify fraudulent transaction by using these machine learning algorithms and can prevent the million dollar loss.

1.1 Credit Card Frauds and Their Issues

The credit card frauds are broadly classified into two major categories: behavioural and application frauds. Application frauds occur when the documents provided are fake, and the card issuer approves the identity. Behavioural frauds occur after the card is issued to customer with real and original identity, but due to mishandling like card lost, there is a fraud. Our focus is on behavioural frauds. There is need of some powerful tool or method which handles such a huge data as manual management is not computationally feasible and machine learning has been a great platform to handle large dataset. But there is a problem in providing original data as it would lead to security breech. So there has been a trade-off between providing original data set or decomposed dataset.

There are some issues or difficulties that are being faced in successful prediction of fraud cases. These are mentioned below:

1. Due to confidentiality issues, the original features of the dataset are not provided, rather PCA decomposed features are provided which gives identical results.
2. The credit card datasets are highly imbalanced in nature. That is, out of the total transactions in a dataset the number of fraudulent transactions is below 1%. So it is a challenge to prevent the model from overfitting.
3. The real-time credit card data is so huge in nature as each day millions of transactions take place, and it is impossible to train on such large real-time data.

1.2 Contribution of Work

In this research work, we have proposed our approach to detect credit card frauds. A standard dataset of a European bank is taken for the implementation work. The proposed approach is tested on four classifiers like random forest, Gaussian Naïve Bayes, logistic regression and K-nearest neighbours. These classifiers are compared by calculating precision, recall, $F1$ score and ROC curve.

The flow of the paper is as follows: Sect. 2 consists the existing literature in this field. In Sect. 3, our proposed approach is given followed by the details of dataset used in Sect. 3.1. Section 4 shows the results that we have obtained on the dataset. The conclusion along with future scope is stated in Sect. 5.

2 Literature Review

We have studied the existing literature in this domain. Different authors have contributed their work.

In [1], the author has carried out the behavioural analysis only on homogeneous transactions in which deep learning algorithm is clubbed along with feature engineering process based on homogeneity-oriented behaviour analysis (HOBA). The authors in [2] have developed a live detection system that permits real-time classification of the fraudulent cases. In [3], authors have compared and discussed three data mining techniques, namely CNN, MPL for fraud detection. In an another work [4], authors have used classification technique for detecting frauds and used tools like Kafka, Spark and Cassandra.

The authors in [5] have extracted features from social interactions upon which feature engineering was used and finally incorporated a neural network to classify fraudulent cases. In [6], authors have examined a comparative analysis between random forest and long short-term memory (LSTM) methods and proved that frauds detected by LSTM were consistently different from Random Forest.

A decision support system for banking fraud detection was developed by authors in [7]. The system examines the spending habits and identifies frauds. The authors have made a comparison between user-centric and system-centric modelling to detect fraudulent transactions. In [8], authors have performed a comparison among 14 different classifiers after a feature selection based on correlation and examined them on the basis of true positives and true negatives.

Another good work is done by the authors in [9]. They have developed a twofold system; firstly, it helps the card users to develop a prediction system to model a delinquency risk and secondly explores the potential of artificial neural networks. The neural network discussed works on client's personal characteristics like spending habits, etc.

Through this paper, we have presented an outlook to detect frauds. We have used four classifiers and compared their performance. We have shown that random forest classifier has performed the best among all four classifiers.

3 Proposed Work

Figure 1 shows our proposed approach.

The training dataset containing transactions data is highly imbalanced in nature. We have taken a standard dataset that contains transactional records of credit card holders of a European bank. It consists of total 30 features out of which, the two features, i.e. ("Amount", "Time"), are taken as it is and other features are reduced by applying principal component analysis method keeping in mind the confidentiality of customers. After that, data pre-processing is done in which scaling of feature "Amount" is done to ensure that it does not overpower other attributes. The feature "Time" need not to be scaled as it is already in the range. The data is then split into train and testing for fitting and testing. After splitting, the training data is balanced by SVM SMOTE balancing technique as the dataset is highly imbalanced in nature. The SVM SMOTE balancing technique ensures that the generated new minority class samples are at the borderlines which help to develop a boundary between different instances. It contains 492 fraudulent cases and 284,315 non-fraudulent cases. The balanced dataset is then passed through four classifiers like random forest, Gaussian Naive Bayes, logistic regression, K-nearest neighbours. A model or a classifier is built. Further testing of the model is done on original data. The performance of classifiers is analysed using standard performance metrics such as precision, recall, $F1$ scores and ROC scores.

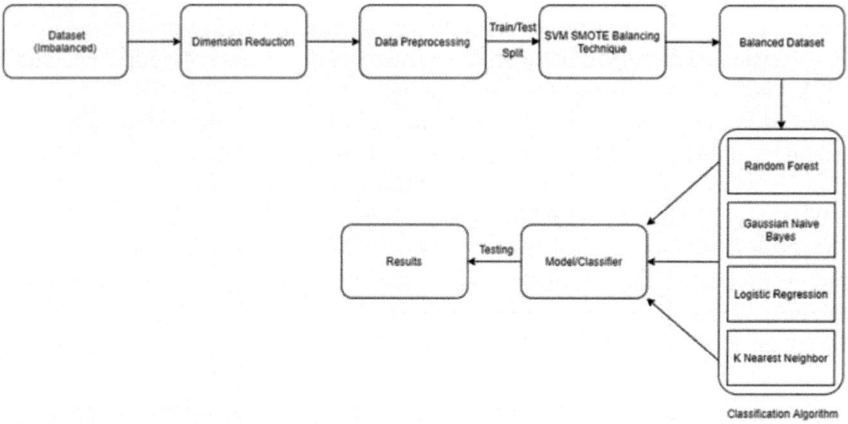

Fig. 1 Proposed approach

3.1 *Dataset Used*

We have referred to Kaggle [10] for the required dataset. It contains transactional records of credit card holders of a European bank. The original features could not be shared due to confidential reasons and PCA decomposed data set consisting of 28 features along with two original features are shared. The feature "Amount" determines the transaction amount and feature "Time" determines the time elapsed between present transaction and first transaction. The data hold the records of two days. There are total 492 fraudulent cases and 284,807 normal transactions which makes the data highly imbalanced. The snapshot of the original dataset is shown in Fig. 2.

4 Results Obtained

We have used four different classifiers for credit card fraud detection, namely random forest, Gaussian Naive Bayes, logistic regression, K-nearest neighbours. Their performance is calculated with the help of standard measures like precision, recall, $F1$ score

```
print(df.tail())
```

```
            Time         V1         V2        V3        V4        V5  \
284802  172786.0 -11.881118  10.071785 -9.834783 -2.066656 -5.364473
284803  172787.0  -0.732789  -0.055080  2.035030 -0.738589  0.868229
284804  172788.0   1.919565  -0.301254 -3.249640 -0.557828  2.630515
284805  172788.0  -0.240440   0.530483  0.702510  0.689799 -0.377961
284806  172792.0  -0.533413  -0.189733  0.703337 -0.506271 -0.012546

              V6        V7        V8        V9  ...       V21       V22  \
284802 -2.606837 -4.918215  7.305334  1.914428  ...  0.213454  0.111864
284803  1.058415  0.024330  0.294869  0.584800  ...  0.214205  0.924384
284804  3.031260 -0.296827  0.708417  0.432454  ...  0.232045  0.578229
284805  0.623708 -0.686180  0.679145  0.392087  ...  0.265245  0.800049
284806 -0.649617  1.577006 -0.414650  0.486180  ...  0.261057  0.643078

             V23       V24       V25       V26       V27       V28  Amount  \
284802  1.014480 -0.509348  1.436807  0.250034  0.943651  0.823731    0.77
284803  0.012463 -1.016226 -0.606624 -0.395255  0.068472 -0.053527   24.79
284804 -0.037501  0.640134  0.265745 -0.087371  0.004455 -0.026561   67.88
284805 -0.163298  0.123205 -0.569159  0.546668  0.108821  0.104533   10.00
284806  0.376777  0.008797 -0.473649 -0.818267 -0.002415  0.013649  217.00

        Class
284802      0
284803      0
284804      0
284805      0
284806      0

[5 rows x 31 columns]
```

Fig. 2 Dataset used

Table 1 Comparative analysis of different classifiers

S. No.	Classifier	Precision	Recall	$F1$ score	ROC score
1	Random forest	0.92	0.91	**0.92**	0.90
2	Gaussian Naive Bayes	0.62	0.81	0.67	0.81
3	Logistic regression	0.57	0.93	0.61	**0.92**
4	K-nearest neighbours	0.60	0.61	0.60	0.61

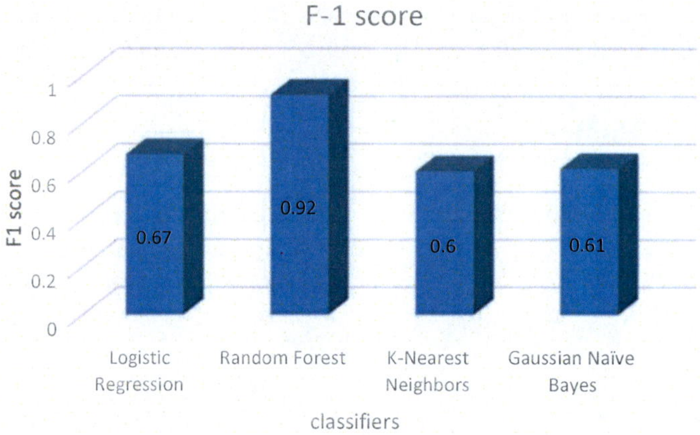

Fig. 3 $F1$ score of four classifiers

and receiver operating curve (ROC) scores. Table 1 shows the values of precision, recall, $F1$ score and ROC curve for all classifiers. The $F1$ score for random forest classifier is best among all others and is equal to 0.92. It is shown graphically in Fig. 3.

The comparison among different classifiers is also shown by ROC and PR curves as shown in Figs. 4 and 5, respectively.

The area enclosed by a ROC curve tells how good a particular parameter classifies the testing data. As depicted by a graph in Fig. 3, ROC score for logistic regression is maximum and is equal to 0.92.

4.1 Discussion

In general, the ROC score consists of multiple threshold values, and thus, it has numerous $F1$ scores in it. On the other hand, $F1$ score consists of single threshold value upon which all results are obtained. For $F1$ score to be high, both precision and recall should be high. But this is not the case with ROC score. This is clearly depicted from Table 1 shown. Random forest has greater $F1$ score as well as greater precision

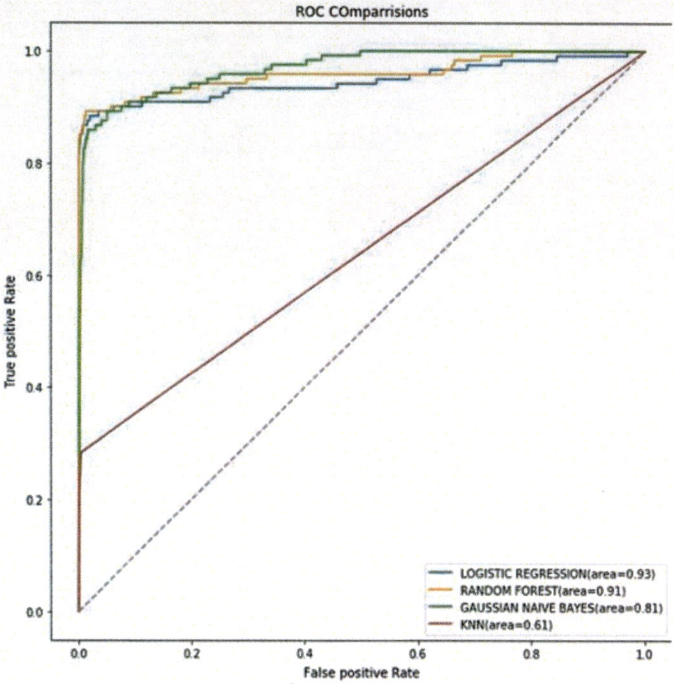

Fig. 4 ROC curves of four classifiers

and recall and despite being has the maximum ROC score logistic regression has pretty bad precision, so we will use $F1$ score to test the ability of classifier. Also when we have imbalanced dataset, we should always use $F1$ score as ROC score averages over all possible thresholds.

5 Conclusion and Future Work

The growing use of Internet and online transactions has led to increase in the number of credit card frauds. Therefore, by this research paper, we have shared an approach to detect credit card frauds. A standard dataset of a European bank is taken for implementation purpose. The proposed approach is tested using four classifiers: random forest, Gaussian Naïve Bayes, logistic regression and K-nearest neighbour. Results have shown that random forest classifier has performed the best among all classifiers.

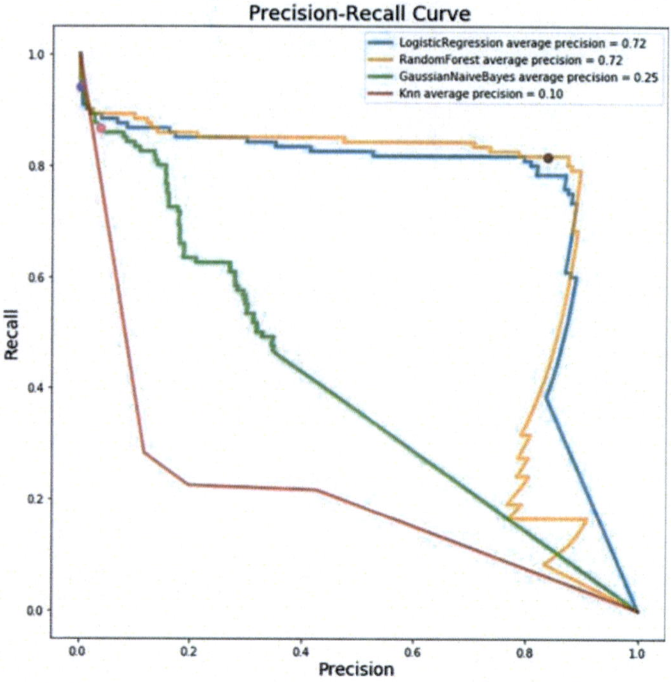

Fig. 5 Precision–recall curves

References

1. Zhang X, Han Y, Xu W, Wang Q (2019) HOBA: a novel feature engineering methodology for credit card fraud detection with a deep learning architecture. Inf Sci
2. Abakarim Y, Lahby M, Attioui A (2018) An efficient real time model for credit card fraud detection based on deep learning. In: Proceedings of the 12th international conference on intelligent systems: theories and applications, pp 1–7, Oct, 2018
3. Sadgali I, Sael N, Benabbou F (2019) Fraud detection in credit card transaction using neural networks. In: Proceedings of the 4th international conference on smart city applications, pp 1–4, Oct, 2019
4. Goswami K, Park Y, Song C (2017) Impact of reviewer social interaction on online consumer review fraud detection. J Big Data 4(1):1–19
5. Gadi MF, Wang X, do Lago AP Credit card fraud detection with artificial immune system. In: International conference on artificial immune systems. Springer: Berlin, pp 119–131, Aug 2008
6. Hajek P, Henriques R (2017) Mining corporate annual reports for intelligent detection of financial statement fraud–a comparative study of machine learning methods. Knowl-Based Syst 128:139–152
7. Carminati M, Polino M, Continella A, Lanzi A, Maggi F, Zanero S (2018) Security evaluation of a banking fraud analysis system. ACM Trans Privacy Secur (TOPS) 21(3):1–31
8. Yusof R, Kasmiran KA, Mustapha A, Mustapha N, Zin M, Asma N (2017) Techniques for handling imbalanced dataset when producing classifier models. J Theor Appl Inf Technol 95(7):1425–1440

9. Sun T, Vasarhelyi MA (2018) Predicting credit card delinquencies: an application of deep neural networks. Intell Syst Account Fin Manage 25(4):174–189
10. https://www.kaggle.com/mlg-ulb/creditcardfraud. Last Accessed on 04 Feb 2020

Voting Ensemble Classifier for Sentiment Analysis

Achin Jain and Vanita Jain

Abstract This paper proposes use of ensemble voting learning classification approach for sentiment classification of movie review dataset collected from IMDB. Sentiment analysis is a technique used to extract opinions from text including reviews, social messaging, etc. Generally, the opinion is classified into positive and negative polarity. The approach has been used in variety of domains including financial, educational, and other areas. Over the years many researchers have worked on sentiment classification to predict the opinion of text using several machine learning algorithms. The work carried out in this paper proposes the use of ensemble voting learning approach in which Naïve Bayes (NB), K-nearest neighbor (KNN), random forest (RF), and decision tree (DT). The results are compared with individual classification algorithm and voting approach. We have also used normalized and denormalized data to compare the accuracy results. The results show that using voting approach decreases root mean square error and increases precision of the classifier. Maximum accuracy of 80.13% is obtained with voting and normalized data.

Keywords Sentiment analysis · Naïve Bayes · Ensemble classification · Random forest · KNN

Abbreviations

AUC	Area Under the Curve
CNN	Convolutional-Neural-Network
DL	Deep-Learning

A. Jain
Communication and Technology, University School of Information, GGSIPU, Sector 16 C, Dwarka, Delhi, India
e-mail: achin.mails@gmail.com

V. Jain (✉)
Bharati Vidyapeeth's College of Engineering, New Delhi, India
e-mail: vanita.jain@bharatividyapeeth.edu

A. Abraham et al. (eds.), *Proceedings of 3rd International Conference on Computing Informatics and Networks*, Lecture Notes in Networks and Systems 167,
https://doi.org/10.1007/978-981-15-9712-1_22

DT	Decision-Tree
KNN	K-Nearest-Neighbor
LSTM	Long-Short-Term-Memory
NB	Naïve-Bayes
NLP	Natural Language Processing
RF	Random-Forest
SVM	Support Vector Machine
TF-IDF	Term Frequency Inverse Document Frequency

1 Introduction

Opinions are views of peoples, society about a topic. In modern world where usage of social media, blogs, review Web sites are increasing at an exponential rate, opinion mining or sentiment analysis becomes a necessary task. Peoples' sentiment about movies, product, or social issues helps in decision making about respective topic. Considering these aspects, the authors in [1] have defined sentiment analysis as "the practice of identification of substantial data from source resources that comes with information about commons perspective". To know about a movie, people usually visit lots of blogs, social networking sites and review sites. But apart from these, IMDB is the major source of data about the movies worldwide [2]. IMDB provides variety of dataset for research work, and for sentiment analysis also IMDB has a large dataset of 50,000 reviews of movies.

Reading each review manually and deciding whether it shows positive or negative opinion about a movie is not possible. To automate this task, concept of natural language processing (NLP) is employed that basically deals with various features extracted from sentence and application of classification algorithm for prediction of review polarity [3]. Sentiment analysis has application in almost every domain such as products, health care, social events, political, financial, and customer services. In this work, we intended to build a reliable sentiment classification model for assessment of movie review dataset. For the experimental work, we have used IMDB database of movie reviews created by Stanford University [4]. Words play very important role in sentiment analysis as they form the features of dataset. Frequency of words present in sentence helps in creation of feature vector set. The basic and most popular used technique is n-gram approach with TF-IDF scheme. In this work, we have used unigram approach where each sentence is tokenized into single words and TF-IDF of each word is calculated. Sentiment score of each review calculated using Vader and boxplot is shown in Fig. 1 displaying the distribution of reviews.

The rest of the paper is organized as follows: Related Work is shown in Sect. 2 while Sect. 3 explains the data preprocessing carried out on the dataset and Experimental Results obtained are presented in Sect. 4. In the end, Conclusion of the work done is presented in Sect. 5.

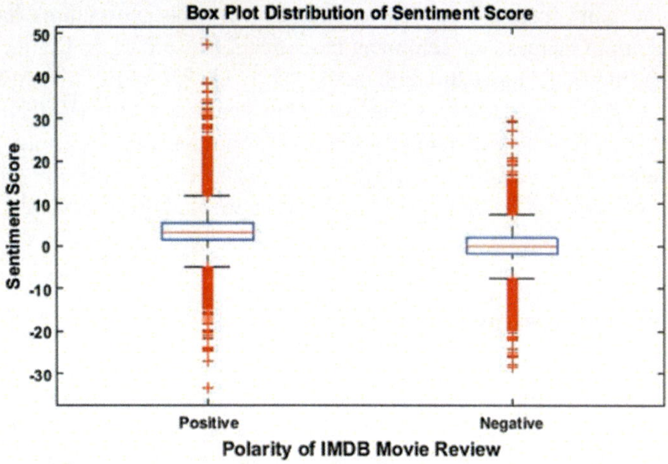

Fig. 1 Boxplot of sentiment score

2 Related Work

IMDB movie review has been one of the most used datasets in sentiment analysis. In this paper, we have also used IMDB movie review dataset for performing classification using various machine learning approaches. This section presents a brief review about previous work carried out on IMDB dataset. In [5], the authors carried out classification of sentiment via n-gram machine learning method. Classification algorithms used to perform experiment work are Naïve Bayes, maximum entropy, stochastic gradient descent, and support vector machine.

The authors in [6] used sentiment analysis techniques to build a hybrid movie recommendation system. The work also showcases the use of "cold start" phenomenon that helps in recommendation of new item with very limited information. In sentiment classification, text features play very important role and in [7] authors have highlighted it with various feature selection and classification algorithms. The authors achieved maximum accuracy with random forest classification algorithm.

Deep learning is an advanced version of artificial neural networks and researchers are already working on sentiment classification using the deep learning. In [8], authors carried out sentiment analysis using neural networks and lexicon. The authors have used bag words, lexicon, and part-of-speech tags for creating feature vector set before applying deep learning algorithms.

In [9], authors have used a hybrid deep learning (DL) approach. The work showcases the use of convolutional neural network (CNN) with long short-term-memory (LSTM) layers. The results showcase that using CNN and LSTM together significantly increases the accuracy of the classification.

There are many feature selection techniques available and authors have used variety of them to increase the sentiment classification accuracy. In [2], the authors have used the hybrid feature extraction methodology in which TF-IDF feature vector set is merged with lexicon features. The dataset is then tested with SVM, NB, KNN, and maximum entropy classification algorithms. The authors in [10] perform comparative evaluation of various feature weighting methods. The work shows that use of Word2Vec with SGD gives best accuracy in sentiment classification task.

3 Data Preprocessing

The IMDB review dataset is first preprocessed using **StringToWordVector** unsupervised filter in WEKA. All the reviews are converted into lower case and each word metric is replaced with TF-IDF. To convert the similar words in one single common word, stemming is done using **LovinsStemmer**. We have limited maximum number of words 1000 and after applying CfsSubsetEval only 66 features are considered for final classification process.

4 Results

In this paper, we have used IMDB dataset containing total of 50,000 rows with 25,000 each for training and testing, respectively. For evaluation of the performance, we have used accuracy, precision, recall, root mean square error and AUC as parameters. Various ML algorithms are applied for training the model and then applied on testing dataset. We have used Naïve Bayes (NB), K-nearest neighbor (KNN), decision tree (DT), random forest (RF), and voting ensemble approach for predicting the sentiment of the movie review. Complete knowledge flow of the process used in this paper is shown in Fig. 2.

Table 1 shows the comparison among various classifiers on testing dataset using recall, precision, and AUC as parameters. Table 2 shows the comparison using normalized dataset with Z-score methodology.

Figure 3 shows the accuracy achieved using all machine learning algorithms with normalized and original dataset used in this paper. From the graph and Tables 1, 2, it is evident that using voting concept helps in increasing the precision value and voting with all four classifiers also increase the accuracy to some extent.

Figure 4 shows the root mean square error comparison graph achieved using all machine learning algorithms with normalized and original dataset used in this paper. From the graph, it is clear that voting concept helps in reducing error value. The results also help in visualizing the improvement using normalized dataset. Error values are decreasing in every classifier algorithm when normalized dataset is used and maximum reduction is achieved in random forest classifier algorithm.

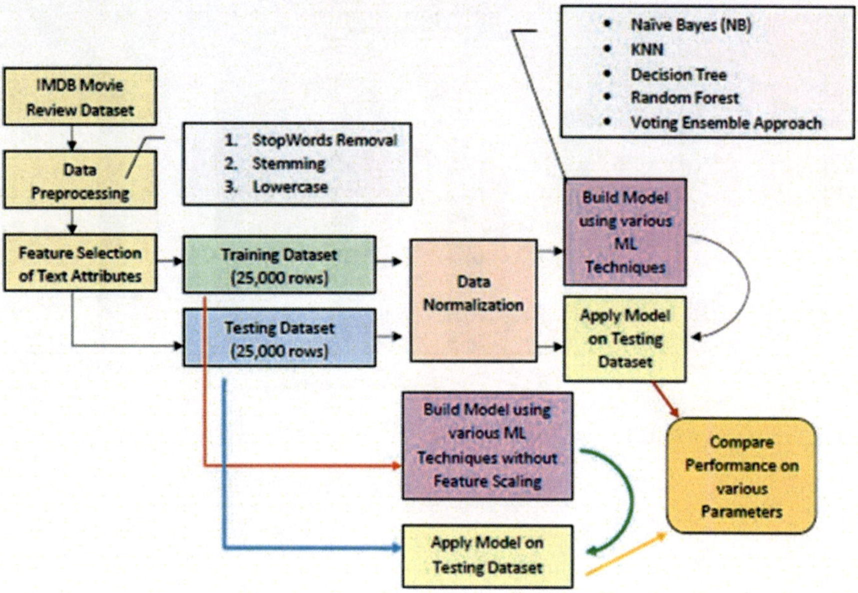

Fig. 2 Knowledge flow of the process

Table 1 Performance evaluation without data normalization

ML technique used	Recall	Precision	AUC
NB	0.79	0.78	0.86
KNN	0.74	0.74	0.78
DT	0.63	0.74	0.64
RF	0.63	0.74	0.72
Voting (NB + KNN + RF + DT)	0.79	0.80	0.81
Voting (NB + KNN + DT)	0.78	0.79	0.80

Table 2 Performance evaluation with data normalization

ML technique used	Recall	Precision	AUC
NB	0.79	0.79	0.86
KNN	0.74	0.74	0.78
DT	0.64	0.74	0.63
RF	0.64	0.74	0.72
Voting (NB + KNN + RF + DT)	0.79	0.79	0.81
Voting (NB + KNN + DT)	0.79	0.80	0.80

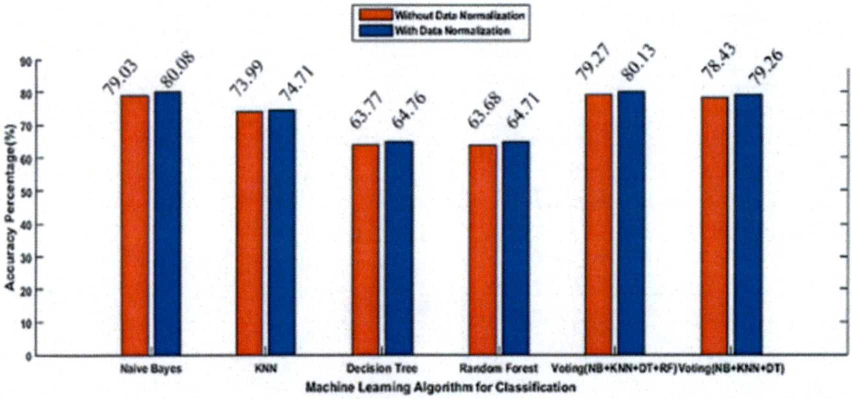

Fig. 3 Accuracy on IMDB movie review dataset

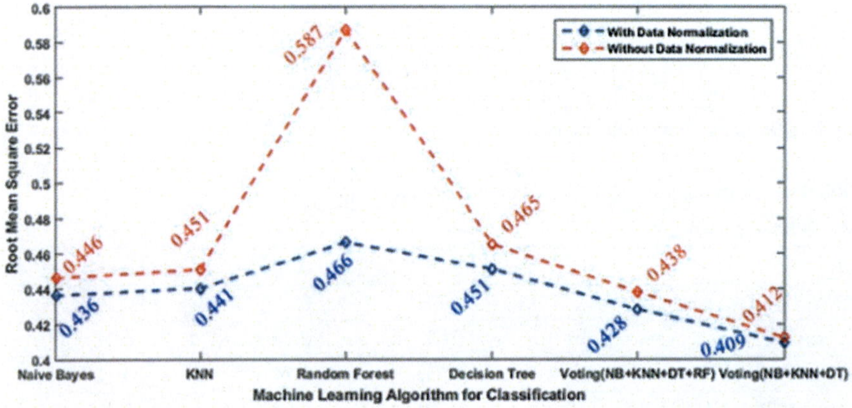

Fig. 4 Root mean square error on IMDB movie review dataset

5 Conclusion

This paper presents the voting ensemble classifier for classification of IMDB movie review polarity. Experiments conducted on normalized and denormalized IMDB movie review dataset shows that voting system with multiple classifiers outperforms single classifier. The results show that voting with NB, DT, RF, and KNN on reduced normalized dataset manages to achieve maximum accuracy of 80.13% with reduced root mean square value 0.428. Further experiment shows that removing one classifier from voting system manages to achieve almost similar accuracy with minimum root mean square error of 0.409. From the experiments, it is clear that voting concept helps in increasing accuracy and precision and reducing error value.

References

1. Giatsoglou M, Vozalis MG, Diamantaras K, Vakali A, Sarigiannidis G, Chatzisavvas KC (2017) Sentiment analysis leveraging emotions and word embedding's. Expert Syst Appl 69:214–224
2. Kumar HM, Harish BS, Darshan HK (2019) Sentiment analysis on IMDb movie reviews using hybrid feature extraction method. Int J Interact Multimedia Artif Intell 5(5)
3. Hussein DMEDM (2018) A survey on sentiment analysis challenges. J King Saud Univ Eng Sci 30(4):330–338
4. Maas AL, Daly RE, Pham PT, Huang D, Ng AY, Potts C (2011) Learning word vectors for sentiment analysis. In: Proceedings of the 49th annual meeting of the association for computational linguistics: human language technologies, vol 1. Association for Computational Linguistics, pp 142–150
5. Tripathy A, Agrawal A, Rath SK (2016) Classification of sentiment reviews using n-gram machine learning approach. Expert Syst Appl 57:117–126
6. Guan Y (2018) IMDB review mining and movie recommendation. In: Conference on statistical learning and data science/nonparametric statistics
7. Sahu TP, Ahuja S (2016, Jan) Sentiment analysis of movie reviews: a study on feature selection and classification algorithms. In: 2016 International conference on microelectronics, computing and communications (MicroCom). IEEE, pp 1–6
8. Shaukat Z, Zulfiqar AA, Xiao C, Azeem M, Mahmood T (2020) Sentiment analysis on IMDB using lexicon and neural networks. SN Appl Sci 2(2):1–10
9. Yenter A, Verma A (2017) Deep CNN-LSTM with combined kernels from multiple branches for IMDB review sentiment analysis. In: IEEE 8th annual ubiquitous computing, electronics and mobile communication conference (UEMCON). IEEE, pp 540–546
10. Sivakumar S, Rajalakshmi R (2019) Comparative evaluation of various feature weighting methods on movie reviews. In: Computational intelligence in data mining. Springer, Singapore, pp 721–730

Dhwani: Sound Generation from Google Drive Images Controlled with Hand Gestures

Tanish Bhola, Shubham Gupta, Satvik Kaul, Harsh Jain, Jatin Gupta, and Rachna Jain

Abstract Dhwani is a project focusing on sound generation from a given image using the content or background of the picture. Audio production from images can have much utilization, especially in reliving the past moment or for exploring the places without even visiting them. With the use of waving hand gesture, the user can open an image and can listen to the sound generated from one photograph at a time. The use of waving hand gesture for accessing the next or previous image has been incorporated into the project for improvizing the user's experience. There is a use of folder linked to Google Drive on which the processing has been performed. So, the user gets an advantage of directly using the images stored on the Google Cloud saving time and hustle to transfer the pictures. Using deep learning, object tracking and computer vision, Dhwani is providing the users with a prominent innovation that comes bundled with high-level user experience.

Keywords SoundNet · MobileNet · Google Drive · Hand gestures · Deep learning · Object tracking · Computer vision · Convolutional neural network

T. Bhola (✉) · S. Gupta · S. Kaul · H. Jain · J. Gupta · R. Jain
Bharati Vidyapeeth's College of Engineering, New Delhi 110063, India
e-mail: tanish.1557@gmail.com

S. Gupta
e-mail: shubhamg0305@gmail.com

S. Kaul
e-mail: satvik.1799@gmail.com

H. Jain
e-mail: harsh.jain96318@gmail.com

J. Gupta
e-mail: jatingpt21@gmail.com

R. Jain
e-mail: rachna.jain@bharatividyapeeth.edu

A. Abraham et al. (eds.), *Proceedings of 3rd International Conference on Computing Informatics and Networks*, Lecture Notes in Networks and Systems 167,
https://doi.org/10.1007/978-981-15-9712-1_23

1 Introduction

The main idea of the proposed model is to develop a mechanism which analyses an image on some pre-decided parameters and give out sounds based on the surrounding appearing in that image so that a person can relive that memory or feel the real presence of that image. The fields such as image recognition, speech recognition and machine translation have already been evolved over the years because of the massive labelled data sets available. But when it comes to natural sound understanding, there has not been much significant development in this area, resulting in small data set of such kind of sounds. People can envision sounds by taking a look at a photograph: The view of a seashore may carry the sound of crashing waves to mind. You may hear hints of horns and road publicizing when you take a look at an image of a bustling intersection of the street. This project is concentrating on this oblivious conduct, where watchers can believe themselves into imaginary soundscapes created with profound learning models. This work depends on the ongoing advancement of the cross-modular data recovery procedure, for instance, picture to sound, content to picture, utilizing deep learning. This project enables us to retrieve the best-coordinated sound document for a scene, out of our enormous ecological sound data set. The soundscapes generated by the AI at times stun us by meeting our desired outcomes; however, occasionally ignore the social and topographical connection. These distinctions and errors lead us to consider how the creative mind functions and how ripe the sound situations encompassing us are.

The proposed model was able to delineate the intended results. It was able to correctly produce the desired sounds based on the objects and surroundings present in an image. For the testing purpose, various pictures of different backgrounds like the beach, stadium and underwater were used, and the results were great. This model can be considered as a great alternative to the present baseline methods in this field.

This paper aims at generating sound from an image based on the objects and surroundings present in an image. This enables the user to relive the moment or experience a place without even visiting it.

This model has tried to take user experience to the next level by allowing the user to control the proposed model with hand gestures. The proposed model is a futuristic and unique approach that can have a large number of implications.

From a broader perspective, this approach can convert a standard image into a live model. This is how the paper tries to benefit the user with best user experience.

The remainder of the paper is composed as follows. The accompanying segment, i.e. Section 2, gives a depiction of different systems and models that have been advanced. Section 3 investigates the philosophy that has been utilized to accomplish the goals of the methodology. Section 4 exhibits the outcomes and examination of the method. It is trailed by the Conclusion and Future Scope in Sect. 5.

2 Related Work

The approach proposed in this paper differs from previous plans and methods followed, and this section will further help distinguish the work done in this paper from the earlier works. The following works have been studied thoroughly, and a new model has been proposed, which has been developed using these studies as a base.

Aytar et al. [1] in 2016—The primary purpose of this paper was to grasp the relations between vision and sound, to deduce a representation of sound from a particular unlabeled video. The objective was to determine if a model could be developed that could determine the objects in the videos using their sounds. The approach was to understand sounds present in nature, rather than focusing on music. The data set chosen for this approach was a collection of two million unlabelled videos, which were beneficial as they contained useful representations of natural sound. Due to the data set being large scale, much deeper networks could be trained, and the sound recognition system was trained through visual discriminative models, built specifically to handle large-scale categorizations. Their experiments left room for more improvement, which suggested that using an even larger data set and training deeper models might help in improving the performance of the model even more. The approach used in this paper is the exact opposite of this method, using image features rather than videos to determine the most suitable sound for that particular image and also comprises a computer vision model for controlling the cursor.

Chen et al. [2] in 2017—This paper centres around cross-modular media observation, attempting to tackle the cross-modular creation issue by making two new data sets, Sub-URMP is comprising of 17,555 pictures matched with a large portion of a second long audio clip from the University of Rochester Musical Performance data set, along with the INIS data set, which contained the ImageNet of five melodic instruments, specifically drums, saxophone, piano, guitar and violin, with each device is combined with a short audio clip of an independent exhibition of particular instruments. This was the first approach which included both grouping and human assessment, and their methodology could create one method from another, i.e. sound from visual and the other way around. This was likewise the primary endeavour to utilize restrictive GANs. Their single restriction was that the exactness was low and that their present strategies were constrained and needed terms of removing excellent portrayals. The method described in this paper only generates sound from image features, but is not limited to only musical instruments and has wider applications than the above approach.

Due to the availability of numerous approaches of user interaction with the keyboard, mouse and other input devices, a new method of interacting with the computer is part of the imminent future. Categorizing the existing approaches into 3D models and appearance-based [3], and giving the advantages and disadvantages of each will shed light on the problems and future development scope of the existing solutions. It was a theoretical approach to determine which of the existing approaches could be optimum. The method reported in this paper combines computer vision techniques with a model that generates sound from images.

Yang and Shi [4] in 2016 centred on improving the presentation of hand signal acknowledgement dependent on Kinect in the human–PC association framework. The plan utilizes depth information to help hand detachment and gets synchronized shading and depth pictures by Kinect. The picture features were investigated and separated to follow the hand zone, with comprehensive element depictions. At last, KNN was employed as the classifier for preparing and acknowledgement for static hand signal, to maintain a strategic distance from the issues in test irregularity. This model depends on Kinect and can be used only in systems which support Kinect. The computer vision model used in this paper can work on all systems with a webcam.

The use of hand gesture recognition is finding increased application in the field of robot control and other related areas as hand gestures can generate enriched information. However, segmenting hand gestures from the background poses to be a complication with using such gestures. A new approach which segments the hands based on skin colour [5] has been proposed. The gesture that is recognized in real time is passed through a neural network for decoding the info, and the results show an accuracy of 98.3%. The model used in this paper uses fingers for acting as a mouse and providing input to the computers, which is combined with the sound generation model.

Pendke et al. [6] in March 2015—The proposed framework was utilized to control the mouse cursor and actualize its capacity utilizing a continuous camera. This framework depended on picture examination and movement recognition innovation to do mouse pointer developments and choice of the symbol. From the outcomes, it was perceived that if the calculations can work in all situations, at that point, the framework will work all the more productively. This framework could be helpful in introductions and to lessen the workspace. This approach is only related to computer vision, whereas the method described in this paper is also composed of a sound generation model which uses image features to generate sounds.

Argyros and Lourakis [7] in 2006 displayed a vision-based interface for controlling a PC mouse utilizing 2D and 3D hand signals. The proposed interface expands upon the past work that allows the identification and following of different hands that can move openly in the field of perspective on a possibly moving camera framework. The proposed interface accomplishes exact mouse situating, smooth cursor development and dependable acknowledgement of signals actuating button occasions. The interface can be utilized as a virtual mouse for controlling any Windows application. Instead of using hand signals, the computer vision model reported in this paper uses fingers for controlling the mouse. This is further added to the sound generation model, which acts upon images to produce sound.

Le and Nguyen [8] in 2014 proposed a continuous fingertip following strategy utilizing a webcamera to empower clients to control their PC mouse by their own uncovered hands remotely. The hand area is right off the bat separated by foundation subtraction and sifted by the morphological opening activities and blob naming. At that point, the convex structure and convexity deformity have utilized the fingers and identify the directions of the fingertip. Next, the instructions of the fingertip are mapped to the screen coordinates and smoothed by the Moving Average. At long last, the occasions relating to the recognized fingers are sent to the PC framework to control the mouse. Exploratory outcomes show that the proposed procedure can effectively tally the finger with the precision of 98.3% and functions admirably continuously. This model, like the one described previously, is again related to only computer vision and does not take into account sound generation, and hence is not as holistic as the approach described in this paper.

Barchiesi et al. [9] in 2015 presented an approach for environment classification based on the corresponding sound they would have. They made use of state-of-the-art technique for the classification of acoustic scenes. They tried to build a framework that is general for acoustic scene classification (ASC). They delineated several algorithms that they submitted for the data challenge focused on ASC. This approach categorizes environments based upon the sounds that they would have. In contrast, the approach described in this paper uses image features to determine the optimum sound related to the particular image.

Salamon and Bello [10] in 2015 deeply examined the k-means algorithm of spherical nature for studying features from audio flags. The audio files they dealt with belonged to urban surroundings. The algorithm they used was simple and precise but equivocal compared to other intricate and time-consuming approaches. They studied the effect of processing different sections of the pipeline shaping the results of their analysis. They tested their method on the most significant public data set available for their task and even differentiated with the available baselines till that time. They demonstrated that feature learning beats other traditional approaches if used to apprehend the specifics of urban surroundings. This approach uses audio files and is only limited to sounds of urban surroundings. The model described in this paper can produce sound for any environment based on features extracted from images.

Serrano et al. [11] in 2002 showcased their approach based on the plot that low-level picture attributes can be made use to generate high-level erudition. They showed that getting an accuracy of around 90% can be easily achieved using the level-level peculiarities. Jumping above this score requires high-level traits and computations, thus preventing these approaches from being used in practical scenarios. The outdoor–indoor classification generally involves the use of high-level peculiarities and calculation, but they achieved comparable results with the use of their presented approach. They used low-level traits and comparatively less-dimensional features with the dual-stage support vector machine (SVM). They were able to achieve an accuracy of 90.2% on a broad set of user's images. The approach in this paper is starkly different from the above approach, as it combines computer vision model with a sound generation model that generates sound from features extracted from an image.

An approach of tracking multiple skins coloured objects in real time by a moving camera has also been put forward [12]. The detection of skin coloured objects is achieved by using a Bayesian classifier, which is further bootstrapped and iterated with supplementary training images. The study was complemented with a prototype, designed for 28 Hz videos with a resolution of 320 × 240 and deployed on Pentium

4 processor. The experimental results have also been provided with the study [13]. This model is related to only computer vision and is not as holistic as the model described in this paper.

Raheja et al. [14] in 2011—Fingertips detection is something which is frequently used for the area of human–computer interaction. This paper presents a new time-efficient procedure that will result in fingertip detection after eliminating the extraneous elements of the input image. Via using skin filter based on HSV colour space and hand cropping based on skin pixels histogram of the hand image, binary silhouette of the image given as input was created. The cropped image would be used to identify or for detecting the fingertips in the image frame. As was with the previous method, this is only a computer vision-based model. The approach described in this paper combines a computer vision model and a sound generation model.

Boutell et al. [15] in 2014—This paper is fixated on the classification mistakes that happen when the class is not fundamentally unrelated by definition. The types are fundamentally irrelevant when there is no crossing point between them; for example, each article has a place just with one class. In typical example acknowledgement issues, types are now fundamentally unrelated. Such problems happen in the semantic scene and report characterization and pharmaceutical conclusions. The proposed framework offers a structure to deal with such issues where a view may contain numerous articles to such an extent that various class marks can depict the scene. The framework presents methodologies for preparing and testing during this circumstance and offers new measurements for evaluating multiple models, with class review and precision, and in general exactness. This model describes a method to overcome classification errors that occur when the classes are not entirely unrelated. It is evident how this approach is different from the one reported in this paper.

In the previous methods, videos have been used to train a classification model for sounds, SoundNet. In this paper, instead of videos, images have been used to generate sound. Image features have been extracted, and they have been related to sound features present in SoundNet, which is an exact reverse of the previous approaches, in which videos and sounds have been used to detect what objects are present in the video. This sound classification model is further combined with computer vision and hand gestures to provide an extraordinary user experience.

3 Research Methodology

The approach adopted has not been trained from scratch; instead, it is about taking help from already generated methods. SoundNet is a deep neural network architecture designed by MIT which was able to recognize the objects and scenes from the sound. MobileNets are a kind of CNN innovated by Google researchers. As the name suggests, they are quick and resource-friendly. They are used for everyday tasks that CNNs are suitable for like detection [16] and classification [17]. The proposed model combines the approaches mentioned above: SoundNet and MobileNet.

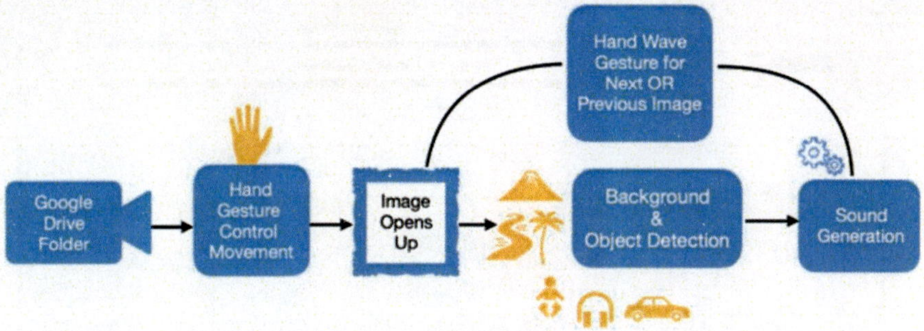

Fig. 1 Workflow diagram of Dhwani

The SoundNet model is used to get object features of all the audio files which are stored in a database. An input image is processed by the MobileNet model, to get the object features of the picture too. Finally, matching them gives a related sound file (Fig. 1).

3.1 SoundNet

MIT presented SoundNet model in 2016. Given a video SoundNet can recognize the scenes of the article utilizing just the sound of that video. The incredible part is that the preparation was done on the unlabelled video data set and, still, the precision accomplished was 92.2%. It introduces a complex auxiliary system that adapts legitimately on crude sound waveforms [18], which is prepared by changing over information from a dream into sound. Even though the system is by vision or by the video, it is not reliant on that during surmising. SoundNet acquires best in class precision on three customary acoustic scene order data sets and recommends that more profound models perform better. It additionally ends up being a significant level locator for minute sounds, for example, winged creature Twitters or groups cheering, even though it was prepared legitimately from a sound without ground truth names (Figs. 2 and 3).

3.2 MobileNet

MobileNets are a class of CNNs designed and introduced by Google researchers used for classification, detection and some other type of general CNN tasks. They are small, fast, accurate and use minimal resources. Unlike the traditional approaches, they use 3×3 depthwise convolution [19] and pointwise convolution of 1×1. The use of depthwise convolutions makes them lightweight [20] (Fig. 4).

Depthwise convolution (1) with 1-filter per channel:

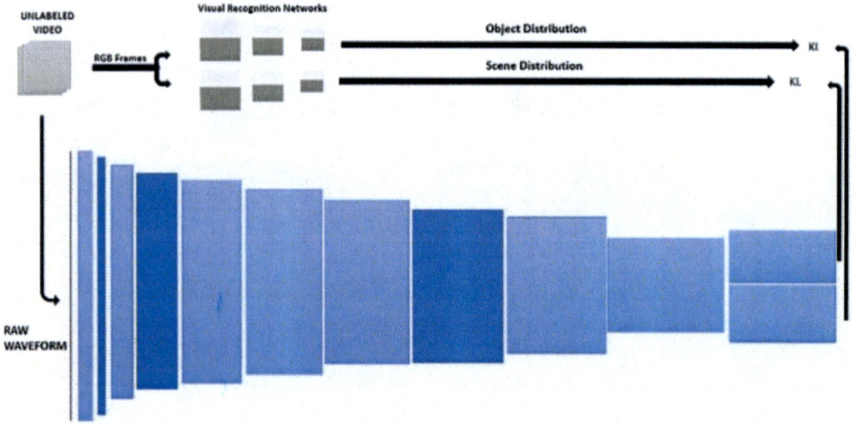

Fig. 2 SoundNet architecture

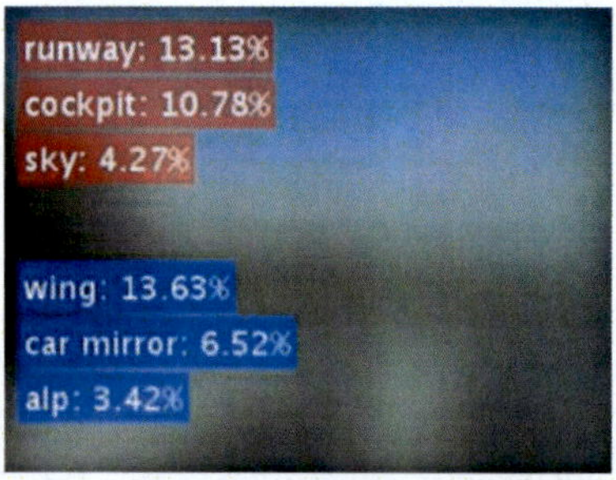

Fig. 3 An example of the output generated by SoundNet

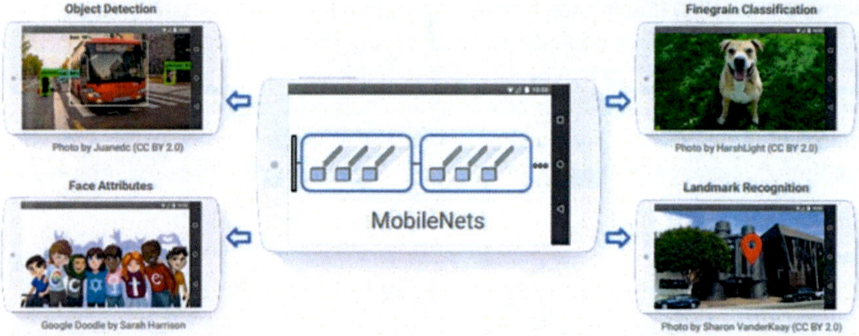

Fig. 4 Various recognition tasks that can be performed by MobileNet

$$\hat{G}_{k,l,m} = \sum_{i,j} \hat{K}_{i,j,m} \cdot F_{k+i-1,l+j-1,m} \quad (1)$$

The computational cost of depthwise convolution (2):

$$D_K \cdot D_K \cdot M \cdot D_F \cdot D_F \tag{2}$$

3.3 Sound Generation

The SoundNet used sounds to predict the objects and acoustic scenes, stopping one step before will reap the features. The MobileNet predicts the purposes from the images, staying one step before will receive the elements. So, image features [21] can be generated using both models. Taking input of the image on which the sound is to be produced then applying MobileNet will give the image features, which can be compared using entropy [22] with the features of the sound files generated via SoundNet. The mp3 file linked to the closest match will be the output audio. That is how the sound is produced with the fusion of the SoundNet and MobileNet models.

3.4 Hand Gestures

The proposed model has attempted to control the mouse cursor development and snap occasions utilizing a camera, in the light of the colour discovery procedure. Here, continuous video has been caught using a webcamera. When the hues are distinguished, the framework performs different activities to follow the cursor and performs control activities, the subtleties of which are given underneath (Fig. 5).

System Description:

- Capturing continuous video utilizing webcamera.
- Processing the individual picture outline.
- Resize and obscuring pictures for better shading identification.
- Colour identification and extraction of various hues (RGB).
- Creating form and making a hover around the identified shading.
- Calculating its centroid.
- Tracking the mouse pointer utilizing the directions acquired from the centroid.

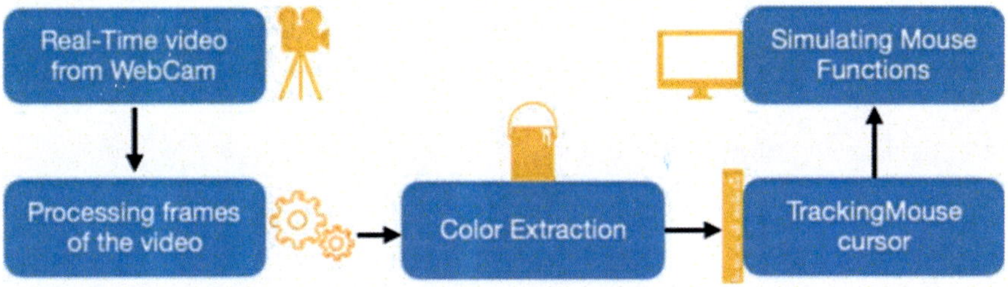

Fig. 5 Hand gestures system

- Simulating the left snap and the correct snap occasions of the mouse by utilizing PyAutoGUI in python IV. Results and Discussions.

4 Results and Discussions

For the experimental purpose, there has been the use of folder linked to Google Drive of the user. The linked folder gives the user more straightforward and convenient access to his/her images saved on Google Cloud. Ten copies of different acoustic scenes and background were placed in that folder. And then there is the execution of hand gesture program, which runs in an infinite loop and detects user's hand gestures. It gives the user the functionality of opening the image and iterating through them with the help of left/right-hand wave gestures [23]. As soon as the user opens up an image, the sound generation program gets executed and produces the sound related to that image. So there is the image displayed on the screen and the related sound being played. Using the wave hand gesture if the user moves to the next or previous image, the sound generation program gets notified and as a result, produces the audio output of the new image being displayed on the screen (Fig. 6).

Comparing with [24] it used a CNN stacked with RNN structure to predict sound features at each time step, and audio samples are reconstructed by example-based retrieval. This project connected the features obtained from MobileNet and SoundNet to generate the sound. Also, it used ImageNet and here the improved version of it, i.e. MobileNet, which is faster is used. In comparison with [25], it made its sound generator with the frame to frame sound generation of max 10 s sound. In contrast, this project used SoundNet in which there is no need to make its generator as it has an audio data set following it. First, the image is passed through MobileNet to get features, then those features are cross-mapped with the SoundNet features and if the features are the same that sound from the SoundNet database is played.

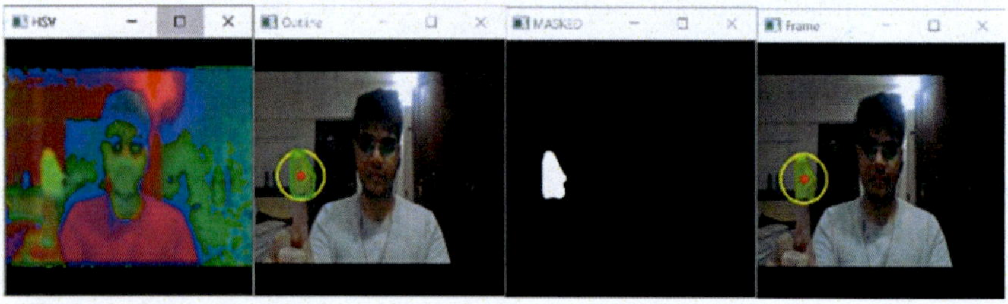

Fig. 6 Hand gesture system simulation

5 Conclusion

The proposed model successfully achieved the aim of the paper, sound generation from images. The outcomes were astonishingly excellent. It was even able to produce accurate distinct sounds for a small child and a crying child. The incorporation of hand gesture movements for control turns out to be a success. The proposed model was able to provide a futuristic and high-level user experience. The applications of the proposed model like reliving a past moment and exploring a place without even visiting it were favourably met. The integration of Google Drive turned out to be useful as the user do not need to get in the hustle of transferring photographs, and the model can quickly locate the user's photographs saved on the Google Cloud. The model was able to delineate the aim of the paper triumphantly.

Till now, the model is making use of an image to produce sound. Being able to play audio from multiple photographs or videos can be one of the future scopes of this project. With this, it might be able to give sound to a silent movie or a GIF. The hand gesture part utilizes a colour, in this case, green colour for its functionality. Making the model achieve this without colour can be another future scope. The user would then be able to operate the model with bare hands. Based on background acoustic scenes and objects, the model produces sound. Every surrounding has a kind of music associated with it like calm music is to mountains. Making the model to recommend songs based on the acoustic surrounding can be an exceptional future scope.

References

1. Aytar Y, Vondrick C, Torralba A (2016) SoundNet: learning sound representations from unlabeled video. NIPS
2. Chen L, Srivastava S, Duan Z, Xu C (2017) Deep cross-modal audio-visual generation. ArXiv: abs/1704.08292
3. Garg P, Aggarwal N, Sofat S (2009) Vision based hand gesture recognition
4. Yang F, Shi H (2016) Research on static hand gesture recognition technology for human computer interaction system. In: International conference on intelligent transportation, big data and smart city (ICITBS), Changsha, pp 459–463
5. Sun J, Ji T, Zhang S, Yang J, Ji G (2018) Research on the hand vision-based interpretation of hand gestures for remote control of a computer mouse gesture recognition based on deep learning. In: 12th International symposium on antennas, propagation and EM theory, pp 1–4, Hangzhou, China
6. Pendke K et al (2015) Int J Comput Sci Mob Comput 4(3):293–300
7. Argyros AA, Lourakis MIA (2006) In: Huang TS et al (eds) Computer vision in human-computer interaction (ECCV 2006). Lecture notes in computer science, vol 3979. Springer, Heidelberg
8. Le PD, Nguyen VH (2014) Remote mouse control using fingertip tracking technique. In: recent advances in electrical engineering and related sciences (201. Lecture notes in electrical engineering, vol 282). Springer, Heidelberg
9. Barchiesi D, Giannoulis D, Stowell D, Plumbley M (2015) Acoustic scene classification: classifying environments from the sounds they produce. SPM

10. Salamon J, Bello J (2015) Unsupervised feature learning for urban sound classification. In ICASSP
11. Serrano N, Savakis AE, Luo J (2002) A computationally efficient approach to indoor/outdoor scene classification. In: Object recognition supported by user interaction for service robots, vol 4, pp 146–149
12. Argyros AA, Lourakis MIA (2004) Real-time tracking of multiple skin-colored objects with a possibly moving camera. In: Pajdla T, Matas JG (eds) ECCV, LNCS, vol 3023. Springer, Heidelberg, pp 368–379
13. Wang RY, Popovic J (2009) Real-time hand-tracking with a color glove. ACM Trans Graph 28(3)
14. Raheja JL, Das K, Chaudhary A (2011) An efficient real time method of fingertip detection. In: 7th international conference on Trends in Industrial Measurements and Automation (TIMA, January 2011), Chennai, India, pp 447–450
15. Boutell M, Luo J, Shen X, Brown C (2004) Learning multi-label scene classification. Pattern Recogn
16. Papageorgiou CP, Oren M, Poggio T (1998) A general framework for object detection. In Sixth international conference on computer vision (IEEE Cat No 98CH36271). IEEE, pp 555–562
17. Cireşan D, Meier U, Schmidhuber J (2012) Multi-column deep neural networks for image classification. arXiv preprint arXiv:1202.2745
18. Oord AVD, Dieleman S, Zen H, Simonyan K, Vinyals O, Graves A, Kavukcuoglu K (2016) Wavenet: A generative model for raw audio. arXiv preprint arXiv:1609.03499
19. Chollet F (2017) Xception: deep learning with depthwise separable convolutions. In: Proceedings of the IEEE conference on computer vision and pattern recognition, pp 1251–1258
20. Howard AG, Zhu M, Chen B, Kalenichenko D, Wang W, Weyand T, Adam H (2017) Mobilenets: efficient convolutional neural networks for mobile vision applications. arXiv preprint arXiv:1704.04861
21. Lowe DG (2004) Distinctive image features from scale-invariant keypoints. Int J Comput Vision 60(2):91–110
22. Xuecheng L (1992) Entropy, distance measure and similarity measure of fuzzy sets and their relations. Fuzzy Sets Syst 52(3):305–318
23. Hong P, Turk M, Huang T (2000) Gesture modeling and recognition using finite state machines. In: Proceedings fourth IEEE international conference on automatic face and gesture recognition (Cat No PR00580). IEEE, pp 410–415
24. Owens A, Isola P, McDermott A (2016) Visually indicated sounds. In: CVPR
25. Zhou Y, Wang Z, Fang C, Bui T, Berg T (2018) Visual to sound: generating natural sound for videos in the wild. In: Proceedings of the IEEE conference on computer vision and pattern recognition, pp 3550–3558

An Empirical Study of Activation Functions for Function Approximation Tasks

Apoorvi Sood, Pravin Chandra, and Udayan Ghose

Abstract Activation functions play a major role in determining the learning speed and generalization capability of feed-forward artificial neural networks. In this paper, an empirical comparison of eight activation functions is reported on 12 function approximation problems. The study allows us to assert that the sigmoidal class of activation functions performed much better than the non-sigmoidal class of activation functions. Out of the six non-sigmoidal activation function, one activation function called the sigmoidal-weighted linear unit is identified as outperforming all other non-sigmoidal activation functions.

Keywords Artificial neural networks · Feed-Forward artificial neural networks · Activation functions

1 Introduction

The role of the activation functions on the capability of feed-forward artificial neural networks (FFANNs) to approximate any continuous function arbitrarily well is well known. The proofs of the universal approximation property of these networks (under different conditions) all impose the following two common requirements on the FFANNs structure:

[1]Henceforth called hidden nodes, for brevity.

A. Sood · P. Chandra (✉) · U. Ghose
University School of Information, Communication & Technology, Guru Gobind Singh Indraprastha University, New Delhi 110075, India
e-mail: pchandra@ipu.ac.in; chandra.pravin@gmail.com

A. Sood
e-mail: soodapoorvi@yahoo.com

U. Ghose
e-mail: udayan@ipu.ac.in

A. Abraham et al. (eds.), *Proceedings of 3rd International Conference on Computing Informatics and Networks*, Lecture Notes in Networks and Systems 167,
https://doi.org/10.1007/978-981-15-9712-1_24

1. The requirement of having at least one hidden layer of nodes.
2. That the activation functions of the hidden layer nodes[1] should be a function that cannot be expressed/represented as a polynomial of finite degree.

For a survey and review, of universal approximation results for FFANNs, may refer to [1] and references therein. The possession of the universal approximation property by these networks may in fact be attributed to the usage of these nonlinear activation functions. Unfortunately, the universal approximation results do not specify or prefer any particular activation function for usage in FFANNs. And, since finite-sized networks are used in practice, it is suggested that some activation functions may be more preferred in practice than others (see [2] for a survey and reasons for exploring new activation functions). This has provided an impetus for research to discover new (and hopefully better) activation functions.

The methodology of learning for FFANNs may be considered as estimation of the function that generates the data available for training. That is, in this paradigm, a set of tuples is available that represents how a particular value for the inputs to a system produces the corresponding output of the system. The dependency between the input(s)–output(s) is to be assessed on the basis of the available data. Or, we may consider all learning tasks (in the supervised learning paradigm) to be a function approximation task. Though it is definitely true that in real-life scenario, the data available may be contaminated by noise, it may be asserted that a methodology that estimates the functional mapping between the inputs and the outputs for non-contaminated (by noise) data may also be preferred for scenarios of estimation from noisy data.

Thus, in this paper, we focus on the approximation capabilities of the networks using some of the commonly used activation functions (including activation functions which are unbounded and have been recently proposed), for the solution of function approximation tasks (wherein neither the input values available nor the corresponding output values are contaminated by noise). Section 2 describes the activation functions used in this work. Section 3 describes the function approximation tasks. Section 4 describes the experiment methodology. The results are presented and discussed in Sect. 5, while the conclusions are presented in Sect. 6.

2 Activation Functions

Traditionally, the commonly used activation functions have been the following two activation functions, namely the logistic/log-sigmoid function (1) and the hyperbolic tangent function (2) defined, respectively, as:

$$\sigma_{01}(x) = \frac{1}{1 + \mathrm{e}^{-x}} \tag{1}$$

$$\sigma_{02}(x) = \tanh(x) = \frac{\mathrm{e}^{x} - \mathrm{e}^{-x}}{\mathrm{e}^{x} + \mathrm{e}^{-x}} \tag{2}$$

These activation functions are bounded that is the limiting value for the functions (toward plus/minus infinity) is a constant [3]. Since the weight update equations of the FFANNs are dependent on the derivative of the activation function values [3], for large net input to the activation functions, the derivative values tend toward zero. This is also called the vanishing gradient problem. In 2000, Hahnloser et al. [4] introduced the rectified linear unit (ReLU) function in the field of dynamical networks. In 2011, Glorot et al. [5] first used the ReLU in the field of deep learning and demonstrated that these activation function performance is better.

The ReLU function is defined as

$$\sigma_{03}(x) = \max(0, x) = \begin{Bmatrix} x; \text{ for } x \geq 0 \\ 0; \text{ for } x < 0 \end{Bmatrix}; \quad x \in \Re \tag{3}$$

where $\Re$ is the set of real numbers. This function has a sparse representation, the function and its derivative are zero for $x < 0$, to rectify this situation, the modified ReLU known as Leaky ReLU (LReLU) function was proposed [6] and is defined as

$$\sigma_{04}(x) = \max(0, x) + \min(ax, 0) = \begin{Bmatrix} x; & \text{for } x \geq 0 \\ ax; & \text{for } x < 0 \end{Bmatrix}; \quad x \in \Re \tag{4}$$

where $a \in \Re$; $1 > a > 0$, the value of $a = 0{:}01$ is used in this work. Both the ReLU and the LReLU are discontinuous functions at 0; moreover, the function LReLU and its derivative are unbounded on both the sides. To alleviate this problem, the exponential linear unit (ELU) was proposed [7] and is defined as

$$\sigma_{05}(x) = \begin{Bmatrix} x; & \text{for } x \geq 0 \\ \alpha(\mathrm{e}^x - 1); & \text{for } x < 0 \end{Bmatrix}; \quad x \in \Re \tag{5}$$

where $\alpha = 1$ (as proposed in [7]). An optimized version of the ELU is the scaled exponential linear unit (SELU) was proposed and is defined as [8]:

$$\sigma_{06}(x) = \lambda \begin{Bmatrix} x; & \text{for } x \geq 0 \\ \alpha(\mathrm{e}^x - 1); & \text{for } x < 0 \end{Bmatrix}; \quad x \in \Re \tag{6}$$

where $\alpha = 1.67326$; $\lambda = 1.0507$ (as proposed in [8]). The functions $\sigma_{3\text{–}6}$ and/or their derivatives are discontinuous at zero, and to remove this discontinuity, the softplus function [5] was proposed and is defined as

$$\sigma_{07}(x) = \ln\left(1 + \mathrm{e}^x\right); \quad x \in \Re \tag{7}$$

This function does not pass through zero or is not zero-centered. A zero-centered, one side unbounded function was proposed by Elfwing et al. [9] and is called the sigmoid-weighted linear unit (SiLU), defined as

$$\sigma_{08}(x) = \frac{x}{1 + \mathrm{e}^{-x}} = x\sigma_{01}(x); \quad x \in \Re \tag{8}$$

3 Function Approximation Tasks

The 12 function approximation tasks taken from literature are

1. G_{01}: This is a one-dimensional function that is available in MATLAB as humps.m:

$$g_{01}(x) = \frac{1}{(x - 0.3)^2 + 0.01} + \frac{1}{(x - 0.9)^2} - 6 \tag{9}$$

   where $x \in (0, 1.1)$ [10, 11].
2. G_{02}: The function is two-dimensional function of MATLAB as mypeaks.m and is

$$g_{02}(x, y) = 3(1 - x)^2\mathrm{e}^{-x^2-(y+1)^2} - 10\left(\frac{x}{5} - x^3 - y^5\right)\mathrm{e}^{-x^2-y^2} - \frac{\mathrm{e}^{-(x+1)^2-y^2}}{3} \tag{10}$$

   where $x, y \in (-3, 3)$ [10, 11].
3. G_{03}: The function is a two input function and has been used in [10–14]:

$$g_{03}(x, y) = \sin(x \times y) \tag{11}$$

   where $x, y \in (-2, 2)$.
4. G_{04}: The function is a two input function and has been used in [10–14]:

$$g_{04}(x, y) = \mathrm{e}^{x\sin(\pi y)} \tag{12}$$

   where $x, y \in (-1, 1)$.
5. G_{05}: This two-dimensional input function has been used in [10, 11, 13, 14]:

$$g_{05}(x, y) = \frac{1 + \sin(2x + 3y)}{3.5 + \sin(x - y)} \tag{13}$$

   where $x, y \in (-2, 2)$.
6. G_{06}: This two-dimensional input function has been used in [10, 11, 13–15]:

$$g_{06}(x, y) = 42.659\big(0.1 + x\big(0.05 + x^4 - 10x^2y^2 + 5y^4\big)\big) \tag{14}$$

   where $x, y \in (-0.5, 0.5)$.

7. G_{07}: This two-dimensional input function has been used in [10, 11, 13–15]:

$$g_{07}(x, y) = 1.3356(1.5(1 - x) + e^{2x-1}\sin\left(3\pi(x - 0.6)^2\right) \\ + e^{3(y-0.5)}\sin\left(4\pi(y - 0.9)^2\right) \tag{15}$$

where $x, y \in (0, 1)$.

8. G_{08}: This two-dimensional input function has been used in [10, 11, 13–15]:

$$g_{08}(x, y) = 1.9\left(1.35 + e^{x}\sin\left(13(x - 0.6)^2\right)e^{-y}\sin(7y)\right) \tag{16}$$

where $x, y \in (0, 1)$.

9. G_{09}: This two-dimensional input function has been used in [13, 14, 16]:

$$g_{09}(x, y) = \sin\left(2\pi\sqrt{x^2 + y^2}\right) \tag{17}$$

where $x, y \in (-1, 1)$.

10. G_{10}: This four-dimensional input function has been used in [13, 14]:

$$g_{10}(x_1, x_2, x_3, x_4) = e^{2x_1\sin(\pi x_4)+\sin(x_2 x_3)} \tag{18}$$

where $x_1, x_2, x_3, x_4 \in (-0.25, 0.25)$.

11. G_{11}: This four-dimensional input function has been used in [10, 11, 13, 14]:

$$g_{11}(x_1, x_2, x_3, x_4) = 4(x_1 - 0.5)(x_4 - 0.5)\sin\left(2\pi\sqrt{x_2^2 + x_3^2}\right) \tag{19}$$

where $x_1, x_2, x_3, x_4 \in (-1, 1)$.

12. G_{12}: This six-dimensional input function has been used in [10, 11, 13, 14, 17]:

$$g_{12}(x_1, \ldots, x_6) = 10\sin(\pi x_1 x_2) + 20(x_3 - 0.5)^2 + 10x_4 + 5x_5 + 0x_6 \tag{20}$$

where $x_1, x_2, x_3, x_4, x_5, x_6 \in (-1, 1)$.

For creation of the training and the test data sets, 1000 samples are generated by uniform random sampling of the input domain, these input samples together with the function values constitute the data set for the learning task. Five hundred such tuples of input–output pairs constitute the training data set, while the other set of such 500 hundred tuples form the test data set.

Table 1 Network architecture summary (I: Inputs, H: Hidden nodes, O: Outputs)

Task	I	H	O	Task	I	H	O
G_{01}	1	20	1	G_{02}	2	15	1
G_{03}	2	10	1	G_{04}	2	10	1
G_{05}	2	64	1	G_{06}	2	17	1
G_{07}	2	11	1	G_{08}	2	18	1
G_{09}	2	24	1	G_{10}	4	10	1
G_{11}	4	30	1	G_{12}	6	25	1

4 Experiment Design

The universal approximation results for the FFANNs impose the condition of minimum one hidden layer. In this paper, therefore we consider the FFANN architecture to have only one hidden layer. The number of nodes in the hidden layer is decided on the basis of exploratory experiments carried out, wherein the number of hidden nodes is varied between 1 and 100. The smallest-sized network that gives a satisfactory error of training is chosen as the network used for detailed experiments. The network architecture for individual tasks is shown in Table 1. The exploratory experiments FFANNs use the logistic activation function (σ_{01}). The training of FFANNs is performed using the weight backtracking variant of the resilient backpropagation algorithm called the improved resilient backpropagation algorithm with weight backtracking (iRPROP^{+}) [18–20]. This algorithm has been demonstrated to compete with second-order optimization algorithms for the training of FFANNs [20].

For the conduction of the experiments, an ensemble of initial weights and thresholds of the FFANNs is created where the weights/thresholds are chosen as uniform random number in the interval $(-1, 1)$. The number of such weights/thresholds created is 50. That is, for FFANNs using any of the five activation functions considered in this paper, 50 networks are trained for each task and for FFANNs using the eight activation functions. Since the number of tasks is 12, the number of FFANNs trained is 12(tasks) $\times$ 8(activation function) $\times$ 50(weight sets) $=$ 4800 networks. Before usage, the data (on a per variable basis), both independent and dependent, are linearly scaled to the interval $[-1; 1]$. The experiments are conducted using these scaled variables, and the results are also reported over the scaled variables. The individual networks are trained for 2000 epochs of training.

For each of the trained network, the mean squared error (MSE) over the training and the test data sets is measured. That is, we have 50 values of the MSEs for each distinct task and activation function (using FFANN). We report the mean of these MSEs (MMSE), the standard deviation of these MSEs (STD), and since median is supposed to be a more robust estimator of the central tendency[2] [21, 22], we also report the median of the MSEs (MeMSE). Since generalization error is the measure

[2]Less influenced by the presence of outliers.

of quality of training, we report the data values and comparison results of activation functions only in terms of the error values obtained on the test set data.

For comparison of the means and the medians of the MSE, we use the Students' t-test and the Wilcoxon rank-sum test, respectively [22].

Experiments are conducted on a Microsoft Windows 10 operating system machine with Intel 17 processor, 8 GB RAM and MATLAB v2018b.

5 Results and Discussions

The summary data of errors/MSE over the test data set is presented in Table 2. For comparison of the performance of the activation function usage in FFANNs, we construct a comparison matrix as below:

1. The comparison matrix can be defined on the basis of comparison of the mean or the median of the MSEs obtained for a particular task and for different activation function using FFANNs (that is the MMSE or the MeMSE values, respectively). For the sake of explanation, the tasks are represented as G_k, where $k \in \{1, 2, \ldots, 12\}$, the activation functions are labeled as σ_{0i} where $i \in \{1, 2, \ldots, 8\}$, and the obtained matrix for the error values (MMSE or MeMSEs) is represented as (for example) $\text{MMSE}_{\sigma_{0i}}^{G_k}$, where $i \in \{1, 2, \ldots, 8\}$ and G_k, where $k \in \{1, 2, \ldots, 12\}$. Then, the following variable is defined that represents if the activation function σ_{0i} performs better than σ_{0j} where $i, j \in \{1, 2, \ldots, 8\}$ using the corresponding MMSE values for the task k:

$$I_{ij}^{(k)} = \begin{cases} 1; \text{ if } \text{MMSE}_{\sigma_{0i}}^{G_k} > \text{ if } \text{MMSE}_{\sigma_{0j}}^{G_k} \\ 0; \text{ otherwise} \end{cases} \tag{21}$$

2. With this, the comparison matrix for the comparison of the activation functions is defined as

$$C_{ij} = \sum_{k=1}^{12} I_{ij}^{(k)} \tag{22}$$

The value C_{ij} represents in how many tasks the FFANNs using the activation function σ_{0i} had better performance as compared to the networks using the activation function σ_{0j}.

3. The matrix can be created by comparison of values or by using any statistical test for comparison of values to check whether the difference between the values is statistically significant or not. We shall use the one-sided Students' t-test for comparison of means of MSEs (MMSE) and the Wilcoxon's rank-sum test for comparison of median of MSEs (MeMSEs). The tests shall be conducted at a level of significance $\alpha = 0{:}05$.

Table 2 Summary of results over the test data set. All values of the statistics are $\times 10^{-3}$

Task	Statistics	σ_{01}	σ_{02}	σ_{03}	σ_{04}	σ_{05}	σ_{06}	σ_{07}	σ_{08}
G_{01}	MMSE	0.151	0.389	21.844	9.559	14.492	7.809	17.391	10.664
	STD	0.129	0.375	52.165	3.117	3.192	4.647	4.121	4.276
	MeMSE	0.105	0.302	9.987	9.961	14.228	6.216	18.046	11.137
G_{02}	MMSE	11.978	13.230	19.918	18.928	22.532	20.240	26.882	22.705
	STD	4.816	4.695	5.061	5.102	2.940	3.301	2.477	2.575
	MeMSE	10.925	11.930	20.053	19.274	22.076	20.372	27.186	22.200
G_{03}	MMSE	9.860	9.009	20.967	13.476	24.865	28.661	31.786	22.328
	STD	2.489	2.368	11.280	8.018	5.609	6.625	11.498	6.430
	MeMSE	9.853	9.120	20.718	12.271	24.682	28.603	30.513	22.788
G_{04}	MMSE	1.880	1.897	13.865	9.899	8.992	8.661	7.046	5.447
	STD	0.858	0.628	13.207	9.565	9.356	3.043	1.551	1.262
	MeMSE	1.640	1.850	10.597	8.608	7.532	7.896	7.014	5.264
G_{05}	MMSE	14.176	9.949	12.223	10.973	30.885	27.021	76.569	30.437
	STD	22.319	2.326	4.675	4.619	4.580	3.463	74.695	11.670
	MeMSE	11.170	9.805	11.868	10.758	30.259	26.725	38.797	26.861
G_{06}	MMSE	8.075	4.293	13.921	10.962	14.694	14.209	22.808	11.619
	STD	4.490	1.505	6.335	4.843	5.692	3.910	3.876	4.801
	MeMSE	6.793	4.035	13.008	9.667	14.762	13.439	23.508	11.368
G_{07}	MMSE	1.646	2.796	11.990	10.269	8.399	8.445	9.723	5.874
	STD	0.467	1.998	10.455	8.178	2.487	5.379	2.319	2.387
	MeMSE	1.559	1.602	9.638	8.665	8.302	7.165	10.181	5.725
G_{08}	MMSE	4.349	3.127	10.089	8.551	9.631	10.997	12.876	6.273
	STD	2.801	0.811	4.438	4.813	3.533	4.450	4.764	2.156
	MeMSE	3.558	3.008	9.003	7.330	8.531	9.597	10.946	6.228
G_{09}	MMSE	48.714	42.892	64.620	55.749	107.383	100.321	151.314	113.071
	STD	14.280	10.414	22.746	15.413	15.787	19.702	51.546	30.981
	MeMSE	49.001	42.809	62.521	56.463	108.758	95.269	135.488	106.872
G_{10}	MMSE	0.342	0.261	1.917	2.103	0.588	2.030	0.363	0.268
	STD	0.340	0.193	0.663	1.509	0.331	1.485	0.281	0.126
	MeMSE	0.240	0.201	1.785	1.665	0.495	1.630	0.324	0.294
G_{11}	MMSE	31.779	28.132	35.446	35.060	31.535	33.065	36.877	31.835
	STD	2.264	2.234	2.940	3.180	1.662	2.616	2.167	2.033
	MeMSE	32.024	27.911	35.325	34.462	31.273	33.132	36.889	31.529
G_{12}	MMSE	0.997	1.750	5.878	6.004	2.947	5.694	2.572	1.686
	STD	0.166	0.390	1.280	1.171	0.586	1.370	0.635	0.346
	MeMSE	0.979	1.789	5.726	5.778	2.783	5.543	2.508	1.702

4. Thus, the matrix C defined has eight rows and columns, we pad this matrix to create an augmented comparison matrix $\bar{C}$ whose 9th row and the 9th column are defined as

$$\bar{C}_{i9} = \sum_{j=1}^{8} C_{ij}; \quad \bar{C}_{9j} = \sum_{i=1}^{8} C_{ij}; \quad \bar{C}_{99} = 0 \tag{23}$$

These entries are the row and column sum of the matrix. We define a activation function to be better if its row sum value is higher.

With these evaluation parameters and from Tables 2, 3, 4, 5 and 6, we may assert the following:

1. The sigmoidal activation functions (σ_{01} and σ_{02}) outperform all other activation functions for these tasks and shallow networks over the test data set.
2. In no case, the sigmoidal functions (σ_{01} and σ_{02}) are significantly worse than the non-sigmoidal activation function, over test data set.

Table 3 Comparison matrix on the basis of value of MMSE (test data set)

Activation	σ_{01}	σ_{02}	σ_{03}	σ_{04}	σ_{05}	σ_{06}	σ_{07}	σ_{08}	Row sum
σ_{01}	00	05	11	11	11	12	12	11	73
σ_{02}	07	00	12	12	12	12	12	11	78
σ_{03}	01	00	00	02	05	07	07	04	26
σ_{04}	01	00	10	00	07	06	08	06	38
σ_{05}	01	00	07	05	00	06	09	03	31
σ_{06}	00	00	05	06	06	00	09	04	30
σ_{07}	00	00	05	04	03	03	00	00	15
σ_{08}	01	01	08	06	09	08	12	00	45
Column sum	11	06	58	46	53	54	69	39	0

Table 4 Comparison matrix on the basis of value of MeMSE (test data set)

Activation	σ_{01}	σ_{02}	σ_{03}	σ_{04}	σ_{05}	σ_{06}	σ_{07}	σ_{08}	Row sum
σ_{01}	00	05	12	11	11	12	12	11	74
σ_{02}	07	00	12	12	12	12	12	11	78
σ_{03}	00	00	00	01	06	06	09	05	27
σ_{04}	01	00	11	00	07	06	09	06	40
σ_{05}	01	00	06	05	00	06	09	02	29
σ_{06}	00	00	06	06	06	00	09	04	31
σ_{07}	00	00	03	03	03	03	00	00	12
σ_{08}	01	01	07	06	10	08	12	00	45
Column sum	10	06	57	44	55	53	72	39	0

Table 5 Comparison matrix over the test data set using the Student's *t*-test

Activation	σ_{01}	σ_{02}	σ_{03}	σ_{04}	σ_{05}	σ_{06}	σ_{07}	σ_{08}	Row sum
σ_{01}	00	03	11	11	11	12	11	10	69
σ_{02}	05	00	12	11	12	12	12	10	74
σ_{03}	00	00	00	00	04	03	07	03	17
σ_{04}	00	00	06	00	06	05	08	04	29
σ_{05}	00	00	05	03	00	05	09	00	22
σ_{06}	00	00	04	02	04	00	08	04	22
σ_{07}	00	00	03	03	02	03	00	00	11
σ_{08}	00	00	07	06	08	08	12	00	41
Column sum	05	03	48	36	47	48	67	31	0

Table 6 Comparison matrix over the test data set using the Wilcoxon's rank-sum test

Activation	σ_{01}	σ_{02}	σ_{03}	σ_{04}	σ_{05}	σ_{06}	σ_{07}	σ_{08}	Row sum
σ_{01}	00	02	11	11	11	12	12	10	69
σ_{02}	06	00	12	11	12	12	12	11	76
σ_{03}	00	00	00	00	05	03	08	03	19
σ_{04}	00	00	05	00	07	06	09	04	31
σ_{05}	00	00	04	03	00	05	09	00	21
σ_{06}	00	00	04	02	04	00	08	03	21
σ_{07}	00	00	03	03	02	03	00	00	11
σ_{08}	00	00	07	06	09	08	12	00	42
Column sum	06	02	46	36	50	49	70	31	0

3. Among the non-sigmoidal activation functions, the function σ_{08} performs the best for generalization (using the results for both MMSE and MeMSE).
4. Out of the functions σ_{01} and σ_{02}, the function σ_{02} may be slightly preferred, but for specific tasks, both the activation functions should be evaluated.

6 Conclusions

In this paper, two sigmoidal functions (σ_{01} and σ_{02}) and six non-sigmoidal function (σ_{03}–σ_{08}) usage as activation functions for FFANNs was evaluated over 12 function approximation tasks. On the basis of the experiments conducted, we may conclude that the sigmoidal activations outperformed the non-sigmoidal functions. Moreover, among the non-sigmoidal functions, σ_{08} performs the best. But, the evaluation has to be conducted over more number of problems before a final conclusion is made.

References

1. Pinkus A (1999) Approximation theory of the MLP model in neural networks. Acta Numerica 8:143–195
2. Duch W, Jankowski N (1999) Survey of neural network transfer functions. Neural Comput Surv 2:163–212. Available http://www.icsi.berkeley.edu/vjagota/NCS
3. Haykin S (1999) Neural networks: a comprehensive foundation. Prentice Hall Inc, New Jersey
4. Hahnloser RHR, Sarpeshkar R, Mahowald MA, Douglas RJ, Seung HS (2000) Digital selection and analogue amplification coexist in a cortex-inspired silicon circuit. Nature 405:947–951
5. Glorot X, Bordes A, Bengio Y (2011) Deep sparse rectifier neural networks. In: Gordon GJ, Dunson DB, Dudk M (eds) AISTATS, ser. JMLR Proceedings, vol 15. JMLR org, pp 315–323
6. Maas A, Hannun A, Ng A (2013) Rectifier nonlinearities improve neural network acoustic models. In: International Conference on Machine Learning (ICML) workshop on deep learning for audio, speech, and language processing, Atlanta, Georgia
7. Clevert D-A, Unterthiner T, Hochreiter S (2015) Fast and accurate deep network learning by Exponential Linear Units (ELUs). arXiv e-prints arXiv:1511.07289
8. Klambauer G, Unterthiner T, Mayr A, Hochreiter S (2017) Self-normalizing neural networks. In: Proceedings of the 31st international conference on neural information processing systems, ser. NIPS'17. USA:Curran Associates Inc., pp 972–981
9. Elfwing S, Uchibe E, Doya K (2018) Sigmoid-weighted linear units for neural network function approximation in reinforcement learning. Neural Networks 107:3–11
10. Sodhi SS, Chandra P (2014) Bi-modal derivative activation function for sigmoidal feedforward networks. Neurocomputing 143:182–196
11. Mishra A, Chandra P, Ghose U, Sodhi SS (2017) Bi-modal derivative adaptive activation function sigmoidal feedforward artificial neural networks. Appl Soft Comput 61:983–994
12. Breiman L (1991) The PI method for estimating multivariate functions from noisy data. Technometrics 3(2):125–160
13. Cherkassky V, Gehring D, Mulier F (1996) Comparison of adaptive methods for function estimation from samples. IEEE Trans Neural Networks 7(4):969–984
14. Cherkassky V, Mulier F (1998) Learning from data—concepts, theory and methods. John Wiley, New York
15. Maechler M, Martin D, Schimert J, Csoppenszky M, Hwang J (1990) Projection pursuit learning networks for regression. In: Proceedings of the 2nd international IEEE conference on tools for artificial intelligence, pp 350–358
16. Cherkassky V, Lee Y, Lari-Najafi H (1991) Self-organizing network for regression: efficient implementation and comparative evaluation, In: Proceedings of the IJCNN-91, vol 1, pp 79–84
17. Friedman JH (1991) Multivariate adaptive regression splines. Ann Stat 19:1–141
18. Riedmiller M, Braum H (1993) A direct adaptive method for faster backpropagation learning: the RPROP algorithm. In: Proceedings of the IJCNN 1993, pp 586–591
19. Riedmiller M (1994) Advanced supervised learning in multi-layer perceptrons—from backpropagation to adaptive learning algorithms. Comput Stand Interfaces 16(5):265–278
20. Igel C, Hüsken M (2003) Empirical evaluation of the improved RPROP learning algorithms. Neurocomputing 50:105–123
21. Huber PJ (1981) Robust statistics. John Wiley & Sons, New York
22. Hettmansperger T, McKean J (1998) Robust nonparametric statistical methods, ser. Kendall's library of statistics: An Arnold Publication no 5, Arnold

A Neoteric Approach for Modeling and Conversion of RDBMS Using UML

Nida Iftekhar, Misbahur Rahman Warsi, Sherin Zafar, Samia Khan, and Siddhartha Sankar Biswas

Abstract RDBMS enables the software systems to manipulate data from a database. Among all the database systems that exist, relational database systems take the major share of the market. Mostly, ER diagrams are used for modeling of RDBMS in design phase. In software applications, application layer is the object-oriented programming language, and RDBMS acts as the backend store. Mostly, UML is being used to model the application layer and ER diagram to model the RDBMS. Modeling of these layers separately creates inconsistency for the developers. Modeling these layers using a single notation, i.e., UML will bridge the gap. Modeling RDBMS using UML to incorporate the database schema as well as the operations to be performed on it. The entire RDBMS information could be scattered across the UML model. Enhancement in the modeling practice will be there using UML as standard for modeling of both layers and proposing new constructs at modeling level for RDBMS. The proposed method identifies different aspects in RDBMS which needs to be captured using UML. The representation of entire relational database system, its components and relationship between them is done using different diagrams in UML. The relationship between entities (tables in RDBMS) is identified based upon some constraints. The proposed method identifies the weaknesses in RDBMS modeling that has been done

N. Iftekhar (✉) · S. Zafar · S. Khan · S. S. Biswas
Department of CSE, SEST, Jamia Hamdard, Hamdard Nagar, New Delhi 110062, India
e-mail: nida.iftekhar@jamiahamdard.ac.in

S. Zafar
e-mail: zafarsherin@gmail.com

S. Khan
e-mail: samia.khan20@gmail.com

S. S. Biswas
e-mail: ssbiswas1984@gmail.com

M. R. Warsi
Aligarh Muslim University, Aligarh, India
e-mail: warsimr@yahoo.com

A. Abraham et al. (eds.), *Proceedings of 3rd International Conference on Computing Informatics and Networks*, Lecture Notes in Networks and Systems 167,
https://doi.org/10.1007/978-981-15-9712-1_25

till now and extracts ways to overcome those weaknesses and propose new constructs at modeling level for RDBMS.

Keywords Unified modeling language (UML) · Relational database management system (RDBMS) · Object oriented · Backend store

1 Introduction

Modeling RDBMS using UML will enhance the understandability and reusability of the database rather than modeling it using ER diagram. The proposed modeling scheme using UML will bring RDBMS modeling close to the object-oriented technology. Modeling both the layers with same technology will bring them closer and increase the easiness for the developer to implement the system.

The proposed method generates a UML profile for modeling RDBMS. Work has been done previously to generate views of the database query operations. Modeling these views and each of the query operations in a system has been done. If modeling of RDBMS is to done using previous approaches, each of the query in the database system needs to be modeled. Rather than modeling each query and increasing the overhead, modeling should be done with an approach to identify the relationships between the relational database tables. This will greatly enhance the understandability of the system for the developer. This paper proposes an approach to identify all the possible relationships in a relational database using the powerful tools provided in UML for identification of the strength and the type of the relationship.

This paper proposes a scheme for modeling RDBMS using unified modeling language (UML). It identifies different aspects in RDBMS which needs to be captured using UML. The representation of entire relational database system, its components and relationship between them is done using different diagrams in UML. The relationship between entities (tables in RDBMS) is identified based upon some constraints. This method proposes new constructs at modeling level for RDBMS. It focuses on structural as well as operational modeling. It will integrate modeling the data schema as well as the operations to be performed on it. It will capture the structure of data and the relationship between data elements. Applying UML to model RDBMS will bridge the gap between the application layer and the backend store.

2 Literature Review

Several database modeling schemes using entity relationship (ER) diagrams and unified modeling language (UML) have been proposed by various researchers to increase the understandability, reusability and maintenance of the relational database systems. It has been defined by the object management group (OMG) and has already become the de facto standard for designing object-oriented software applications [1].

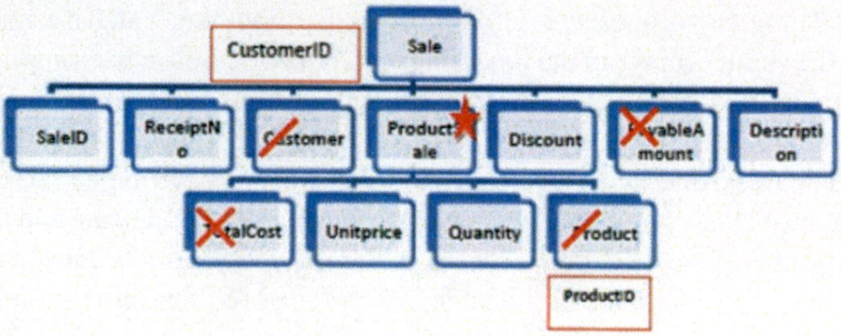

Fig. 1 Hierarchical structure of table sale

Most of the work related to modeling of relational database management system focus on either the structure or the operations on the RDBMS.

Shuxin Yin et al. [2] proposed a technique to model database operations using UML. They have proposed a technique to model the atomic operations in SQL DML (data manipulation language), i.e., INSERT, UPDATE, DELETE and SELECT. They have worked toward providing a unified view of the entire system to the users and developers. They have given an approach to model database operations so that it becomes easier for the developer to understand and implement the system in lesser amount of time.

Attaran et al. [3] proposed a technique to directly derive the database and the relational tables in it out of the systems' classes. It aims at reduction of phases in the database design. They have represented the interface in a tree outward appearance. Their method follows a hierarchical format for representation and gives the resemblance of a decision tree. They have used different shapes for representation of the tables in the database, its properties and attributes, need for foreign key, dependent attribute, etc., represented in Fig. 1.

Sinha et al. [4] analyzed the entity relationship schema (ERS) and proposed an augmented schema called object relationship schema (ORS) having additional constructs to the entity relationship schema. ORS can be used to measure the database development effort in an objective and quantitative manner. It gives the privilege to apply required resources toward development, testing and maintenance of the RDBMS. The ORS clearly represents the relationship between the objects of the system.

Al-Fedaghi [5] worked toward the incompatibility of the object-oriented paradigm and relational database systems. There is an inherent mismatch between these two models due to which the developers face problem. They have proposed a technique to integrate the object-oriented and relational database model into a unified conceptual model called the Flowthing Model. It enhances the understanding of the required mapping and provides a high-level framework for development of a unified model for object-oriented and relational database technologies. They have proposed a diagrammatic representation rather than the tree-based object-oriented graphs.

Lo and Hung [6] proposed a UML profile for modeling database retrieval to overcome the shortcomings in the modeling techniques that have been applied previously to model relational database management systems. They have modeled the query operations so as to generate the code automatically. They have defined a set of stereotyped classes, one stereotyped attribute, and a set of stereotyped relationships. The stereotyped classes provide the view of the database table schema and the views of the query operations performed on the tables. Stereotyped attributes are for the representation of the columns of the table. Stereotyped relationships are being used to denote the calculation of set operations between tuples.

Lee [7] have developed a methodology using UML as a standard notation for computer applications. Nikose et al. [8] have generated operation specification by utilizing class diagrams. Albert et al. [9] have compared ER and UML class diagrams representing comprehension tasks on data models. De Lucia [10] have utilized Unified Modeling Language to model the operations in relational databases. Song et al. [11] have proposed an approach to derive a data model from object model. Fakhar and Muhammad [12] have modeled relational database views using OCL (version 2.0). Akhtar et al. [13] have generated profiles of academicians using data from different websites. Iftekhar et al. [14] have reverse engineered relational database to generate its corresponding UML class diagram. Demuth and Hussmann [15] have utilized the UML/OCL constraints for relational database design.

3 Problem Formulation

In the previous researches, they are separately modeling each query. Separately modeling each query and then using it in code is time consuming. Modeling should be done at an abstract level. Above methods focus either on the operations modeling or the representation of the structure of the RDBMS tables. None of them considers the relationship that exists between the tables based upon the operations performed on them.

Any of the above method is not considering the logical grouping that may exist between the tables. They are not making any use of the powerful tools like association, aggregation and generalization provided by UML which represent the kind and strength of relationship that may exist between the relational database tables. Moreover, in the above-mentioned approach by Lo and Hung [6], modeling each query will increase the overhead. Modeling should be done at an abstract level to increase the understandability rather than increasing the overhead of modeling each query and then implementing it in the code.

View modeling needs to be given attention as those views which are used again and again should be modeled and then used further wherever required. User operations modeling should also be considered. The operations performed by the user should also be taken into consideration. The user operations affect the relational database tables. So, their effect must be captured using the user operations modeling.

None of the above methods have identified the relationship between relational database tables based upon logical relations. Logical grouping of tables should be taken into account. There must be many tables in the relational database tables which are logically related to each other. It must be taken into account so as to capture a strong relationship between the tables.

Operations to be performed on groups of tables and their issues have not been considered in any of the above methods. The tables which are not logically related must have some common operations that cannot be modeled as the operations of their aggregate class. These kinds of issues and their corresponding solutions must be considered.

4 UML Profile for RDBMS

4.1 Entity Representation

Entities in the proposed work are represented using the notation similar to that of the class in UML, i.e., a rectangular box consisting of three compartments. Description of the content of these three compartments is as given below:

- First compartment in the class includes the name and the stereotype of the class. Stereotype is the variation on an existing modeling element with modified intent which represents the further classification of the elements in UML. The class is itself a representation of a type of entity, so it has been classified with a stereotype ≪Entity≫.
  (≪……≫ is the syntax to specify the stereotype).
- The second compartment represents the attributes of the entity with their types and visibility. It may also contain additional information about the attributes of class like initialization value and stereotype.
- The third compartment represents the behavior of the entity. Every entity has a behavior associated with it. The operations applied on the entity which represent the behavior of entity according to different constraints are added to the third compartment. Figure 2 represents ER model of entity person and its representation in UML.

4.2 Attributes in UML

The attributes of a class in UML represent the status and information about an entity. The columns in a table are represented as attributes of the entity. There must be visibility defined for each of the attributes. The name, visibility and type of each of the attributes are specified in the class specification. The attributes are placed in the second compartment of the class declaration in UML. The visibility of the attribute

Fig. 2 Representation of an entity using ER diagram and UML

can be private, public or protected. The type of the attributes is the analysis type. It may change during the design phase. The name of the attribute is separated from type by a colon.

4.3 Key

Keys in RDBMS are used to access the table. Primary keys uniquely identify row in a table, while other keys access the data in other tables which are related to the table. The primary key is unique which is either declared by the user as primary key or often generated by the database to ease the updates of data and uniquely identify tuples in a table. Foreign key is always derived as a relationship of the base table to other tables. Keys are implementation of the key constraints, which specify content of the key (which column in base table builds the key), as well as physical implementation of the key. For identification of key representing columns in table, they are tagged with stereotypes in UML class. For representation of foreign key constraint, the stereotype ≪FK≫ is used and for foreign key ≪FK≫ is used.

4.4 Semantics of Association

In RDBMS, there are several operations which affect multiple tables at a time. These operations either retrieve data or modify data present in the tables. So, there may exist a relationship between tables which are affected by the same operation at a time. Relationship may exist between two tables based upon the common operations applied to them. The tables which are related through their common operations are linked to each other via association class which has the common operations of both the tables. The association class does not contain any attributes but has the common operations of both the tables which are linked via association class. In Fig. 3, account and customer class are associated through the association class named holds which

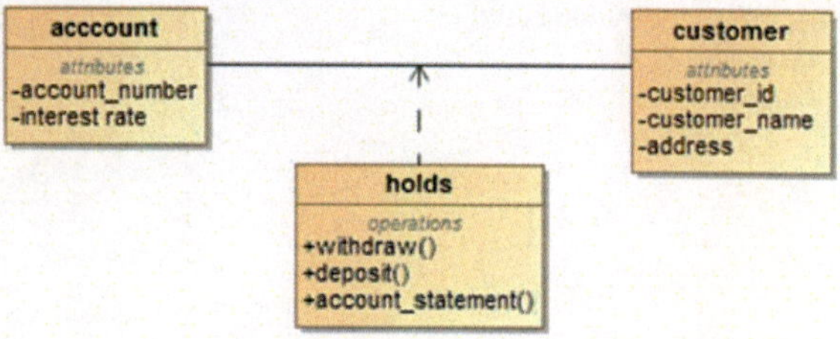

Fig. 3 Relationship between account and customer class with association class holds

has the operations withdraw, deposit and account_statement. In all these operations, both the account class and the customer class are affected.

4.5 Aggregation Based upon Logical Grouping

The classes which are logically related are grouped together by aggregation to constitute their aggregate class. This way the classes which are belonging to a similar area are kept as the constituent classes of the aggregate class. The operations which are common to the constituent classes are modeled as the operations of the aggregate class, i.e., the operations which affect the constituent classes of the aggregate class are kept under the third compartment of the aggregate class. Let us consider an example of a hospital database. In hospital management system, the doctors, nurses and the staff belong to a particular group, i.e., hospital. So, they can be modeled as logically grouped represented by aggregation. Figure 4 is the representation of logical grouping in hospital management system.

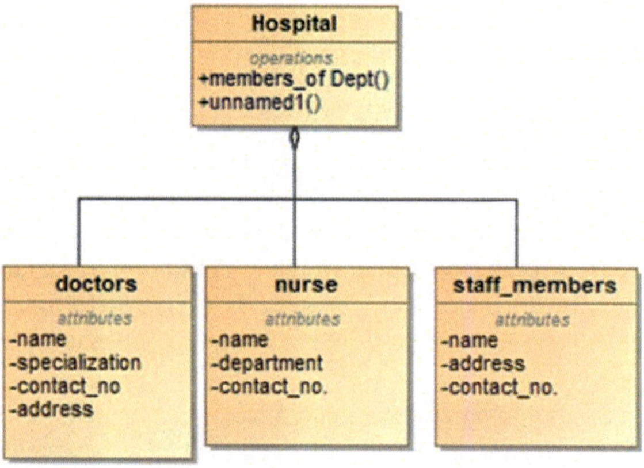

Fig. 4 Logical grouping of classes using aggregation

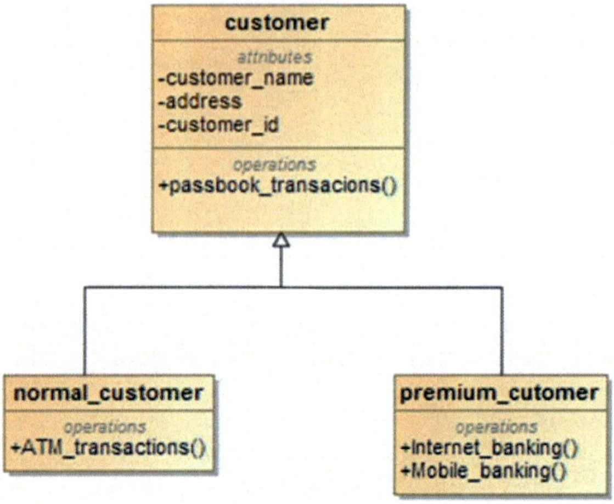

Fig. 5 Generalization based upon attributes and operations

4.6 Generalization Based upon Attributes and Operations

If the customer table in the bank database is considered, some of the attributes are applicable to all the customers like customer_id, customer_name, account_number, address, etc., but there are some differences between the customers. Some will do banking through passbook and ATM only, but some of the customers will opt for advanced services like Internet banking and mobile banking. So, there can be many categories of the customers depending upon the services they opt. These services can be represented by the operations of the class which is representing the entity corresponding to a table in the bank database. Figure 5 is the representation of such a relationship.

4.7 Reverse Generalization

Whenever generalization is done for classes in UML, super class is created first, and then, based upon the additional attributes, its subclasses are created. We always proceed from top to bottom while creating a hierarchy. The superclass having the common attributes of all the subclasses is created first. After that, subclasses having the additional attributes are created. They inherit the common attributes from their superclass.

As the name reverse generalization indicates, generalization is done in it in a reverse manner. In this paper, the proposed technique is to first create a subclass and then its superclass. Based upon the operations applied on the subclass, the operations result in a class with lesser attributes as compared to the class on which operations are applied. In RDBMS, ALTER with DROP command is an operation which gives the resultant table with lesser attributes in comparison with the base table, and the

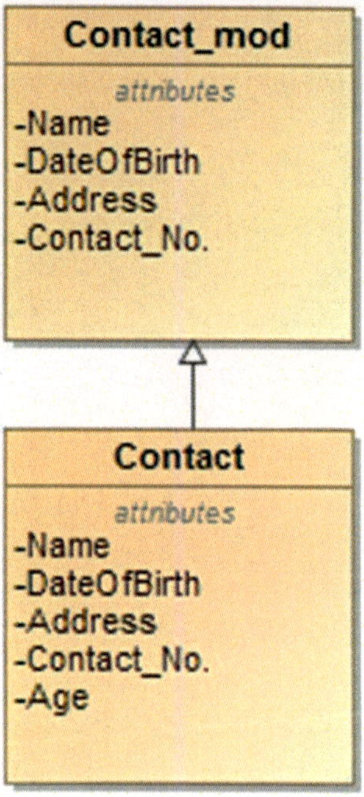

Fig. 6 Reverse generalization applied to class contact

attributes of the resultant are common with the base table. This paper proposes a technique to model this kind of operation in RDBMS using reverse generalization. The following command and its modeling in UML using our technique of reverse generalization are depicted in Fig. 6.

ALTER table Contact DROP Age.

4.8 Aggregation for UNION, EXCEPT aND INTERSECT

The UNION operator in SQL returns the combined results from the base tables on which it is applied and removes the duplicates. The derived table or unit is the aggregate of the base tables on which the UNION operator is applied. The proposed approach will model this kind of SQL statements using aggregation. The base tables act as the parts of the derived table. For example: If the following SQL query is applied to two tables namely customer and supplier:

SELECT ContactName, Address, City FROM Customer
UNION
SELECT ContactName, Address, City FROM Supplier ORDER BY Contact-Name.

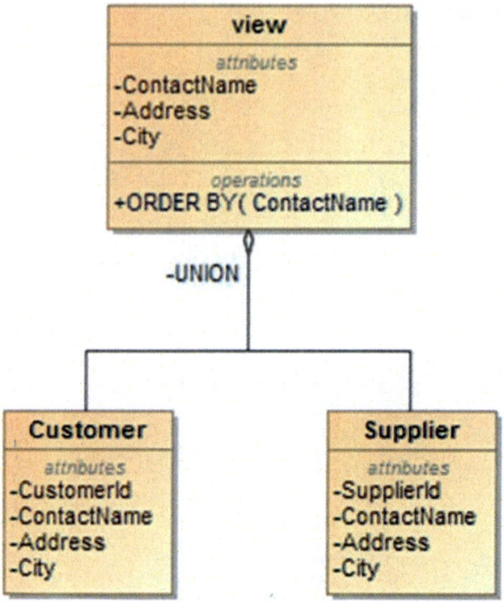

Fig. 7 Modeling of operator UNION using proposed approach

Figure 7 shows the modeling of UNION operator applied to customer and supplier table.

When EXCEPT clause is used between two SELECT statements, it returns the tuples that are outcomes of the first SELECT statements and removes the tuples which are the outcome of the second SELECT statement. The following SQL query is being considered for modeling which uses EXCEPT clause:

SELECT id, name, amount, date FROM Customer
EXCEPT
SELECT id, name, amount, date FROM Supplier.

Figure 8 depicts the modeling of EXCEPT clause applied to customer and supplier table.

In the similar manner as UNION and EXCEPT operator, INTERSECT can be modeled.

4.9 *Aggregation for JOIN—NATURAL JOIN, OUTER JOIN aND INNER JOIN*

JOIN operation when applied to two tables gives the resultant table with the Cartesian product of the base tables, having attributes of both the tables. The JOIN in which comparison is done between attributes using equality condition is an EQUIJOIN. When a condition is specified on the JOIN such that there is no repetition of attributes in the resultant table, it is the NATURAL JOIN.

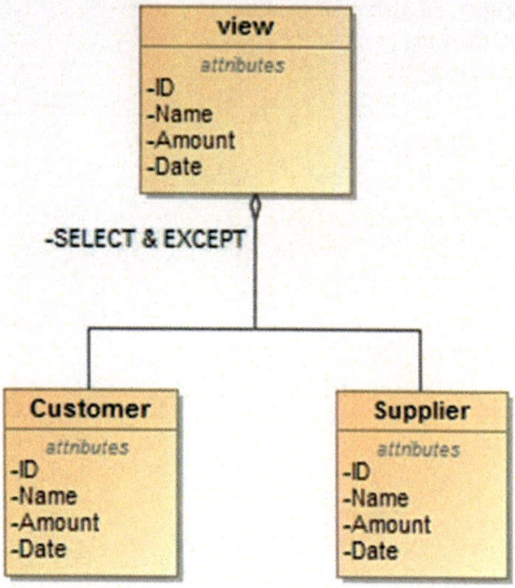

Fig. 8 Modeling of query having EXCEPT clause using aggregation

JOINS can be categorized to INNER JOIN and OUTER JOIN. The join operation in which the tuples that do not satisfy the join condition are removed from the resultant table is INNER JOIN. The JOIN in which all the tuples of either the first table or the second table are kept in the resultant table whether they satisfy the join condition or not is OUTER JOIN. If they do not satisfy the join condition, their other attributes are marked as NULL in resultant table. The selection of which table's attributes must be kept in result is based upon whether it is LEFT OUTER JOIN or RIGHT OUTER JOIN. If it is LEFT OUTER JOIN, all the attributes of first table will be kept, and if it is RIGHT OUTER JOIN, all attributes of second table will be kept.

The JOIN will be modeled using aggregation operator applied to the classes corresponding to the base tables. The resultant class will be aggregate of both the base classes having attributes of both classes as its attributes.

Considering the following query with INNER JOIN:

SELECT Order.OrderID, Customer.CustomerName, Order.OrderDate
FROM Order
INNER JOIN Customer
ON Order.CustomerId = Customer.CustomerId.

Above query modeled as shown in Fig. 9.

Considering the following query, the proposed approach will model it in following way:

SELECT ID, Name, Amount, Date FROM Customer
LEFT JOIN ORDER ON Customer.ID = Order.OID
EXCEPT
SELECT ID, Name, Amount, Date FROM Customer
RIGHT JOIN ORDER ON Customer.ID = Order.OID.

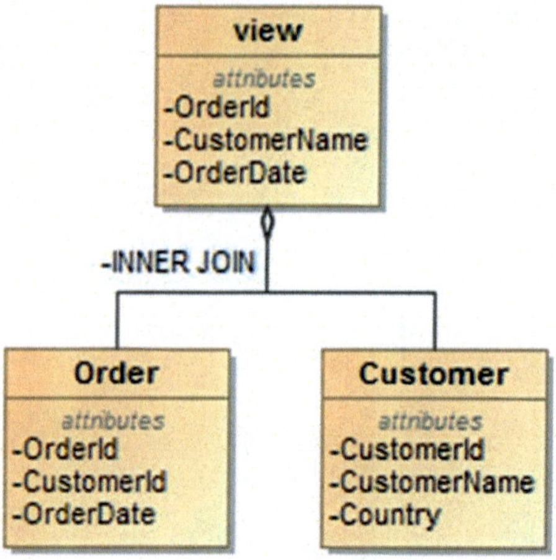

Fig. 9 Modeling of INNER JOIN using proposed approach

Modeling this query with multiple JOIN conditions is depicted in Fig. 10.

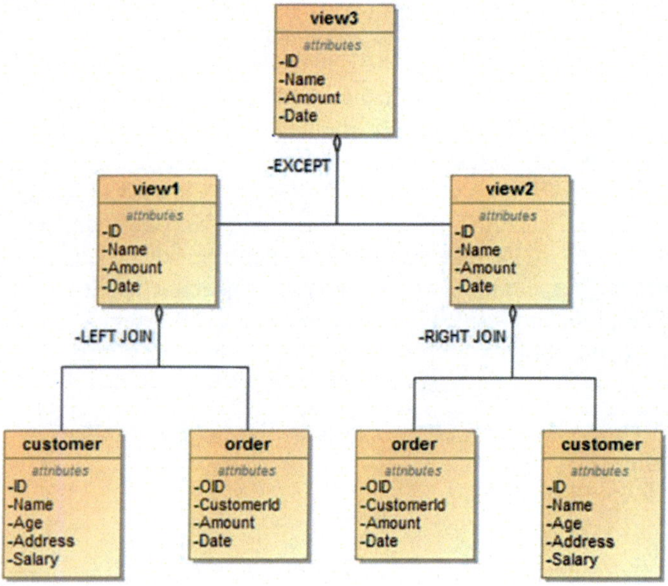

Fig. 10 Modeling of EXCEPT clause with RIGHT and LEFT JOIN using proposed approach

4.10 Semantics of Generalization

The generalization is a relationship in which the subclass extends the behavior of superclass with additional attributes or the superclass generalizes the behavior and properties of its subclasses. Now, the issue is how to apply generalization to the tables in RDBMS. Generalization to tables is being applied based upon the following two conditions:

i. The subclass must contain additional attributes as compared to the superclass
ii. The subclass must have additional operations in comparison with the superclass.

Explaining the first scenario using the table of an account in a bank database. It may have attributes namely account_number, amount, interest rate, minimum balance, etc. Account can be of type savings_account and current_account. Both these accounts have additional properties. Current_account has overdraft_amount attribute and different minimum balance limit. Savings_account has interest rate attribute and some other minimum balance limit. Classes corresponding to the tables savings_account and current_account in RDBMS can be generalized to class corresponding to account. The superclass account will be modeled as abstract because no table corresponding to it will exist in the bank database. The tables savings_account and current_account will be the subclasses having all their attributes. It will indicate that the superclass account is the generalization of its subclasses savings_account and current_account, and corresponding to the subclasses, there exists tables in the database. Figure 11 is the representation of account class.

In the second scenario, the subclass contains the additional operations as compared to the superclass. Some of the operations of the subclass are unique and are not applicable to the superclass and some inherited from the superclass. If the customer table in the bank database is considered, some of the attributes are applicable to all the customers like customer_id, customer_name, account_number, address, etc., but there some differences between the customers. Some will do banking through passbook and ATM only, but some of the customers will opt for advanced services like Internet banking and mobile banking. So, there can be many categories of the

Fig. 11 Generalization based upon attributes applied to account class

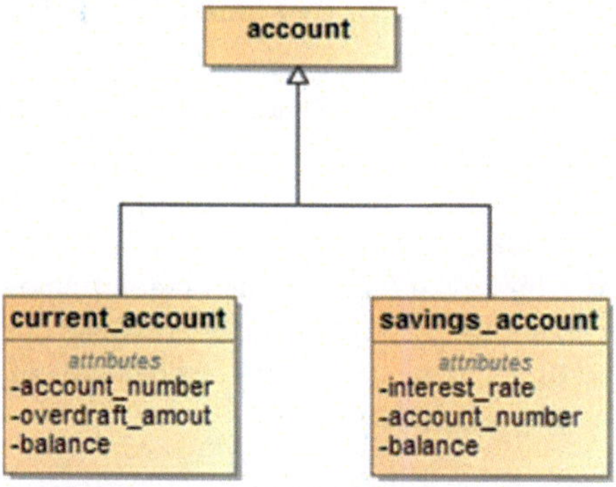

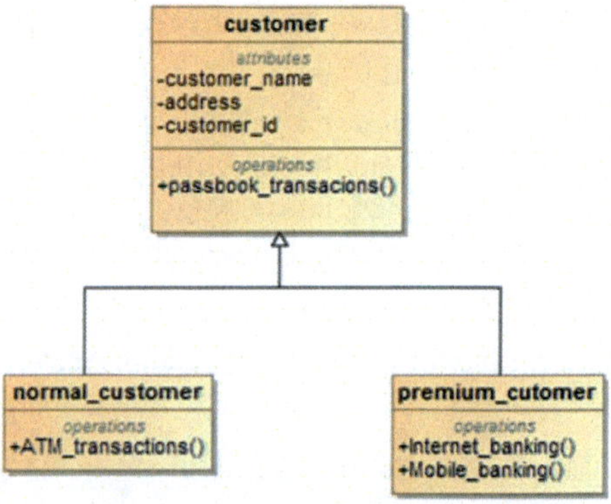

Fig. 12 Generalization based upon operations applied to customer class

customers depending upon the services they opt. These services can be represented by the operations of the class which is representing the entity corresponding to a table in the bank database. Figure 12 is the representation of such a relationship.

4.11 Generalization for ALTER Command Using ADD

The attributes and the operations applied on them can be changed by using the ALTER command. Either attributes can be to the base table or some attributes can be removed from the base table. To model the base table and the resultant table after applying ALTER command, generalization will be used. If attributes are added to the base table then, the base table will be considered as superclass, and the resultant table after addition of attributes will be considered as the subclass. For example: If the SQL statement using ALTER command is considered on table Person initially having attributes P_id, Name, Address: ALTER TABLE Person ADD DOB. Figure 13 represents the modeling of the ALTER command.

4.12 Modeling of Operators in Queries Using UML

Whenever a view is generated in RDBMS, several constraints are applied to the SQL query using the AND and OR operators as per the desired results. AND operator in SQL behaves similar to the logical AND in which all the conditions must be true. This will be modeled as the operation which is applied to the class representing the table in the RDBMS, and the attributes which are to be compared using AND operator are passed as parameters to the operation. If the following query on a table PROJECT is modeled with attributes Pname, Pnum, Dnum, Fname and Lname. It will be modeled as in Fig. 14.

SELECT * FROM PROJECT WHERE Dnum = 20 AND Lname = 'Smith';

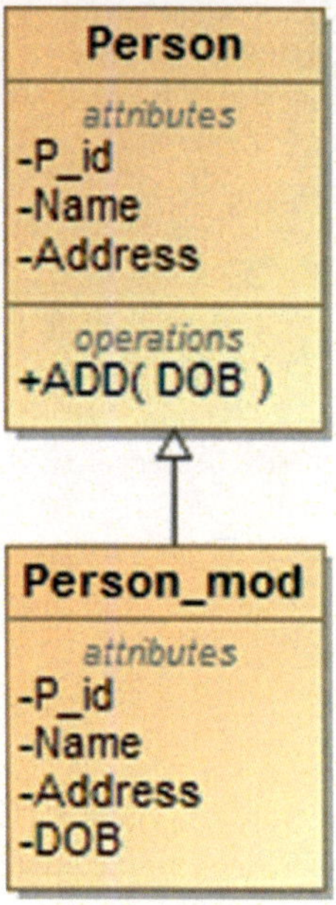

Fig. 13 Modeling of ALTER with ADD using generalization

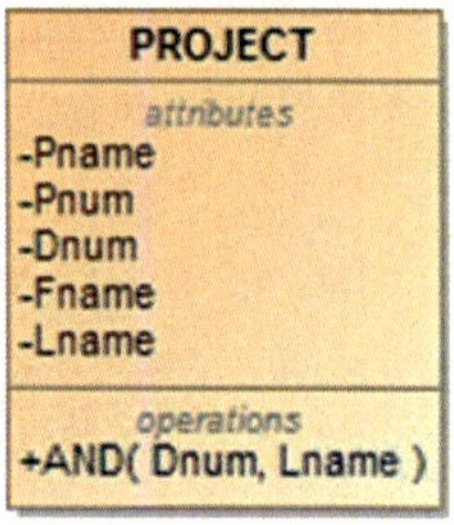

Fig. 14 AND operator applied to table PROJECT

Similarly, when OR operator is used in SQL query, the results are returned when either of the conditions with OR operator are true. OR operator is modeled in the same manner as AND.

In SQL, ORDER BY clause is used to sort the results. It provides the feature of sorting in ascending order by default and in descending order if the keyword DESC mentioned with it. The proposed approach models ORDER BY clause as an operation of the class which contains the attributes of the base table called in the SELECT clause as the attributes of the class representing the view and ORDER BY clause as the operation of the view class with the sorting attribute as its parameter. Modeling the following SQL statement is represented in Fig. 15:

SELECT Pnumber, Pname, Plocation FROM PROJECT
ORDER BY Pname.

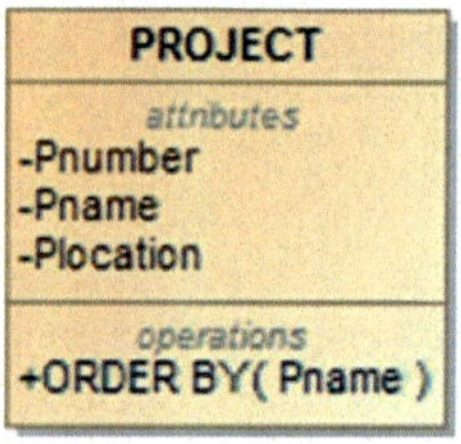

Fig. 15 Modeling of ORDER BY operator applied to table PROJECT

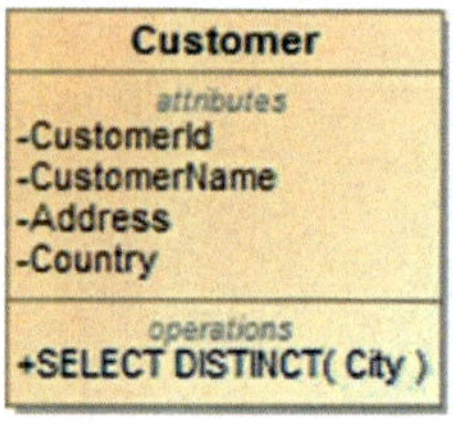

Fig. 16 Modeling of DISTINCT operator

Similarly, GROUP BY clause in SQL is used to group the results obtained based upon one or more columns. GROUP BY clause will be modeled in the similar manner as ORDER BY explained above with GROUP BY replacing ORDER BY in query and class as well.

DISTINCT operator is used when one wants to retrieve unique, i.e., non-repeating results in the result set. The proposed approach will model it as an operation of the class which is representing the result obtained by the SELECT statement on the base table, and the retrieved columns of the table representing the attributes of the class. It will be modeled in the following way:

SELECT DISTINCT City FROM Customer.

Figure 16 depicts the modeling of a query having DISTINCT operator.

4.13 Interface Modeling for Classes not Logically Grouped

The user interacts with the system through interfaces. The system provides various services to the user, and the user interacts with the system in different ways performing the operations according to the needs. Modeling these operations will provide a better view of the functionalities to be implemented in the system to the developer. It will increase the easiness of the developer to understand the system in a better way.

Till now, the operations applied to the tables have been modeled as the operations of the classes corresponding to the tables. If an operation is applied to multiple tables, it cannot be modeled as an operation of the tables because execution of that operation once will affect multiple tables. This is a major issue with operation modeling till now. The proposed approach will model it with use case diagram. The operations will correspond to the use cases in the system and the user to actors. The tables affected during the execution of one use case will be modeled using the sequence diagram. In the sequence diagram, the tables will correspond to classes.

5 Implementation and Results

In our implementation, a relational database management system is modeled using UML with the approach that has been proposed in this paper. Then, using the model for development of a software system using RDBMS as a backend store. For this purpose, MagicDraw 18.1 (Tool to develop models in UML) has been used to create a UML model of the relational database management system. For the development of a software system, java servlets to create a web application which provides interface to the user for various operations to be performed. For the storage of data, MySQL is being used which acts as a backend store for the Web application. Figure 17 represents the flow of data:

First of all, the proposed modeling scheme is being used for modeling the RDBMS using MagicDraw 18.1 tool. The modeling constructs are being applied that have been proposed in this paper. This approach models the RDBMS using the approaches that have been proposed in this paper namely aggregation by logical grouping, generalization based upon attributes, generalization based upon operations and reverse generalization. Also, the tables are modeled which are not logically related but have some common operations applied on them using interface modeling, i.e., through use case diagrams and sequence diagrams for representation of each use case so as to represent the tables affected in a particular use case. To verify the effectiveness of proposed modeling scheme, a software system has been implemented using the developed model in UML. A Web application has been created based upon the modeling scheme using Java Servlets in NetBeans IDE 7.3.1 and using MySQL which acts as a backend store for storing the corresponding data.

Bank management system is taken as a case study to illustrate the difference between the previous approaches and our proposed approach for relational database management system modeling. As the previous approaches were analyzed for RDBMS modeling, it was discovered that more work needs to be done to bring

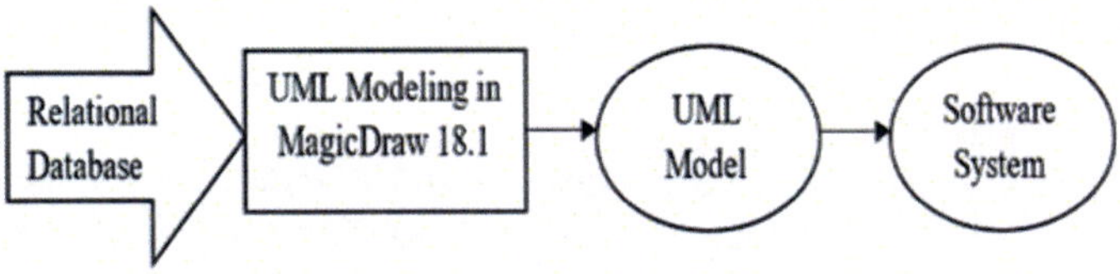

Fig. 17 Workflow to model and develop software system using proposed approach

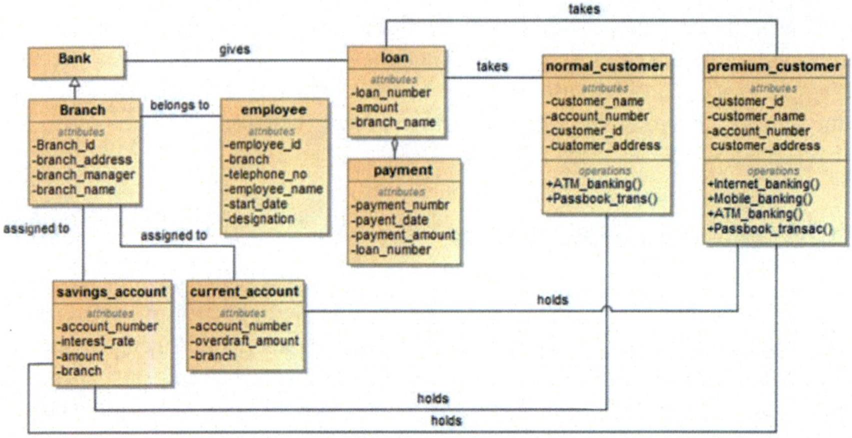

Fig. 18 Generated UML model for bank management system using previous approaches

the application layer modeling closer to the RDBMS modeling. Figure 18 shows the UML model for bank management system using previous approaches.

A number of limitations were found in previous approaches if relational database management system is modeled using them. Following are the limitations in previous approaches that have been managed to overcome in proposed approach:

- No criteria defined to model the complex operations. In Fig. 19, account and the employees that belong to the same branch may have some common operations. There is no way out in previous approach to manage this issue.
- If the classes are related by some common operations that affect them simultaneously, there is no method to model those operations. In Fig. 19 when customer performs transactions, both account class and customer class are affected. These transactions need to be modeled which is not there is previous approach.
  Repetition of operations in classes. In Fig. 19, normal_customer and premium_customer have some operations common which are modeled two times. There is no way out to avoid this repetition in previous approach.
- Multiple associations of same kind. In Fig. 19, we can see holds association is there three times. If both types of customers are grouped using some criteria, it will lead to avoidance of repetition of associations.

Bank management system is being modeled using the modeling scheme that has been proposed in this paper using UML in MagicDraw 18.1. There are eight tables in bank database schema corresponding to each of which a class exists in the model. Other classes in the model are created either to represent aggregation by logical grouping or generalization and are marked as abstract. Tables corresponding to the abstract classes does not exist. Figure 19 shows the UML model for bank management system using the proposed approach.

The proposed scheme also manages the issue when the tables are not logically related, but there exists some operations which are applied to multiple tables. To

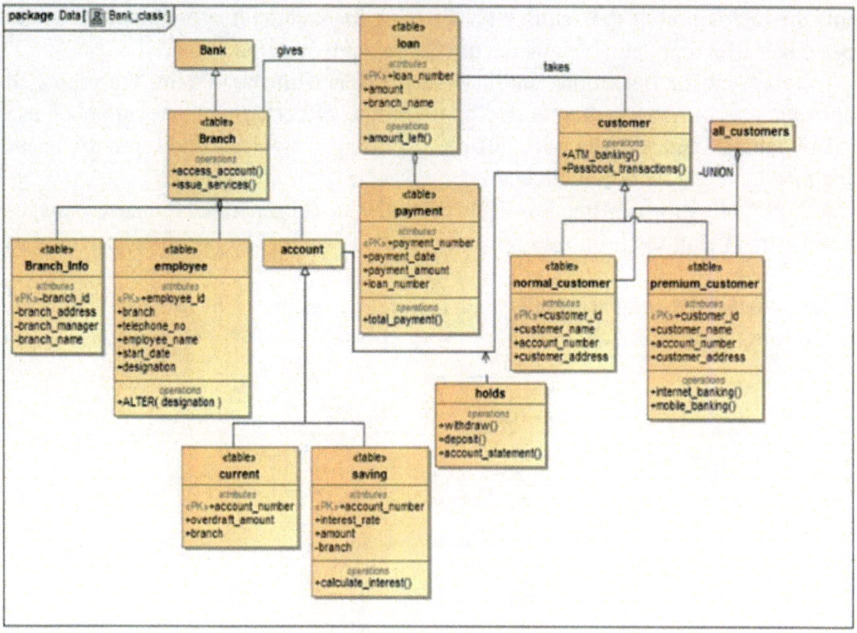

Fig. 19 UML model for bank management system using proposed approach

show the effect of those operations, use case diagrams have been used in the proposed approach. Whenever a use case is executed, it effects multiple tables. To represent the changes occurring in tables, use case diagrams have been used. Figure 20 repre-

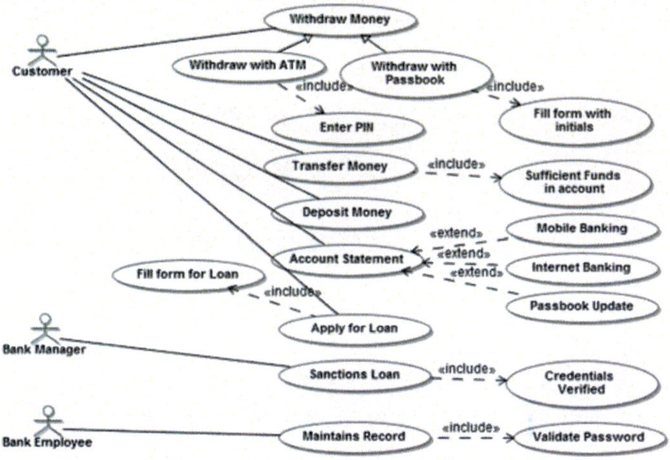

Fig. 20 Use case diagram for tables not logically related

sents the interaction of the actors with the bank management system as well as the operations affecting the tables which are not logically related.

To represent the operations and user interaction with the system, the proposed approach uses sequence diagrams. To model the effects of the execution of use case on tables, sequence diagrams are shown in Fig. 21 corresponding to use cases sanctions loan and withdraw money.

A Web application is being developed using the model generated for bank management system using the proposed scheme. Figure 22 is for the interface provided to users.

Following is the comparison between features incorporated by previous approach and our proposed approach.

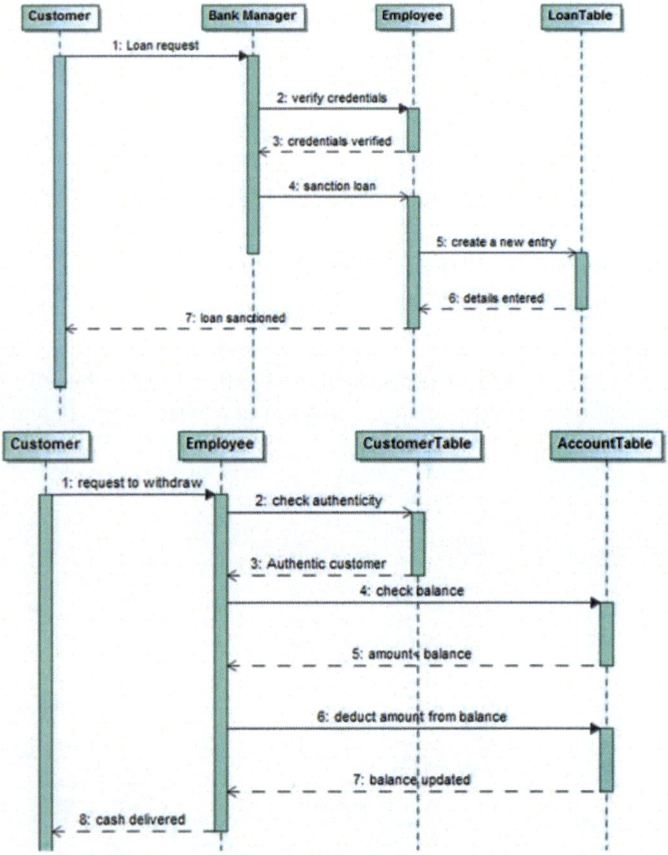

Fig. 21 Sequence diagram for use case sanctions loan and withdraw money

Fig. 22 Banking application developed using the proposed approach

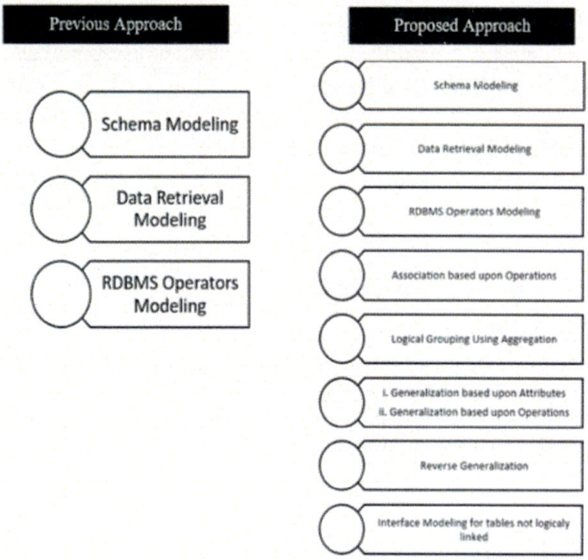

6 Conclusion and Future Work

In this paper, a modeling scheme is proposed for RDBMS using unified modeling language. To model relational database management system, MagicDraw 18.1 tool is being used.

Application modeling and data modeling using the same object-oriented approach not only increase the easiness for the developer but also increase the understandability and reusability. In this paper, logical grouping using aggregation is proposed which will ensure that the entities which have a strong logical relation are always grouped.

Generalization based upon attributes and generalization based upon operations are proposed which ensure that the entities are categorized based upon the attributes and operations.

The proposed modeling scheme ensures that the maximum RDBMS information is scattered across the UML model. The proposed approach uses the richness of UML in the best possible manner so as to increase the understandability of the software system for the developer bridging the gap between the application layer and the backend store, i.e., RDBMS.

The proposed approach ensures that the maximum RDBMS information is modeled with UML using our modeling scheme. This approach has narrowed the gap between application layer and relational database management system modeling, more work can be done to bring them close further. Identification of stronger relationships between RDBMS tables brings the RDBMS layer modeling closer to the application layer modeling. Identification of strong relationship between the tables of relational database management system can be improved if the dependency among the tables is considered. Cardinality of associations can also be taken into account. Cardinality can be specified at each end of relationships so as to identify how many instances of one class are related to the instances of other class. Inclusion of cardinality and dependency between classes can be taken into account for identification of more relationships between classes.

References

1. Balsters H (2003) Modelling database views with derived classes in the UML/OCL-Framework. In: UML, LNCS, vol 2863, pp 295–309
2. Ray I, Shuxin Yin (2005) Relational database operations modeling with UML. In 2013 IEEE 27th international conference on advanced information networking and applications (AINA). Taipei, Taiwan, pp 927–932
3. Attaran H, Far AH (2011) A novel technique for object oriented relational database design. In: IEEE 10th international conference on Cybernetic Intelligent Systems (CIS), IEEE
4. Sinha BR et al (2013) Database modeling with object relationship schema. In: International conference on Information Technology Based Higher Education and Training (ITHET), IEEE
5. Al-Fedaghi S (2014) Unified representation for object oriented and relational models. In: 11th International conference on electrical engineering/electronics, computer, telecommunications and information technology (ECTI-CON), IEEE
6. Lo C-M, Hung H-Y (2014) Towards a UML profile to relational database modeling. Appl Math 8(2):733–743
7. Lee S (2012) Unified Modeling Language (UML) for database systems and computer applications. Int J Database Theory Appl 5(1)
8. Nikose Ms MC, Dhande SS, Bamnote GR (2012) Query optimization in object oriented databases through detecting independent subqueries. Int J Adv Res Comput Sci Software Eng 2(2)
9. Albert M, Cabot J, Gómez C, Pelechano V (2011) Generating operation specifications from UML class diagrams: a model transformation approach. Data Knowl Eng 70:365–389
10. De Lucia A et al (2008) Data model comprehension: an empirical comparison of ER and UML class diagrams. In: The 16th IEEE International Conference on Program Comprehension (ICPC), IEEE

11. Song E, Yin S, Ray I (2007) Using UML to model relational database operations. Comput Stand Interfaces 29:343–354
12. Fakhar L, Muhammad AG (2007) Design of a simple and effective object-to-relational mapping technique. In: Proceedings of the 2007 ACM symposium on applied computing, pp 1445–1449
13. Akhtar N, Iftekhar N, Varshneya S (2015) Online profile maker. In: IEEE International Conference on Electrical, Computer and Communication Technologies (ICECCT), IEEE
14. Iftekhar N et al (2019) Reverse engineering of relational database schema to UML Model. In: International conference on electrical, electronics and computer engineering (UPCON), IEEE
15. Demuth B, Hussmann H (1999) Using UML/OCL constraints for relational database design. In: «UML»'99—The unified modeling language. Springer, Berlin, pp 598–613

A New Activation Function Validated on Function Approximation Tasks

Akash Mishra, Pravin Chandra, and Udayan Ghose

Abstract In this paper, a bounded, continuous, differentiable and non-monotonically increasing function is proposed (σ_{05}). The efficiency and efficacy of using this function as an activation function for hidden layer nodes of FFANNs is demonstrated on five function approximation problems. The proposed function is compared against four bounded, continuous, differentiable, and non-monotonically increasing function (also called sigmoid functions). The results demonstrate that the usage of the proposed activation function as an activation function leads to creation of networks that generalize better than FFANNs using other (sigmoidal) activation function(s).

Keywords Artificial neural networks · Activation functions · Feed-forward artificial neural networks · Non-monotonic activation function

1 Introduction

Feed-forward artificial neural networks (FFANNs) implement the supervised learning paradigm. In this paradigm, the generator of data (available) is estimated from the data itself.[1] The universality of approximation by FFANNs is established by the universal approximation results for these networks (Refer [1] and references therein for the details and survey of the proofs of universal approximation

[1]Thus, the paradigm may also be called learning from data.

A. Mishra · P. Chandra (✉) · U. Ghose
University School of Information, Communication and Technology, Guru Gobind Singh Indraprastha University, New Delhi 110075, India
e-mail: pchandra@ipu.ac.in; chandra.pravin@gmail.com

A. Mishra
e-mail: akash.mishra@ipu.ac.in

U. Ghose
e-mail: udayan@ipu.ac.in

A. Abraham et al. (eds.), *Proceedings of 3rd International Conference on Computing Informatics and Networks*, Lecture Notes in Networks and Systems 167,
https://doi.org/10.1007/978-981-15-9712-1_26

by FFANNs). The conditions under which the proofs of universal approximation by these networks (of any arbitrary function or mapping) is obtained may be summarized as:

1. There should be at least one hidden layer of nonlinear nodes in the network.
2. The nonlinearity at the hidden layer nodes[2] should be a function that cannot be represented as a polynomial of finite degree.

The universal approximation results only indicate the type of activation function, that if used at the hidden nodes allow universal approximation by FFANNs. It does not provide any indication whether any particular activation is to be preferred over others. The set of functions satisfying the condition imposed (of being a non-polynomial) is infinite. It has been asserted that for infinite sized networks (including infinite training time, that is for infinite resources), the usage of any of these functions as activation functions at hidden nodes may be equivalent. But, for finite sized (finite resource) networks, specific activation functions may demonstrate faster training as well as better generalization capability [2] (and references therein).

The generally used activation functions belong to the class of functions called sigmoidal functions. The generally used sigmoidal activation functions are a class of functions which are bounded, continuous, differentiable, and monotonically increasing functions which are also not representable by polynomials of finite degree. The conditions of being continuous and differentiable is a requirement of the training algorithms used for training these networks to solve a particular learning task (as the weight update rules require the derivative of activation functions). The condition of boundedness is more a stability requirement for the weights updates while monotonicity is a historical feature as the seminal paper of Rumelhart et al. [3] used the logistic activation function as a continuous and differentiable approximation to the McCulloch–Pitts function [4].

Since, the universal approximations results do not require sigmoidality, in this paper, we propose an activation function that is continuous, bounded, and differentiable everywhere, but is not a monotonically increasing function. The efficiency and efficacy of using the proposed activation function is demonstrated on function approximation problems as all problems of learning from data (supervised learning) may be treated as problems of estimating the function that generated the output data for the given input data. The proposed function is compared against the commonly used sigmoidal activation functions.

The paper is organized as follows: In Sect. 2, we describe the commonly used activation function and also describe and discuss the proposed activation function. In Sect. 3, we describe the function approximation tasks used to compare the behavior of the activation functions. The experiment design is described in Sect. 4. Results of the experiments are presented and discussed in Sect. 5 while the conclusions are presented in Sect. 6.

[2]Also called hidden nodes for the sake of brevity.

2 Activation Functions

The generally used activation functions belong to the sigmoidal class of functions and are [5] the logistic function defined as:

$$\sigma_{01}(x) = \frac{1}{1 + \mathrm{e}^{-x}} \tag{1}$$

With the derivative given by:

$$\frac{\mathrm{d}\sigma_{01}(x)}{\mathrm{d}x} = \sigma_{01}(x)(1 - \sigma_{01}(x)) \tag{2}$$

The hyperbolic-tangent function is defined as:

$$\sigma_{02}(x) = \tanh(x) = \frac{\mathrm{e}^{x} - \mathrm{e}^{-x}}{\mathrm{e}^{x} + \mathrm{e}^{-x}} \tag{3}$$

With the derivative given by:

$$\frac{\mathrm{d}\sigma_{02}(x)}{\mathrm{d}x} = 1 - (\sigma_{01}(x))^2 \tag{4}$$

The arc-tangent activation function is defined as:

$$\sigma_{03}(x) = \frac{2}{\pi} \tan^{-1}\left(\frac{\pi x}{2}\right) \tag{5}$$

With the derivative given by:

$$\frac{\mathrm{d}\sigma_{03}(x)}{\mathrm{d}x} = \frac{1}{1 + \left(\frac{\pi x}{2}\right)^2} \tag{6}$$

The σ_{03} constants are chosen, so that the output of the function is between ± 1. And, a rational function proposed in [6] and defined as:

$$\sigma_{04}(x) = \frac{x}{1 + |x|} \tag{7}$$

With the derivative given by:

$$\frac{\mathrm{d}\sigma_{04}(x)}{\mathrm{d}x} = \frac{1}{(1 + |x|)^2} \tag{8}$$

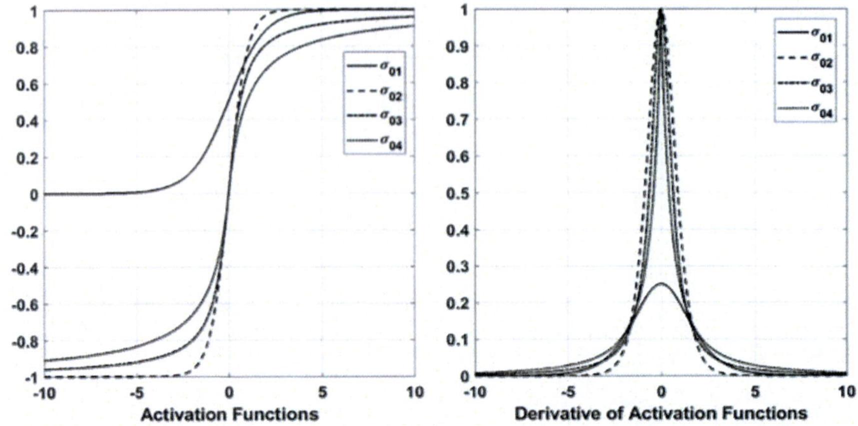

Fig. 1 Commonly used activation functions σ_{01}–σ_{04} and their derivatives

The variation of these functions (σ_{01}–σ_{04}) and their derivatives with the independent argument is shown in Fig. 1. The following may be easily established as properties of these commonly activation functions:

- These functions are continuous.
- These functions are bounded.
- These functions are differentiable everywhere.
- These functions are monotonically increasing.

In this paper, a new activation function is proposed, and the activation function is defined as:

$$\sigma_{05}(x) = \frac{1}{2}\left(\frac{x}{1+|x|}\right)^2 + \frac{1}{2}\frac{x}{1+|x|} \tag{9}$$

With the derivative given by:

$$\frac{\mathrm{d}\sigma_{05}(x)}{\mathrm{d}x} = \frac{1}{2}\frac{1}{(1+|x|)^2}\left(\frac{2x}{1+|x|}+1\right) \tag{10}$$

The variation of the proposed activation function and its derivative is shown in Fig. 2. The following properties of the proposed activation function can be easily established:

- The function σ_{05} is continuous.
- The function σ_{05} is bounded.
- The function σ_{05} is differentiable everywhere.
- The function σ_{05} is not monotonically increasing. In fact, for $x < -1$, it is a decreasing function, while for $x > -1$, it is an increasing function.

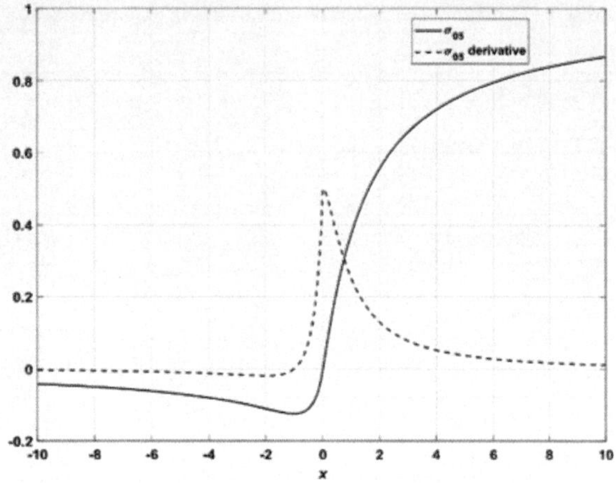

Fig. 2 Proposed activation functions σ_{05} and its derivative

The efficiency and efficacy of using the proposed function as activation function for hidden nodes of FFANNs are demonstrated on function approximation tasks by comparing the generalization errors with FFANNs using the commonly used activation functions generalization errors.

3 Function Approximation Tasks

The function approximation tasks used in this work has been collected from literature and is summarized in Table 1. For each task, the data set is generated by uniform random sampling of the input domains. These input samples together with the function value at these inputs define the tuples that constitute the data set. 1000 such tuples are generated and partitioned into two (sub)sets of 500 tuples each. One of these subsets is designated the training data set while the other is used as the test data set.

4 Experiment Design

For the design of the architecture of the FFANNs used for conduction of the experiments, the following defines the methodology:

1. *Initial weights*: All initial weights and thresholds are chosen as uniform random values in the range $(-1,1)$.
2. *Architecture*: The universal approximation results for FFANNs assert that one hidden layer network is minimally required. Thus, in this paper, we consider only one hidden layer network. The number of nodes in the hidden layer is fixed

Table 1 Function used as learning tasks in the experiments

Task	Function	Input domain	Reference(s)
T_{01}	$g_{01}(x, y) = \begin{cases} 3(1-x)^2 e^{-x^2-(y+1)^2} \\ -10\left(\frac{x}{5} - x^3 - y^5\right) e^{-x^2-y^2} \\ -\frac{e^{-(x+1)^2-y^2}}{3} \end{cases}$	$x, y \in (-3,3)$	Sodhi and Chandra [7], Mishra et al. [8]
T_{02}	$g_{02}(x, y) = \sin(x \times y)$	$x, y \in (-2,2)$	Sodhi and Chandra [7], Mishra et al. [8], Breiman [9], Cherkassky et al. [10], Cherkassky and Mulier [11]
T_{03}	$g_{03}(x, y) = \frac{1+\sin(2x+3y)}{3.5+\sin(x-y)}$	$x, y \in (-2,2)$	Sodhi and Chandra [7], Mishra et al. [8], Cherkassky et al. [10], Cherkassky and Mulier [11], Maechler et al. [12]
T_{04}	$g_{04}(x, y) = 1.9\left(1.35 + e^x \sin\left(13(x-0.6)^2\right) e^{-y} \sin(7y)\right)$	$x, y \in (0,1)$	Sodhi and Chandra [7], Mishra et al. [8], Cherkassky et al. [10], Cherkassky and Mulier [11], Maechler et al. [12]
T_{05}	$g_{05}(x, y) = \sin\left(2\pi\sqrt{x^2+y^2}\right)$	$x, y \in (-1,1)$	Cherkassky et al. [10], Cherkassky and Mulier [11], Cherkassky et al. [13]

by conducting exploratory experiments. In these experiments, the number of nodes in the hidden layer is varied. The smallest sized network (on the basis of number of hidden nodes), that achieves a satisfactory error, is chosen as the network architecture. The network architectures used is summarized in Table 2.

3. *Error function*: The mean squared error function is used.
4. *Data normalization*: All inputs and outputs are normalized to the interval [−1,1]. It is these scaled variables that are used for experimentation.

Table 2 Network architecture summary (I: Inputs, H: Hidden nodes, O: Outputs)

Task	I	H	O
T_{01}	2	15	1
T_{02}	2	10	1
T_{03}	2	64	1
T_{04}	2	18	1
T_{05}	2	24	1

5. *Training algorithm*: The training algorithm used is a first-order (gradient-based) algorithm called improved resilient backpropagation with weight back-tracking algorithm [14–16]. The networks are trained for 2000 epochs of training.
6. *Design of experiments*: For each learning task, an ensemble of 50 sets of weights and thresholds is created. FFANNs using different activation functions are all initialized by using weights/thresholds from this ensemble of weights for the task. That is, networks that use different activation functions all start from the same configuration in the weight/threshold space. Thus, in all, 5(task) $\times$ 50(weight sets) $\times$ 5(activations) = 1250 networks are trained. For each network trained, we measure the mean squared error (MSE) on the data sets. These measurements, for each activation functions, are a set of 50 values of the MSE on the training and test data sets, separately. These results are summarized by the following statistical measures: (1) mean of the MSEs (MMSE), (2) the standard deviation of these ensemble of values of MSEs (STD), (3) the median of these MSE values (MeMSE), (4) the maximum of the MSEs (MaMSE), and the minimum MSE value (MiMSE), for each activation function and task pair. To test the statistically significance of the difference between the MMSEs and the MeMSEs obtained for different activation and task pairs, we use the Students' t-test and the Wilcoxon rank-sum test [17, 18], respectively. Experiments are conducted on a Microsoft Windows 10 operating system machine with Intel 17 processor, 8 GB RAM, and MATLAB v2018b.

5 Results and Discussions

The generalization error (error over the test data set) of a model to solve a particular learning task is a measure of the quality of the solution. Tables 3, 4, 5, 6 and 7 summarize the results obtained in the experiments. From these tables, the following is observed:

1. The comparison of the MMSE values given in Tables 3 and 4 allows us to conclude that the FFANNs using the proposed activation function σ_{05} outperforms all other networks (using activation functions σ_{01}–σ_{04}).
2. The comparison of the MeMSE values given in Tables 3 and 5 allows us to conclude that the FFANNs using the proposed activation function σ_{05} outperforms all other networks (using activation functions σ_{01}–σ_{04}). Though for one task, each σ_{02} and σ_{04} performs better.
3. The statistical significance of the mean values reported in Table 3 and summarized in Table 4 is only demonstrating the difference in the MMSE values. To test whether these differences are statistically significant or not, a one-sided Students' t-test with significance level of $\alpha = 0.5$ was conducted. The results of this test are summarized in Table 6. From this table, we infer that in no case the FFANNs using the proposed activation function (σ_{05}) performs worse than the FFANNs using the other activation functions. And, for most of the tasks, it outperformed the other networks.

Table 3 Summary of results over the test data set. All values reported $\times 10^{-3}$

Task	Statistics	σ_{01}	σ_{02}	σ_{03}	σ_{04}	σ_{05}
T_{01}	MMSE	12.251	12.869	10.690	9.280	9.078
	STD	4.542	4.681	3.944	3.420	3.462
	MiMSE	6.182	6.107	6.257	3.838	4.224
	MaMSE	22.315	22.984	20.875	19.475	20.395
	MeMSE	11.095	11.834	9.348	8.508	8.712
T_{02}	MMSE	9.958	8.559	12.694	9.729	8.534
	STD	2.116	2.037	2.504	2.275	3.439
	MiMSE	5.445	5.233	7.805	5.226	2.036
	MaMSE	15.183	13.690	17.838	16.711	15.053
	MeMSE	9.887	8.252	12.664	9.480	8.769
T_{03}	MMSE	10.080	8.659	8.206	5.868	2.773
	STD	3.211	2.089	0.888	1.207	1.494
	MiMSE	4.895	4.704	6.493	3.781	1.133
	MaMSE	18.606	14.669	10.000	9.060	8.367
	MeMSE	10.039	8.547	8.152	5.634	2.403
T_{04}	MMSE	4.138	3.416	3.985	3.089	2.998
	STD	1.930	1.302	1.218	0.848	0.974
	MiMSE	2.507	2.014	2.173	1.900	1.772
	MaMSE	10.150	9.463	7.376	5.887	6.489
	MeMSE	3.308	3.101	3.807	2.936	2.834
T_{05}	MMSE	44.212	43.170	48.278	36.504	27.636
	STD	19.052	10.562	8.231	9.019	8.578
	MiMSE	14.565	19.719	35.862	18.574	16.319
	MaMSE	99.299	74.297	76.240	62.071	50.704
	MeMSE	39.442	43.274	46.867	34.398	24.635

Table 4 Comparison results of MMSE over test data

Activation	σ_{01}	σ_{02}	σ_{03}	σ_{04}	σ_{05}	Row sum
σ_{01}	00	01	02	00	00	03
σ_{02}	04	00	03	01	00	08
σ_{03}	03	02	00	00	00	05
σ_{04}	05	04	05	00	00	14
σ_{05}	05	05	05	05	00	20
Column sum	17	12	15	06	00	00

Values specify the number of tasks in which the row activation functions provided better result than the column activation functions

Table 5 Comparison results of MeMSE over test data

Activation	σ_{01}	σ_{02}	σ_{03}	σ_{04}	σ_{05}	Row sum
σ_{01}	00	02	03	00	00	05
σ_{02}	03	00	03	01	01	08
σ_{03}	02	02	00	00	00	04
σ_{04}	05	04	05	00	01	15
σ_{05}	05	04	05	04	00	18
Column sum	15	12	16	05	02	00

Values specify the number of tasks in which the row activation functions provided better result than the column activation functions

Table 6 Comparison results of MMSE over test data

Activation	σ_{01}	σ_{02}	σ_{03}	σ_{04}	σ_{05}	Row sum
σ_{01}	00	00	01	00	00	01
σ_{02}	03	00	03	01	00	07
σ_{03}	02	01	00	00	00	03
σ_{04}	04	03	05	00	00	12
σ_{05}	05	04	05	03	00	17
Column sum	14	08	14	04	00	00

Values specify the number of tasks in which the row activation functions provided better result than the column activation functions. The comparison is using the one-sided Student's t-test at significance level of $\alpha = 0.05$

Table 7 Comparison results of MeMSE over test data

Activation	σ_{01}	σ_{02}	σ_{03}	σ_{04}	σ_{05}	Row sum
σ_{01}	00	00	02	00	00	02
σ_{02}	03	00	03	01	00	07
σ_{03}	02	01	00	00	00	03
σ_{04}	04	03	05	00	00	12
σ_{05}	05	04	05	03	00	17
Column sum	14	08	15	04	00	00

Values specify the number of tasks in which the row activation functions provided better result than the column activation functions. The comparison is using the one-sided Wilcoxon's rank-sum test at significance level of $\alpha = .05$

4. The statistical significance of the median values reported in Table 3 and summarized in Table 5 is only demonstrating the difference in the MMSE values. To test whether these differences are statistically significant or not, a one-sided Wilcoxon rank-sum test with significance level of $\alpha = 0{:}5$ was conducted. The results of this test are summarized in Table 7. From this table, we infer that in no case

the FFANNs using the proposed activation function (σ_{05}) performs worse than the FFANNs using the other activation functions. And, for most of the tasks it outperformed the other networks.

5. Among the existing activation function that are used in this work σ_{04} is better performing than σ_{01}, σ_{02}, and σ_{03}.

6 Conclusions

In this paper, a bounded, continuous, differentiable, and non-monotonically increasing function is proposed (σ_{05}). The efficiency and efficacy of using this function as an activation function for hidden layer nodes of FFANNs are demonstrated on five function approximation problems. The proposed function is compared against four bounded, continuous, differentiable, and non-monotonically increasing function (also called sigmoid functions). The results demonstrate that the usage of the proposed activation function as an activation function leads to creation of networks that generalize better than FFANNs using other (sigmoidal) activation function(s).

References

1. Pinkus A (1999) Approximation theory of the MLP model in neural networks. Acta Numerica 8:143–195
2. Duch W, Jankowski N (1999) Survey of neural network transfer functions. Neural Comput Surv 2:163–212. Available on online http://www.icsi.berkeley.edu/vjagota/NCS
3. Rumelhart DE, McClelland JL, PDP Research Group (eds) (1986) Parallel distributed processing: explorations in the microstructure of cognition. MIT Press, Cambridge
4. McCulloch W, Pitts W (1943) A logical calculus of the ideas immanent in nervous activity. Bull Math Biophys 5:115–133
5. Haykin S (1999) Neural networks: a comprehensive foundation. Prentice Hall Inc, New Jersey
6. Eliott OL (1993) A better activation function for artificial neural networks. Institute for Systems Research, University of Maryland, College Park, Maryland, Technical Report ISR TR 93–8
7. Sodhi SS, Chandra P (2014) Bi-modal derivative activation function for sigmoidal feedforward networks. Neurocomputing 143:182–196
8. Mishra A, Chandra P, Ghose U, Sodhi SS (2017) Bi-modal derivative adaptive activation function sigmoidal feedforward artificial neural networks. Appl Soft Comput 61:983–994
9. Breiman L (1991) The PI method for estimating multivariate functions from noisy data. Technometrics 3(2):125–160
10. Cherkassky V, Gehring D, Mulier F (1996) Comparison of adaptive methods for function estimation from samples. IEEE Trans Neural Networks 7(4):969–984
11. Cherkassky V, Mulier F (1998) Learning from data—concepts, theory and methods. John Wiley, New York
12. Maechler M, Martin D, Schimert J, Csoppenszky M, Hwang J (1990) Projection pursuit learning networks for regression. In: Proceedings of the 2nd international IEEE conference on tools for artificial intelligence, pp 350–358
13. Cherkassky V, Lee Y, Lari-Najafi H (1991) Self-organizing network for regression: efficient implementation and comparative evaluation. In: Proceedings of the IJCNN-91, vol 1, pp 79–84

14. Riedmiller M, Braum H (1993) A direct adaptive method for faster backpropagation learning: the RPROP algorithm. In: Proceedings of the IJCNN, pp 586–591
15. Riedmiller M (1994) Advanced supervised learning in multi-layer perceptrons—from backpropagation to adaptive learning algorithms. Comput Stand Interfaces 16(5):265–278
16. Igel C, Hüsken M (2003) Empirical evaluation of the improved RPROP learning algorithms. Neurocomputing 50:105–123
17. Huber PJ (1981) Robust statistics. John Wiley & Sons, New York
18. Hettmansperger T, McKean J (1998) Robust nonparametric statistical methods, ser. Kendall's library of statistics: an arnold publication No 5. Arnold

Determining the Predictive Accuracy of Loan Defaulters Using R

Mugdha Sharma, Aarnav Madan, Akshat Shakarwal, Abhay Pratap Singh, and Nitin Kumar

Abstract The banking sector shares a great contribution in maintaining the economy of the country. Default in bank loans shares vital role in risk management of various bank institutions. Bank helps the customer by giving many loans, credit cards, investment, mortgage, and others. Bank loan is also an important part of banking institutions which define as the probability of returning money to the bank by its users. Although, with increase in the participant of banking loan user, the number of defaulters is increasing with the minute and that is leading to huge losses for the banking industry. Machine learning has been used to tackle this issue. This research paper proposes a more accurate way to predict loan defaulters. The model proposed in this paper predicts defaulters using a logistic regression area under Roc curve of 77% which beats the earlier accuracy of 75%. Similarly, in this paper, decision trees, random forest, and the SVM models have been able to achieve better accuracy than the models proposed in the past.

Keywords Exploratory data analysis · ML · Data science

M. Sharma (✉) · A. Madan · A. Shakarwal · A. P. Singh · N. Kumar
Department of Computer Science and Engineering, Bhagwan Parshuram Institute of Technology, Delhi, India
e-mail: mugdha.sharma145@gmail.com

A. Madan
e-mail: Madanm.aarnav@gmail.com

A. Shakarwal
e-mail: akshatshekhnew@gmail.com

A. P. Singh
e-mail: abhaytrekk@gmail.com

N. Kumar
e-mail: starknitin@gmail.com

A. Abraham et al. (eds.), *Proceedings of 3rd International Conference on Computing Informatics and Networks*, Lecture Notes in Networks and Systems 167,
https://doi.org/10.1007/978-981-15-9712-1_27

Abbreviations

ROC curve	Receiver Operating Characteristic curve
SVM	Support Vector Machine
ML	Machine Learning
K-NN	K-Nearest Neighbors

1 Introduction

With the increase in the amount of data being generated day by day, machine learning algorithms have got stronger, and data analytics have become an integral part of the industry. Various machine learning methods are being applied to solve serious business problems [6].

This research paper is designed to pay attention to one of the greatest challenges that the banking sector is facing currently. Non-repayment of loans has caused major losses to the banking sector. This is a major concern due to which banks have started to invest more and more in developing bank loan risk models that help them in reducing the risk factor of providing the loan to the customer. This is done using machine learning and predictive modeling.

Machine learning techniques that this research paper is using to find the loan prediction defaulter are logistic regression, Rpart decision tree, random forest, and SVM. Now, let us talk about logistic regression—statistic logistic model is used to find the probability of particular class or we can say event. For example, passed/failed, win the event/loss the event, alive/dead or healthy/sick. This may be applied to several event to find whether an image contains cow, bird, or any animal. Each object is detected in the image would be assigned a probability of between 0 and 1 and sum adding to one.

The decision tree is the most efficient and most favored tool which used to classify and predict dataset. A decision tree is like tree structure where each internal node denotes a test on an attribute, each branch represents an outcome of the test, and each leaf node (terminal) holds a class label [6].

Random forest or can be said as random decision forest is an assemble learning method for classification, regression, and other tasks that operate by constructing a multitude of decision trees at training time and outputting the class, i.e., mode of classes (classification) or mean prediction (regression) of individual trees.

The input/contribution of this research paper.

The overall research process design for the proposed study is shown in Fig. 1.The objectives of the study are as follows:

To get to know which factors affect the person to default on his payment

To conduct in-depth exploratory data analysis to get insights into the data available

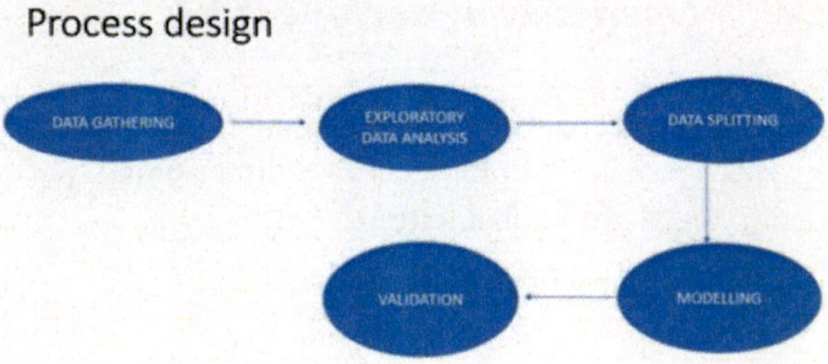

Fig. 1 Research process

To build a predictive model that accurately predicts whether the person would default or not
To improve the accuracy of the models implemented in the past work.

2 Machine Learning

2.1 Logistic Regression

Logistic regression is a classification technique used to solve a classification problem that involves predictions of a factor variable. It comes under a supervised learning algorithm that means the target variable should be known beforehand to use this algorithm.

2.2 Decision Trees

The decision tree is an unsupervised machine learning algorithm used to predict a variable by finding out the most important variables and then creating a tree-based structure using them [1]. It can be used for both regressions as well as classification problems. It can cause a problem of overfitting, but the ease of its implementation is a big factor of its being used in the industry.

2.3 Random Forest

Random forest is an unsupervised machine learning technique that uses an ensemble method to create multiple decision trees and come up with the best model using those. A random forest can also be used for both classification and regression problems. The random forest takes a lot of time to train since the generation of multiple decision trees takes time.

2.4 Support Vector Machine (SVM)

Support vector machine is used to solve regression and classification problems. In this, each data item is plotted in an n-dimensional space. Vectors are used to uniquely identify each group distinctly.

3 Literature Review

Based on the past literature, we have seen many different types of machine learning techniques have been used like logistic regression, decision trees, random forest, and K-NN [2, 11, 12].

The most used technique that we observed was that of decision trees. This is because of the ease of which it can be implemented. It is a technique where we find the most significant variables and make a tree concerned about that [3]. Radom forest is another such technique that was used quite often. It generates many decision trees and ensembles them to create a model with the best accuracy. The best accuracy was found out to be of Sayjdhah [6]. Nowadays, the banking sector uses efficient use of machine learning techniques with several classification techniques to split up the customers to predict the trends [8, 10]. They want to keep the all details of the customers to understand the behavior of payment data which is added to the loan scoring literature to anticipate their defaults [4]. Some researcher used the Bayesian network used for the graphical representation model showing the probability of correlation of variables [7]. Few researchers have proposed the hybrid approaches also such as merging genetic algorithms with neural network approaches to detect the financial frauds [5].

4 Methodology

4.1 Datasets

In this project, the datasets are used, and it is generated by the banking loan operations by the user. The datasets consist of 25 variables with 30,000 samples. This dataset has been used in various researches previously too [9]. So, it is not unsigned yet, the dataset includes a binary variable of Yes equal to 1 and No is equal to 0, for example, default payment outcome [6]. Table 1 shows the basic description of dataset.

Default_Payment Next_Month → This is the target variable that has to be predicted. It tells whether the person would default or not. It is a factor variable with 2 levels "Yes" and "No" as shown in Fig. 2.

Table 1 Explanation of dataset

Attribute-name	Descriptions
ID	UserID
Limit__Bal	Amount of the given credit (NT dollar)
Sex	Gender (1 = male, 2 = female)
Education	Education (1 = graduate school, 2 = university, 3 = others)
Marriage	Martial status (1 is married, 2 is unmarried)
Age	Age (year)
Pay-1_Pay-5	Status of repayment from April to September 2005
Bill_Amt1 to Bill_Amt-6	Bill statement amount from April to September 2005
pay_Amt1 to pay_Amt6	Amount paid or compensate in September 2005
default_payment_Next_Month	Amount which has to be compensated in next month

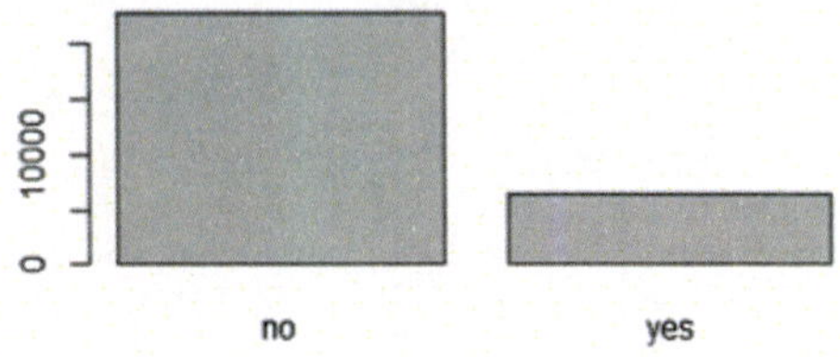

Fig. 2 Snapshot of the dependent variable

4.2 *Dataset Pre-Processing and Feature Selection*

Firstly, the data was cleaned to build the model. Then, all the NA values were removed. This is done so that the model can run smoothly. The next step is feature selection where only the important variables are kept and all the obsolete variables are removed. In this paper, multicollinearity and correlation have been considered to observe feature importance. Then, the outliers were treated. The interquartile range has been used to remove outliers. Outliers are values that do not follow the pattern, and these values make the model deviate from the correct predictions. Figure 3 shows the variable of bill_amt2 before and after the treatment of outlier. Feature scaling technique is also used to scale the features to a certain range to make the logistic model work fast and efficiently. Z score normalization was used to scale the features.

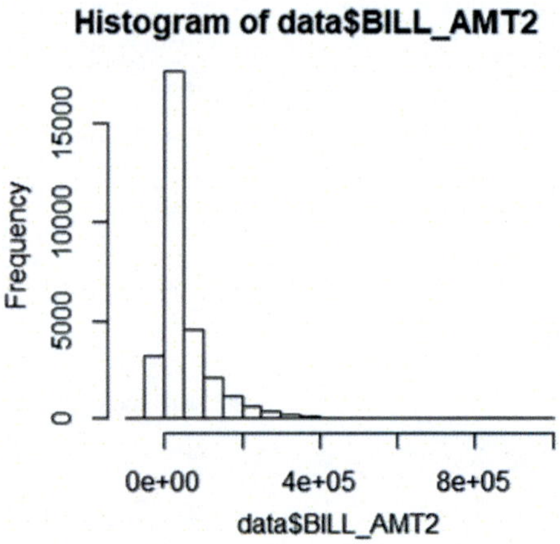

Fig. 3 Before outlier treatment

4.3 Data Visualization

This section provides interesting insights into the data that helps in understanding the relationship between different variables, all the nitty-gritty of the data are understood using different visualization techniques. Also, the demographics about the people using the credit card can be figured out from Fig. 4. Figure 5 shows the split of our dependent variable gender-wise.

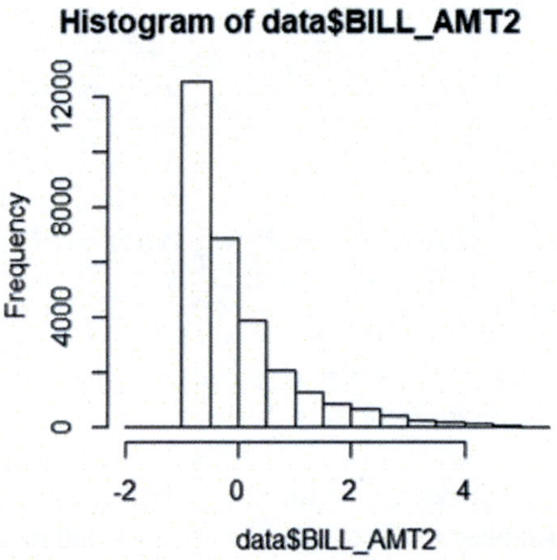

Fig. 4 After outlier treatment

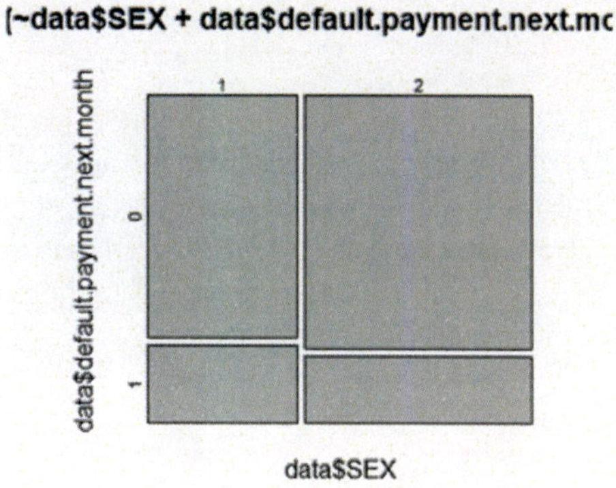

Fig. 5 Gender demographics

4.4 Data Partitioning

Data was split into two parts, train and test data. This is done so that first we can train the model on train data and then we can do the error analysis on test data. This helps us prevent overfitting and also makes the model more flexible to new data points. The data is split in the ratio of 70:30 to ensure we have sufficient data points in the test data to train the model well.

5 Performance and Evaluation

As seen in Fig. 6, it can be observed that the model used in this paper gives better results from the model being used in previous papers. Even though the accuracy of logistics reduces, but the gain in the ROC curve area shows that the model being used in the paper is much more stable and equipped to handle new incoming data.

Fig. 6 Model outcome. 1 Yashna Sayjdhah, 2 model used in this paper

	ROC(1)	ROC(2)	ACCURACY(1)	ACCURACY(2)
LOGISTIC	75%	77%	82%	79.5%
DECISION TREE	64%	64.3%	82%	82.3%
RANDOM FOREST	77%	77%	81.8	81.75%

1->Yashna Sayjdhah
2->model used in this paper

6 Model Comparison and Discussion

Four models have been used in this paper. Out of the four, random forest provides the best accuracy and a comparable area under the curve for our ROC curve making it the most stable and best-equipped model to predict the defaulters.

Decision tree was also considered, but since its area under the curve for the ROC curve is way too less, so we choose random forest as the best and most suited model for this paper.

7 Conclusion

At this stage, according to the predicting model, we have concluded that almost half of the population is married. Most of the people are graduate and from the university. Male and female have equal percentages. Bill amounts are skewed which need to be treated. Most of the people are aged between 20 and 60. More men tend to default than women in terms of ratio. Married people and others tend to default more than single. People having school or university education tend to default more. The techniques which are used in this model are random forest, decision tree, and logistic regression. Random forest has the best accuracy in this model with 81.75%. Based model had (Yashna Sayjhda 2018) the accuracy 81%. Main objective of this model is to detect the defaulters who take the loan from the bank and refuse/fail to pay within the given time which was provided by the bank itself. This paper checks on different parameters which customers likely to default more.

References

1. Rising credit card delinquencies to add to U.S. banks' worries. Accessed 13 Nov 2017
2. Venkatesh A, Gracia S (2016) Prediction of credit-card defaulters: a comparative study on performance of classifiers. Int J Comput Appl 145(7):36–41
3. AghaeiRad A, Chen N, Ribeiro B (2016) Improve credit scoring using transfer of learned knowledge from self-organizing map. Neural Comput Appl 28(6):1329–1342
4. Bakoben M, Bellotti T, Adams N (2017) Identification of credit risk based on cluster analysis of account behavior. Department of Mathematics, Imperial College London, London SW72AZ, United Kingdom
5. Azimi A, Hosseini M (2017) The hybrid approach based on genetic algorithm and neural network to predict financial fraud in banks. Int J Inf Secur Syst Manage 6(1):657–667
6. Yashna S, Kasmiran KA, Hashem IAT, Alotaibi F (2018) Credit card default prediction using ML techniques, Malaysia
7. Xia Y, Liu C, Li Y, Liu N (2017) A boosted decision tree approach using Bayesian hyper-parameter optimization for credit scoring. Expert Syst Appl 78:225–241
8. Yap B, Ong S, Husain N (2011) Using data mining to improve assessment of creditworthiness via credit scoring models. Expert Syst Appl 38(10):13274–13283

9. Yeh I (2017) UCI machine learning repository: default credit card clients data set. Online Archive.ics.edu. Available at. https://archive.ics.uci.edu/ml/datasets/default+of+credit+card+clients. Accessed 13 Nov 2017
10. Ghasemi A, Motahari AS, Khandani AK (2010) Interference alignment for the K user MIMO interference channel. In: IEEE International Symposium on Information theory proceedings (ISIT), pp 360–364
11. Bellotti T, Crook J (2013) Forecasting and stress testing credit card default using dynamic models. Int J Forecast 29(4):563–574
12. Harrell F (2015) Regression modeling strategies: with application to linear models, logistic and ordinal regression, and survival analysis. Springer International Publishing, Berlin

A Review on Human Behavior Using Machine Learning for Ambient Assisted Living

Vanita Jain, Nishant Khurana, and Sameer Bhardwaj

Abstract With advances in machine learning, the evaluation and analysis of human behavior continue to attract large number of researchers around the globe. In this paper, we furnish an extensive overview of ways to identifying, analyzing and assessing human behavior, taking into account various behavioral characteristics. Most promising attributes and recognition techniques for vision and sensor-based approaches have been detailed. Most prominently used datasets for both vision- and sensor-based approaches have also been studied, keeping in mind the nature, source and applications of the same in the field of human behavior and activity detection. The study indicates that sensor-based approaches tend to have an upper hand because of the privacy breach caused by vision-based approaches, which accounts for the evolving usage of sensor-based monitoring for real-time behavior detection. Various other deep learning methods and their applications in the field of behavioral recognition have also been stated.

Keywords Human behavior · Ambient living · Sensory · Visual · Attributes

Abbreviations

AAL	Ambient Assisted Living
ADL	Activities of Daily Living
AmI	Ambient Intelligence
DBN	Deep Belief Network

V. Jain (✉) · N. Khurana · S. Bhardwaj
Bharati Vidyapeeth's College of Engineering, New Delhi, India
e-mail: vanita.jain@bharatividyapeeth.edu

N. Khurana
e-mail: khurananishant74@gmail.com

S. Bhardwaj
e-mail: sameerbhardwaj16@gmail.com

A. Abraham et al. (eds.), *Proceedings of 3rd International Conference on Computing Informatics and Networks*, Lecture Notes in Networks and Systems 167,
https://doi.org/10.1007/978-981-15-9712-1_28

EFS Evolution Fuzzy Systems
HBA Human Behaviour Analysis
HBU Human Behavioral awareness and Understanding
HMM Hidden Markov Model
HOHA Hollywood Human Action
SCC Source Code Control
SRBM Social Restricted Boltzmann Machine
SVM Support Vector Machines
TLN Temporary Logic Network

1 Introduction

Human behavior essentially means the response of an individual to any kind of stimuli, whether internal or external. It refers to all the physical actions and observable emotions associated with any individual. Although some traits of an individual's personality remain consistent throughout the life, but lot of the behavioral pattern changes as one moves from childhood to adulthood. Behavior is not only determined by age and genetics, but is also influenced significantly by thoughts and feelings [1]. This gives us great insight into an individual's psyche, attitude and values.

Human behavioral awareness and understanding (HBU) has become one of the most promising areas of research in health care. Healthy lifestyle behaviors improve a person's overall health and thereby increase lifespan, while physical inactivity, improper nutrition, anxiety and sleep disturbance can reduce the life expectancy of an individual [2]. Not only this, psychological, physical and social changes can also affect healthy and active aging. Therefore, analyzing and monitoring human behavioral changes can help prolong an individual's life [3].

There are two main mechanisms for monitoring and analyzing the human behavior, namely visual and sensory monitoring. Several vision-based approaches have been discussed in [4]. However, most visual procedures are a privacy issue for consumers because they are subject to constant visual monitoring [5]. Therefore, sensor-based behavior evaluation is the most widely accepted solution for capturing human behavior and movement in a smart environment [6]. Sensor technology is the basic idea behind sensor-based approaches for behavioral and activity evaluation. Sensory data is mainly state transition of different parameter values, which is usually processed through probability, statistical analysis techniques, data fusion and previous knowledge of the activity. The two major methods to sensor-based behavior and activity identification are namely: data-and knowledge-dependent methods. Learning the activity and behavioral patterns through machine learning, data analysis and data mining is the idea behind the data-driven approach. Because data-driven approaches require large amounts of data to train different classifiers, they are usually presented as a supervised and deep learning model. Knowledge-based approaches rely on previous information available in this area for modeling without the use of collected data [7].

Human behavioral models play a very critical part in the context of smart environments, wherein a person's long-term motivations are determined by his or her current actions [8].

Ambient intelligence (AmI) addresses the prevailing and intelligent environment that continually helps people in carrying out their routine activities [9]. AmI presents new ideologies where smart devices are present ubiquitously to enable intuitive and intelligent interactions between an individual and his/her respective environment.

Ambient assisted living (AAL) [10] is considered to be one of the most prominent concepts, derived from AmI. AAL's mission is to apply AmI's idea and its techniques to help people with special needs, particularly senior citizens, to live longer in their natural settings. Various examples of such smart solutions range from simple devices like intelligent drug dispensers, drop detection sensors or bed sensors to complicated systems such as interactive systems and connected homes.

In the AAL environment, the process can be viewed as a cycle of analyzing and monitoring the environmental status, achieving a particular goal, or assessing the consequences of all the actions possible, and ultimately working on the environment to achieve a desired state [11]. Sensing refers to getting accurate information about the users and the environment. Communication means that all parts of the AAL environment must be interconnected to establish a connection. The final stage of acting means that any type of AAL environment must be capable of working with variety of actuators to achieve its goal [12, 13].

The paper is divided into seven sections. Section 2 describes the various terminologies used in human behavior. Vision-based approaches are explained in Sect. 3. In Sect. 4, different databases used in vision-based approaches are discussed. Sensor-based approaches are explained in Sect. 5. Section 6 narrates the different databases used in sensor-based approaches. Finally, Sect. 7 concludes the paper.

2 Human Behavior Terminologies

In this section, various HBA terminologies from the most relevant and recent research works have been detailed in order to understand the classification and differences by various authors.

In [14], the author has defined an action terminology that most researchers agree on. Three levels from bottom to top are as follows:

- Basic motion is referred to as motion or action primitive and refers to the nuclear unit that produces action. Therefore, as indicated in [15], an action primitive is a nuclear motion, which can be described at the organ level.
- The middle level of abstraction is called action which is nothing but a combination of different actions primitives.

- The highest level of abstraction involves a large number of events, environments and interacting entities.

The three levels of abstraction can be clearly illustrated using the food preparation example. When preparing meals, individual movements of the palms are considered an action primitive. The process of putting ingredients on the cabinet or mixing them together would be called actions, and finally, the full procedure of preparing meals involving various actions interacting in different combinations with different objects and environment would be called an activity.

This terminology is clear and is often used by various authors who analyze human behavior. Because this is the type of classification which is exclusively concentrated on actions, it becomes hard to use this terminology in higher-level methodologies, where the main goals are behavioral analysis and activities of daily living (ADLs). For this reason, a lot of researchers prefer to use their own terminology to suit their goals.

In [16], author states activity as a mix of objects and actions. Here, actions are identified by series of verbs, and positions are identified by series of nouns that are nothing but results of actions. Rather than identifying actions, recognition of objects is done to address human activity. In [17], the authors have distinguished amongst activities and actions, by stating that integrated actions amongst limited number of individuals are called actions. Behaviors are defined as models of human movement with a comprehensive description of behaviors and movements in [18]. In contrast to [14], the authors consider the level of dependence on the environment, human interaction and objects.

In [19], behaviors are considered to be patterns in a series of inspections of events or activities. Smart homes equipped with binary sensors are helpful in detecting routine activities such as eating, cooking, watching TV and are also helpful in detecting repeatable patterns and anomalies.

User behavior is a complex structure composed of large number of defining elements. To properly describe them, various elements are defined: actions, activities and behaviors [20]. Actions are normal conscious and voluntary movements, while behaviors form the most complicated arrangement. The model described in [20] divides the behavior into two major categories: Interactivity behaviors and intraactivity behaviors. This enables better modeling of different aspects of user behavior. It identifies two types of behavior: intraactivity (describing how a person performs the same activity differently) and interactivity (describing the everyday life of occurring actions and activity sequences). Action sequences are used in interactivity behavior as they provide a detailed description of the user behavior.

A much detailed description of the user behavior is categorized by defining three attributes:

- Actions are transient, conscious and voluntary muscle movements by the user (e.g., pushing/pulling, etc.).
- Activities are long but limited and involve many simultaneous activities (e.g., meal preparation, taking a bath, walking, etc.).

- Behavior describes how the user demonstrates these various activities at different time intervals.

3 Vision-Based Approaches

Behavior is defined as the highest level of complexity according to the categorical classification. It seems to be a series of long-term activities that take place in a specific order. It can be considered as the daily routine of the observed person. Deviations from routine can be considered exceptional and provide information about the development, health or independence of the individual living alone [19].

Public behavior in the scene seems to be a circadian rhythm of activity in some of these tasks. That is the development of the activities of daily living (ADL) [19]. By learning a person's circadian activity rhythm every week or five days, deviations in system behavior can be detected from observed normal behavior.

Context awareness systems can be found in [21, 22], while video comment systems can be found in [23–25]. These systems are dedicated to "behavioral awareness" because they are not limited to identifying the activities they perform at all times, but to understand and extract more information from the identified activities.

In [26], the author deals with the situation recognition. They have a system that captures a person in three dimensions with cameras and gathers information about that person's speed, posture and interactions with other people, furniture or accessories, depending on the distance between them. This setting uses the headset to determine if a person is speaking. Ambient noise can be detected using microphones. The number code is assigned to every arrangement possible (single or multiple mass, with or without ambient noise). By merging the information extracted from the codes, various conditions can be learned and identified using the left-right hidden Markov model.

In [27], the author summarizes the interpretation of tennis sequences. Location, speed and action details are combined and given in an action classifier, which is given to hidden Markov model (HMM). The top level of this HMM is behavior, where the sensitive HMM output recognizes a number of events that are more common. The future scope is to develop an abnormality detection application.

In [28], the author proposes a system of low-level information and combines with two different contexts: spatial and temporal. Vision sources are used to learn and identify behaviors using the hidden Markov model for hierarchical forecasting. The AAL environment uses its methods in the view of a house, where different behaviors are continuously tracked. This routine includes activities such as sitting, going to the toilet, sleeping on bed, having breakfast and walking. To monitor for deviations in the expected routine (duration, time or location), people are monitored to know the normal sequence of activities.

The results were compared with those of other methods in [29, 30]. In [29], the authors propose the switching hidden semi-Markov model that makes room for detecting the abnormality based on the duration of the activity. The order of operations and the number of operations are limited to six. The space for activities is also limited as the space is split.

In [31], the authors present a method for modeling, identification and interaction. The modeling action uses threads (associated with each actor/agent) and time constraints. These parameters are fed to the temporary logic network (TLN) for identification.

In [32], the authors identify behaviors in pairs or groups of individuals by extending the methods presented in [28]. A switchover module called SCC control (SC) is used for switching between two distinct hidden Markov model-based systems that works by identifying the number of people in the scene. Methods apply in the AAL environment.

4 Datasets Used in Vision-Based Approaches

- HOHA—Hollywood Human Action [33]: This record contains eight types of action notes in 32 video clips. An alternate version is available, which along with existing data, contains approximately four new features and additional 1200 min of video. The video clips are collected from various movies, because of which the main focus is on the people in the foreground and background changes are common. So, this record can be very challenging and useful.
- KTH Human Motion Dataset [34]: This dataset contains six types of human actions in four scenarios with 25 subjects. These scenes are displayed in over 2000 videos. Videos are categorized by actions, so it is easy to exclude unwanted tasks. In contrast to the HOHA dataset [33], the backgrounds are homogeneous and clear, which facilitates background separation.
- Weizmann Human Action Dataset [35]: Single human motion capture of ten people using fixed cameras in the foreground in different environments. There are about 340 MB video sequences available. The system relies on space-time properties and can detect complex actions, such as ballet movements.
- INRIA Xmas [36]: In this dataset, 390×291 pixel video images were taken from five different angles. There are 13 different acts of 11 actors. These activities took place three times in the arbitrarily chosen perspective of the viewpoint. The background and brightness settings are consistent and clear.
- TUM Kitchen Record [37]: This dataset targets everyday activities in the kitchen landscape with minimal activity. Most people do table operations in different ways. Some pick items individually; and others behave spontaneously and pick many items at the same time. Video images are recorded at 25 fps and have a resolution of 384×288 pixels. Markers collect data with an entire body tracker. The frames are labeled separately for the person's right hand, left hand and torso.
- MuHAVI Dataset [38]: This dataset contains video data from multiple cameras targeting silhouette methods for detecting human actions. Continuous but uneven background images are taken at night with streetlights. A camera is installed in every corner and on each side of the rectangular platform. These cameras capture

Table 1 Vision datasets and their respective accuracies

Name of the paper	Dataset	Methodology	Results and conclusions
"Learning realistic human actions from movies"	HOHA	Support vector machines	Automatic annotation achieves 60% precision
"Recognizing human actions: a local SVM approach"	KTH human motion dataset	Support vector machines	An accuracy of 62.4% was achieved on all the subsets of the dataset
"Free viewpoint action recognition using motion history volumes"	INRIA Xmas motion acquisition sequences	K-means clustering	Recognition rate of 82.79% was achieved for 78.79% correctly classified motions and 14.08% false positives
"The TUM kitchen dataset of everyday manipulation activities for motion tracking and action recognition"	TUM kitchen dataset	Conditional random fields	The overall classification accuracy reaches 82.9%
"MuHAVi: a multicamera human action video dataset for the evaluation of action recognition methods"	MuHAVI dataset	Cross validation	An average classification accuracy of 82.4% was achieved
"Action MACH a spatio-temporal maximum average correlation height filter for action recognition"	UCF sport action datasets	Cross validation	The overall accuracy achieved on the dataset was 69.2%

16 different actions from seven actors three times. Each frame is recorded at 25 fps with a resolution of 720 × 576 pixels.

- UCF Sport Action Dataset [39]: This dataset is approx. 200 video sequences with a resolution of 720 × 480 pixels. To get realistic training data, images are deliberately taken from real scenes (usually TV channels). Pictures from sports broadcasts or from YouTube introduce significant changes in settings for motion, object display and scaling, perspective, opacity and brightness. That is why they are so challenging. This record contains both actions and activities (Table 1).

5 Sensor-Based Approaches

The learning methods used in the literature to assess human behavior are very broad and range from simple naive Bayes classification [40, 41] to support vector machines

[42], dynamic Bayesian networks [43, 44], online classifiers (or ascent) [45] and hidden Markov models [46]. Knowledge-based approaches aim to increase existing domain knowledge to avoid using labeled records for training.

In [47], the authors studied physical motion using data obtained by making the user wear five small biaxial accelerometers simultaneously on different parts of the body. Twenty subjects were observed in order to collect the acceleration data. Subjects were asked to complete their daily tasks but were not specifically told how to manage them. The correlations of the data were calculated for the acceleration, power, mean, frequency domain entropy, and many classifiers were tested using these properties. The decision tree classifiers [48, 49] showed the best performance.

In [50], in order to analyze ADL from sensor readings, the authors used two evolving classifiers [51] from in the ambient assisted living environment. They introduced the evolution fuzzy systems (EFS)-based approach, where each ADL is defined by a model that adapts to changes in ADL.

In [52], the authors present a simple, accurate and inexpensive way to install a sensor network. A record containing 28 days of sensor data and annotations is made available to the community. Markov models and conditional random fields show through a series of experiments how recognition functions work.

In [53], DBN-R and DBN-Artificial Neural Networks are proposed by the authors for predicting human behavior, along with the widely used prediction algorithm DBN-Support Vector Machines. They also present an online learning algorithm for the restricted Boltzmann machine. After being tested on the MIT dataset, the results generated by the algorithm were more accurate as compared to the traditional algorithms such as SVM, to predict when the user will indulge in a new activity. DBN-R displayed the most promising results.

In [54], the authors introduce SRBM, an in-depth learning model for examining human behavior and meanings in communication networks. The model incorporates all aspects of human behavior such as motivation, perceived social impacts and environmental events. The model outperforms classical methods while predicting future activities. This method uses unique techniques to manage constitutional area knowledge and design of human behavior. Testing on a real social network provides important information:

- Oncology-based user presentations improve the accuracy of deep learning methods for assessing human performance.
- The SRBM model clearly states all decision makers.
- The behavioral educators learned in the model are dependable, and their perceptible accoutrement serves as good explanations for human behavior with individual characteristics.

In [55], the authors proposed a model for deep neural networks, which uses evolutionary algorithms based on neural folding networks [56] to detect human actions. The combination of models created with evolutionary algorithms overcomes the limitations of individual models by combining classes of integral information.

In [57], the authors proposed a multilevel linear regression model that describes consumer performance using action, tasks, internal performance and interaction behavior, using short-term memory. Intensive learning is demonstrated through architecture [58, 59], which models different operating behavior. Architecture provides a likelihood model that allows the user to predict subsequent actions and identify inconsistent user behavior. A number of constructs have been analyzed, each analyzing how they behave for different numbers. The proposed behavioral model is used in many new projects to detect mild cognitive impairment in elderly. Most of the studies discussed in the domain of behavioral study and AAL [60–67] are based on other sensory devices [62, 68].

6 Datasets Used in Sensor-Based Approaches

- MIT Home Dataset [69]: MIT's home data contains two independent databases collected from two homes in two weeks. There were 84 sensors in the first apartment and 77 sensors in the second apartment. Sensors were incorporated into everyday objects, such as cabinets, refrigerators and light switches.
- Kasteran dataset [70]: The dataset was collected by putting 14 binary sensors in a three-room residence. The sensors were put in different places such as cabinets, kitchen, freezers or washrooms. A 26-year-old subject was observed for 28 days, and sensor data was collected. The activities described general tasks such as getting out of the house, using the toilet, bathing, sleeping, cooking breakfast, preparing dinner, getting a drink, using dishwasher, flushing the toilet, opening the fridge and using the microwave (Table 2).

Table 2 Sensor datasets and their respective accuracies

Name of the paper	Dataset used	Technique used	Results and conclusions
"Predicting human behaviour with recurrent neural networks"	Dataset provided by T.L.M. van Kasteren, published on Google	Recurrent neural networks	Higher predictions gave higher accuracies, with five predictions giving an accuracy of 85%
"Online activity recognition using evolving classifiers"	Dataset provided by T.L.M. van Kasteren, published on Google	Evolving classifiers	The evolving classifiers achieved an accuracy of 85% on the dataset
"Accurate activity recognition in a home setting"	Dataset provided by T.L.M. van Kasteren, published on Google	Conditional random fields	The classifiers achieved a class accuracy of 79.4%
"Human behaviour prediction for smart homes using deep learning"	MIT home activity dataset	Restricted Boltzmann machines	The method used shows an accuracy of 51.89% as compared to the traditional methods which show a 43%

7 Conclusion

The paper describes the different levels of human behavioral analysis in the ambient assisted environment. Most promising attributes and recognition techniques for vision- and sensor-based approaches have been detailed. Vision techniques clearly show that many practices produce strong and high success rates but also cause privacy problems for the user. Deep learning models for sensor-based approaches can be very successful and provide an opportunity for future scale and development. Progress in the AAL field can prove to be very valuable as it can help in improving the quality of life. It can also increase life expectancy for cognitively impaired and elderly.

References

1. Affendy AH, Nurilia A (2014) The influence of subconscious mind on human behavior. J Postgrad Curr Bus Res 2(2)
2. Salah AA, Gevers T, Sebe N, Vinciarelli A (2010) Challenges in human behavior understanding. In: Proceedings first international workshop, HBU, Springer, Istanbul, pp 1–12
3. Wang L, Hu W, Tan T (2003) Recent developments in human motion analysis. Pattern Recogn 36(3):585–601
4. Weinland D, Ronfard R, Boyer E (2011) A survey of vision-based methods for action representation, segmentation and recognition. Comput Vis Image Underst 115(2):224–241
5. Yilmaz A, Javed O, Shah M (2006) Object tracking: a survey. ACM Comput Surv 38(4):13
6. Chen L, Hoey J, Nugent CD, Cook DJ, Yu Z (2012) Sensor-based activity recognition. IEEE Trans Syst Man Cybern Part C Appl Rev 42(6):790–808
7. Almeida A, Azkune G (2018) Predicting human behaviour with recurrent neural networks. Appl Sci 8(2):305
8. Yordanova K (2011) Human behaviour modelling approach for intention recognition in ambient assisted living. In: Novais P, Preuveneers D, Corchado JM (eds) Ambient intelligence—software and applications. Advances in intelligent and soft computing, vol 92. Springer, Berlin, pp 247–257
9. Ramos C (2007) Ambient intelligence—a state of the art from artificial intelligence perspective. In: Neves J, Santos MF, Machado JM (eds) Progress in artificial intelligence. EPIA 2007. Lecture notes in computer science, vol 4874. Springer, Berlin, pp 285–295
10. Sun H, Florio VD, Gui N, Blondia C (2009) Promises and challenges of ambient assisted living systems. In: Sixth international conference on information technology, ITNG, IEEE, Las Vegas, Nevada, US, pp 1201–1207
11. Cook D, Das S (2007) How smart are our environments? An updated look at the state of the art. Pervasive Mob Comput 3(2):53–73
12. Camarinha-Matos L, Vieira W (1999) Intelligent mobile agents in elderly care. Robot Auton Syst 27(2):59–75
13. Costa R, Carneiro D, Novais P, Lima L, Machado J, Marques A, Neves J (2008) Ambient assisted living. In: 3rd symposium of ubiquitous computing and ambient intelligence, Springer, Berlin
14. Moeslund TB, Hilton A, Krüger V (2006) A survey of advances in vision-based human motion capture and analysis. Comput Vis Image Underst 104(2):90–126
15. Poppe R (2010) A survey on vision-based human action recognition. Image Vis Comput 28(6):976–990
16. Wu J, Osuntogun A, Choudhury T, Philipose M, Rehg J (2007) A scalable approach to activity recognition based on object use. In: IEEE 11th International Conference on Computer Vision (ICCV), vol 200. IEEE, Rio de Janeiro, pp 1–8

17. Turaga P, Chellappa R, Subrahmanian V, Udrea O (2008) Machine recognition of human activities: a survey. IEEE Trans Circuits Syst Video Technol 18(11):1473–1488
18. Ji X, Liu H, Li Y, Brown D (2008) Visual-based view-invariant human motion analysis: a review. In: Lovrek I, Howlett R, Jain L (eds) Knowledge-based intelligent information and engineering systems, vol 5177. Lecture notes in computer science. Springer, Berlin, pp 741–748
19. Monekosso DN, Remagnino P (2010) Behavior analysis for assisted living. IEEE Trans Autom Sci Eng 7:879–886
20. Chaaraoui AA, Climent-Pérez P, Flórez-Revuelta F (2012) A review on vision techniques applied to human behaviour analysis for ambient-assisted living. Expert Syst Appl 39:10873–10888
21. Brdiczka O, Crowley J, Reignier P (2009) Learning situation models in a smart home. IEEE Trans Syst Man Cybern B Cybern 39:56–63
22. Chung P-C, Liu C-D (2008) A daily behavior enabled hidden markov model for human behavior understanding. Pattern Recogn 41:1572–1580
23. Robertson N, Reid I (2006) A general method for human activity recognition in video. Comput Vis Image Underst 104:232–248
24. Robertson N, Reid I, Brady M (2006) Behaviour recognition and explanation for video surveillance. In: The institution of engineering and technology conference on crime and security, pp 458–463
25. Yamamoto M, Mitomi H, Fujiwara F, Sato T (2006) Bayesian classification of task-oriented actions based on stochastic context-free grammar. In: 7th international conference on automatic Face and Gesture Recognition (FGR). IEEE, Southampton, UK, pp 317–322
26. Lafferty JD, McCallum A, Pereira FCN (2001) Conditional random fields: probabilistic models for segmenting and labeling sequence data. In: Proceedings of the eighteenth International Conference on Machine Learning (ICML '01). Morgan Kaufmann Publishers Inc., San Francisco, pp 282–289
27. Fischer A (2015) Training restricted boltzmann machines. KI - KünstlicheIntelligenz 29(4):441–444
28. Van Kasteren T, Noulas A, Englebienne G, Krose B (2008) Accurate activity recognition in a home setting. In: UbiComp—proceedings of the 10th international conference on ubiquitous computing. ACM, Seoul, pp 1–9
29. Duong T, Bui H, Phung D, Venkatesh S (2005) Activity recognition and abnormality detection with the switching hidden semi-markov model. In: IEEE computer society conference on computer vision and pattern recognition CVPR1. IEEE, San Diego, pp 838–845
30. Nguyen N, Phung D, Venkatesh S, Bui H (2005) Learning and detecting activities from movement trajectories using the hierarchical hidden markov model. In: IEEE computer society conference on computer vision and pattern recognition, CVPR 2. IEEE, San Diego, pp 955–960
31. Hongeng S, Nevatia R (2001) Multi-agent event recognition. In: IEEE international conference on computer vision, eighth ICCV. IEEE, Vancouver, pp 84–91
32. Liu C-D, Chung Y-N, Chung P-C (2010) An interaction-embedded hmm framework for human behavior understanding: with nursing environments as examples. IEEE Trans Inf Technol Biomed 14:1236–1246
33. https://www.di.ens.fr/~laptev/actions/. Accessed 23 March 2020
34. https://www.csc.kth.se/cvap/actions/. Accessed 23 March 2020
35. http://4drepository.inrialpes.fr/public/viewgroup/6. Accessed 23 March 2020
36. https://ias.in.tum.de/dokuwiki/software/kitchen-activity-data. Accessed 23 March 2020
37. https://ias.in.tum.de/dokuwiki/software/kitchen-activity-data. Accessed 24 March 2020
38. http://velastin.dynu.com/MuHAVi-MAS. Accessed 24 March 2020
39. https://www.crcv.ucf.edu/data/UCF_Sports_Action.php. Accessed 24 March 2020
40. Bao L, Intille SS (2004) Activity recognition from user-annotated acceleration data. In: Proceedings of the international conference on pervasive computing. Springer, Berlin, pp 1–17
41. Pratama BY, Sarno R (2015) Personality classification based on Twitter text using Naive Bayes, KNN and SVM. In: International conference on data and software engineering, IEEE, Yogyakarta, Indonesia

42. Fatima I, Fahim M, Lee YK, Lee S (2013) A unified framework for activity recognition-based behavior analysis and action prediction in smart homes. Sensors 13(2):2682–2699
43. Oliver N, Garg A, Horvitz E (2004) Layered representations for learning and inferring office activity from multiple sensory channels. Comput Vis Image Underst 96:163–180
44. Liao W, Zhang W, Zhu Z, Ji Q (2005) A real-time human stress monitoring system using dynamic bayesian network. In: IEEE computer society conference on Computer Vision and Pattern Recognition (CVPR'05), vol 3. IEEE, San Diego, pp 70
45. Ordóñez FJ, Iglesias JA, De Toledo P, Ledezma A, Sanchis A (2013) Online activity recognition using evolving classifiers. Expert Syst Appl 40:1248–1255
46. Van Kasteren T, Noulas A, Englebienne G, Kröse B (2008) Accurate activity recognition in a home setting. In: Proceedings of the 10th international conference on ubiquitous computing. ACM, New York, pp 1–9
47. Bao L, Intille S (2004) Activity recognition from user-annotated acceleration data. Pervasive Comput 3001:17
48. Podgorelec V, Kokol P, Stiglic B et al (2002) J Med Syst 26(5):445–463
49. Safavian SR, Landgrebe D (1991) A survey of decision tree classifier methodology. IEEE Trans Syst Man Cybern 21(3):660–674
50. Ordonez F, Iglesias J, De Toledo P, Ledezma Espino A, de Miguel Araceli S (2013) Online activity recognition using evolving classifiers. Int J Expert Syst Appl 1248–1255
51. Lones M, Smith S, Alty J, Lacy S, Possin K, Jamieson D, Tyrrell A (2013) Evolving classifiers to recognise the movement characteristics of Parkinson's disease patients. IEEE Trans Evol Comput 18:559–576
52. Sutton C, Mccallum A (2012) An introduction to conditional random fields. Foundations and trends in machine learning, pp 267–373 (Now Publishers)
53. Choi S, Kim E, Oh S (2013) Human behavior prediction for smart homes using deep learning. In: Proceedings—IEEE International workshop on robot and human interactive communication. IEEE, Gyeongju, pp 173–179
54. Phan N, Dou D, Brigitte P, Kil D (2016) A deep learning approach for human behavior prediction with explanations in health social networks: social restricted Boltzmann machine (SRBM+). Social network analysis and mining. Springer-Verlag Wien, Austria
55. Paul E, Krishna Mohan C (2017) Human behavioral analysis using evolutionary algorithms and deep learning. Hybrid intelligence for image analysis and understanding. Wiley, Hoboken
56. Aghdam H, Heravi E (2017) Guide to convolutional neural networks: a practical application to traffic-sign detection and classification. Springer, Berlin
57. Du K-L, Swamy MNS (2019) Recurrent neural networks. Neural networks and statistical learning. Springer, Heidelberg
58. Hochreiter S, Schmidhuber J (1997) Long short-term memory. Neural Comput 1735–1780
59. Gers FA, Schmidhuber J, Cummins F (1999) Learning to forget: continual prediction with LSTM. In: 9th International Conference on Artificial Neural Networks (ICANN), IET, Edinburgh, UK
60. Cardinaux F, Brownsell S, Hawley M, Bradley D (2008) Modelling of behavioural patterns for abnormality detection in the context of lifestyle reassurance. Progress in pattern recognition, image analysis and applications. Lecture notes in computer science, vol 5197. Springer, Heidelberg, pp 243–251
61. Hara K, Omori T, Ueno R (2002) Detection of unusual human behavior in intelligent house. In: Proceedings of the 2002 12th IEEE workshop on neural networks for signal processing. IEEE, Martigny, pp 697–706
62. Hayes T, Pavel M, Kaye J (2008) An approach for deriving continuous health assessment indicators from in-home sensor data. In: Technology and aging: selected papers from the 2007 international conference on technology and aging, vol 21. Ios Press, Amsterdam, pp 130–137
63. Jain G, Cook D, Jakkula V (2006) Monitoring health by detecting drifts and outliers for a smart environment inhabitant. In: Proceedings of the international conference on smart homes and health telematics. Springer, Berlin, pp 1–8

64. Mahmoud S, Akhlaghinia M, Lotfi A, Langensiepen C (2011) Trend modelling of elderly lifestyle within an occupancy simulator. In: UkSim 13th international conference on computer modelling and simulation. UkSim, Cambridge, pp 156–161
65. Park K, Lin Y, Metsis V, Le Z, Makedon F (2010) Abnormal human behavioral pattern detection in assisted living environments. In: Proceedings of the 3rd international conference on pervasive technologies related to assistive environments. ACM, New York, pp 9:1–9:8
66. Virone G, Sixsmith A (2008) Monitoring activity patterns and trends of older adults. In: 30th Annual international conference of the IEEE engineering in medicine and biology society. IEEE, pp 2071–2074
67. Wood A, Stankovic J, Virone G, Selavo L, He Z, Cao Q et al (2008) Contextaware wireless sensor networks for assisted living and residential monitoring. IEEE Network 22:26–33
68. Hara K, Omori T, Ueno R (2002) Detection of unusual human behavior in intelligent house. In: Proceedings of the 12th IEEE workshop on neural networks for signal processing. IEEE, Martigny, pp 697–706
69. https://sites.google.com/site/tim0306/datasets
70. https://courses.media.mit.edu/2004fall/mas622j/04.projects/home

Voice Speech and Recognition—An Overview

Ayush Anand and Raju Shanmugam

Abstract In the present scenario, when the automation has gain so much strength, all the support and machinery to human conversation are getting linked with voice controls. In all the field be it Internet of things, artificial intelligence or communication. In order to reduce the efforts on giving instructions and implementing them, the world has changed its behavior toward voice recognition. The voice speech recognition allows the user to understand the spoken voice commands and generate the result on the basis of that. This deals with the voice received and generating outputs based on the hearing capability of the machine. The result is generated and is made available for the further. Voice-based devices or applications are growing a lot. It uses state art process in speech-to-text, natural language understanding, deep learning and text-to-speech. The first step to build a voice application is to listen for user voice constantly and then understand the voice to understand and implement things. This deals with understanding in various languages under the section and reflects the work assigned to it. Speech recognition which is also known as automatic speech recognition (ASR) and voice recognition recognizes the spoken words and phrases and converts them to a machine-readable text, and speech recognition technology let users control digital devices by speaking instead of using conventional tools such as keystrokes, buttons or keyboards. From automated phone systems to Google voice to digital assistant, i.e., voice recognition, ASR helps in daily life activities. Even the bluetooth used in cars uses ASR. Voice recognition has taken a complete scan all over the arena. The tool of automation in voice further deals with the communication among the machines without the code access and suggestively talking with the help of compatible recognizable words. This would strengthen the compatibility and speed up the automation process by decreasing the amount of efforts made. Such work would overcome the error and would possibly give a better way of transformation.

A. Anand (✉) · R. Shanmugam (✉)
Unitedworld School of Computational Intelligence, Karnavati University, Gandhinagar, Gujarat, India
e-mail: ayushananad636@gmail.com

R. Shanmugam
e-mail: srajuhere@gmail.com

A. Abraham et al. (eds.), *Proceedings of 3rd International Conference on Computing Informatics and Networks*, Lecture Notes in Networks and Systems 167,
https://doi.org/10.1007/978-981-15-9712-1_29

As Edmund L. Andrews researcher from Stanford recently on March 23, 2020 said that the possibility of error could be minimized to a great extent with the evolution under this process.

Keywords Speech to text · Text to speech · ASR · HMM · Neural network

Abbreviations

ASR	Automatic Speech Recognition
HMM	Hidden Markov Model
NN	Neural Network
WER	Word Error Rate

1 Introduction

In voice speech recognition, the users are made to make the voices in the customized language which has to be understood by the system, and interpretation is made what was said. This enables the microphone and stresses on the recognition factor of the different voices generated by different people by picking up the word from the speech made. The search can be performed on any device with input as voice. Language is the most important factor for a system to understand and give the most accurate results of what the user search as output. This covers across languages, dialects and accents, as users want a voice assistant that understands both of them and speaks to them with a better understanding. The search method is similar to performing normal search on the Web site; the only difference is the search that is conducted using speech, rather than text.

This covers the ability of a machine to decode the voices made by human as an input signal to the system. The main purpose includes operating a device, performing the commands assigned by the most common is text-to-speech format wherein the system understands whatever the human is saying, and after understanding by differencing it on the language basis and various pitch and difference in voice, it recognizes and writes the suitable text which is further ready for other various activities. The result can be searched each time by calling out the program and is limited to ten words. Further, chunk size and sampling rate are adjusted accordingly in order to get the complete out of the system.

It is expected to meet the demands by hearing out the same sound with different frequency and pitch with different voice, and it has been made understandable where the system understands the voice easily and shows quick implementation and the result of it from the voice made for the system (Fig. 1).

Fig. 1 Hierarchy automation model

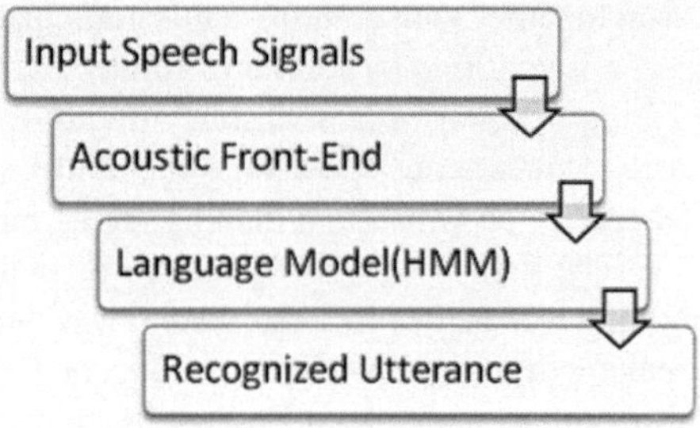

Among the technology gaining the market, in 2018, Amazon alone sold around ten million of echo devices marking as one of the biggest selling devices. Further, the paper suggests the development in the automation field where the system could communicate among each other and deal with the solution to the problem in real quick time. This became very problematic and a difficult scenario to deal with, yet with the inducement in technology, the program was called off. Wherein it was seen that the system earlier started communicating in their own encrypted format and the project had to shut down, this being the challenge where the encryption should be held with the developer so that they can auto tune the faults. Various fields such as call centers and IVR systems use speech recognition tool for many organizations and for self-service as well. The load of the customer and client is cleared off with the help of this. Further, it has a lot of practical deployment in the field of dictation solutions. The speech-to-text technology has given the opportunity to express the ideal without typing anything. Even the field of education is entirely covered up with the recognition technology, where the students are creating and documenting the word files without actually typing. This has made the work and load for the students much easy and has reduced a great amount of load from them. For the disabled, the applications serve as a new hand where they are able to learn things more quickly and adapt the learning habitat as fast as any other people. This paper deals with the growth of automatic voice recognition in the field of automation where the system communicates within them creating a sentimental analysis for them and leading to the easier work.

2 Literature Survey

Voice speech recognition roots can be traced on from an early time and could be majorly understood by understanding how we as human are able to recognize the statements made. These happen under a set of parameters of speech classification which includes the isolated words which are the single words that are interpreted one at a time, and then, according to that, it is interpreted. Following are the connected and continuous words. Under connected words, it allows the users to speak naturally, and the computers are made to examine the words, but it allows the separate words to run together with creating a minimum pause in between. Another classification of speech is continuous, similar to connect words, it allows under to speak naturally by creating boundaries and various difficulties aroused, while speech is implemented. Lastly, spontaneous speech has the ability to take control of the various voice tools and connected words and making them work altogether.

As the first speech recognition was mainly focused on numerology rather than building words. In 1952, the system was developed which recognized a dingle voice speaking digits that too in loud frequency. From that, it took 10 more years for IBM (Shoebox) to understand English language and responded to 16 words. Much later came the concept of hidden Markov model (HMM), and it took as a storm by the 80s. Further with due course of time, it was made to consumer availability in the later 90s where ASR was introduced known as automatic speech recognition.

The best extract for speech recognition is ASR. However, other there were other recognition systems which were speaker-dependent system, under which it requires training before any word or sentence is made to system. Speaker-independent system is the software recognizes voice with minimal training. In discrete speech recognition, the user has to take a pause before speaking the other word in order to let the system understand. In continuous speech recognition, the system understands in the normal speaking flow, and lastly, in natural language, not only understanding is there but an automated response is also generated based on that.

With the involvement of recognition among the automation for making the workload easier, it was to be stated the condition of work and the related work in this field. In the year 2009, Anusuya [1] stated the field of acoustic–phonetic approach to the recognition along with artificial intelligence and discriminative learning—HMM-ANN. In the year 2012, Singh [2] stated the explanation of hidden Markov model involvement in the recognition field to reduce the load. It showed the development of a speech model which was created on the platform of machine interface. Sahu [3] in May 2013 further added to the field the concept of connected words where the different lengths of the word took different time for the system to apt the voice. Lleida [4] stated that the utterance of the similar would be taken as single word to reduce the load on the machine and the paper dealt with the making the machine more optimized to get only the required data and ignore the rest of the body. In October 2017, Cambria [5] under his article on the review of research paper developed a theory which stated that about reduction of 30% in the work would be done if the sentiment analysis of the systems is done. Burt [6] in October 2019 stated the quality configuration in speech where the receiver was given the priority as the interference from the source might cause problem. Shobha Dey [7] in April 2014 did the work in the intelligent recognition of the words where the predictive tests were converted into the closest similar word so that better and accurate results could be achieved. Sharma and Hakar [8] in 2012 kept the main focus on speech denoising using different types of filters which could help in the clarity of the tone and volume of the content. Acharjee et al. [9] presented a paper on voice recognition system: Speech-to-text, where the continuous format of speaking style was directly converted into speech with intellectually calculating the pause between the words and at the same time prediction analysis was performed on the words to get the accurate sentence. Paliwal [10] in the year

2003 developed the use if the spectral co-bands of centroids where it displayed the dynamic approach to SSC coefficients revealing that they are stronger enough to deal with the noise than the MFCC specifications.

Over the year's speech recognition has gained a physical importance and in the technological field, the advancements are seen much more in information recovery where the information is recovered using the concept of speech recognition tool. Artificial intelligence comes into the picture which is a very important field in relation to the work done in speech-to-formatting or text-to-speech, the ultimate goal is to translate and judge the automation just like a human would do, and the automation here just increases the capabilities with it. Further, with the artificial intelligence as base, it is done on the basis of bottom up, top down and the blackboard ways. In the case of bottom up, the cases with the less priority are induced first, and then, the cases with the higher priority are instantiated. The reverse of this takes place in the top down approach. This sums up the involvement of artificial intelligence in speech recognition tool. However, the blackboard approach takes the inclusion of various approaches among the acoustic, semantic, surrounding and the easy methodology process to act.

3 Problem Statement

In the early voice recognition tool, it had capabilities of hearing out only numerical values and further extends to hearing a sentence that too with a limited training to the system and mainly focused in one language. Then came the advancement where multiple language support was possible and automated response could be generated.

Whenever we say a word or call out a sentence that is understood by the system, it interprets in its own way, by generating the various results, as different users have different voices. Filtering out all the disturbances and noise cancelation of the voice, the sentence is interpreted as a natural language, and based on that, result is generated, with the system having a functioning microphone and for an automated response a speaker. The search is made on the various search engines by making the speech.

Further, the utterances are the area to be dealt of, and it got grouped into short and long utterances where more than 10 words say up to 100 words can be taken and least specific which means ignoring the basic pronunciation errors in favor of fast customer experience with taking less interruptions in the code. While under short, it is focused on very specific variation of speech. Various training data of different voices are assigned to overcome this and in order to provide an accurate result.

4 Proposed System

The basic working of the voice speech recognition is the intake of speech signals which goes under preprocessing, and language modeling is performed on it with the help of various models such as uniform model and stochastic Model. Speech signals from sources are captured with the help of microphone, where generally sampling of frequency ranges between 8 and 20 kHz. In the need of calling signals of voice, it is advised for the signals to have a sampling rate of 8 kHz, and the general microphones are rated to be at the frequency of 16 kHz. Modeling of language is done in order to spot the right word sequence by predicting the nth words using preceding word technique.

Our proposed system is automatic sound recognition where it uses the concept of hidden Markov model, in which an alternative approach speech recognition is to construct a model of each word to recognize the vocabulary and make the system understand working of it. This is a modeling approach which consists of three major parameters namely the model, the method of computing the probability of the model and the methods of computing the parameters of the model. Our system here deals with listening to the user and understanding it to the different levels of connected, isolated and continuous word. On taking these values as input, the result is generated which shows the result in the form search made. The model thus suggests the capabilities of using the automation search recognition techniques in order to communicate between the systems to remove the chances of error and help to grow it faster and makes it more reliable and data is accurate. The workload is also reduced with the inducement of such act. The recognition saves time, and it helps in understanding the speech in different voice and sentiments to judge them as a question or an imperative remark. These studies are done by making the system learn by providing them a different set if training data. The length of words further creates an another set of test cases for the system to understand. All these tasks are performed on a unit testing basis to increase the readability and decrement in chances of error by the system.

The tool of speech recognition in machinery helps in the automation among the machines wherein the system will help in communicating with each other and reduce the human effort of hard coding. The system understands the concept of sentiment analysis and further contributes to each other with the medium of communication, thus helping resolving the queries system is developed in such a way that it understands the sentiment of other devices working along and takes the input signals from them and regulates on the basis of that making it much easier, cost effective and reliable to the system. The idea has been taken from the tech system developed by Facebook, where the system designed to work for the designated work started to talk among themselves and further the project had to call off as the securities issues lacked because in contrast to the work done the outcome showed that they were talking in an encrypted format making it harder for the system to understand the working procedure of the system and thus it turned out to be serious threat as the security concern was not taken. The cause of such an event could be drawn to the sentiment analysis where the understanding is developed and it further takes care of the methodology of the system. Speech recognition is divided into four subcategories on the basis of its properties, and they are speech mode, speaker mode, speaking style, word size and spoken styles. All these deals with the way the voice is enabled in order the system to understand the given data. Under speech mode isolated where the words are separated by a good pause and other mode is continuous where the speaking is

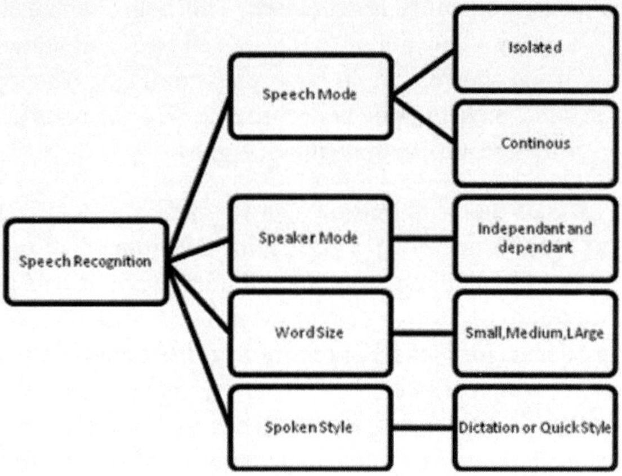

Fig. 2 Speech recognition classification

done in the flow for the system to understand the data. Speaker modes are further categorized as independent and dependent parts. The type of words that are given to the system to analyze may deal with problem with the long, short and medium words. Here, the complexity increases with the increment in the length of the words. The spoken style is however dependent of the type of style which is dictation, the continuous format or the quick style (Fig. 2).

5 Methodology

The proposed method includes the involvement of few classifiers and those are

1. Hidden Markov Model (HMM)
2. Neural Network (NN)

1. **Hidden Markov Model**: The hidden Markov model is a model basically a statistical model where the system is modeled and the modeled system is assumed to be a process called the Markov process where the parameters seem to be unknown as the wholesome tasks lie in generating the hidden parameters from the given data which is also known as observable data. It is the heart of the voice recognition facility and provides a natural framework for connecting such models to it. Further, it helps in finding the unknown variables from the list of known variables.
2. **Neural Networks**: The neural network is a concept of analyzing the data based on certain flow and list of algorithms that explains the flow very similar to as that of a human brain has the capabilities of interacting to the system. The network has an adaptability behavior where it plays as the smartest role in which it can change its behavior according to the condition and mark itself as that to help the situation and paradigm. A neural network is trained by adjusting the inputs taken and based on the network performance. The network classifies the data correctly, which helps increases the probability of correct answers while the other similar

data probability is decreased. This helps in canceling out the other close words that are very similar to the spoken ones and allows the system to make the best of the choices, and it chooses accordingly. The signals sent make a connection sets which compile in accordance with the units and with help of the training set which results in the output (Fig. 3).

Under the proposed path, it requires a general set of skill and proficiency for the user to make the right choice of pronunciation and clear voice as the voice received by the system is filtered under various categories of modeling concepts in the underlying design of the model. The system receives the voice interpretations and on subcategory levels classifies it to the statement received and then is interpreted which is then answered by an automated response on the basic understanding of the system. However, it currently deals with the text-to-speech interpretation, and the link is forwarded to the search engine where the search iOS made and the result is displayed. Pattern recognition deals with acquiring the details from the machine and dealing with the understanding and producing a result dataset. The input speech signal is sent where the speech signals are learned and understood and is followed upon to pattern comparison which takes help from the pattern of unknown signals for the accuracy, and then, the output is generated. The performance of the voice recognition system is solely based on the swiftness and how précised the data is going to be. The accuracy measured in terms of the performance accuracy is usually rated with word error rate (WER), wherein the swiftness measured in the single time factor. Other two factors are important to the contribution and that includes the word error and command success. However, the performance basis of judgment was mainly dependent on the error occurrence of the words connected to it, and the system translation deals with the accomplishment of the recognition in voice. The difficulty was in understanding the difficult words coming through the way as different words have different length. The problem was resolved by first aligning recognized word sequence with the speaker and taking help of the dynamic alignment of the string that is supposed to be the word.

Word error rate can formulate into

$$\mathrm{WER} = (X + Y + Z)/C$$

Where

X Count of substitutions

Y Count of deletions

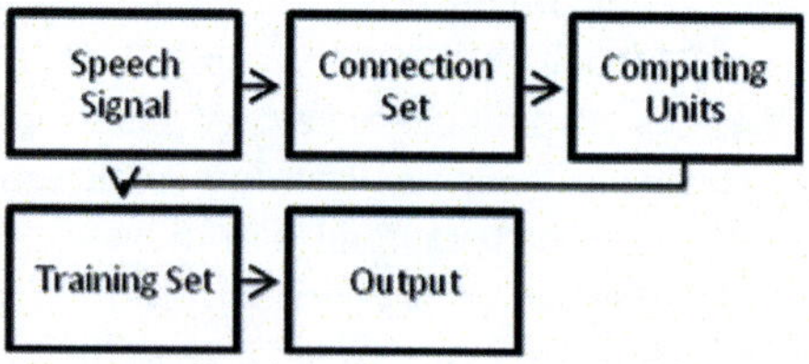

Fig. 3 Neural network approach

Z Count of insertions
C Count of words in the reference.

6 Conclusion and Future Research Directions

Automatic speech recognition has become a strong tool for the system communications to be made automated. The limited availability of the resources could not stop the growth anyway. But with the right techniques and extensive use, it can be made more productive for the customer experience as well as the automation levels in the production machineries as well. The future of the speech recognition for various uses in the various fields will get wider and wider and is going to be one of most vast arenas ahead.

The training data and recognition can be speed up by making it more prone to the environment conditions. Seeking voice of different personalities is tough. However, a major work is being done on that by teaching the system through many dataset available. The platform is going to support the demands of customer as well as automation needs too. The current usage of speech in consumer fields has increased for fold. According to the survey analysis of get elastic in 2019, it was found that around 20% of queries are submitted through voice on Google. 72% of people use voice recognition to manipulate and use devices. The global market has boomed to 187%. Considering the same if this automation is reached in the field of automation design, then the market would grow faster, and the workload can be reduced at a great level.

However, the work has been done on this level as Ford and Toyota have already been introduced with Alexa over few years and Hyundai has teamed up with Google to make it as a virtual assistant. Around 64% of the customers have shown their interest and investment to the technology. Predictions suggest that by the end of the decade around half of the search will be done through voice search. In USA, the rise would be from 13 to 55% in upcoming two years. The field of automation among the machines to communicate themselves has started, and with the proper encrypts voice enabled partnership, they could work as fine as ever.

Not only in the field of automation but the speech recognition has been a boon to impaired patients with difficulty in making out the sentence or words to the people. With the help of tools, these are made possible some of which includes HTK, Julius and Sphinx which are open source toolkits.

References

1. Anusuya MA (2009) Speech recognition by machine. IJCSFS 6(3)
2. Singh B, Kapur N, Kaur P (2012, March) Speech recognition with Hidden Markov Model
3. Prabhakar O, Sahu NK (2013) A survey on voice command recognition technique. IJAR 3(5)

4. Lleida E et al (2000) Utterance verification in decoding and training procedures. IEEE Trans. Speech Audio Process 8(2)
5. Cambria E (2017) A practical guide to sentimental analysis
6. Burt C (2019, October) Ensuring quality in voice biometrics data collection
7. Dey S (2014, April) Intelligent system speech recognition
8. Sharma K, Hakar P (2012) Speech denoising using different types of filters. Int J Eng Res Appl 2(1)
9. Acharjee K, Das P, Prasad V (2015, November) Voice recognition system: speech-to-text
10. Paliwal KK, Gajic B (2003) Conference: acoustics, speech, and signal processing. In: Proceedings. (ICASSP '03). 2003 IEEE international conference
11. Spratt EL (2002) Computers and art in the age of machine learning
12. Yu D, Deng L (2015, March) Automatic speech recognition—a deep learning approach
13. Wankar P (2014) Research paper on basic of neural network
14. Damodaran S (2015) Guidance for hearing impaired people 4(2)
15. Sharma R (2004) Design and implement multi-language converter using machine learning techniques. IJCSMC J
16. Bajorek JP (2000) Guiding principles of voice user research 1(1)
17. Doyle S (2006) Automatically improving a voice recognition system
18. Science in Me (2020, April) Voice and speech recognition market analysis by 2020
19. Magrath D (2020, March) Listening and speaking: teaching hints
20. Singh K (2012) A review of speech literature

Prediction of Kyphosis Disease Using Machine Learning

Pavithra Chatter, D. V. Swetha Ramana, S. Suzain, and P. V. Suma Latha

Abstract The word kyphosis refers to the spinal curve that reflects an aberrantly rounded back that might happen as a result of distress, contagious disease, developmental anomalies, inherited and also sometimes iatrogenic disease. Kyphosis can occur to people of all ages and is commonly found in adolescents and children affected by malnutrition which are at times very difficult to predict or to confirm as they come in different postures and sizes. The main purpose of this paper is to predict and suggest the necessary treatment to be given to the diseased based on the data given by the user. The admin creates the data-chart of the patient based on the data given by the patients and prepares the treatment form for the patients where the details of the severity of the disease are also mentioned. The admin provides information about the hospitals suitable for the treatment of the patients. Depending upon the present condition of the patient and his/her needs, he/she should be treated. Our experiment data is collected from government hospital Vijayanagara Institute of Medical Sciences, Ballari, and the treatments were discussed and suggested by Doctors in VIMS, Ballari. The main goal is to help them provide the treatment for their recovery.

Keywords Kyphosis · Scheuermann's kyphosis machine learning

P. Chatter (✉) · D. V. Swetha Ramana · S. Suzain · P. V. Suma Latha
Rao Bahadur Y Mahabaleswarappa Engineering College, Ballari, Karnataka, India
e-mail: pavithracse.rymec@gmail.com

D. V. Swetha Ramana
e-mail: swethasrihari@gmail.com

S. Suzain
e-mail: suzainsamms0026@gmail.com

P. V. Suma Latha
e-mail: sumalathapv1@gmail.com

A. Abraham et al. (eds.), *Proceedings of 3rd International Conference on Computing Informatics and Networks*, Lecture Notes in Networks and Systems 167,
https://doi.org/10.1007/978-981-15-9712-1_30

1 Introduction

Kyphosis in general terms can be described as a spinal curve that results in the hunched back as shown in Figs. 1 and 2.

1.1 Kyphosis Disease

Kyphosis disease is classified into three types:

- Postural
- Scheuermann's
- Congenital

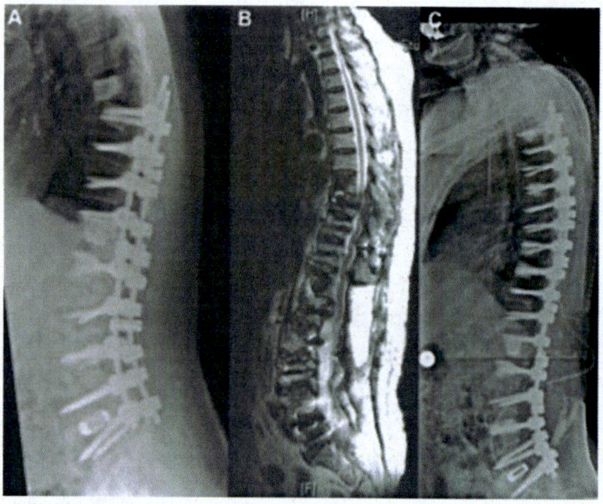

Fig. 1 Images of a person with kyphosis

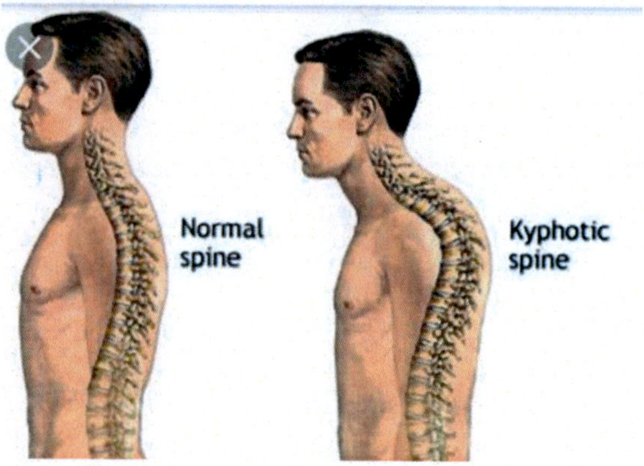

Fig. 2 Images of a person with kyphosis

Postural Kyphosis
As a consequence of bad sitting and standing postures, humans have round back or postural kyphosis. This round back problem is very usual in the age groups from 12 to 18 years, as the youngsters usually hunch resulting in the bent spine [1, 2]. When the patient lies on a flat surface, these abnormalities are not seen on x-rays properly. These postural problems need to be corrected immediately by doctors with proper education, and treatment is suggested by doctors in the form of exercises which strengthen the back muscles.

Scheuermann's
The thoracic curve is generally between 45 and 75° in this type of kyphosis disease [3]. This might cause usually when more than three side by side placed vertebrae have vertical wedging of more than 5° and have a triangular look, which causes in cut down of the space between those vertebrae. With Scheuermann's disease, the patient will have many problems like Schmorl's nodes, herniated disk, thickened anterior longitudinal ligaments and spinal deformity [4]. It was first seen in Scheuermann, the Danish radiologist after whose name the disease is named. It was understood by the radiologist that this was because of avascular necrosis, which stops the growth of the bones [5].

Congenital
Congenital kyphosis is due to the inherited aberrant development of the spine. This disease is most severe and common and might be because of incomplete formation of the spine in the mother's womb [6, 7]. This congenital kyphosis is usually accompanied by an abnormal urinary collecting system. The doctor usually suggests an x-ray of kidneys and a spine MRI. Treatment for this is surgery which and can stop the progression of the curve. Non-surgical treatments with close observation and follow up are not successful.

The main purpose of this paper is to predict whether the person is affected by kyphosis based on the data given by the user. Through this method, it is easier to predict the disease in early stages and provide them with necessary treatment as at times if delayed it could become fatal.

2 Literature Survey

In [8], the authors studied the relation between forwarding head posture, rounded shoulders and added kyphosis disease. These three problems may occur either in the blend or alone. Cervical lordosis values and thoracic kyphosis values were strongly correlated.

In [9], authors studied adult kyphosis and discussed different surgical and non-surgical treatments. The non-surgical treatment, which is not improved from years, was done to reduce the symptoms and involves physical therapy. The surgical treatment, however, has evolved drastically to improve the results. These surgical treatments for different conditions were explained by presenting three cases. In [10], authors to detect and predict kyphosis disease applied machine learning algorithms Random Forest, Support Vector Machines, Artificial Neural Network and found that Artificial Neural Networks is more accurate than other two algorithms.

3 Working

The registered user's data-chart is formed by the admin by the details which were the outcome of his test. The admin performs the further process by scheduling the patient's treatment process and the type of the physician to be consulted to treat the diseased.

- The required algorithms and data must be fed into the systems in advance.
- The admin latter initiates the finding, extracting and summarizing the relevant data of the diseased.
- The latter makes predictions based on data analysis.
- Further calculates the probabilities for specific results based on the data of how worse the situation of the patient is.
- The patients are made to adapt certain developmental processes autonomously optimizing the process based on recognized patterns of the spine.

3.1 Algorithm

Start

a. User logs in after registration.
b. Enters details of the X-ray report like spine curvature degree, age and address.
c. Classify a patient based on two features degree and age.
d. Check thoracic curve degree
e. if (degree >45).
f. The risk of that person having kyphosis is high.
g. Check age feature.
h. if (age >10) and (age <15).
i. Label the patient with kyphosis = 1.
j. Depending upon the start and number features, the patient is suggested treatment.
k. View the treatment.
l. Shows the kyphosis prediction graph.

End.

4 Experimental Setup and Results

In this section, we discuss an experiment done to predict and treat kyphosis disease based on four features, out of which two features age and degree are used for prediction and two features start and number are used to suggest treatment

4.1 Data Sets

The data sets which are used for predicting kyphosis disease in this paper are collected from postgraduate students, Vijayanagar Institute of Medical Sciences, Ballari, Karnataka, India. Data is collected based on patient history. The data set collected consists of 200 instances and characterized by four attributes age, number, start and degree. The data set further is divided into training data and test data according to the ratio 75–25, which is 75% data is considered as training set data. And the rest 25% of the data is considered as test set data.

4.2 Results

Figure 6 presents the sample results of this work. It labels accurately whether a patient is suffering from kyphosis disease based on two features age and thoracic curve degree, if so, our software suggests treatment to the patient and also nearby hospitals that treat kyphosis disease. To evaluate our work, we cross verified with treatment sheets manually which were discussed and collected from VIMS, Ballari.

Figures 3, 4, 5, and 6 which explain how this particular software developed by us works to predict and suggest kyphosis disease.

5 Conclusion

A person suffering from kyphosis, depending upon his present condition and requirements, has to be treated, as some might be congenital and some iatrogenic. Those patients suffering from kyphosis at early ages in some situations can be treated easily by counseling and exercises, in some situations might need to undergo surgery, or handled by making them wear braces based on the severity of the patterns. The data given by the patients plays a vital role in predicting the disease and providing information about the necessary treatment to be taken by the diseased. Our software predicted accurately the disease and suggested nearby hospitals and treatments, which was appreciated by doctors in the government hospital in our city. Our work

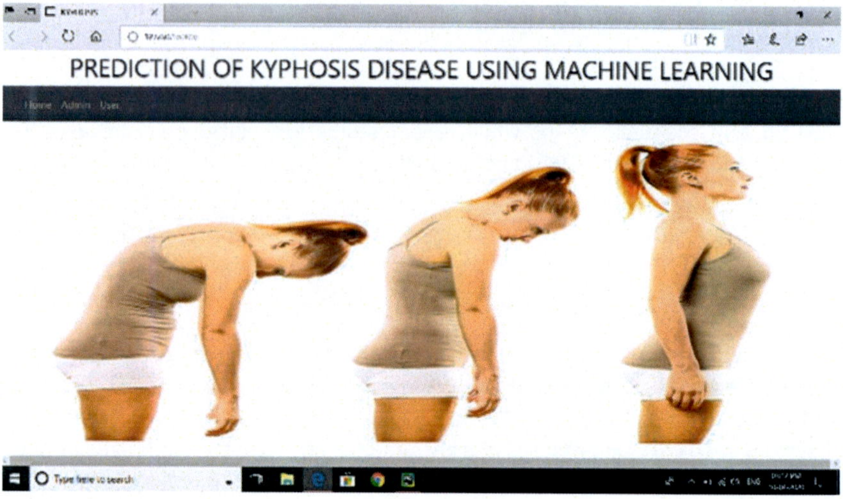

Fig. 3 Home page showing the process of the treatment. It consists of the admin and user page

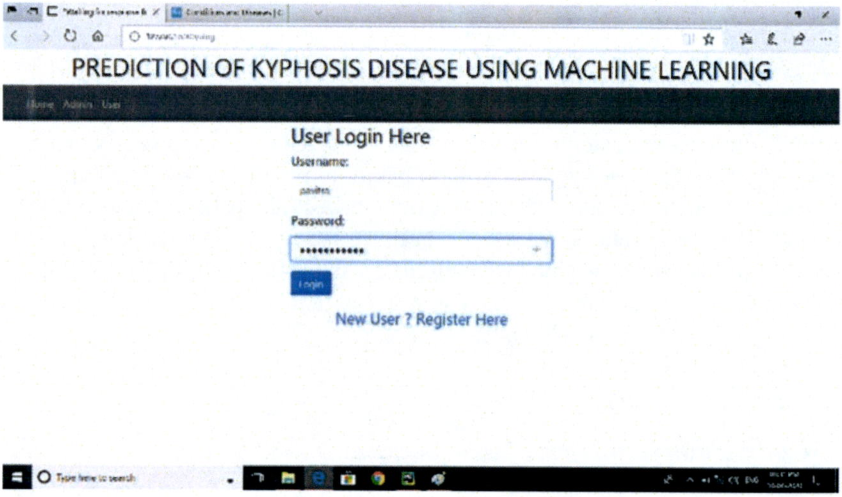

Fig. 4 User login page where the user login by entering the credentials required. After entering into the page, the user fills the form which contains the details about his/her disease, thus making it available for further treatment

did not consider pulmonary condition which is also an important attribute of kyphosis disease.

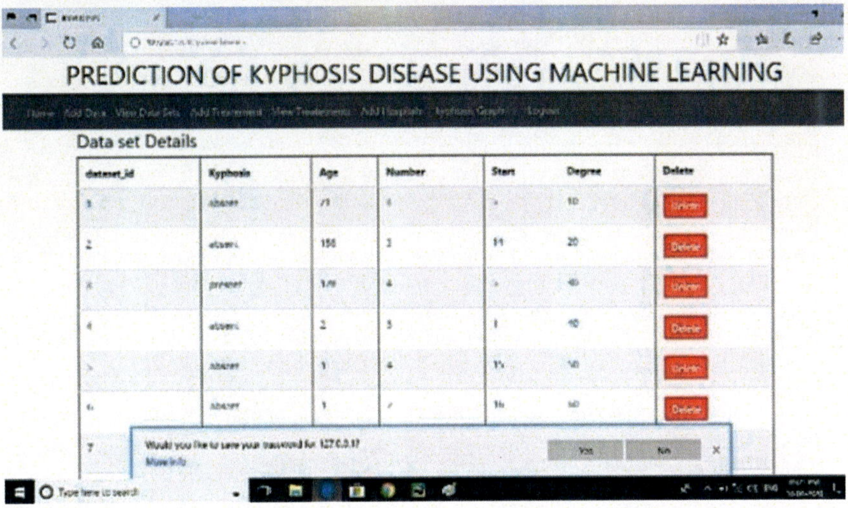

Fig. 5 Data set used for this paper which consists of attributes id, a label named kyphosis, age, point number, start number and degree at which angle is the thoracic curve

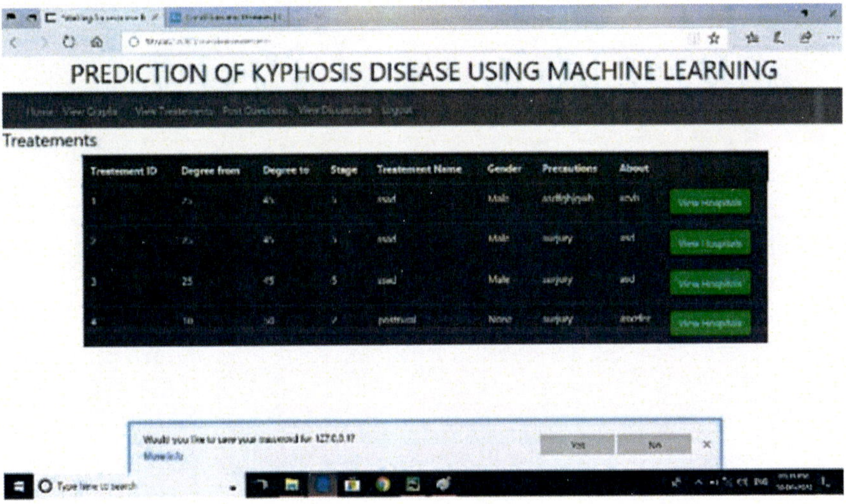

Fig. 6 Treatment suggested by our software for the patient

References

1. Been HD, Poolman RW, Ubags LH (2004) Clinical outcome and radiographic results after surgical treatment of posttraumatic thoracolumbar kyphosis following simple type A fractures. Eur Spine J 13:101–107
2. Bell OF, Walker JL, O'Connor (1989) Spinal deformity following multiple laminectomy in children (abs). Presented at the 56th annual meeting of the American academy of orthopaedic

surgeons, Las Vegas
3. Bernhardt M, Bridwell KH (1989) Segmental analysis of the sagittal plane alignment of the normal thoracic and lumbar spines and thoracolumbar junction. Spine 14:717–721
4. Booth KC, Bridwell KH, Lenke LG, Baldus CR, Blanke KM (1999) Complications and predictive factors for the successful treatment of flatback deformity (fixed sagittal imbalance). Spine 24:1712–1720
5. Blumenthal SL, Roach J, Herring JA (1987) Lumbar Scheuermann's. A clinical series and classification. Spine 9:929–932
6. Benson ER, Thomson JD, Smith BG, Banta JV (1998) Results and morbidity in a consecutive series of patients undergoing spinal fusion for neuromuscular scoliosis. Spine 23:2308–2318
7. Bick EM, Copel JW (1950) Longitudinal growth of the human vertebra. A contribution to human osteogeny. J Bone and Joint Surg 32A:803–814
8. Singla D, Veqar Z (2017) Association between forward head, rounded shoulders, and increased thoracic kyphosis: a review of the literature. J Chiropractic Med 16(3):220–229
9. Zhang P, Peng W, Wang X, Luo C, Xu Z, Zeng H, … Ge L (2016) Minimum 5-year follow-up outcomes for single-stage transpedicular debridement, posterior instrumentation and fusion in the management of thoracic and thoracolumbar spinal tuberculosis in adults. Br J Neurosurg 30(6):666–671
10. Dankwa S, Zheng W (2019) Special issue on using machine learning algorithms in the prediction of kyphosis disease: a comparative study. Appl Sci 9(16):3322
11. Galbusera F, Casaroli G, Bassani T (2019) Artificial intelligence and machine learning in spine research. JOR Spine 2(1):e1044

Natural Language Processing: History, Evolution, Application, and Future Work

Prashant Johri, Sunil K. Khatri, Ahmad T. Al-Taani, Munish Sabharwal, Shakhzod Suvanov, and Avneesh Kumar

Abstract It is quite hard to imagine a smart system like a voice assistant or a chatbot or a recommender system without natural language processing (NLP). It all starts with an initial unit that first interprets the data (audio or text) provided and then start making sense of the data, and after proper processing of the data, the actual steps are followed by the machine to throw some replies or get the work done. NLP does not fall under a discipline; rather it is a part of several different disciplines, i.e., computer science, information engineering, artificial intelligence (AI), and linguistics. The concern of NLP is the interaction between a computer and human languages. NLP areas include speech recognition, machine translation, automatic text summarization, part-of-speech tagging, etc. Generally, NLP is used in many real-time applications like smart homes, smart offices like Alexa, Cortana, Siri, and Google Assistant. The history of NLP generally started in the 1950s and has come a long way from then and improved a lot. This paper discusses the history of NLP, its evolution, its tools and techniques, and its applications in different fields. The paper also discusses the role of machine learning and artificial neural networks (ANNs) to improve NLP.

P. Johri (✉) · M. Sabharwal · A. Kumar
Department of Computer Science, Galgotias University, Greater Noida, India
e-mail: johri.prashant@gmail.com

M. Sabharwal
e-mail: mscheckmail@yahoo.com

A. Kumar
e-mail: avneesh.avn119@gmail.com

S. K. Khatri
Amity University, Tashkent, Uzbekistan
e-mail: sunilkkhatri@gmail.com

A. T. Al-Taani
Department of Computer Science, Yarmouk University, Irbid, Jordan
e-mail: ahmadta@yu.edu.jo

S. Suvanov
Faculty of Mathematics and Informatics, Samarkand State University, Samarkand, Uzbekistan
e-mail: shakhzod_suvanov@yahoo.com

A. Abraham et al. (eds.), *Proceedings of 3rd International Conference on Computing Informatics and Networks*, Lecture Notes in Networks and Systems 167,
https://doi.org/10.1007/978-981-15-9712-1_31

Keywords NLP · Machine learning · ANNs · Deep learning · ATNs

Abbreviations

AI	Artificial Intelligence
ANNs	Artificial neural networks
ATNs	Augmented Transition Networks
DL	Deep Learning
DMMH-SVM	Dual-Margin Multi-Class Hypersphere Support Vector Machine
ML	Machine Learning
NLP	Natural Language Processing
RNNs	Recurrent Neural Networks

1 Introduction

In today's fast-moving world, almost everything is just a one click away. Hey Google, what the weather would be like next Monday? You may want sunglasses; it would be 18°C (Sunny). It is a quite basic command for the voice assistant, and it is a way be low its capability. We have come a long way in AI and life without it will be difficult. This came as a result of the continuous efforts of researchers in the field of AI. NLP was proposed to bridge the gap between the human language and computer's understanding. Researchers proposed different approaches to enable the computer to process and understand the human language. One of the early researches in NLP was in machine translation. The goal of machine translation is to develop automated programs to translate text or speech from one natural language to another.

The reach of AI is not confined in a field rather it is expanded in every field that one can think of; this is somewhat the same for NLP. Several fields of study are incorporating the concept of NLP to make systems robust and automated to meet the needs of the future. Apple integration of Siri voice assistant on their iPhones in 2011 was the first groundbreaking achievement of AI especially NLP that was very visible even with little experience in technology. The assistance was not too much from experienced staff, whereas it opened the door for other manufacturers to do their experiments in the technology and came up with more accurate and advanced products. Later, different companies started proposing new approaches in NLP in different fields such as medical field.

Recent NLP approaches are based on deep learning. Earlier approaches that employed deep learning did not gave good results since the processing power needed for deep learning implementation is very high. Nowadays, we have very efficient computers that can perform complex tasks within a fraction of a seconds, and the data required for training the machine learning model is abundant too. Most of the AI technologies use NLP as a crucial part. During the past decade, many NLP approaches have been proposed. Some of the ongoing research in NLP investigates various ways for improving deep learning approaches used in NLP, such as the use of recurrent neural networks (RNNs) to guess the theme of the article and suggesting the upcoming word in a sentence.

The primary objective of this paper is to provide an understanding of the rise of NLP, its evolution, recent applications and suggest future applications that can take advantage of this technology.

The paper is organized as follows. Section 2 discusses the history and evolution of NLP from merely a word translation to a real-time language processing. Section 3 discusses the implementation of NLP approaches. Section 4 explains the tools and techniques that NLP incorporates and the use of different components for improving its effectiveness. Section 5 presents the applications of NLP and how NLP is contributing in different fields. The last section discusses the gaps in NLP research and the areas where NLP still needs refinements for future work.

2 History and Evolution

Language is a medium of delivering thoughts, information, and ideas along with emotions, imperfections, and ambiguity. It is difficult to think of a language as a combination of mathematical rules that works simultaneously to form a sentence or a phrase. Thus, making it difficult to form logical combination that can be applied to the machine, to make it understand the language and return the results in the same way.

The first time when the world came across the term "translating machine" was in the mid-1930s when the first patent for "translating machine" was appeared. There were two patents for the same technology. First was for Georges Artsrouni, who used a bilingual dictionary to map the words of one language directly to another using paper tape. It was basic approach and did not provide a way to deal with grammar for a language. The second approach was given by a Russian, named Peter Troyanskii, who gave a detailed strategy to tackle the grammar of a language. He used the bilingual dictionary along with a method to deal with grammar of the language established over Esperanto (originally titled, the international language). Both were the disruptive approaches toward the technical domain, but they fail to provide the working model and remain the conceptual approaches only.

The first attempt of using NLP was by the Germans in World War II. Enigma, the name of the machine, was used for encrypting the secret message of Germans' into the secret code and is used also to transfer the message to the field commanders and military units of Germans placed in Europe, to further arrange the attacks and met the requirements of the army. It was one of the great achievements of Germans to be able to communicate secretly even in the most precarious situations. Later in 1946, Britain comes up with, Colossus, a machine that was capable of successfully decrypting the secret code generated by Tunny (code name given to Enigma, by the

British). It was the turning point of World War II since the British was able to know the position of the German military, their strategies, army conditions, strengths and could take prior action accordingly.

During World War II, the British government cryptographic establishment was placed in Bletchley Park. Bletchley Park was the place where Alan Turing, along with other intelligence agents, decoded the German message and provides insights about the German military. This was indeed the groundbreaking achievement in World War II, but it was also the first wave for the advancement in the field of modern-day computing.

In 1950, a Turing test (imitation game) is proposed by Alan Turing in order to determine whether a computer can think like a human or not. The test includes three players: a man (player-1), a woman (player-2), and an interrogator (player-3). Player-3 tries to find out the gender of player-1 and player-2, with the help of written conversation only. But the catch of the game was, player B will lead the interrogator to the correct solution, and player A will try to trick the interrogator and lead it to the wrong solution. Now, Turing suggested replacing player A by a machine. If player C successfully able to identify the gender of both the players, the machine fails the test, otherwise it wins the test [1]. This test was not to make a machine solve the problem but to identify whether the machine can perform tasks that humans can do indistinguishably or in other words, can machine think the way humans think.

Any language is incomplete without the use of proper grammar. The true sense of a sentence is made only when the proper utilization of grammar is done. Until 1957, there was no way to incorporate grammar into the machines and make them understand its true sense. The true evolution in the field of NLP comes in the year 1957 when Noam Chomsky introduced the syntactic structures. Chomsky emphasis on “formalized theory of linguistic structure” [2]. He works on refining the set of rules to form vigorous linguistics based on universal grammar. But later, Charles Hockett found out several drawbacks to Chomsky’s approach. The major one was, Chomsky, considered language as a well-defined, stable structure, and a formal system, which was possible only in an exceptionally ideal condition [3].

One of the early areas of NLP was machine translation. The goal of machine translation is to develop automated programs to translate text or speech from one natural language to another. For example, an automatic text translation program was developed through a joint project between Georgetown University and IBM Company in 1954. Experimental results showed that the proposed program was successfully translated sixty Russian sentences into English.

It was done by directly mapping the sentence and make use of the dictionary, which was explicitly maintained by the author, and the author claimed that within the coming few years, the machine would become capable of translating consistently. But the progress comes way slower than expected which hit the funding of the project and it reduces dramatically.

In the late 1960s, a machine was developed by Terry Winograd at Massachusetts Institute of Technology (MIT) named SHRDLU. It was the first NLP computer program that was able to perform the tasks like moving objects, determining the current state, and remembering names. It performs all these tasks in the “blocks world" environment. It became the first major achievement of AI and lures the attention of the researchers to optimize this technology. But the success of the program gets halted when it comes to interpreting the more real-world situations which were ambiguous and complex [4].

In the year 1969, Roger Schank introduced the concept of tokens that provide a better grasp of the meaning of a sentence. These tokens include real-world objects, real-world actions, time, and locations. For every sentence, these tokens provide the machine with better insights into the things happening and the object involves in it [5].

Till now, all the rules that were used to make sure that machine could understand a sentence are based on the phrase structuring. One way or another, researchers try to make well-defined set of rules that a machine could follow to deduce the meaning of the phrase. In the year 1970, William Woods introduced the concept of augmented transition networks (ATNs) for the representation of natural languages. He used finite set automata with recursion to reach the final state and conclude the meaning of the phrase according to the information available. The concept provides the result possible for information available (part of a sentence available) and makes changes to the meaning as per the information further provided. It comes up with a guess when the phrase provided is leading to an ambiguous situation. ATNs use recursion to solve the problem where enough information is not provided and postpone the decision until more information is provided [6].

Most of the approaches used until now to make the machine understand natural languages were based on the handwritten rules that machine used to match. But in the 1980s, another emerging field of computer science was holding its grasp in the area of computing, that was, machine learning. Machine learning was the major shift from the handwritten rules to the concept making. Machine learning algorithms provide more advanced way to interpret the ambiguity and further provide acceptable evidence for the consideration of a decision. Algorithms like decision trees use if-then rules to better deduce the optimal result and probabilistic algorithms back up the decision opted by the machine by providing enough confidence.

Currently, a shift has been made to deep learning. This was due to the dominance of deep learning to perform tasks that are difficult to perform by simply using rules and fixed criteria. When it comes to NLP, there is no way to solve the ambiguity of the language. Single word can have multiple meanings according to its neighboring situation, and it is not possible to write a rule or to use a decision tree to represent every possible meaning. Deep learning solves this problem efficiently as it does not require a programmer to provide the rules for deciding rather an algorithm itself deduces the process of mapping an input to an output.

3 Language Implementation

Processing human language is one of the toughest tasks for a machine to tackle. It is not a single form or concept that contributes to sentence formulation, rather there are several forms like nouns, verbs, adjectives, determinants, etc., that are used in association with each other to come up with the meaning of the sentence. Moreover, the ambiguity in the sentence meanings, like in the sentence: "he will be running a marathon soon", and in a sentence: "the software is running faster than expected," adds another layer of complexity. In both phrases, "running" is used, but the meaning of the running is different. The meaning of a word depends on the reference to which it is used. Parsing these types of sentences containing the ambiguous words is difficult for computers. The task becomes more complex when the grammar is considered. The grammar can be extended to handle sub-categorization by proposing additional rules or constraints (e.g., would is often used in conditional sentences with a clause beginning with "If"). But the use of rule-based approaches may give inefficient results due to the ambiguity and complexity of the language.

Addressing all these problems is an essential task to move forward, and it gave rise to statistical NLP [7]. Parsing rule is addressed with the probabilistic approach in statistical parsing. Probability is assigned to an individual rule, which is determined through machine learning algorithms. Numerous detailed rules are collected, and broader rules are defined. These broader rules help in building decision trees, and statistical parser always parses a sentence with the maximum likelihood. Thus, statistical approach produces better results.

NLP can be comprised of five major components: morphological analysis, syntactic analysis, semantic analysis, discourse analysis, and pragmatic analysis [8]. Parsing is the main step in syntactic analysis; it is the process of analyzing the constituent words in a text, based on an underlying grammar, to determine its syntactic structure.

The five components include some sub-tasks like: text decomposition, spelling, morphological parsing, and stemming. The parsing process results in a parse tree having sentence as root, noun-phrase, verb-phrase, etc., as intermediate nodes, and individual words as leaves.

Semantics in NLP is a process of deducing the meaning of the text. It performs the process like semantic analysis (check meaningfulness of a text), word sense disambiguation (determining the correct sense of the word, in case of ambiguity in meaning), lexical semantics (checks for the synonyms, antonyms, homonyms, etc.)

Pragmatic deals with the extraction of information from a piece of text [9]. It is further divided into three sub-fields: reference resolution (detecting reference), discourse analysis (determining the structure of the text), dialog interpretation (interpreting the information from the text).

4 NLP Tools and Techniques

NLP is concerned with making computers to interpret, understand, and to manipulate human languages. Traditionally, the interaction of humans to computers is done through a programming language. When it comes to human language interaction with the machine, achieving this interaction is quite challenging, as human language is highly ambiguous, contains slangs with unusual meanings, and contains social contexts. The task becomes more challenging when the accent is taken into consideration, as people from different regions have a different accent.

NLP incorporates two major tasks: syntax analysis and semantic analysis. Syntax analysis is used to make the arrangement of the words in the sentence in such a way that it starts to make grammatical sense. It helps NLP to assess the meaning of the sentence based on the grammatical rules. Semantic analysis is done to discover the meaning behind the words and their use in a sentence. It is applied by NLP for understanding the structure and meaning of a sentence.

Recently, deep learning has received big attention by researchers in NLP due to the availability of huge (big) amount of texts (data) for natural languages. Standard NLP tools and techniques are fine to use, but deep learning has altogether revolutionized the entire process. The massive success of deep learning is due to two major reasons:

The increase of the amount of data available:

Today, the data is producing at a tremendous rate, and the pace is only accelerating with the growth of the Internet of things (IoT). With this amount of data, the training process of the deep learning model can be done with greater precision, and the model can work on a variety of examples. Previously, deep learning fails to get this much exposer because of the limited amount of data available.

The high processing power available:

Deep learning model requires high-end hardware with faster processing capabilities to perform the training and testing phases. Earlier, the machine was not that capable of providing this high processing power to meet the requirement. But now, the hardware is easily available to carry out the operation of deep learning appropriately.

Deep Learning Challenges

Deep learning approach requires a huge amount of labeled data for training the model, and getting this amount of labeled data is one of the main hurdles to NLP. Labeled data is not readily available, and to overcome this hurdle, the data needs to be labeled explicitly which is a highly expensive and time-consuming process.

To make this process faster, deep learning engineers came up with a suggestion of using a semi-supervised approach. Semi-supervised learning is a combination of supervised and unsupervised learning. In this learning approach, a small amount of labelled data is first used to train the model to label the remaining unlabeled data and make the labeling process much faster. It makes the labeling process more efficient and less expensive.

Earlier NLP research was based on rule-based approach, i.e., the machine was not trained to identify the meaning of the sentence or phrase, but it is trained to look for certain words or collection of words in certain patterns and respond with a specific action when the word or pattern is encountered. Deep learning has the advantage because it uses a more intuitive and flexible approaches in which many examples were used to identify the speakers' intent and then provide its own response.

NLP Tools

There are several open-source tools available for NLP. These include

1. Python Tools: Natural Language Toolkit (NLTK), TextBlob, PyTorch-NLP, Textacy, SpaCy.
2. Java Tools: OpenNLP, StanfordNLP, CogCompNLP.
3. Node Tools: Retext, Compromise, Natural, NlP.js.

5 NLP Applications

NLP has many applications in various fields. Deep learning has opened the door for the research in the field of NLP. NLP is widely used in the IoT devices. Voice assistant like Cortana Alexa, Siri, and Google Assistant have made their own market lately. It can be seen in homes and offices quite often. Voice assistant is not confined to these areas only rather it has also started implementing in automobiles too.

There are tones load of comments, ideas, suggestions been posted on the Internet from social media. It became quite hard for an enterprise to go through all these comments to monitor the performance of their business. That is where the sentimental analysis, another primary use case of NLP comes to rescue. Sentiment analysis helps businesses to filter out the essential information from a massive amount of text data and make an optimal decision in the right direction in order to max out their profits.

NLP can be used to analyze a large amount of textual data and make the process faster and efficient. A large amount of information is stored in the form of text, and it is time-consuming to go through this text manually and come up with a solution. Before NLP, this was done manually, but the entire process gets revolutionized once deep learning-based NLP arises. Take the medical records of patients, for example, there are a huge amount of medical reports generated, and specialized doctors need to go through all these reports to suggest a diagnose. This is a slow process, but it can be made efficient by using NLP for text analysis.

Twitter is a platform where every day millions of opinions are exchanged. It is important to filter out the essential data from all the tweets. Here, NLP provides the support via sentiment analysis which helps to classify tweets in three categories, namely positive, negative, and neutral. Various techniques are incorporated for the classification like the unigram model, a feature-based model, and a tree kernel-based model [10]. But sometimes, before actual sentiment analysis of the product for its positivity or negativity among the audience, it is important to classify the product itself from the other. Hardly anyone is interested in the opinion of an individual for a product, topic, etc. So instead of classifying the opinion, quantification is needed. Quantification gives a broader perspective on a particular topic [11].

Sentiment analysis can be done using various techniques. One of the techniques is phrase-level sentiment analysis for recognizing contextual polarity. This technique uses to classify the contextual data using a top-down approach which first detects if the data is neutral or polarizing and afterward checks the polarity of the data [12].

Another area that NLP has revolutionized is spam filtering. With the rise in Internet availability throughout the world, the data generated is also increased. But the downside is many of the organizations are misusing the technology by misleading people. Spam filtering in e-mailing application is quite popular. Spam filtering uses various machine learning algorithms to differentiate between fake and legit data. Some of the techniques are decision tree, logistic regression, random forest, Bernoulli's Naive Bayes, linear and Gaussian SVMs, etc.

Apart from e-mail applications, YouTube is also using spam filtering to filter out the spam data from the comment section of videos. One such technique is using a tool called TubeSpam. It uses Bernoulli's Naive Bayes for the same. In this technique, the comparison is done between already classified comments by TubeSpam and the actual ones on YouTube. It initially uses the default classifier for all the videos and then improves by using suggestions is something is wrongly classified [13].

Apart from YouTube and e-mail spam filtering, Web pages also require a mechanism to filter out spam. Web pages are also not prone to spams. There is a continuous effort by Black Hat SEOs to inject spams. To deal with the same, the approach that is made is Dual-Margin Multi-Class Hypersphere Support Vector Machine (DMMH-SVM). This approach provides a high precision and recall rate and reduces the false positive [14].

Radiology is a medical discipline concerned with diagnosis and treatment of diseases in animals and human bodies using medical imaging. The report generated is highly specific and requires specialized lookup for further diagnosis and treatment. NLP can be used with machine learning in medical applications to medical reports with commendable accuracy. NLP can be applied to extract real-life concepts from unstructured data, sometimes leveraging advancement in machine learning to perform classification tasks [15].

Classification is one of the broadly used techniques in different business markets. Classification is used to filter the important e-mails from the spam ones. It is easy to classify a mail as spam mail or not because the size of the data is large. For big data, the chance of detection increases. But when the data is small, like in reviews, it is difficult to differentiate between the real ones and the fake ones, which hinders the credibility of the entire review collection. Identification and removal of such fake review requires automation since it is a difficult task for humans to perform manually. Machine learning algorithms can be used with NLP to produce good results [16].

NLP is an important discipline to make electronic health records a supportive source of data to meet needs and helping in main activities for researchers and clinicians while reducing their needs for data and charts review. NLP can be used to extract and process clinical data or information for both structured and unstructured forms [17]. NLP also could be used for patient's classification and supporting critical clinical tasks like clinical decision-making and producing quality reports [18].

A clinical decision-making system can be developed using patient's behavior toward a product, medicine, or treatment using NLP techniques. For instance, aspect-based sentiment analysis can be used in clinical decision-making for backing personalized therapy analytically [19].

NLP makes it easy for people to write down essays, e-mails, articles, etc. For example, the software Grammarly can detect the theme of the article once it is written, can check the grammar, and can suggest better ways to write a sentence.

As most of the information deals with digital documents and contracts nowadays, there is a need for a mechanism that protects the privacy of an organization. Named entity recognition can be used to monitor and detect privacy violations in online contracts by automatically monitoring personally identifiable information [20].

Search engines like Google, Bing, DuckDuckGo, etc., use machine translation tools to translate the content of a Web page into another language while holding the meaning and message intact.

Many organizations use chat-bots in their Web sites as a primary source of interaction between site and the user. Chat-bots process the query that the user enters and shows the results and suggestions accordingly. Chat-bots are considered much faster than humans to process the information and come up with a result of suggestions, thus, reducing the friction between the problems and the solutions.

6 Future Work

NLP came a long way when it comes to its implementations and applications. From bilingual dictionaries to handwritten rules, then from phrase structuring to ATNs, and finally machine learning, NLP has evolved in different stages with better and more advanced processing capabilities. It is hard to deny the rise of AI in the future to come. There are a lot of applications of NLP through the door of deep learning, and it is performing as per human expectations, and in some fields, it is outperforming humans too. Today, in our surroundings, most of the things we use are automated, as machine learning has overwhelmed the market with advanced devices, but there is still a continuous need for improvement. One may think that NLP had its effect on every area which can take advantage of, but still several areas can utilize this technical attainment and help in the progress toward a better future for everyone.

Nowadays, NLP is performing quite well in textual and audio data. NLP has achieved a commendable speed to process these data. But still, there is a setback for NLP, while processing the sarcasm, irony, and idiom in data. More research is needed in NLP to process various clinical data. This could help in the identification of early symptoms and targeting the root cause by suggesting the potential drug in the pharmaceutical field. Further, the extension can be made to this field using deep learning methods to train the machine over various structured and unstructured data.

References

1. Turing AM (1950) Computing machinery and intelligence. Mind 49:433–460
2. Lees Robert (1957) Review od syntactic structures. Language 33(3):375–408. https://doi.org/10.2307/411160 JSTOR 411160
3. Hockett C (1966) Language, mathematics and linguistics, current trends in linguistics. In: Theoretical Foundations (Chapter 3). Mouton, The Hague, pp 155–304
4. Winograd Terry (1980) What does it mean to understand language? Cogn. Sci 4:209–241
5. Schank R (1969) A conceptual dependency parser for natural language proceedings. In: Conference on computational linguistics, Sång-Säby, Sweden pages, pp 1–3
6. Wanner E (1980) The ATN and the sausage machine: which one is baloney? Cognition 8(2):209–225
7. Nadkarni PM, Ohno-Manchado L, Chapman WW (2011) Natural language processing: an introduction. J Am Med Inform Assoc 2011(18):544–551
8. Manning C, Schuetze H (1999) Foundation of statistical natural language processing. MIT Press, Cambridge, MA
9. Jurafsky D, Martin JH (2018) Speech and Language Processing, 2nd edn. Prentice-Hall, Englewood Cliffs, NJ
10. Agarwal A, Xie B, Vovsha I, Rambow O, Passonneau R (2011) Sentiment analysis of twitter data. In: Proceedings of the workshop on language in social media (LSM 2011), pp 30–38, Portland, Oregon
11. Nakov P, Ritter A, Rosenthal S, Sebastiani F, Stoyanov V (2019) Sentiment Analysis in Twitter. Cornell University, SemEval-2016. arXiv:1912.01973vl[cs.CL]
12. Wilson T, Wiebe J, Hoffmann P (2005) Recognizing contextual polarity in pharse-level sentiment analysis. In: Proceedings of Human Language Technology Conference and Conference on Empirical Methods in Natural Language Processing (HLT/EMNLP), pp. 347– 354, Vancouver, October
13. Alberto TC, Lochter JV, Almeida TA (2015) TubeSpam: comment spam filtering on youtube. In: 2015 IEEE 14th International conference on machine learning and applications (ICMLA)
14. Kumar S, Gao X, Welch I, Mansoori M (2016) A machine learning based web spam filtering approach. In: 2016 IEEE 30th international conference on advanced information networking and application
15. Chen P-H (2019) Essential elements of natural language processing: what radiology should know. The Association of University Radiologists
16. Vijayakumar B, Muhammad Fuad MM (2019) A new method to identify short-text authors using combinations of machine learning and natural language processing techniques. In: 23rd international conference on knowledge-based and intelligent information & engineering systems
17. Alemzadeh H, Devarakonda M (2017) An NLP-based cognitive system for disease status identification in electronic health records. In: 2017 IEEE EMBS International Conference on Biomedical & Health Informatics (BHI)
18. Juhu Y, Liu H (2019) Artificial intelligence approaches using natural language processing to advance EHR-based clinical research. American Academy of Allergy, Astama & Immunology
19. Silva P, Goncalves C, Godinho C, Antunes N, Curado M (2020) Using natural language processing to detect privacy voilation in online contracts. The 35th ACM/SIGAPP Symposium on Applied Computing, 2020
20. Hiremath BN, Patil MM (2020) Enhancing optimized personalized therapy in clinical decision support system using natural language processing. J. King Saud Univ

High Performance Computing

CNN-LSTM-Based Facial Expression Recognition

Puneet Singh Lamba and Deepali Virmani

Abstract Expressions play an imperative part in human interactions as it let humans express their intentions and feelings without words. Mental state of a subject can be mined from the expression extracted. In recent times, deep neural network has outperformed traditional handcrafted descriptors including spatiotemporal local binary pattern (LBP), LBP-TOP, HOG, etc., when it comes to feature extraction for facial expression recognition (FER). In this paper, an amalgam of convolution neural network (CNN) and long short-term memory (LSTM) [recurrent neural network (RNN)] is employed to extract essential features for recognizing the expression from the target frame. To increase the performance, transfer learning concept is engaged to get learned parameters (weight/bias). To accomplish transfer learning, leading layers of ResNet-50 (trained on thousands of image frames) are used. Further, a LSTM layer (time distributed) is affixed to the existing model. The model is further trained (CK+ database) with different activation functions, and a relative analysis is performed. Maximum accuracy of 94% is attained with the hybrid model (CNN-LSTM with SELU and ELU).

Keywords Transfer-Learning · RNN · ResNet50 · ImageNet · LSTM

P. S. Lamba (✉)
University School of Information, Communication and Technology, GGSIPU, Sector 16 C, Dwarka, New Delhi, India
e-mail: singhs.puneet@gmail.com

Bharati Vidyapeeth's College of Engineering, Paschim Vihar, New Delhi, India

D. Virmani
Department of Computer Science, Bhagwan Parshuram Institute of Technology, Delhi, New Delhi, India
e-mail: deepalivirmani@gmail.com

A. Abraham et al. (eds.), *Proceedings of 3rd International Conference on Computing Informatics and Networks*, Lecture Notes in Networks and Systems 167,
https://doi.org/10.1007/978-981-15-9712-1_32

Abbreviations

LBP	Local Binary Pattern
FER	Facial Expression Recognition
CNN	Convolution Neural-Network
LSTM	Long Short-Term Memory
RNN	Recurrent Neural Network
CK+	Cohn-Kanade+
ROC	Receiver Operating Characteristic
AUC	Area Under Curve
MTCNN	Multitask Cascaded Neural Network

1 Introduction

Emotions/expressions play a substantial role in human interactions as it let humans express their intentions and feelings without words. Human facial expressions are elusive and instinctive. Understanding facial expressions assists us to recognize and understand the true mental state of a subject. Initially, the expressions were recognized by the matching some traditional handcrafted expression descriptors (including spatiotemporal pattern (LBP) [3–5], LBP-TOP [6], directional mean optical flow feature [7]) mined from the target frames [1, 2]. However, the weakness with these approaches is lack of essential information required for a worthy feature map. In recent time, deep learning-based methods for expression recognition are getting popular due to its capability in extracting minute features and maintaining relationship among frame pixels. Deep learning-based approaches especially CNN, LSTM (RNN) are widely being utilized to resolve various computer vision problems [8] including frame classification [9], driver drowsiness, medical reporting and many more. In almost all the specified computer vision areas, deep learning approaches have outperformed the traditional techniques by a margin. A number of CNN, RNN-based models are proposed for FER [10–13]. In this paper, a model with an amalgamation of CNN and LSTM (RNN) is presented to solve the problem of facial expression recognition. A pre-trained image classification CNN model ResNet50 is used as a part of transfer learning along with LSTM layer to predict the expression.

Key contributions of the paper:

1. A hybrid model (CNN-LSTM) is presented which extracts essential features from the frames for classification.
2. For preprocessing, dataset faces are allied in accordance with eye position to condense the variance.
3. Leading layers of a pre-trained image classification CNN model ResNet50 are used to extract the feature map.
4. Time distributed LSTM layers are utilized to study the variation in frames.

5. Various activation functions are incorporated, and the performances are compared via ROC graphs.

Section 2 expounds the related study. Methodology along with experimental setup is explained in Sect. 3. Results obtained from the experimental setup are expounded in Sect. 4. Section 5 clinches the entire study.

2 Related Study

In recent years, CNN has evolved into most conventional approach as far as deep learning techniques are concerned. AlexNet [9] is based on conventional layered engineering consisting of convolution, max-pooling layers with rectified linear units (ReLUs) as activation function. Szegedy et al. [14] presented GoogLeNet, a 22-layer deep network, the quality of which is assessed in the context of classification and detection. In [15], inception layer is utilized to accomplish best in class results. Succeeding the success of inception layers, many variants have also been offered [16]. When it comes to handling sequential data (speech [17], NLP [18, 19], gait recognition [20]), RNN has performed at par CNNs. RNN with its ability to recollect information from past instances and learn relative dependencies with frames has been proved expedient over CNN. Further, the concoction of RNN and CNN has achieved better results specifically in computer vision problems [21, 22]. Author in [23] "proposed a novel framework (spatial temporal RNN) to fuse the learned information from multiple signal sources into a spatial temporal dependency model". Expressions are also recognized from video sequences using a fusion of CNN and RNN [24–26]. Similar works are being proposed for videos as well as speech recognition [27, 28]. A novel CNN model to reduce redundancy of features already learned providing quality feature set for the succeeding layer is proposed by Xie and Hu [29]. Ouellet [30] proposed a real-time emotion detection application using CNN. In this work, the initial feature set is extracted using a pre-trained image classification model based on CNN which is further flattened (time distributed) and fed into RNN (LSTM) layer which helps in recollecting the learned features (as required in CK+ dataset), resulting in accurate prediction of facial expressions.

3 Methodology Used

This section discourses the various concepts being used to build up the final expression recognition model. Section 3.1 explains the ResNet-50 image classification model. Transfer learning concepts and dataset used are explained in Sects. 3.2 and 3.3. Performance parameters used to analyze the results and experimental setup are elaborated in further sections.

3.1 Image Classification Models

ResNet-50 is a well-known CNN architecture trained on thousands of images (Image-Net database). It has 50 CNN layers which make it possible to learn minute feature representation from image frames (size: 224*224), thus equipping capability to categorize the image frame into 1000 classes. Inception-v3 is an another CNN trained on images from Image-Net database. In comparison with ResNet-50, it has 48 layers having the capability to categorize frames (size: 299*299) into 1000 object categories. In this paper, leading layers of ResNet-50 except last few layers (layers used for classification) are incorporated for feature extraction.

3.2 Transfer Learning

To train an extremely large dataset, a deep convolution network may take hours or even days. To cut off the training period, pre-trained weights/bias from already trained models (developed/tested for standard computer vision datasets) can be used. In deep learning, transfer learning is a technique whereby a network (trained to solve a particular problem) is reused to solve a different problem with minute modifications. Layers from the already trained model are then incorporated within the new model trained on the problem of interest resulting in decreased training time and low generalization error. In this paper, transfer learning is applied to recognize facial expressions using a pre-trained image classification model (ResNet-50) as explained in Sect. 3.1.

3.3 Dataset

In this work, a state-of-the-art dataset: Cohn-Kanade+ (CK+) (extensively used to assess facial expressions) as précised in Table 1 is used. Figure 1 displays a snapshot

Table 1 CK+ dataset

Properties	Descriptions
Subjects/sequences	123/593
Face	Single face
Gray/colored	Gray
Resolution	640*480
View	Anterior
Face expression	From 593 sequences, 327 are labeled with seven expressions
Sequence length	Variable lengths (at least ten frames)

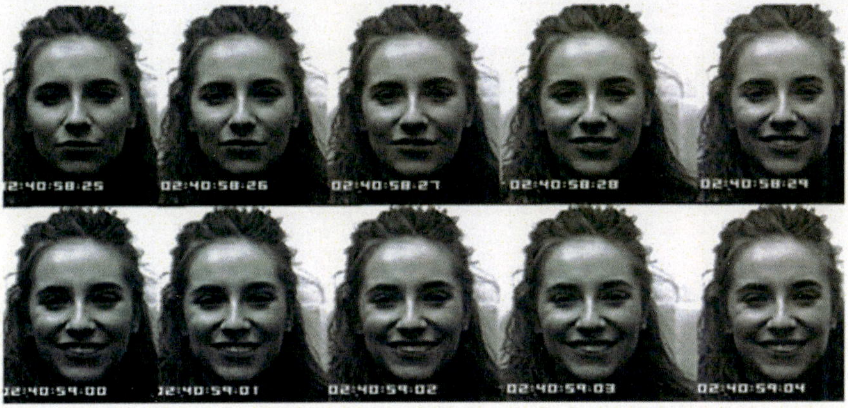

Fig. 1 Snapshot of the CK+ dataset. Expression transition from neutral to happy through ten frames [31]

of the CK+ database.

3.4 *Performance Parameters*

Human facial expressions are recognized using transfer learning [using learned weights from pre-trained model (mentioned in Sect. 3.1)] and LSTM (an artificial RNN). The performance of the model is analyzed in terms of various performance parameters (accuracy, precision, recall, receiver operating characteristic (ROC) curve, F1-score and area under the curve (AUC) score).

3.5 *Experimental Setup*

This section explains the experimental setup employed for the hybrid model (CNN + LSTM) which is used to foresee the expression from the input image frame. Preprocessing is a vital step in image-related processing. This section further explains the steps employed for the preprocessing and feature extraction using transfer learning.

Preprocessing. Before feature extraction, preprocessing is performed to improve the performance of the FER system. In CK+ database, there are multiple folders for each emotion, and each folder contains at least ten frames showing a transition from neutral to the actual emotion. In our case, we have selected last six frames from every folder whose labels were available. Further as far as preprocessing step is concerned, dataset faces are allied in accordance with eye position to condense the variance. Multitask cascaded neural network (MTCNN) is used to locate the accurate eye coordinates. Figure 2 explains the preprocessing algorithm in detail. The input

```
Algo. Face_align
function FaceAlign(image)
        left_eye_middle, right_eye_middle:= MTCNN(image)
        angles:= angles between left and right eye middles
        desired_rght_eye_x:= 1-0.35
        distance:= square_root(dx**2+dy**2)
        desired_distance:= desired_rght_eye_x-0.35
        desired_distance*:=256
        scaled:= desired_distance/distance
        eyes_middle:= middle of line joining eyes
        N:= Affine matrix for 2d rotation
        ttx:= .5*256
        tty:= 0.35*256
        N[0,2]+:= (ttx-eyes_middle_x)
        N[1,2]+:= (ttx-eyes_middle_y)
        results:= affine transformation to image using N and face width and height
        return results
End function
```

Fig. 2 Algorithm for preprocessing

to output transition of the preprocessing step can be seen from Fig. 3.

Feature Extraction using Transfer Learning. For feature extraction, leading layers of ResNet-50 architecture (used for learning the features of the image frame) are utilized. The layers removed are the ones used for classification. The preprocessed image frames are fed as input in batches to the ResNet-50 model, and the resultant features maps ($7 \times 7 \times 2048$) are saved as shown in Fig. 4a. These resulting feature maps [six frames of same emotion ($6 \times 7 \times 7 \times 2048$)] are flattened (time distributed) and are fed into a LSTM layer (with 256 neurons) as shown in Fig. 4b followed by a dense layer (512 neurons) for classification. The loss used for training is categorical cross entropy with sgd optimizer.

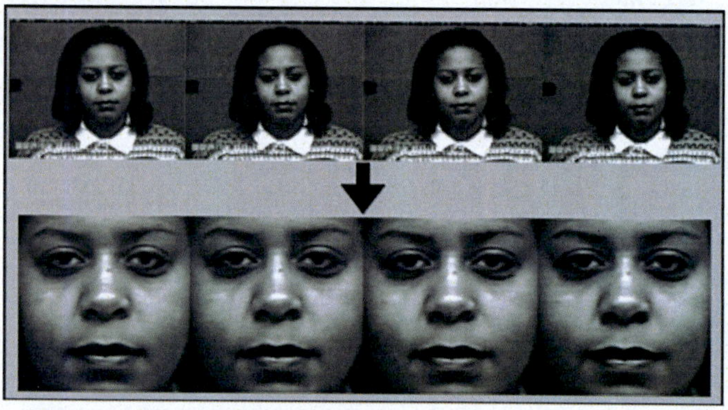

Fig. 3 Faces aligned using the preprocessing algorithm

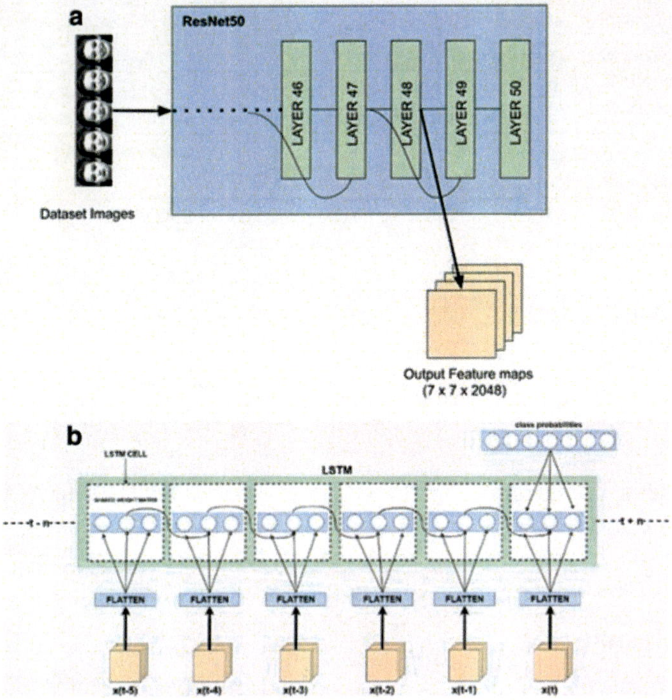

Fig. 4 **a** Feature extraction using a pre-trained model. **b** Time distributed LSTM layer (taking six frames at a time)

4 Experimental Results

A state-of-the-art dataset (CK+) is processed to investigate the performance of the model. The model is instigated distinctly with five activation functions and is graded as per the performance parameters. Performance parameters (precision, recall, ROC graphs, etc.) are used to measure and examine the performance of hybrid system under observation.

4.1 Result Description and Analysis

The comparative analysis of different activation functions is presented in Table 2, where macro average is the normal average calculated and weighted average depends on the size of each class. It can be easily perceived that with scaled exponential linear unit (SELU) and exponential linear unit (ELU) as activation function, the model performs with utmost accuracy, whereas with sigmoid activation function, the performance has degraded profusely. Further, other performance parameters follow

Table 2 Comparative examination with different activation functions

S. No.	Activation function	Precision		Recall		F1-score		Accuracy
		Weighted average	Macro average	Weighted average	Macro average	Weighted average	Macro average	
a	Exponential linear unit	0.94	0.92	0.94	0.89	0.94	0.90	0.9400
b	Rectified linear unit	0.92	0.90	0.92	0.86	0.92	0.87	0.9200
c	Sigmoid	0.85	0.87	0.74	0.56	0.69	0.52	0.7400
d	Scaled exponential linear unit	0.94	0.92	0.94	0.89	0.94	0.90	0.9400
e	Exponential	0.93	0.92	0.93	0.88	0.93	0.89	0.9300

the same curve. Comparative ROC curve along with their AUC for each expression is shown in Fig. 5.

5 Conclusion

The paper presents a fusion model (CNN + LSTM) to sense the expression with an amalgam of CNN and RNN using transfer learning. For feature extraction, leading layers of a pre-trained model-ResNet-50 with learned weights/bias are used. Further, LSTM layer is added to enhance the performance. The model is evaluated on CK+ dataset. For preprocessing, faces are allied in accordance with eye position to reduce the variance. Further, performance parameters are used to evaluate and compare the model with five activation functions. Among the activation functions, SELU and ELU (acc: 94%) have outperformed its equivalents by a margin.

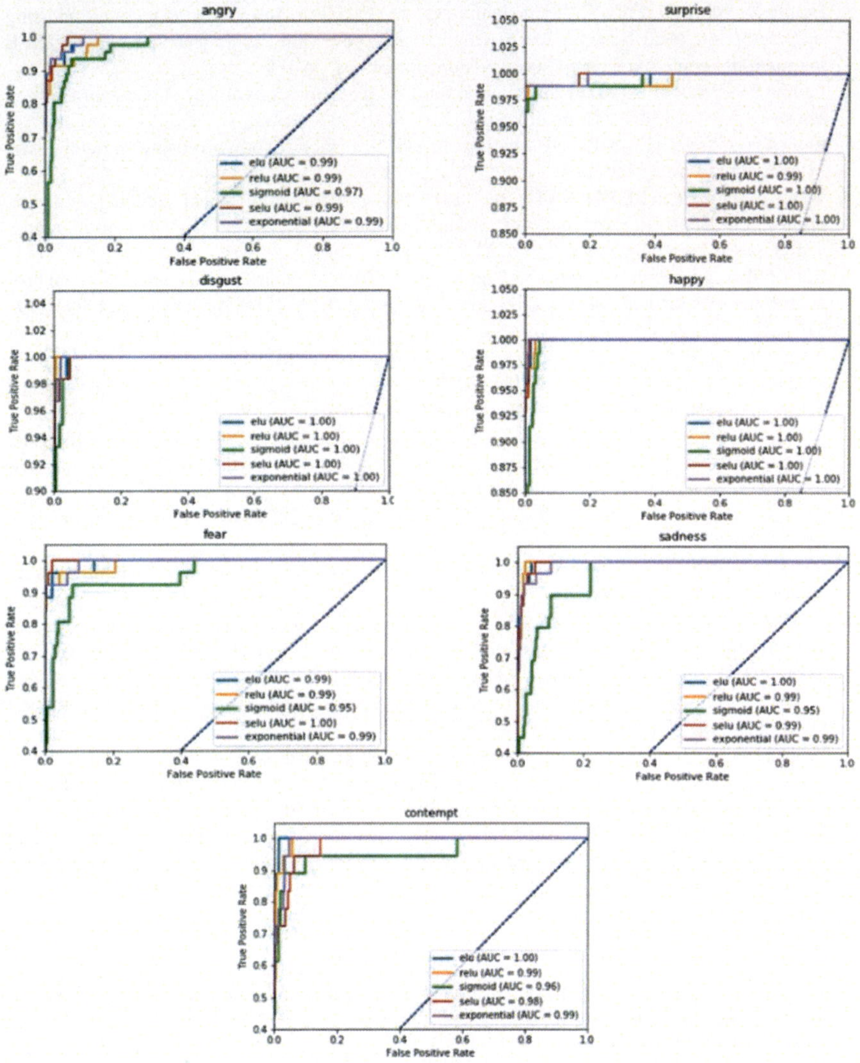

Fig. 5 Comparative ROC curve for each expression

References

1. Mukherjee S, Vamshi B, Reddy KSVK, Krishna RV, Harish SVS (2016) Recognizing facial expressions using novel motion based features. In: Proceedings of the tenth Indian conference on computer vision, graphics and image processing. ACM, pp 1–8
2. Li X, Hong X, Moilanen A, Huang X, Pfister T, Zhao G, Pietikäinen M (2017) Towards reading hidden emotions: a comparative study of spontaneous micro-expression spotting and recognition methods. IEEE Trans Affect Comput 9(4):563–577

3. Huang X, Wang SJ, Zhao G, Piteikainen M (2015) Facial micro-expression recognition using spatiotemporal local binary pattern with integral projection. In: Proceedings of the IEEE international conference on computer vision workshops, pp 1–9
4. Lamba PS, Virmani D (2018) Information retrieval from emotions and eye blinks with help of sensor nodes. Int J Electr Comput Eng (IJECE) 8(4):2433–2441
5. Lamba PS, Virmani D (2019) Information retrieval from facial expression using voting to assert exigency. J Discrete Math Sci Crypt 22(2):177–190
6. Wang Y, See J, Phan RCW, Oh YH (2014) LBP with six intersection points: reducing redundant information in LBP-top for micro-expression recognition. In: Asian conference on computer vision, pp 525–537. Springer, Cham
7. Liu YJ, Zhang JK, Yan WJ, Wang SJ, Zhao G, Fu X (2015) A main directional mean optical flow feature for spontaneous micro-expression recognition. IEEE Trans Affect Comput 7(4):299–310
8. Goodfellow I, Bengio Y, Courville A (2016) Deep learning. MIT Press
9. Krizhevsky A, Sutskever I, Hinton GE (2012) Imagenet classification with deep convolutional neural networks. In: Advances in neural information processing systems, pp 1097–1105
10. Guo J, Zhou S, Wu J, Wan J, Zhu X, Lei Z, Li SZ (2017) Multi-modality network with visual and geometrical information for micro emotion recognition. In: 2017 12th IEEE international conference on automatic face & gesture recognition (FG 2017), pp 814–819
11. Gan YS, Liong ST, Yau WC, Huang YC, Tan LK (2019) Off-apexnet on micro-expression recognition system. Sig Process Image Commun 74:129–139
12. Satya V (2018) Revealing true emotions through micro-expressions: a machine learning approach. CMU SEI Insights
13. Patel D, Hong X, Zhao G (2016) Selective deep features for micro-expression recognition. In: 2016 23rd international conference on pattern recognition (ICPR), pp 2258–2263. IEEE
14. Szegedy C, Liu W, Jia Y, Sermanet P, Reed S, Anguelov D, Erhan D, Vanhoucke V, Rabinovich A (2015) Going deeper with convolutions. In: Proceedings of the IEEE conference on computer vision and pattern recognition, pp 1–9
15. Mollahosseini A, Hasani B, Salvador MJ, Abdollahi H, Chan D, Mahoor MH (2016) Facial expression recognition from world wild web. In: Proceedings of the IEEE conference on computer vision and pattern recognition workshops, pp 58–65
16. Ioffe S, Szegedy C (2015) Batch normalization: accelerating deep network training by reducing internal covariate shift. In: Proceedings of the 32nd international conference on machine learning, pp 448–456
17. Graves A, Mohamed AR, Hinton G (2013) Speech recognition with deep recurrent neural networks. In: 2013 IEEE international conference on acoustics, speech and signal processing, pp 6645–6649. IEEE
18. Graves A, Jaitly N (2014) Towards end-to-end speech recognition with recurrent neural networks. In: International conference on machine learning, pp 1764–1772
19. Mikolov T, Karafiat M, Burget L, Cernocky J, Khudanpur S (2010) Recurrent neural network based language model. In: Proceedings of INTERSPEECH, pp 1045–1048
20. Sanin A., Sanderson C, Harandi MT, Lovell BC (2013) Spatio-temporal covariance descriptors for action and gesture recognition. In: 2013 IEEE workshop on applications of computer vision (WACV), pp 103–110. IEEE
21. Visin F, Kastner K, Cho K, Matteucci M, Courville A, Bengio Y (2015) Renet: a recurrent neural network based alternative to convolutional networks. arXiv preprint arXiv:1505.00393
22. Visin F, Kastner K, Courville A, Bengio Y, Matteucci M, Cho K (2015) Reseg: a recurrent neural network for object segmentation. arXiv preprint arXiv:1511.07053
23. Zhang T, Zheng W, Cui Z, Zong Y, Li Y (2018) Spatial–temporal recurrent neural network for emotion recognition. IEEE Trans Cybern 49(3):839–847
24. Khorrami P, Le Paine T, Brady K, Dagli C, Huang TS (2016) How deep neural networks can improve emotion recognition on video data. In: 2016 IEEE international conference on image processing (ICIP), pp 619–623. IEEE

25. Jain DK, Kumar R, Jain N (2017) Decision-based spectral embedding approach for identifying facial behaviour on RGB-D images. In: Proceedings of international conference on communication and networks, pp 677–687. Springer, Singapore
26. Jain DK, Zhang Z, Huang K (2016) Hybrid patch based diagonal pattern geometric appearance model for facial expression recognition. In: Chinese conference on intelligent visual surveillance, pp 107–113. Springer, Singapore
27. Chernykh V, Prikhodko P (2017) Emotion recognition from speech with recurrent neural networks. arXiv preprint arXiv:1701.08071
28. Fan Y, Lu X, Li D, Liu Y (2016) Video-based emotion recognition using CNN-RNN and C3D hybrid networks. In: Proceedings of the 18th ACM international conference on multimodal interaction, pp. 445–450
29. Xie S, Hu H (2017) Facial expression recognition with FRR-CNN. Electron Lett 53(4):235–237
30. Ouellet S (2014) Real-time emotion recognition for gaming using deep convolutional network features. arXiv preprint arXiv:1408.3750
31. Lucey P, Cohn JF, Kanade T, Saragih J, Ambadar Z, Matthews I (2010) The extended Cohn-Kanade dataset (CK+): a complete dataset for action unit and emotion-specified expression. In: 2010 IEEE computer society conference on computer vision and pattern recognition-workshops, pp 94–101. IEEE

Delve into the Ambience of Anonymity: Deep Web

Kritika, Harjas Kalsi, Himanshi, Rajneesh Dubey, and Yash Mittal

Abstract The determination of this research is to analyze the darknet Web sites. This analysis is done on the basis of the number of users accessing it, active servers hosting darknet Web sites anonymously and traffic on the network. A survey was conducted in the community on dark web and Mariana's web; after that, Tor services are used to collect the data and interpret it. Skillful comparison graphs are plotted between years, and other parameters (like users, servers, traffic, performance) have been used, and variations are also indicated that the number of users is increased, and it is seen that most of the relays are run by Linux. A Google form was also disseminated through social media to different sections of our society. Responses are interpreted and collected due to which it is seen that a good proportion of society is aware of dark web but not Mariana's web. This survey spreads awareness about the dark web and its services.

Keywords Relay · Bridge · Berkeley software distribution (BSD) · Onionoo · Descriptors

---

Kritika (✉) · H. Kalsi · Himanshi · R. Dubey · Y. Mittal
BPIT, GGSIPU, New Delhi, India
e-mail: kriti27.balihar@gmail.com

H. Kalsi
e-mail: harjakalsi12@gmail.com

Himanshi
e-mail: himanshi0875@gmail.com

R. Dubey
e-mail: rkrajnees3@gmail.com

Y. Mittal
e-mail: yashmittal157@gmail.com

A. Abraham et al. (eds.), *Proceedings of 3rd International Conference on Computing Informatics and Networks*, Lecture Notes in Networks and Systems 167,
https://doi.org/10.1007/978-981-15-9712-1_33

Abbreviations

BSD	Berkeley Software Distribution
DDOS	Distributed Denial Of Service
FBI	Federal Bureau of Investigation
Fig.	Figure
Govt.	Government
TOR	The Onion Router
URL	Uniform Resource Locator

1 Introduction

This paper presents the modern web, dark web which is famed as "darknet," but it is necessary to know the fact that dark web and darknet are not identical. Darknet is a term used in the 1970s, for networks which are secluded from the ARPANET, generally for security purposes, such as compartmentalization [1]. Over time, the term was also used for overlay networks, which are essentially networks that utilize software and hardware to create multiple levels of abstraction. Dark web encloses the hidden online information or documentation. These documentations are not indexed on the usual search engines like Yahoo!, Bing, Google, and others. The dark web is browsed with the help of a specific browser TOR. It is an acronym for the software "The Onion Router." It is a non-profitable organization. TOR makes us innocuous on the Internet. It precludes somebody inspecting our Internet connection, and it also evades the sites we access from detecting our physical location or other personal information. Tor is the version of Firefox with patched various privacy anxieties [2]. Tor was released for public on Sept 20, 2002 and accessible in almost 32 languages [3]. This research also includes the Marianas web, where Marianas web is the deepest part of the Internet. Only some percentage of people have heard about it, and accessing it is almost impossible. It can only be accessed when the address of the site and its key is available, unlike the dark web where we only need the address of the site. It is named Marianas from Mariana Trench which is known as the deepest part of the oceans. It contains the secret data of the government and some other agencies. And it has the highest level of security that hacking it is almost impossible. But, yes, it can be hacked by using a quantum computer also known as a supercomputer [3].

The paper is organized as follows: In Sect. 2, we describe dark web, how it was established and how the TOR browser works. In Sect. 3, we describe information flow in TOR. Services of dark web are categorized, and how much percentage of web are used by the sites are described in Sect. 4. Analysis and comparison of users according to countries and operation systems are described in Sect. 5. In Sect. 6, the result of survey is conducted in our society while the conclusion is presented based on our analysis which is described in Sect. 7.

2 Literature Review

It is well known about the dark web and its various activities—from legal to illegal and free to paid—but people may not be aware of its origin and how it came into the digital world. Many think that it originated only a few years ago, but the truth is that it has been there since the beginning of the Internet.

On October 29, 1969, a student from the University of California used ARPANET to transmit the first electronic message from one computer to another. The first transaction over darknet was cannabis. In the early 70s, students at Stanford used ARPANET for selling drugs. During the 80s, storing illegal and sensitive data became difficult. To tackle this problem, users created "data havens" where sensitive data was stored and these data havens were on the Caribbean islands [4].

After the mid of 1990, data storing became faster with improvement in compression algorithms. So, various services were made available to provide music and other files for free. In 2000, Lan Clarke developed a Web browser named "Freenet" which provides users complete anonymity. Later, the US Naval Research Laboratory released "The Onion Router" or TOR. It attracted a huge section of people who wanted to use illegal services of darknet. Even though the US govt. created it, it became one of the top ways to buy illegal content. There came a problem in money exchange which was later resolved. On January 3, 2009, Satoshi Nakamoto mines the first bitcoin, an untraceable cryptocurrency. It boosts the process of money laundering and other criminal activities [5].

The first modern e-commerce black market Web site was launched in February 2013, Silk Road, and later, it was banned. After a few months, its versions were released.

Now, everyday, there is some new Web site in the dark web which provides various illegal services like child pornography, hitman hiring, etc.

3 Proposed Work

3.1 Process Flow of Tor Services

Due to the anonymity of Tor, it is the prevalent nesting ground for crooks and illegal activities. Tor is a powerful tool whose network is similar to a Virtual Private Network (VPN), but Tor is not 100% anonymous; it has many legitimate uses; with the help of Tor Metrics, different facts are analyzed about the dark web. Tor Metrics is the crucial mechanism which Tor requires to estimate the working and ongoing application of its technologies. Tor Metrics comprises several services that work together to accumulate, aggregate, and present data from the Tor net and other coupled services. It is Onionoo applying a protocol to upkeep JSON documents over HTTP [6].

Fig. 1 Information flow in TOR

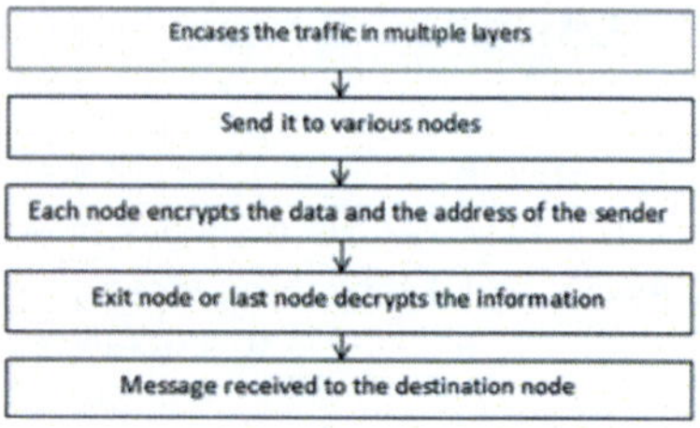

Tor bridges and relays gather cumulative data of their norm along with bandwidth and linking users per country. With the help of basic aggregation, it protects the privacy of connected users.

To deliver the ancient data archived, the Tor Metrics Web site encloses the customizable strategies [7] to represent the traffic, relay users, bridge users, and the other downloaded statistics in the claimed time and delivered to the certain country.

Figure 1 outlines the working of TOR. It anonymizes our identity or online activity by encasing our traffic in multiple layers. This makes Tor secure. Tor picks a random path from source to destination service every time.

Categorization of services on the deep web was done by accessing hidden wiki Web site and surfing some famous Web site and collecting data. Reference was also taken from [8] for knowing the most popular services on deep web. Darknet sites were accessed through Tor browser on KALI LINUX operating system on a virtual machine for better anonymity and protection. Surfing of darknet sites was done by darknet search engines named Candle and Torch. Services on the darknet sites were not used. However, there was a lot of information to be collected such as price details of services [9] and payment methods and URLs of popular onion services.

Dark web provides numerous services, and this paper categorizes them with the help of some case studies.

AlphaBay is one of the largest marketplaces on the darknet after Silk Road where thousands of criminals anonymously sold or bought weapons, drugs, stolen identities, and many offensive products. The operator of this Web site, a 25-year-old Canadian boy living in Thailand, was arrested by FBI in 2017. He operated AlphaBay for more than two years and had transactions exceeding $1 billion in digital currency or bitcoins. Now, AlphaBay is officially down [10].

Doctor X is a Spanish physician named Fernando Caudevilla. He was active on many Web sites like Silk Road, Silk Road 2.0, Evolution marketplace, etc., from 2011 to give advice on reduction of usage of drugs. Dr. X contributed all his free time on projects related to usage and reduction of drugs [11].

Exploring and analysis of dark web: This paper was studied to get insight into the type of sites hosted on the deep web and get their approximate count on the dark web. This included all the types of dark Web sites being hosted which also includes a number of sites.

4 Categorization of Services

1. Drug market: Most of the hidden web constitutes sites which are involved in selling illegal drugs such as cannabis, cocaine, heroin, ecstasy, and speed. About 5.1% sites on deep web are related to drugs [12].
2. Political secrets: Services that posts articles and blogs on political secrets and current political environment of a country. About 0.9% darknet sites are related to politics.
3. Malware/ransomware providers: Services that offer malware and ransomware. Malware includes mostly for Windows and UNIX operating systems. Malware providing darknet sites constitutes about 3.2% of deep web.
4. Forged documents: Duplicate documents such as passports, license, and social security numbers are provided for illegal use. There are about 1.1% of darknet Web sites related to forged documents.
5. Counterfeit currencies: These sites supply counterfeit currency heavily focusing on the dollar. Counterfeit related services are small part of darknet sites constituting about 1.1% [12].
6. Market place: These are the sites which provide all types of above-mentioned services. About 10.2% of onion sites on deep web fall in this category.
7. Violent content: This content includes selling of weapons, hiring hitmen for individuals, torture, terrorism, child molestation. About 2.8% of sites are related to violent services.
8. Doxing: Posting personal information or images online to malign a given individual
9. Pirated software: Distribution of versions of licensed software at highly reduced prices.
10. Adult content: This section includes pornography including illegal child pornography, bestiality (animal sex). Adult content-related services have a greater number of darknet sites on the deep web [12]. Adult content-related services constitute about 4.7% on deep web (Table 1).

All transactions are done in cryptocurrencies mainly bitcoin. However, some drug-selling Web sites also use dollars. Usage of bitcoin and other cryptocurrencies are increasing day by day on the dark web.

5 Analysis and Comparison

Tor services are used to find the increase in number of users (relay and bridge), onion sites, and different operating systems hosting servers of the dark web. The following plots are made from the data collected.

Figure 2 represents the percentage of relay users according to different countries. This data is estimated by directly connected users. It also determines that the

Table 1 Various darknet sites

Name of site	Link
Drug sites	
Eucana	rso4hutlerfirefqp.onion
Cannabis UK	fzqnrlcvhkgbdwx5.onion
Peoples drug store	newspdsuslmzqazvr.onion
Political sites	
Political profile	6sgjmi53igmg7fm7.onion
Political secrets	onion53ehmf4q75.onion
Malware sites	
Dark rift	darkrift2iywe544.onion
Cs Trezor	nzy5csztrezor5zx.onion
Fake documents	
Fake passport	fakepass5me6dw3g.onion
Fake id	fakeidskhfik46ux.onion
Counterfiet currency sites	
Fake dollar	qkj4drgvpm7eec1.onion
Billsman	z5uz2t7emaueuj25.onion
Market sites	
Midnight market	midnightkt43f3fk.onion
Tennebera market	vzmvmmaldx3h7vx6.onion
Voilent content sites	
Red room	redroomfing27toi.onion
Hitmen hire	killersp4h67dntp.onion

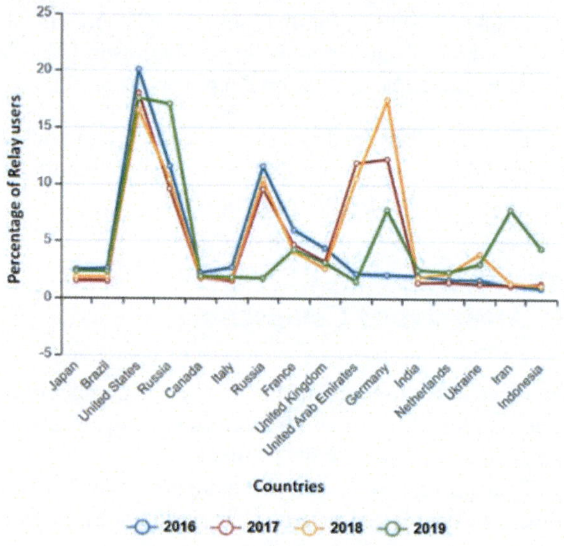

Fig. 2 Country-wise relay users

maximum percentage of relay users are from the USA. States except in 2018 where Germany has the maximum percentage of relay [13, 14].

Figure 3 represents the percentage of bridge users according to different countries. This data is estimated by users connected via bridges. This graph determines that the maximum percentage of bridge users are from the USA in 2016, the United Arab Emirates in 2017, Russia in 2018, and Iran in 2019 [13, 14].

Figure 4 shows the evolution of dark web in terms of number of onion sites over the years. It can be see that there is no general trend of increasing or decreasing of onion sites. However, 2018 was the year which hosted most number of sites. At the time of writing of this paper, there are about 65,000 (1000 give or take) onion sites on the dark web [11, 14].

Figure 5 represents the number of relays running on different operating systems in different years. It can be seen that most of the relays are run by Linux followed by BSD and then Windows. The reason behind this is the flexibility provided by Linux to customize it according to usage. Also, running relays remain almost the same every year [13, 14].

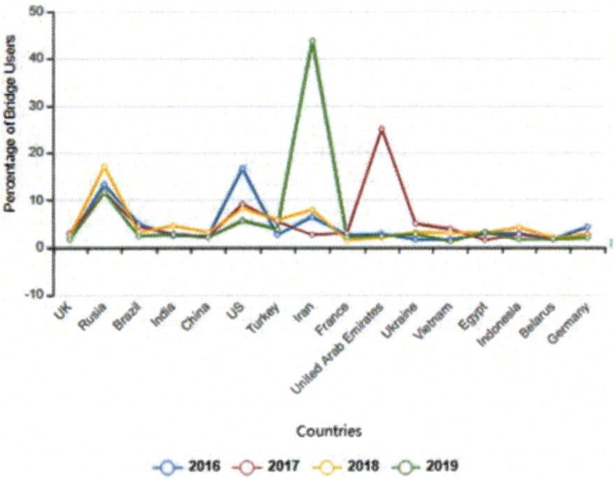

Fig. 3 Country-wise bridge users

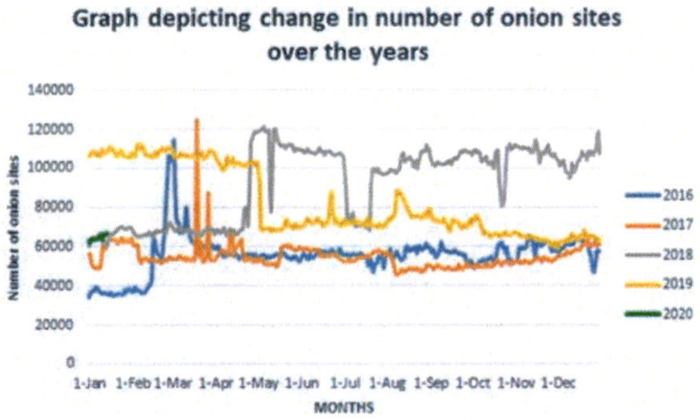

Fig. 4 Change in number of onion sites over the years

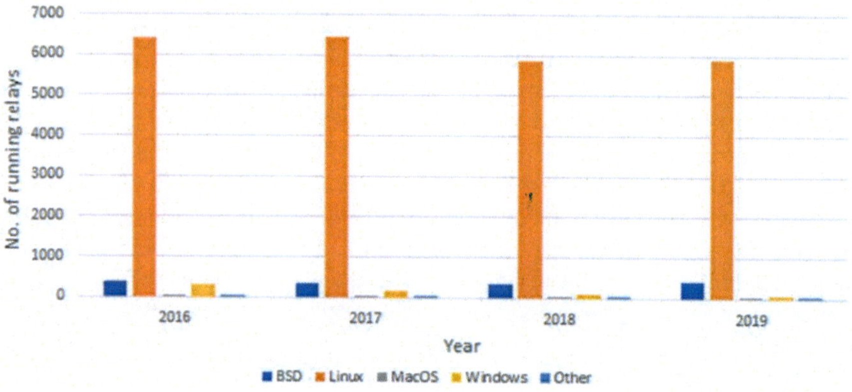

Fig. 5 Running relays by different operating systems

Table 2 Response of questions asked

People know about Dark web	75%
People visited Dark web	18.5%
People know about Marianas Web	31%
People visited Marianas Web	2.2%

Table 3 Types of dark web sites used

Type of site	Percentage users
Black market	16.4%
Hitman hire	5.9%
Pornography	4.4%
Human torture	5.9%
Hidden Wiki	7.4%

6 Survey of Community on Dark Web

A Google form is circulated through social media to different sections of our society. Responses are interpreted and collected in tabular form (Tables 2 and 3).

7 Conclusion

The conclusion of this paper includes that the growth of the deep web has been stagnant over the last two years. There have been several instances of a sudden rise in the number of onion sites in the year 2017 and 2018. In India, approximately, 1370 mean relay users were added in the past four years. This paper explores that the number of dark web users increases day by day. 2018 was the year in which most number of onion sites were active. The USA has the maximum number of relay users. Russia is the second only to the USA in terms of average number of users, as over 345,428 (17.32%) access onion services daily. Before 2017, Germany was among top countries in relay users count, but due to DDOS attack on Tor browser, its relay users' number declined in subsequent years. Outside Europe and USA, Iran has the highest number of Tor users, which counts for 50% more users than UK, despite having one-third of its Internet population. Iran has the maximum number of bridge users due to strict censorship policies of the Iranian government. However, on June 2019, relays in Iran were reported to be unblocked once again, and user numbers recovered to about 25% of the maximum they had reached during the previous period of unblocking. They then dropped precipitously, and bridge users increased. Tor is also popular in Israel, and it had more Tor users than India in 2014; however, after 2015, there was a gradual increase in India's relay users count. Europe constitutes over half of daily relay users, having the highest penetration, as the service is used by an average of 80 per 100,000 European Internet users. Germany in particular accounts for over 159,217 users a day, which is about one-fifth of the entire European Tor daily user base. UK, Netherlands, France, and Ukraine are other European regions with a high penetration. Despite a very small Internet population, in regions like Monaco, Andorra, and Liechtenstein, hundreds of people use Tor services everyday. This research paper also shows that Linux is the most used operating system for accessing the deep web using the Tor browser. In 2019, there are more than 5000 relays running on Linux as compared to the Windows operating system while BSD (UNIX based operating system) is the second-most preferred operating system. The use of Windows for accessing dark web is in gradual decline from 2016. The conclusion of a small community survey in New Delhi, India, shows that most of the people have heard about the deep web irrespective of their profession, but very few people have accessed and none have used any service provided on the dark web. The main motive of this survey was to create awareness among people and provide them the information about good and bad aspects of the deep web. Categorization of services provided on the dark web was also done which shows that drug sites, bitcoin market, and adult content hosting sites are major contributors to deep web networks.

References

1. Retzkin S: Hands-on dark web analysis, 1st edn. Packt Publishing
2. Tor Documentation. https://2019.www.torproject.org/docs/faq#WhatIsTor. Last accessed 4 Mar 2020
3. Tor-Wikipedia Homepage. https://en.m.wikipedia.org/wiki/Tor_(anonymity_network. Last accessed 15 Feb 2020
4. Ranker Homepage. https://www.ranker.com/list/history-of-the-dark-web/jordan-breeding. Last accessed 11 Feb 2020
5. TechNadu Homepage. https://www.technadu.com/dark-web-history/52017/. Last accessed 11 Feb 2020

6. Tor Blogpage. https://blog.torproject.org/collecting-aggregating-and-presenting-data-tor-network. Last accessed 28 Feb 2020
7. Loesing K, Murdoch SJ, Dingledine R (2010) A case study on measuring statistical data in the tor anonymity network. In: Proceedings of the workshop on ethics in computer security research (WECSR 2010), Tenerife, Canary Islands, Spain, January 2010
8. Faizan M, Khan RA (2019) Exploring and analyzing the dark web: a new alchemy. First Monday 24, 5–6 May 2019
9. Winter P, Edmundson A, Roberts LM, Dutkowska-Zuk A, Chetty M, Feamster N (2018) How do tor users interact with onion services? In: Proceedings of the 27th usenix security symposium
10. Baravalle A, Lee SW (2018) Dark web markets: turning the lights on AlphaBay. In: Conference 2018, LNCS, vol 11234, pp 502–514. Springer, Heidelberg
11. SAPAS-Interview with Fernando. https://sapas.sk/en/doctorx/. Last accessed 20 Feb 2020
12. Gulati S, Sharma S, Agarwal G (2018) The hidden truth anonymity in cyberspace: deep web. In: Conference 2018, AISC, vol 673, pp 719–730. Springer, Heidelberg
13. Bartlett J (2014) The dark net: inside the digital underworld, 1st edn. Cornerstone Digital Publishing
14. Tor Metrics. https://metrics.torproject.org/. Last accessed 2 Feb 2020
15. TechnicalSagar-What is Marianas web. https://youtu.be/J2XoTLpsPwY. Last accessed 28 Feb 2020

CNN Performance Evaluation for Image Classification

Pooja Mudgil, Arpit Choudhary, Bhavya Sethi, and Garima Chhabra

Abstract Predictions based on chest X-ray are usually a complex and time-taking process. Due to its complicated nature, it is often hard for radiologists to detect the disease correctly in their first examination only. Diagnosing X-ray to find out about the potential disease requires careful observation along with appropriate knowledge of anatomical, physiology and pathological principles. In this performance evaluation paper, we talk about convolutional neural network (CNN) that is undoubtedly one among widely trusted and effective deep learning algorithms for the purpose of image classification which is subsequently used in disease detection using chest X-ray. The basic working technique involves various machine learning techniques applied to automate the predictions of numerous thoracic diseases, namely hernia, pneumonia, edema, fibrosis, emphysema, pneumothorax and cardiomegaly by analyzing the chest X-ray images. Along with these, advantages and shortcomings of CNN are also included in the respective manner.

Keywords Thoracic diseases · Lung diseases · Independent binary classifier · Chest X-ray · Deep learning

P. Mudgil (✉) · A. Choudhary · B. Sethi · G. Chhabra
Department of Information Technology, Bhagwan Parshuram Institute of Technology, Delhi, India
e-mail: engineer.pooja90@gmail.com

A. Choudhary
e-mail: choudharyarpit99@gmail.com

B. Sethi
e-mail: sethibhavya98@gmail.com

G. Chhabra
e-mail: chhabragarima11@gmail.com

A. Abraham et al. (eds.), *Proceedings of 3rd International Conference on Computing Informatics and Networks*, Lecture Notes in Networks and Systems 167,
https://doi.org/10.1007/978-981-15-9712-1_34

1 Introduction

From the past few years due to the rapid changes in the field of Digital Content recognition, the field of computer vision has also evolved which has increased the complexity of the task of image classification which plays a vital role in data storage and retrieval systems.

Artificial intelligence is a field that emphasizes the development of intelligent machines, thinking and working like humans and a subfield of which is computer vision.

An image consists of a set of pixels in which each pixel is assigned a particular value and we need a significant amount of calculations to achieve the goal of image classification. Due to this, advanced computer system with increased processing power and higher configuration is required. Also, it is not humanly possible to make decisions based on these inputs as the time required for computation for providing accurate results is high.

Observing images through human vision provides a better edge over the analysis of the same image by various algorithms implemented by the system due to which there exist some complex problems in the current classification algorithms and considerable efforts have been made to solve these issues.

As a result, we have implemented deep learning algorithms in our system in order to achieve as accurate results as possible.

2 Literature Review

Various project and research papers were taken into consideration while planning and executing this project. Some of them were completely based on the techniques of image processing, while others use the artificial neural networks (*ANN*) or connectionist systems for making prediction or forecast of the diseases based on the chest X-ray images.

Khobragade et al. [1] proposed an automated smart system to detect lung diseases like lung cancer, tuberculosis and pneumonia using chest films or radiographs. Based on the results obtained, it was inferred that techniques used for image preprocessing like image segmentation and histogram equalization give better results for the chest radiographs. Pattern recognition techniques like the feedforward artificial neural networks provide better results.

Chen et al. [2] developed a new methodology or framework to dramatically supplement the dataset. Further, this augmented or amplified dataset was used to train the CNN model for the diagnosis of thoracic diseases, and this helped to improve and enhance the performance and efficiency of the model quite substantially. The future scope was to collect millions of different images without any labels from local hospitals to improvise the operation of the existing CNN model.

Wang et al. [3] made an effort to create a "human–machine glossed" compendious database of chest X-ray which provides the pragmatic methodological and clinical concerns of handling a minimum of ten thousands of patients. They carried out a large quantified operation based on the general eight thorax pathology categorizations and feebly monitored positions using database of chest X- ray.

Chan et al. [4] suggested an approach to partition and divide the lungs in different peculiar regions using the concept of several overlaying tiles or blocks. These regions of abnormalities are traced by patterns which are transformed mainly through computation of numerous overlaying levels of block. Eventually, the technique was able to effectively and efficiently analyze the specific regions of lung ailments in the images of chest X-ray and helps to improve the conceivable examination of the lacking concerns of the emphysema area.

Paredes et al. [5] devised a new algorithm in order to increase the efficiency of the usual image classification algorithm and thus, was successful in achieving the state-of-the-art accuracy by using small chunks of medical images under consideration. These small patches acted as the local features. The algorithm used for the classification along with the categorization of the image was K-NN which is the K-nearest neighbor algorithm (k-NN). This algorithm greatly helped in achieving a high level of accuracy in the process.

Li et al. [6] used a similar approach of using CNN for image classification. They modified the usual convolutional neural network algorithm according to their needs and then trained this customized CNN for the purpose of classifying and categorizing the patches which were present in the images of lungs in their dataset. Their model had only one single layer of convolutional neural network as opposed to the multi-layer CNN usually used for image classification. The deep features of the data set were extracted using this single layer of the CNN, and it also aided in overcoming the problem of over-fitting. Along with it, their model also proved to be successful in obtaining the best performance in classification when compared with other different types of feature extraction algorithms which include SIFT-scale-invariant feature transform (SIFT), LBP-rotation-invariant local binary pattern (LBP) features and the restricted Boltzmann machine used for unsupervised feature learning.

3 Proposed Work

The framework of human visual productively perceives and identifies objects within disordered sights. However, for the synthetic systems, which is yet quite troublesome due to the dependency on perspective, self-generated variability in object variability, and the major class variation of various object types. Deep progressive neurotic models usually imitate the idea of the vertebral visible crusts, and by the network, an accord is considered as the most encouraging structures for similar undertakings. The major thriving hierarchic object gratitude frameworks concentrate on extracting topical characteristics from the available set of intake images, convoluting the spots of image with screening. Now, these filtered reflexes are further progressively down-converted and re-filtrated, which results in a profound positive feedback and profound mesh structure whose succumb highlights the carriers which are finally organized.

The existing approaches to deal with object identification generally take vital utilization of machine intelligence techniques. In order to enhance the level of accuracy and strengthen the exhibition, we can gather bigger and complex datasets, adapt potential and enhanced exemplars, and utilize improvised systems to avert the problem of cloying [7]. To study about the several thousands of objects present in millions of clichés, we require a paradigm with a better training ability.

Although the intense intricacy and immense variation of the object perception work conclude that this issue cannot be handled even with a complex collection as voluminous as the NIH chest X-ray dataset, the paradigm must also have enough precedent awareness about how to indemnify for all the information that we actually fail to have. Convolutional neural networks (CNNs) comprise one similar type of the paradigm. Their ability can be constrained by changing the breadth and length and also make persistent and primarily accurate assumptions regarding the characteristics of images like fortitude of details and vicinity of pixelated dependence. Accordingly, contrasted with general positive feedback neurotic systems with similar calibrated tiers, CNNs may have many less associations and benchmarks; thus, they are quite simpler to be trained, whereas their technically finest execution is possibly to be the only somewhat worse.

In spite of appealing characteristics of convolutional neural network, and regardless of the general proficiency of their local structure, they still have been narrowly costlier to be applied in bulk to larger goals pictures. Fortunately, the existing GPUs, combined along with profoundly upgraded usage of 2D intricacy, are amazing enough to encourage the learning of engagingly larger CNNs, and reset dataset, like NIH chest X-ray dataset, contains adequate labeled instances to train such paradigms using the training dataset without severe overfitting.

3.1 Working of the CNN Algorithm

This part explains the working methodology of the algorithm in a brief manner. The input is given to the network in the form of a 2D image. The network consists of an input or initial layer that accepts the image of X-ray as an input and the output or final layer from where we obtain the proficient output, while the intermediary layers are termed as hidden layers [8]. As already stated, the mesh comprises a succession of convolving and subsampling layers. Together, these layers provide an approximation of the input image data. CNNs actually exploit the spatial local reciprocation by implying a local linked prototype among various neurons of adjoining layers. Nerve cells in "x" layer say are linked to a localized subgroup of nerve cells from the former layer of (x-1), where the nerve cells of the (x-1) layer have contiguous receptive fields. Diagrammatic course of various layers in sequence depicts distributed weights in the CNN algorithm, and each sparse sieve is repeated through the complete visible sphere. Such units then compose a feature map, and these share weight vectors and bias. It represents three such units of hidden layers of the similar characteristic map. Same colored weights are kept mutual, thus strained to be similar. The gradients of the shared weights are calculated as the total of all gradient of aggregated specifications being communal. Such replication in some fashion permits characteristics to be traced and identified disregarding of the fact where they are located or positioned in the visible range. Moreover, shared weights also allow to reduce the statistics of unfettered training specifications. Because of the controlled execution, CNN is able to attain improved conjecture to visibility issues. CNN also makes use of the idea of maxpooling that is a type of nonlinear down-sampling. In this technique, the input image is partitioned into non-overlying rectangular frames. The output for all the subregions is the maximal value.

3.1.1 Convolution Layer

The primary layer of the CNN network which is the convolution layer comprises a convolution mask, function expression and a bias term. Combining them, they produce result for that particular layer. The example depicts a masked 5 $\times$ 5 matrix which implements convolution aloft a 32 $\times$ 32 characteristics map as input. The resulting output is a 28 $\times$ 28 matrix. Then, bias is added, and sigmoid function is applied on the matrix. The convolution layer has been applied four times for various filters, i.e., 16, 32, 64 and 128.

3.1.2 Subsampling Layer

The subsampling layer comes after the convolutional layer. It has the exact number of frames as present in the convolutional layers. The intent of the layer is to decrease the extent and complexity of the characteristics map. This breaks the input figure into chunks of 2 $\times$ 2 and implements averaging. Subsampled layers maintain the related data among various features and not to the exact relations. We have then applied maxpooling which is used to reduce the dimensions and assumed features for better feature extraction; then to convert the matrix format of images to a vector, we have used flattening through a fully connected layer. To pass the data from one layer to another layer, we have to put the data into the neurons using dense function layer.

3.2 Preparing Database

The input data is provided as an image itself. These images are then converted to grayscale images as data information is more important for the network as compared to the color information [9]. Also, the images are reduced to 64 $\times$ 64. Since the images present in the datasets are larger, pyramid reduction is used to resize them

64 × 64. The edifice of the image is the kind of structure of information which is devised to provide uphold for decisive and effective calibrated contortion using these diminished representation of images. This comprises a sequence of copies of an initial image which are decreased in resolution as well as in sample density in regular steps.

3.3 Working Model

The assessment matrix applied here contains the F-beta scores, precision and recall. In binary classification where the answer is either yes or no (if disease is found or not), F-score is considered of more use as compared to accuracy level of the model. This is due to the fact that in bipartite denominations that if any disease is present or not, then divisions are not balanced. Consider an illustration that we are having a frivolous segregator which makes guess to the major class, this would achieve 70% accuracy when there is an 70/30 division and 50% accuracy when that has a 50/50 division.

Recall and precision majorly target on the count of peoples that are predicted to be influenced, thus surmounting the quality of slant input and the benefit of forecasting the ailment of a person. Precision defines ratio of peoples that have been accurately forecasting the ailment to all those that were anticipated to be ill. Recall defines ratio of peoples that have been envisioned as ill to all those who were infected actually. These indicating parameters are very useful in making prediction the illness, so we require an index which can be aggregate of recall and precision both.

Paradigms are required to be assured if the person is expected to suffer from any sickness that signifies what all expectations of disease are majorly satisfied as when a person is analyzed to be suffering from an illness, and it is very distressing. Hence, it becomes really critical, because it is a procedure to help doctors in diagnosing patients. Therefore, we require higher level of accuracy, i.e., more precision value and less recall value, in reference to smaller value of β, assuming β to be 0.5 for calculating F-beta score. It requires to avert garbling ill persons to shun their sickness, bypassing the actual patient at higher level of exposure. Such cases would delectate lower precision value and higher recall value, in reference to larger β, assuming β to be 2 for F-beta score calculation. The objective of this strategy is to justify the advancement in technology and provide helping hand to bolster the doctor in forecasting the illness as to identify the disease which would require various examinations of the patient. This will make the person anxious before getting an insight of further result, so we choose F-score for $\beta = 0.5$.

The order of steps is as follows:

- Load data has been processed into ram, and data is processed as before. Images are in RGB format.
- Implement network architecture as designed by the architect.

- Implement metric function including binary accuracy, precision score with threshold, recall score with threshold, Fbeta score with beta and threshold.
- Implement data generator, model checkpoint and model loss function.
- Training model with training parameters, logging training validation loss and training validation accuracy.
- Save the best model and test model on the testing set.

4 Experimental Results

Differences of gender in the Determinants and Consequences of Thorax Diseases

Gender refers to "the categories of roles and relationships that are socially constructed, attitudes, relative power, behaviors, personality traits, values, and influence that society assigns to the two genders on a differential basis. This is related, characteristics and gender roles do not exist in isolation, but are defined in relation to each other and by the relationships between men and women."

Various surveys on health based on different genders have been performed by community scientists across the world, and it has been observed that only botanical differences are not sufficient to explain the variation in health. In contrast, the results also depend on various socioeconomic factors along with the prevailing conditions of the society; hence, it can be concluded that gender is an important factor for any study related to human health.

With the health system developed, i.e., Chester primarily used for the detection of 14 thorax diseases based on the X-ray images of the patient along with other factors such as age, gender and patient data, we were able to analyze the trends of diseases among men and women. The result for the same has been presented in the form of bar graphs. Figure 1 Disease distribution based on patient gender (Fig. 2).

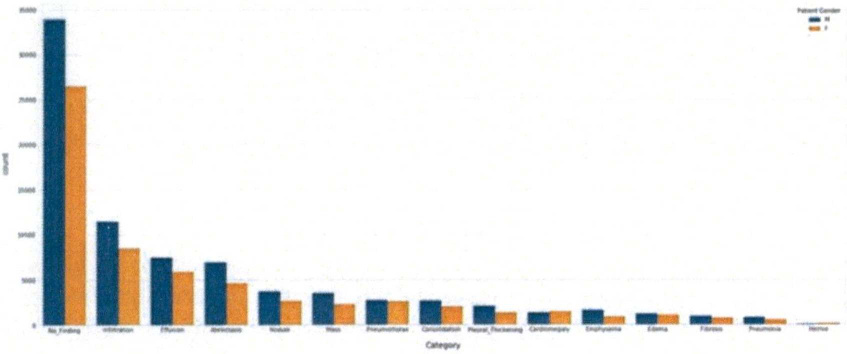

Fig. 1 Disease distribution based on patient gender

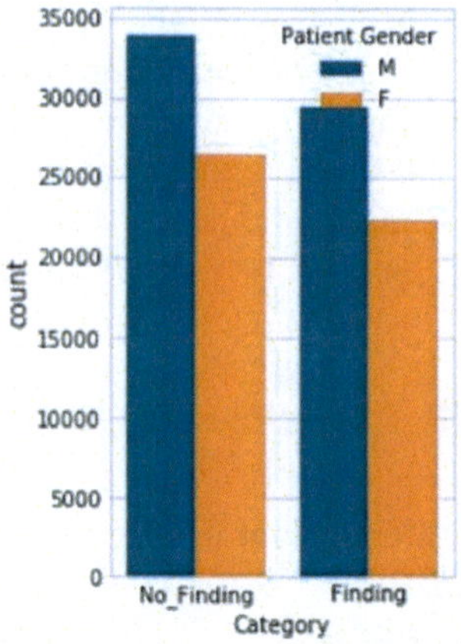

Fig. 2 Disease finding based on patient gender

The above bar graphs reveal that cases of diseases like fibrosis, pneumonia, hernia and other popular lung diseases like atelectasis, effusion and infiltration are very low. Dispersal of diseases is not regular. With the help of the graph, it can be observed that the cases for men for a particular disease are higher as compared to that of women. In contrast where there is uncertainty about the disease, the difference is observed to be quite low.

In pneumonia that is quite a common disease primarily caused due to some bacterial infection, it leads to the inflammation of the lungs. It has been observed that women have a greater chance of surviving if they get infected. Study of these variations can help in designing better treatments which can lessen the number of people getting infected and also help the patients in a faster recovery.

Bacteria coming from the nose which also include the pneumonia causing germs enter into the body through the respiratory system. Although the body has quite strong resistance against these attacks, special cells that are known as macrophages are deployed as a "first response" to kill these harmful bacteria when they enter into the lungs. But this system will not work if the immunity system of the body is weak or if the individual is already suffering from any infection. This is the reason many patients with respiratory infection need to be hospitalized and taken care of. Females have a stronger resistance against pneumonia because of production of estrogen which is the primary female sex hormone, and it increases the ability of the immunity system to fight against the bacteria.

Age Variation in Prevalence of Thorax Diseases

Age is a vital factor in the data analysis needed for the detection of thorax diseases. Thus, taking into consideration various age groups, graphs have been plotted showing the trends of diseases by categorizing people into young- and middle-aged people (Fig. 3).

From the given graph, it is observed that middle-aged people are at more risk of getting thorax disease and are frequently sent for medical examinations. Nowadays, youngsters are also giving importance to early diagnosis.

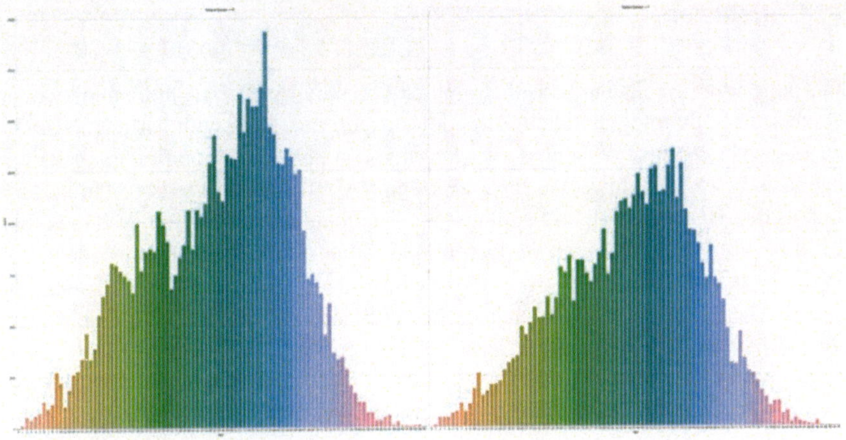

Fig. 3 Disease distribution based on patient age

The immunity system of the human body is a collection of various types of cells and organs that work in collaboration to protect the body from attacks by various infection causing organisms like fungi, bacteria and parasites. White blood cells are an important part of this immune system which fight against foreign substances that attack the body known as antigens.

When respect to germs aging has both pros and cons. As we grow older, we are able to fight against antigens faster. The human body gains antibodies for the germs defeated by the immunity system in the past; due to this phenomenon, adult is at less risk of getting cold and other respiratory infections.

It can be deduced from the graph that thorax disease is more common in middle age people which are due to the following reasons:

- There is a scarcity of white blood cells in the human body which are capable of fighting against the new antigens; due to this, when an older human body encounters a new type of bacteria or antigen, the body is not able to defend against it.
- The number of antibodies built for defending against antigens is less in older people and is less able to associate with the antigen. Due to this, common diseases like influenza, pneumonia and tetanus are the causes of death of older people as they are more likely to get infected.
- As we grow old, the immunity system loses its tolerance against some of the infections, and sometimes, the immune cells can also attack particular organs or tissues of the body that is known as autoimmune disorder. Some of the common autoimmune disorder are arthritis, rheumatoid, scleroderma and lupus.

5 Conclusion

This paper can be considered as the research for analyses of convolutional networks which are deeply integrated and also allows experimentation with various ideas. In our research, we have presented a system that can help doctor and radiologists for diagnosis of chest X-rays. We have also emphasized various needs of the system such as disease prediction and explanation. The main purpose of the system is that it can be used as an assistant to doctor. We also ensure the privacy of patients by designing the system in such a way that the information is processed locally and not leaked. The system is easily scalable and can be used for the needs of 1-1 Million users with little overhead.

Previously, there have been a lot of studies that focus on chest diseases and use the structure of artificial neural network. But all these studies have used a different approach to neural network along with the different datasets, and thus, the direct comparison of our research was not possible.

In the end, it can be deduced that:

- The structure of various neural network can be used in the diagnosis of thorax diseases.
- CNN was used for chest disease diagnosis, and it provided the best results with a fair amount of accuracy.
- The results we got by using two hidden layers in our CNN structure were better as compared to those with only one hidden layer.
- Also, it can be concluded that these neural network structures can prove to be helpful as a support system that enhances learning-based decisions and contribute to the medical world.

References

1. Khobragade S, Tiwari A, Patil C, Narke V (2015) Automatic detection of major lung diseases using Chest Radiographs and classification by feed-forward artificial neural network. In: 2016 IEEE 1st international conference on power electronics, intelligent control and energy systems (ICPEICES)
2. Chen J, Qi X, Tervonen O, Silven O, Zhao G, Pietikainen M (2016) Thorax disease diagnosis using deep convolutional neural network. In: 2016 38th annual international conference of the IEEE engineering in medicine and biology society (EMBC)
3. Wang X, Peng Y, Lu L, Lu Z, Bagheri M, Summers RM (2017) Chest X-ray8: hospital-scale Chest X-ray database and benchmarks on weakly-supervised classification and localization of common thorax diseases. In: 2017 IEEE conference on computer vision and pattern recognition (CVPR)
4. Chan Y-H, Zeng Y-Z, Wu H-C, Wu M-C, Sun H-M (2018) Effective pneumothorax detection for Chest X-ray images using local binary pattern and support vector machine. J Healthcare Eng 2018:1–11

5. Sharma A, Raju D, Ranjan S (2017) Detection of pneumonia clouds in chest X-ray using image processing approach. In: 2017 Nirma University international conference on engineering (NUiCONE)
6. Study on air pollution and control investment based on environmental theory model. https://www.researchgate.net/publication/326000772_Study_on_Air_Pollution_and_Control_Investment_from_the_Perspective_of_the_Environmental_Theory_Model_A_Case_Study_in_China_2005-2015
7. Learning vector quantization for machine learning by Jason Brownlee. https://machinelearningmastery.com/learning-vector-quantization-for-machine-learning/
8. Understanding of convolutional neural network (CNN)—deep learning. https://medium.com/@RaghavPrabhu/understanding-of-convolutional-neural-network-cnn-deep-learning-99760835f148
9. NIH Chest X-ray dataset. https://www.kaggle.com/nih-chest-xrays/data

Indexing-Based Peculiarity Extrication from Biomedical Images Using Deep Learning

Tanvi Chaudhary, Rajat Mudgil, Neha Agarwal, and Deepika Sandil

Abstract Biomedical image classification is an important task to generate classification maps for deadly diseases like cancer, tumor, etc. In this paper, a novel technique for classification of biomedical images is proposed. The proposed technique takes raw biomedical image as input and applies indexing to retrieve the parameters used to classify different diseases based on deep learning. The proposed Indexing-Based Peculiarity Extrication from biomedical images using deep learning (IBPE) is evaluated using GLCM, Gabor filter, fuzzy clustering means and SVM. To achieve an effective biomedical image classification, this framework system isolates its work in various stages; these phases are important to give the better classification accuracy. Evaluation proves the efficiency of the proposed IBPE with an achieved accuracy of 96.7 over 92.5% of the existing method.

Keywords Indexing-Based peculiarity extrication from biomedical images using deep learning · Gray-level co-occurrence matrix · Support vector machine · Open access series of imaging studies · Artificial neural network

Abbreviations

IBPE Indexing based peculiarity extrication from biomedical images

T. Chaudhary (✉) · N. Agarwal
Amity University, Noida, Uttar Pradesh 201313, India
e-mail: tanvicy1234@gmail.com

N. Agarwal
e-mail: agarwalnehajain@gmail.com

R. Mudgil · D. Sandil
BPIT, Rohini 110089, I.P University, India
e-mail: rajatmudgil24@gmail.com

D. Sandil
e-mail: deepikasandil@bpitindia.com

A. Abraham et al. (eds.), *Proceedings of 3rd International Conference on Computing Informatics and Networks*, Lecture Notes in Networks and Systems 167,
https://doi.org/10.1007/978-981-15-9712-1_35

GLCM	Grey level co-occurrence matrix
SVM	Support vector machine
ANN	Artificial neural network
LBP	Local binary pattern
AI	Artificial intelligence
ML	Machine learning
OASIS	Open access series of imaging studios
SDAE	Stacked denoising autoencoder
HOG	Histogram of arranged angles
NN	Neural network

1 Introduction

Images are the most critical resources in the biomedical sector for analyzing, diagnosing and examining. Doctors, clinical specialists and pertinent scientists from emergency clinics and clinical foundations contemplate and look at previews best. Because of this, we have huge database of images and they are delivered in the form of ultrasound, MRI, X-ray and so forth. As a result, it is very tough and impractical for the physicians to observe and interpret all the images from this big amalgam of image database manually. But we cannot forget about the use and importance of these images, so our project is going to discover a more practical manner to deal and make the most use of them for our expecting diseases with extraordinary accuracy. Tools including a retrieval device for the biomedical images perform a key role in the era of digital health.

This device can be used to detect, observe and cure complex illnesses with minute differences between them resulting in dealing with and improving human health. There are positive traditional techniques that have been used on this subject or for widespread retrieval of pixels, methods like color, histograms, edges, texture, shape and all forms of other variations. It is quite evident from the earlier studies that accurate extraction function and to remove away the redundant statistics are the two most important fulfilling elements in the indexing and retrieval of biomedical image system. It is carried out by way of reducing the number of dimensions of the given input image. For that, we have got visual function descriptors named as global visual feature descriptor and local visual feature descriptor. They both are broadly proposed and used in the field of biomedical image retrieval and analysis. Many researchers have worked on the part of improvement and use of world visual capabilities for the retrieval of biomedical images [1]. They proposed the concept which says that wavelet may be transformed to use pixel values and build an image signature. Then, the photo texture at various scales is represented which is achieved by way of describing the decomposition [2, 3] of worldwide distribution from wavelet coefficients including each sub-band.

But, in current nearby years, visual features have taken a top position and are being extensively used. For example, one among the maximum extensively adopted descriptor in applications of biomedical photo retrieval is local binary pattern [3, 4]. Due to its promising performance, LBP and its numerous variants are utilized in biomedical pictures and then came the local mesh pattern [4, 5] descriptor presented by Murala and Wu. Even though after diverse discoveries and theories, traditional characteristic descriptors had weaker discriminative and representative strength especially for same images but from precise categories IBPE and the usage of deep learning. To increase the efficiency and outcome, deep learning came into existence for indexing and then retrieving the classified image. Although it is an interesting area for the researchers, still a lot of development is in progress. In our framework IBPE, we are proposing a technique which is feasible and have the capability to classify images with the help of queried image which is having only minor or minute differences between them.

2 Review of Related Work

2.1 Machine Learning

Logical study of factual models and calculations that are used by PC framework (also shown in Fig. 1 [6]) can be used to play out a particular task without utilizing express guidelines; the following instance is called as machine learning. It is a part of artificial intelligence (Al).

Numerical model on sample data, called as "training data," is built in machine learning. This machine has the ability to make assumptions or conclusions without being separately programmed to do the work. There are endless types of applications where machine learning is used. One of them can be filtering of email and vision of computer; in such tasks, it is impractical to make a standard algorithm to perform the task effectively. There is one more part of ML which is the field of study called as data mining [7].

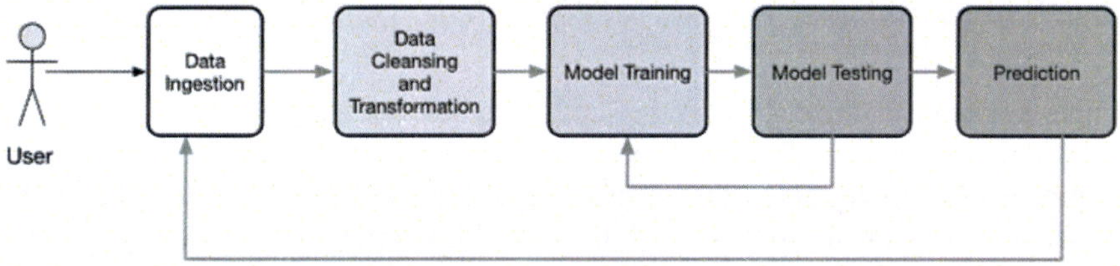

Fig. 1 Machine learning system

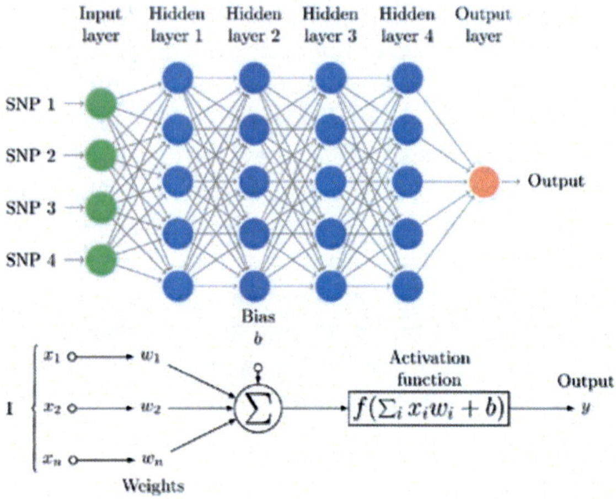

Fig. 2 Deep learning tasks

2.2 *Deep Learning*

Wide collection of AI techniques on ANN comprises of deep learning also called as deep structured learning. Learning can be grouped in three categories—supervised, semi-supervised or unsupervised learning. Areas having vision of PC, acknowledgment of discourse characteristic language preparing, acknowledgment of sound, community which is informal separating, interpreting machines, bioinformatics, tranquilize plan, examination of pictures.

It is effective in the health sector for delivering results similar to and even always better than specialists that are human. GLCM and Gabor filter will be used for the image feature extraction [8]. In biological systems, we have information processing and distributed communication nodes inspired ANN. There is a lot of difference between ANNs and biological brains. On one hand, biological brain of most living organisms is dynamic (plastic) and analog, while on the other hand neural networks (as per shown in Fig. 2 [1–9]) tend to be symbolic and static [7].

2.3 *Supervised and Semi-supervised Learning*

Ample amount of information having both data sources and the perfect results is utilized to construct a scientific model in supervised learning algorithm. Making information for an algorithm that is supervised learning algorithm would consider images without or with that article (info), and every image will be having a patch depicting that it contains the item when we somehow managed to decide if an image contained a specific object [1]. In exceptional scenario, the information may be just reachable halfway bound to extraordinary feedback advancement of mathematical models from deficient making information a segment of the example information that does not have names is done in semi-supervised learning algorithms [6].

3 Proposed IBPE Using Deep Learning

We will be going according to the flowchart shown in Fig. 3. Biomedical image classification is an important task to generate classification maps as no. of world observation diseases like cancer, tumor,. etc., are increasing day by day, and these diseases contain different tools capable of capturing imagery time to time and utilized for a wide range of application. Thus, classification of cancer imagery has the current area of researches, and classification results can be used for different real-time applications. This system proposed a novel approach for classification of six different types: actinic keratosis, basel cell carcinoma, cherry nevus, dermatofibroma, melanocytic nevus and melanoma by utilizing disease imagery. To achieve an effective disease image classification framework, this system isolates its work in various stages; these phases are important to give the better classification accuracy, and in the next page, we have described these phases in detail. To filter the unnecessary information, we use noise filtering and make our images free of different types of noises using image processing toolbox. Fuzzy clustering means is used for image segmentation and clustering. The proposed IBPE classifies works or process in six phases that

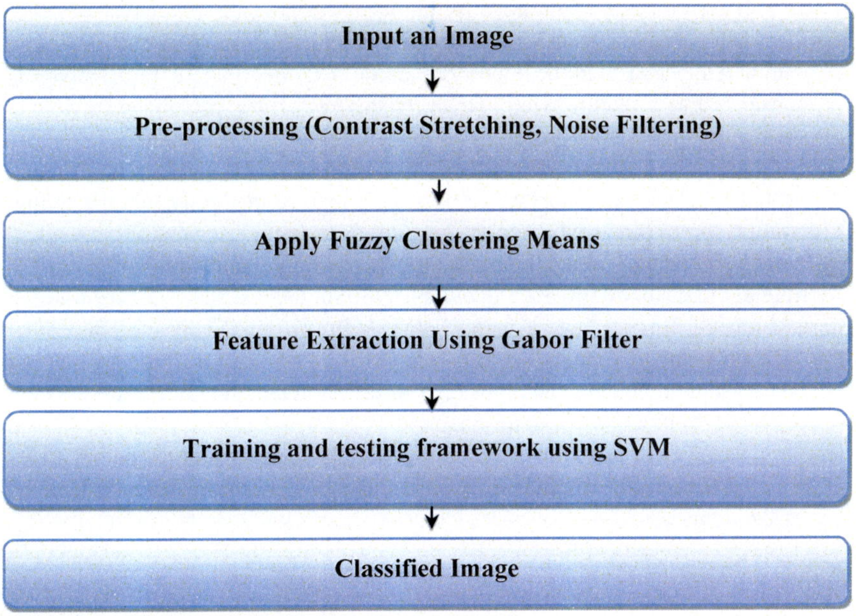

Fig. 3 Proposed framework working

is taking raw biomedical images as input for pre-processing. Indexing the usage of GLCM, Gabor filter, fuzzy clustering approach and SVM is implemented to acquire a powerful classification. IBPE is evaluated on Open Access Series of Imaging Studies (OASIS) dataset.

Here are the steps defined as per the proposed IBPE framework using deep learning:

(i) We will start by taking an input image from our image database.

(ii) The second step is taking input from the raw input image so that we can pre-process the image. So, pre-processing of raw input image is basically done with the help of noise filtering and contrast stretching (also known as normalization). By these techniques, we are attempting to endeavor and enhance to improve the complexity in a picture by extending the scope or range of intensity values of force esteems it contains to traverse an ideal scope of qualities. Also, the unwanted noise is filtered from the image preserving the details of the same.

(iii) After we get filtered and clean image, we apply fuzzy clustering means algorithm (also called as k-means algorithm) which is an unsupervised machine learning algorithm. Here, we find out the centroid of the data points and then calculate the distance of each data point from the given centroids until the clusters formed become constant. And because it works on fuzzy logic, so that is why each data point can belong to more than one cluster. These all significant steps are required to classify our image and extract the essential features in order to determine the correct disease with its accurate type and stage.

(iv) These steps will be followed by the last step of feature extraction with the help of Gabor filter. Right now, we are trying to examine or analyze whether there is a particular recurrence content in the picture in explicit bearings in a confined locale around the point or district of examination. Texture analysis is considered as a feature while we are trying to analyze any image or we can say that feature extraction additionally incorporates surface analysis or texture analysis using the gray-level co-occurrence matrix (GLCM). It is a factual strategy for analyzing surface or analyzing texture that considers the spatial relationship of pixels which is the gray-level co-occurrence matrix.

(v) Above all the steps make the first phase of our implementation, and now, it is time to inculcate intelligence in our system. So, the last step goes for training and testing our framework using SVM. It is surrounded by the concept of decision plane also called as hyperplane for classification. It is a plane that gives a separation between object sets of different class memberships and can be a straight line or a curve in shape.

(vi) This way we classify our image and group them according to their features which is then used to map them down and identify the correct disease depicted by the image.

4 Important Algorithms

4.1 *Fuzzy Clustering Means (Soft Clustering Or Soft K-Means) Algorithm*

It is a technique for grouping which permits single piece of data to take a position with at least a couple of bunches. This strategy is as often as possible utilized in design acknowledgement.

It is dependent on the minimization of the accompanying target work [1–9]:

$$J_m = \sum_{i=1}^{N} \sum_{j=1}^{C} u_{ij}^m \left\| x_i - c_j \right\|^2 \quad 1 \le m < \infty \tag{1}$$

In above Eq. 1, we have m that stands for any number real more frequent as compared to 1, u_{ij} is the extent for indulgence of x_i in the j group, the ith of d-dimensional estimated information is represented by x_i, c_j is the focus of d-measurement of the bunch, and ||*|| is any standard that can be used to communicate the similitude between any wanting information and the inside. Fluffy dividing is done with the update of cluster centers c_j and the membership u_{ij} which is done via an optimization that is iterative, of the objective function depicted earlier:

$$u_{ij} \frac{1}{\sum_{k=1}^{C} \left(\frac{\|x_i - C_j\|}{\|x_i - C_k\|} \right)^{\frac{2}{m-1}}} \quad c_j = \frac{\sum_{i=1}^{N} u_{ij}^m . x_i}{\sum_{i=1}^{N} u_{ij}^m} \tag{2}$$

This condition will end when:

$$\max_{ij} \left\{ \left| u_{ij}^{k+1} - u_{ij}^{k} \right| \right\} < \delta \tag{3}$$

end foundation in the range of 0 and 1 somewhere while k denotes cycle steps. This strategy meets to a neighborhood least or a seat purpose of Jm. The calculation is made out of the accompanying advance as per shown in Fig. 4 [1].

4.2 *Support Vector Machine (SVM)*

A decision plane is utilized to define the decision boundaries, and this comes under the concept of support vector machine. A plane that gives a separation between object sets of different class memberships is called a decision plane. A diagram sample is shown in the figures below. In the below case, there are two dot groups GREEN or RED to which the objects belong. Right side we have all the objects GREEN and on

Fig. 4 Fuzzy clustering means algorithm

1. Initialize U equal to[uij] matrix, $U^{(0)}$
2. At k- step: calculate the centers vectors $C^{(k)}=[cj]$ with $U^{(k)}$
3. $c_j = \frac{\sum_{i=1}^{N} u_{ij}^m \cdot x_i}{\sum_{i=1}^{N} u_{ij}^m}$
4. Update $U^{(k)}$, $U^{(k+1)}$
5. $u_{ij} = \frac{1}{\sum_{k=1}^{C} \left(\frac{\|x_i - c_j\|}{\|x_i - c_k\|} \right)^{\frac{2}{m-1}}}$
6. If $\|U^{(k+1)} - U^{(k)}\| < \delta$ then STOP; or go to Step Number 2.

the left side, all the objects RED which is separated by the line that defines boundary as per shown in Figs. 5 and 6 [1]. Any white circle (new object) depicting on the right side is denoted, i.e., as GREEN.

Set of items into their individual gatherings is finished by a classifier, an ideal case of linear classifier. Generally, characterization assignments are not unreasonably straightforward and frequently require progressively structures that are complex which help in making an exact division. It is depicted in the outline beneath. It is very evident that we do need a curve more complex than a line for complete differentiation of GREEN and RED objects as per shown in Fig. 7 [1]. Hyperplane classifiers are called when we had to draw separate lines to differentiate different class membership objects. And we usually have support vector machine to take care of these cases.

Classification SVM Type 1: Support Vector Machine Type 1:

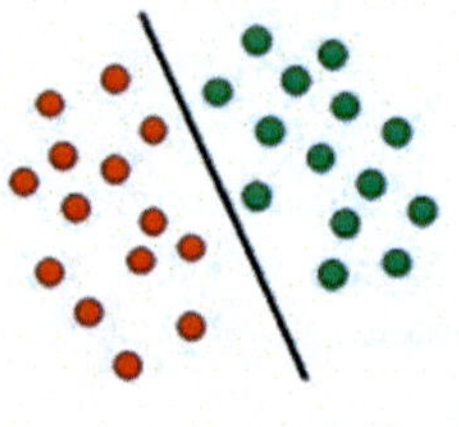

Fig. 5 Straight line hyperplane

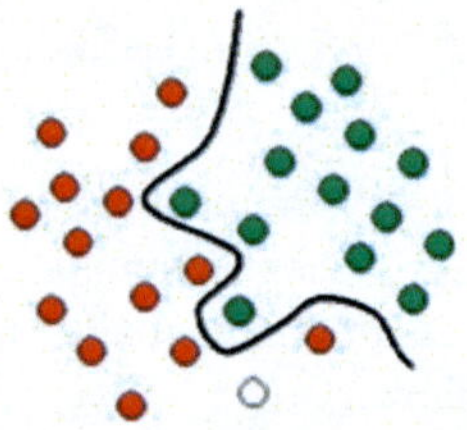

Fig. 6 Curve hyperplane

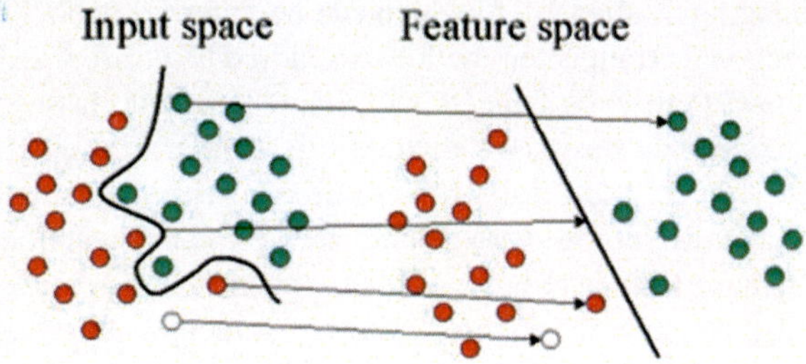

Fig. 7 Hyperplane classifier

$$\frac{1}{2}w^T w + C\sum_{i=1}^{N} \xi_i y_i\left(w^T \emptyset(x_i) + b\right) \geq 1 - \xi_i \text{ and } \xi_i \geq 0, i = 1, \ldots N \quad (4)$$

In the above Eq. 4, C stands for constant capacity, w stands for coefficient vectors, b shows a constant, and shows epsilon parameters. The index denoted by i is used to label the N number of training cases.

Classification SVM Type 2: Rather than classification support vector machine type 1, we have classification support vector machine type 2 model:

$$\frac{1}{2}w^T w - \upsilon\rho + \frac{1}{N}\sum_{i=1}^{N} \xi_i y_i\left(w^T \emptyset(x_i) + b\right) \geq \rho - \xi_i, \xi_i \geq 0, i = 1, \ldots N \text{ and } \rho \geq 0 \quad (5)$$

5 Result Analysis of the Proposed IBPE

We contrast our methodology and the regular and renowned deep preference learning algorithm (convolutional neural network [11, 12], stacked denoising autoencoder [13], GLCM, NN [12], Discrete wavelet transform [14] and SVM) set of rules which has been broadly used in biomedical images recovery or for classifying and retrieving them. What is more, we analyze it with various general photo or image recovery systems with color histograms and the histogram of arranged angles (HOG). A basic factor is that we additionally contrast it with those types of pictures or image recovery methodologies which utilize the profound form as a deep learning field of capacity extraction. This observation and analysis show that it is not just the profound trademark improving the ordering execution, anyway our proposed system. To extensively inspect the presentation of the proposed structure, we pick a few open biomedical photograph datasets (OASIS database [15–17]) as measure for our analysis or examining for our tests.

The OASIS database contains an understanding set with biomedical pictures of complete cerebrum neuroimaging made with the guide of the Open Access Series of

Imaging Studies. [5, 6] which can be accessed publicly and freely having the cross-sectional arrangement with two hundred fifty-three (253) tumor subjects and ninety-one (91) disease subjects with unprecedented classes. For biomedical photograph recovery purposes, we divided this dataset into nine classifications (155, 98, 14, 16, 10, 12, 14, 17 and 8 pics) in step with the state of the ventricular from each image. Right now, varieties between pics are exceptionally little having only minor changes which are not feasible to recognize and recover. To process our framework on this database, we break down the recuperation results between our significant tendency acing estimation and different procedures over the OASIS database in Figs. 7 and 8 (inquiry photograph sooner than applying Gabor filter) and subsequent to utilizing the Gabor filter, the utilization of explicit recovery execution assessment criteria through changing the wide assortment of top coordinating pics. All in all, the shaggy bunching strategy calculation has an across the board improvement over the conventional capacity descriptors in biomedical photograph recovery. In many words, Fig. 9 and Table 1 additionally recommend that the propelled ordering generally

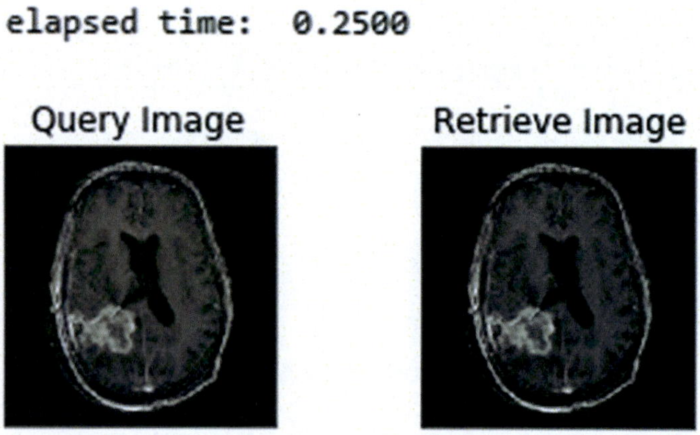

Fig. 8 Image depicting results

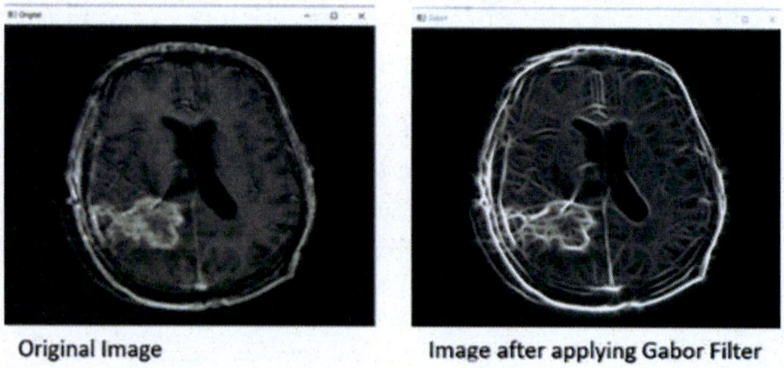

Fig. 9 Gabor filter, GLCM application

Table 1 Accuracy between the existing approach and proposed approach

Techniques	No. of queried images	Classification accuracy
GLCM, Gabor filter, fuzzy clustering means and SVM (Proposed)	100	96.7%
GLCM, NN, Discrete wavelet transform, stacked denoising autoencoder and SVM (Existing)	100	92.5%

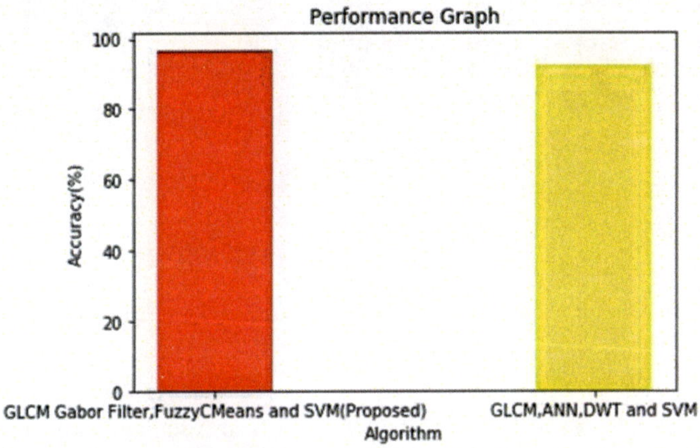

Fig. 10 Performance bar graph of the proposed algorithm over the existing technique

speaking execution of our proposed structure does not just originate from fluffy grouping calculation, anyway our total system. The proposed IBPE classifies works or process in six phases that are taking raw biomedical images as input for pre-processing. Indexing the usage of GLCM, Gabor filter, fuzzy clustering approach and SVM is implemented to acquire a powerful classification. IBPE is evaluated on Open Access Series of Imaging Studies (OASIS) dataset, and evaluation shows the efficiency of the proposed IBPE with a finished accuracy of 96.7 over 92.5% of the existing technique proven in Table 1 and Fig. 9 (Figs. 10 and 11).

6 Conclusion

Technique Indexing-Based Peculiarity Extrication (IBPE) from biomedical images using deep learning is proposed in this paper for classification of deadly diseases like cancer and tumor by utilizing cancer and tumor imagery classification. The proposed IBPE classifies symptoms in six phases. Raw biomedical image is taken as input for pre-processing. Indexing using GLCM, Gabor filter, fuzzy clustering means and SVM is applied to achieve an effective classification. IBPE is evaluated

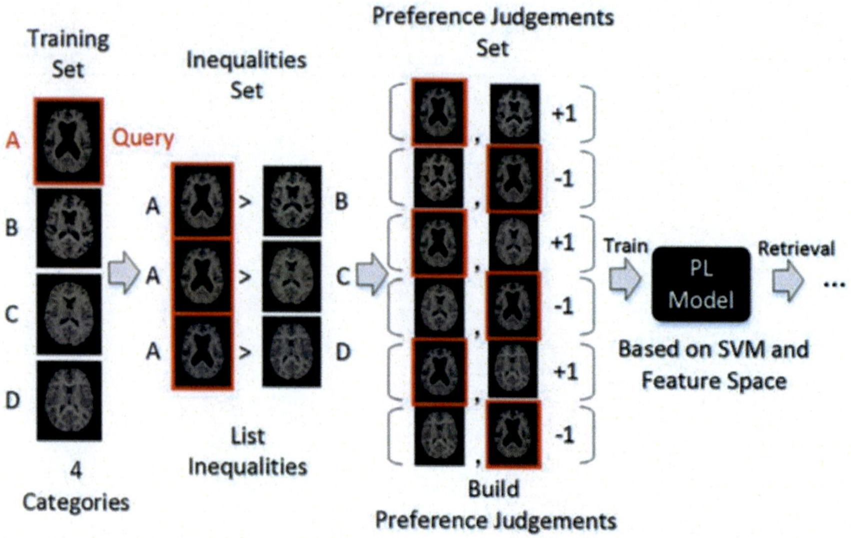

Fig. 11 Preference learning model technique to retrieve or classify the images which are comparable with a queried image

on Open Access Series of Imaging Studies (OASIS) dataset, and evaluation proves the efficiency of the proposed IBPE with an achieved accuracy of 96.7 over 92.5% of the existing technique.

References

1. Pang S, Orgun MA, Du A, Yu Z (2017) Leveraging deep preference learning for indexing and retrieval of biomedical images. In: 2017 8th international IEEE/EMBS conference on neural engineering (NER)
2. Shinde A, Rahulkar A, Patil C (2019) Biomedical image indexing and retrieval based on new efficient hybrid approach using directional decomposition and a novel local directional frequency encoded pattern: a post feature descriptor. Springer Science + Business Media, LLC, Part of Springer Nature 2019
3. Ojala T, Pietikainen M, Maenpaa T (2002) Multiresolution gray-scale and rotation invariant texture classification with local binary patterns. IEEE Trans PAMI 24(7):971–987
4. Murala S, Wu QJ (2014) Local mesh patterns versus local binary patterns: biomedical image indexing and retrieval. IEEE JBHI 18(3):929-938
5. Dubey SR, Singh S, Singh R (2015) Local bit-plane decoded pattern: a novel feature descriptor for biomedical image retrieval. IEEE JBHI
6. Quellec G, Lamard M, Cazuguel G, Cochener B, Roux C (2010) Wavelet optimization for content-based image retrieval in medical databases. Med Image Anal 14(2):227–241
7. Smith B, Arabandi S, Brochhausen M, Calhoun M, Ciccarese P, Doyle S, Gibaud B, Goldberg I, Kahn Jr CE, Overton J, Tomaszewski J (2015) Biomedical imaging ontologies: a survey and proposal for future work. J Pathol Inf 6

8. Quddus A, Basir O (2012) Semantic image retrieval in magnetic resonance brain volumes. IEEE Trans Inf Tech Biomed 16(3):348–355
9. Panga S, Orgunb MA, Yu Z (2018) A novel biomedical image indexing and retrieval system via deep preference learning. Elsevier Comput Methods Programs Biomed 158:53–69
10. Masood A, Al-Jumaily A, Anam K (2015) Self-supervised learning model for skin cancer diagnosis. In: 7th international IEEE/EMBS conference on NER, pp 1012–1015
11. Pei X (2015) Emphysema classification using convolutional neural networks. In: International conference on intelligent robotics and applications. Springer International Publishing, pp 455–461
12. Krizhevsky A, Sutskever I, Hinton GE (2012) Imagenet classification with deep convolutional neural networks. In: Advances in neural information processing systems, pp 1097–1105
13. Su H, Xing F, Kong X, Xie Y, Zhang S, Yang L (2015) Robust cell detection and segmentation in histopathological images using sparse reconstruction and stacked denoising autoencoders. In: International conference on medical image computing and computer—assisted intervention, pp 383–390
14. Belić J, Halje P, Richter U, Petersson P, Kotaleski JH (2015) Behavior discrimination using a discrete wavelet based approach for feature extraction on local field potentials in the cortex and striatum. In: 7th international IEEE/EMBS conference on NER, pp 964–967
15. Marcus DS, Wang TH, Parker J, Csernansky JG, Morris JC, Buckner RL (2007) Open access series of imaging studies (OASIS): cross-sectional MRI data in young, middle aged, nondemented, and demented older adults. J Cogn Neurosci 19(9):1498–1507
16. Noor AM, Holmberg L, Gillett C, Grigoriadis A (2015) Big data: the challenge for small research groups in the era of cancer genomics. British J Cancer
17. Andreu-Perez J, Poon CC, Merrifield RD, Wong ST, Yang GZ (2015) Big data for health. IEEE JBHI 19(4):1193–1208
18. Chakrabarty N (2019) Brain MRI Images for Brain Tumor Detection (Version 1) [Data Files]. Retrieved from https://www.kaggle.com/navoneel/brain-mri-images-for-brain-tumor-detection
19. Carranza D (2019) Skin Cancer Classification (Version 1) [Datasets], Oncology and Cancer. Retrieved from https://www.kaggle.com/tags/oncology-and-cancer, https://www.kaggle.com/ingbiodanielh/skin-cancer-classification-with-resnet-50-fastai

Calculating the Proof of Work Using Volunteer Computing

Rishab Lamba, Vanita Jain, and Dharmender Saini

Abstract In this era of abundance of computing devices like smartphones, tablets, and laptops, their computation power is yet to be utilized efficiently. To fully utilize their computation power, we have made use of a JavaScript implementation of WebRTC and WebSockets to generate a grid network of connected devices. One of the core features of this platform is its ease of use, where the users can tap into the power by just opening their Chrome browsers, be it on their laptop, computers, or mobile devices. The workload is divided into these connected devices, and the output of each sub-problem is generated in parallel. The core objective of this paper is to illustrate how this approach will significantly reduce the time required to calculate the nonce for a transaction in the crypto-mining process and demonstrate the results of this grid network by comparing the performance in a combination of multiple daily drivers like mobiles and laptops.

Keywords Volunteer computing · Blockchain transactions · Nonce calculation

Abbreviatios

WebRTC	Web real-time communication
BOINC	Berkeley open infrastructure for network computing

R. Lamba · V. Jain (✉) · D. Saini
Bharati Vidyapeeth's College of Engineering University, New Delhi, India
e-mail: vanita.jain@bharatividyapeeth.edu

R. Lamba
e-mail: rishablamba407@gmail.com

D. Saini
e-mail: dsaini77@gmail.com

A. Abraham et al. (eds.), *Proceedings of 3rd International Conference on Computing Informatics and Networks*, Lecture Notes in Networks and Systems 167,
https://doi.org/10.1007/978-981-15-9712-1_36

1 Introduction

According to a study in 2018, more than one billion mobile devices were sold, where the computation power of these smartphones rivals that of desktop devices [1]. With the highest-end mobile devices, RAM is going as far as 8 GB. Collectively, they are a repository of unlimited computational potential. While there have been significant advancements in distributed computing, they have failed to include the daily drivers of the common people. The installation of the existing software can be a tedious process; accessing the cloud platforms requires financial infrastructure and financial plans; moreover, the grid platforms require admin-level privileges. These are some of the barriers which restrict the inclusion of billions of devices that collectively have a potential equal or larger than these pre-existing networks [2]. BOINC [3] is one of the most popular volunteer computing platforms, but the installation of the software on the host computer requires a strenuous effort from the user. And again BOINC does not provide a solution to incorporate the abundant computation power of mobile devices. Because of these existing limitations and paradigms, the paper proposes to use the V8 engine of the Google Chrome Web browser with the help of WebRTC [4] and WebSockets [5] to make a seamless yet easy to use, a distributed network of parallelly running online devices. A rather abstract implementation which uses the existing tools and technologies, this platform takes the input of the problems in the form of a simple JavaScript [6] function and relays the tasks to a task scheduler which organizes and handles the distribution of the tasks to each of its nodes, which are the devices connected to the network. The application of one such network can deal with any sort of computer program. In theory, the throughput will continue to rise with the increment of such devices in a linear fashion. For a simple demonstration of the speeding of data processing, the paper compares the result of hashes produced per second for proof-of-work calculation in a blockchain transaction with rising computer nodes. Overall, this implementation provides a simple deployment interface with an easy to use distributed network.

The paper is divided into six sections; of which the current Sect. 1 gives the general introduction of the problem statement. Section 2 of the paper describes various approaches to use distributed computation that have been sought after following which Sect. 3 explains one of our major supported real-world use cases, i.e., crypto-mining, in which the nonce for each transaction is to be calculated. Through this use case, it will become pretty evident how our platform makes it easier and efficient to use other remote devices by making use of their computation power. After which, the paper is continued to discuss the design principles of our platform and present in a comprehensive manner the implementation and its analysis in Sect. 4. In Sect. 5, the output and its implications are discussed. Finally, Sect. 6 concludes the paper.

2 Traditional Decentralized Approaches for Computation

2.1 Parallel Computing

Parallel computing is a method of processing in which multiple calculations or process execution are conducted simultaneously [7]. Problems are subdivided and addressed simultaneously since each resource applied to the job is running simultaneously [7]. It saves money and time by operating together multiple resources to minimize time and reduce potential expenditure. If the local resources are scarce, it may take advantage of non-local services. Serial computation wastes the available computational capacity, and thus, parallel processing allows computers to work better. Parallel computing is used in the industry for operations on databases, advanced graphics, real-time simulation of systems, data mining, augmented reality, and virtual reality.

2.2 Distributed Computing

Distributed computing is a much broader technology that has been around for more than three decades now. Simply stated that distributed computing is computing over distributed autonomous computers that communicate only over a network. Distributed computing systems are usually treated differently from parallel computing systems or shared-memory systems, where multiple computers share a common memory pool that is used for communication between the processors [8]. Distributed memory systems use multiple computers to solve a common problem, with computation distributed among the connected computers (nodes) and using message passing to communicate between the nodes. For example, grid computing is a form of distributed computing where the nodes may belong to different administrative domains. Another example is the network-based storage virtualization solution which can use distributed computing between data and metadata servers [9].

2.3 Volunteer Computing

Volunteer computing is a process in which individuals (volunteers) provide programs with computational services and use the methods of distributed computations to do collaborative computation and/or storage. Volunteers are usually members of the general population, having mobile devices or low-end devices connected to the Internet. Organizations including colleges and businesses can also offer to use their devices [3]. Volunteer-based systems have to account for systems that are geographically dispersed; a node may join and leave at any moment, so it has to balance load effectively since resources are not dedicated. Moreover, collectively, people now have more computer power than any organization via their devices, like desktops, laptop computers, tablets, smartphones, etc. While work has been done to expand the use of mobile devices for voluntary computing [10, 11], the new collaborative personal computing solution is the first to build unique platforms enabling public programmers to easily leverage the computational resources of their mobile computers and social networks. Right now, Berkeley Open Infrastructure for Network Computing (BOINC) is the middleware program used most commonly, and it provides client software for Android, Linux, Mac OS X, Windows, and other Unix variants.

3 Use Case for Volunteer Computing in Crypto-Mining

In any public blockchain [12], blocks are needed to be verified before adding. Each transaction is validated by a miner who runs a brute force implemented computer program to verify these transactions [13]. This process of verifying transactions is called crypto-currency mining or crypto-mining. It is after this verification process that the transaction is added to the blockchain with other verified transaction as a block. Each transaction is timestamped, and a 256 Hex code is generated for them. A special nonce is calculated upon adding which the final 256-bit code will contain a certain fixed number of zeros in it. The only way to calculate this typical nonce value is to brute-forcing through each and every number combination until the conditions for the required pattern are met [14]. As with increasing the number of zeros in the final 256-bit Hex code, the process becomes more tedious; thus, the need for powerful computing becomes substantial.

Currently, large mining farms need to be established to compute this particular nonce, which makes the infrastructure cost considerably very high for a rather simple process of counting. Thus, an alternative to this approach can be realized through the introduction of volunteer computing, through which users as volunteers could just open their Web browsers to expedite this process efficiently and economically. Our work illustrates how a "proof-of-work" for a crypto-currency can be distributed on multiple volunteers. We assume volunteers willfully donate the processing time for the benefit of a third party, as in a fundraising effort for some causes or organizations. The proof-of-work algorithm corresponds to that of bitcoin: a miner searches for a nonce such that

$$sha256(\text{nonce} + \text{block}) <= \text{target}$$

where *sha256* is a hashing function, a nonce is a number, and the block is an arbitrary string. The lower the target, the lower the probability to find a valid nonce, and therefore, the harder it is to "mine" a new block. The work is distributed by making every volunteer mine the same block by trying different nonces, i.e., the work is divided into multiple mining attempts that consist of testing all nonces in a range of values. An attempt results in a "success" when a nonce is found, or a "failure" if none of the nonces were valid. The mining process lazily creates as many mining attempts as there are processors. There are therefore no upper bound on the number of volunteers that may join. Miners monitor the results, and as soon as a valid nonce is found, the new attempt descriptions are updated to reflect the next block to mine.

4 Implementation

4.1 Simple Interface for Programming

A simple implementation of a JavaScript function takes in a sequence of input and returns the corresponding output values. The function takes an input of the particular number and a callback function to determine the type of return values. Web browsers of mobile devices and computers are used in parallel to execute the JavaScript code. The file to be used can be saved as by any name with the ".js" extension. Moreover, a time out, let us say of 1000 ms, can be added, so that the output is not to be confused with the throughput of the distributed network.

4.2 Simple Interface for Programming

Deployment of a distributed server on this platform is a simple process. For this implementation, they should have all the node modules install and should have the latest installation of NPM and node. Deployment can be done on a command-line interface with simple configurations. By default, a server will be set up at port 5000, from where the nodes can be further connected to share the workload. Port 5001 is for administrative functions, like viewing the analytics and performance of each node present in the system.

The user just needs to open a Chrome browser and type in the port address to connect with the distributed network as a volunteer. When a new volunteer is added to the network, the sub-tasks are distributed to the nodes which then, respectively, process them and returns the value. The processing is done on their open Google Chrome tabs of their machine. The simplicity of this deployment is that the user just needs to open the tabs in the background and can go on to continue doing other work. The process does not take much of the bandwidth since the actual computation is done over the V8 engine, pre-installed on every machine including smartphones. The nodes do not even have to be in close proximity to the serving station, and they just need to have access to the URL and a good Internet connection. For communication, the information can be relayed by using WebRTC and WebSocket. With this kind of deployment, there is no need to install any kind of software application or to buy some high-end hardware to support parallel processing. A simple browser tab is enough to do the job.

4.3 Programming Model

JavaScript functions. It applies a function, let us say $f(x)$, on a series of input variables. Our model is revolved around streaming inputs and outputs through relaying simple and then map them to their respective outputs. This sort of application can process the input in any order necessary, but the output needs to be processed with respect to their inputs. In other words, the sub-task may be distributed in any order and processed without sequence, but the output must be displayed in sequence in which the input sequence was made. This kind of model makes the whole underlying process rather simple but robust. It is powerful enough to parallelly coordinate multiple devices to produce a sequential output. A declarative concurrent program results in the same outcome, irrespective of how many threads that make up the performance perform their tasks. This makes the model easy to program with one partaking processor in a sequential environment and a dozen parallel cases. It is important to order results to maintain the syntactic sequential property. The relative processor speed might otherwise affect the orientation of the outcomes and allow the non-determinism to be observed. Initially, we chose a streaming map scheduling design as it aligns more problems than the typical voluntary computer problems model with stand-alone inputs, which do not usually require orders. Problems involving un-ordered parameters could be simplified to a streaming form by switching values increasingly in a more general order. This streaming variant also allows a number of values to be used.

4.4 Technology and Design Choices

For a variety of reasons, our design was centered on Web technologies.

1. A wide range of personalized devices are compatible (mobile devices, embedded devices, tablets, desktops, and laptops).
2. In modern browsers, virtual machines run numerical JavaScript applications at speeds within 3 times C [15]. Performance levels are close enough to C to improve the output of tasks across multiple simultaneous browsing tabs.
3. Browsers offer a good sandbox to avoid software execution from disrupting the operating system of the host within a web page.
4. WebRTC [4] allows for direct client contact.
5. Social networking links allow your users to organize their social networks rapidly.
6. Both WebRTC and WebSocket can detect disconnections at any instance.

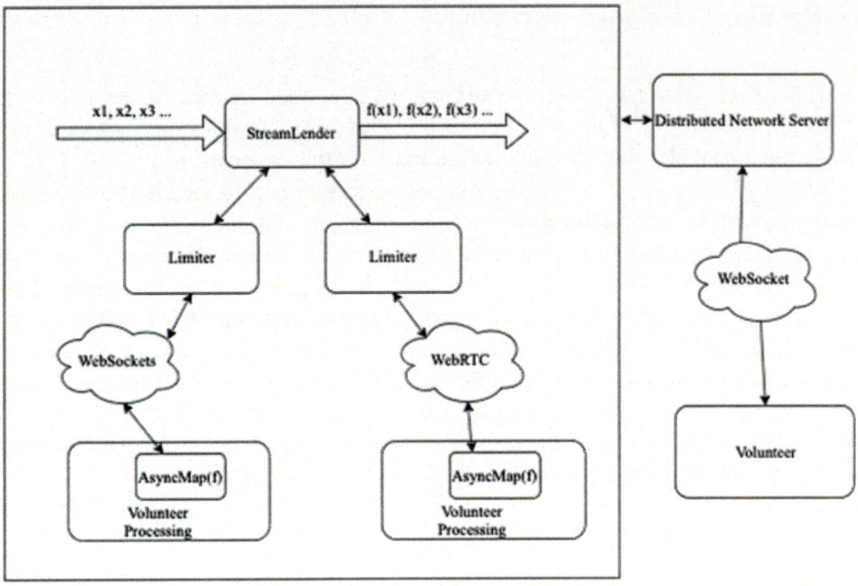

Fig. 1 Architecture of the volunteer network

4.5 Architecture

Figure 1 shows the main modules as well as the manner in which they are interconnected. They collaborate to create a map that services a value stream with a function. We use Node.js but may also function as just a host Web application. Deployment involves running the command-line tool that winds up the process. HTTP links can then be made solely from volunteer devices to the host. The user code such as the function f is obtained by using the HTTP link, and eventually, a WebSocket and a WebRTC connection is created. A public server that uses a separate network socket interface is used to bootstrap the WebRTC connection, which includes signage of potential connecting endpoints between peers. This connection is closed following the establishment of the WebRTC link. Building on a small basic server like the Raspberry Pi board [16] or public category of a cloud such as Heroku is possible since the signals need little resources [17].

Since its implementation reads all of the variables accessible at the transmitting side, we have connected the overall number of variables to be borrowed through our new module Limiter: A limited number of outputs are initially allowed to pass until the threshold is reached, and then, new results are allowed per each new data. With a sufficiently large range, packet forwarding in both directions occurs parallel to the measurements, thereby hiding the transmission delay. The individual values are processed within workers using the current component AsyncMap which relates the function $f(x)$ for the inputs.

5 Results

To demonstrate how our model is working, we have used in total six devices; two MacBook Air, two Acer One SW110, one OnePlus 5 (Mobile device), one MacBook Pro. Table 1 provides the relevant configuration for this experiment.

In Table 3, we can see that as the number of devices is increased in the network, the number of hashes produced per second increases with them. We can also see that the number of hashes produced is slightly less if a single device is taken into perspective. For example, MacBook Air individually produced 60,157 hashes/s in Table 2 which is slightly less than the individual contribution of MacBook Air in Table 3, i.e., 106,282/2 = 53,141 hashes/s. This can be accounted for network latency and time delay in the scheduling process. But overall, we do get a significant increase of hashes produced per second as the devices add on. Moreover, this experiment also demonstrates that lower-end devices like MacBook Air, if work collectively, can produce better results than higher-end devices like MacBook Pro.

Table 1 Device configurations

Devices	RAM	Processor	Approx price (Rs)
MacBook Air	8 GB	Intel Core i5	65,000
Acer One SW110	2 GB	Intel Atom X5	16,000
One Plus 5	8 GB	Qualcomm Snapdragon 835	30,000
Macbook Pro	16 GB	i7-2nd gen	173,000

Table 2 Hashes produced/second by individual devices

Devices	Hashes/s
MacBook Air	60,157
Acer One SW110	8719
One Plus 5	43,251
Macbook Pro	96,834

Table 3 Individual device's throughputs of hashes produced per second in combinations

Devices (number of devices)	Hashes/s	%
Macbook Air (2)	106,282	41
Acer One SW110 (2)	16,835	6.5
One Plus 5 (1)	41,927	16
Macbook Pro (1)	94,319	36.5
Total	259,363	100

Table 4 Comparison of a high-end device with a couple of lower-end devices

Devices (number of devices)	Hashes/s	%
Macbook Air (1)	60,157	28
One Plus 5 (1)	8719	22
Macbook Pro (1)	43,251	50
Total	96,834	100

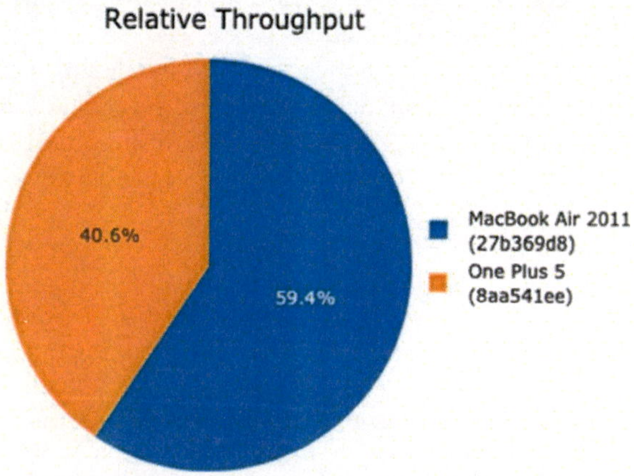

Fig. 2 Relative throughput of MacBook Air and OnePlus 5

From Table 4, it can be observed that two low-end devices, like the MacBook Air 2011 model and a OnePlus 5, can give almost the same throughput as that of a high-end device like MacBook pro. The difference in the total price of the devices, where the two lower-end devices are essentially half the cost of the MacBook Pro, further substantiates our intended result to reduce the overall cost of the infrastructure for computation.

Moreover, if the throughputs of a mobile device like OnePlus 5 and a laptop such as MacBook Air (2011) are studied in Fig. 2, the correlation between different devices and their optimization with Chrome's V8 engine comes into view. Despite having the same RAM specifications, the relative throughput of the laptop is more, since the Chrome browser app is well optimized to make full usage of the RAM than that of the mobile chrome app. It is not unusual since laptops generally have more powerful processing chips inside them. But the relative throughput in Fig. 2 of both devices is only off by a factor of 9%, and this will continue to go down with more and more advancements in the processing power in smart phones. Hence, it can be said that our platform uses almost the full potential even of low-end devices to give quality-driven results.

6 Conclusion

This paper shows the development of our system, a new tool for personal volunteer computing for the people who can use the browser of their mobile device to coordinate the implementation of a feature in a stream of values, by a range of fault-prone personal devices contributed by volunteers. We explained how the declarative competitor model simplified its programming and how it decomposed its application in reusable modules. We discussed how this implementation can help to expedite the blockchain nonce creation process, and how it is theoretically more efficient and robust. The simplicity and versatility of our deployment will allow more developers to exploit the computational prowess of local and remote devices. Moreover, with the ongoing research, the feedback that we are getting from our system suggests that a cluster of low-power personal devices can give a comparable throughput to that of high-end devices.

References

1. Herrera D, Chen H, Lavoie E, Hendren L (2018) Numerical computing on the web: benchmarking for the future. In: Proceedings of the 14th ACM SIGPLAN international symposium on dynamic languages. ACM, pp 88–100
2. Lavoie E, Hendren L (2019) Personal volunteer computing. In: Proceedings of the 16th ACM international conference on computing frontiers. ACM, pp 240–246
3. Anderson DP (2019) Boinc: a platform for volunteer computing. J Grid Comput 1–24
4. Nurminen JK, Meyn AJ, Jalonen E, Raivio Y, Marrero RG (2013) P2p media streaming with html5 and webrtc. In: 2013 IEEE conference on computer communications workshops (INFOCOM WKSHPS). IEEE, pp 63–64
5. Bullotta R, Canosa J, DeRemer B, Mahoney M (2015) System and method of using dynamic rest messages with web-sockets. US Patent App. 14/222,083
6. Harthi OAA, Alalfi MH, Dean T (2019) Detection of feature interaction in dynamic scripting languages. In: Proceedings of the 29th annual international conference on computer science and software engineering. IBM Corp, pp 130–137
7. Gottlieb A, Almasi G (1989) Highly parallel computing. Benjamin/Cummings Redwood City, CA
8. Sitaram D, Manjunath G (2012) Related technologies. In: Moving to the cloud. Elsevier, pp 351–387
9. van Steen M, Tanenbaum AS (2016) A brief introduction to distributed systems. Computing 98(10):967–1009
10. Pramanik PKD, Choudhury P, Saha A (2017) Economical supercomputing thru smartphone crowd computing: An assessment of opportunities, benefits, deterrents, and applications from India's perspective. In: 2017 4th international conference on advanced computing and communication systems (ICACCS). IEEE, pp 1–7
11. Soyata T (2015) Enabling real-time mobile cloud computing through emerging technologies. IGI Global
12. Crosby M, Pattanayak P, Verma S, Kalyanaraman V et al (2016) Blockchain technology: beyond bitcoin. Appl Innov 2(6–10):71
13. Mukhopadhyay U, Skjellum A, Hambolu O, Oakley J, Yu L, Brooks R (2016) A brief survey of cryptocurrency systems. In: 2016 14th annual conference on privacy, security and trust (PST). IEEE, pp 745–752

14. Nakamoto S et al (2008) Bitcoin: a peer-to-peer electronic cash system
15. Khan F, Foley-Bourgon V, Kathrotia S, Lavoie E, Hendren L (2015) Using javascript and webcl for numerical computations: a comparative study of native and web technologies. ACM SIGPLAN Notices 50(2):91–102
16. Brock JD, Bruce RF, Cameron ME (2013) Changing the world with a raspberry pi. J Comput Sci Coll 29(2):151–153
17. Middleton N, Schneeman R (2013) Heroku: up and running: effortless application deployment and scaling. O'Reilly Media, Inc

PV Emulator Model Design Using AI-Based Controllers

Simmi Sharma and Dheeraj Joshi

Abstract With exponential increase in solar applications, the use of PV emulator for testing of solar PV panels before installation is considered as an important facet. Many researchers have examined PV panel characteristics using resistive load in PV emulators. However, applicability of fuzzy logic and particle swarm optimization-based controllers with load variation is a distant approach pertained in this paper for attaining parameters at maximum power point. The aim of this paper is to design a PV emulator model by employing AI-based controllers with varying resistive load and to assess its performance. Stability analysis is done to check the validity of the model.

Keywords Photovoltaic emulator · Fuzzy logic controller · Particle swarm optimization · Resistive load variation · Simulation using MATLAB

Abbreviations

V_{in}	Input voltage
S	Switch
D	Duty cycle
L	Inductor
C	Capacitor
R_s, R, R_1, R_2	Load resistors
V_o	Output voltage
SSA	Small-signal analysis

S. Sharma (✉)
BPIT, Delhi, India
e-mail: simnit08@gmail.com

D. Joshi
EE Department, DTU, Delhi, India
e-mail: ee.dheeraj@gmail.com

A. Abraham et al. (eds.), *Proceedings of 3rd International Conference on Computing Informatics and Networks*, Lecture Notes in Networks and Systems 167,
https://doi.org/10.1007/978-981-15-9712-1_37

PV	Photovoltaic
P	Proportional
I	Integral
FLC	Fuzzy logic controller
AI	Artificial intelligence
PSO	Particle swarm optimization
TF	Transfer function
MPP	Maximum power point

1 Introduction

Worldwide exponential demand and interest in solar-based systems require fast and efficient designing and installation of solar panels. The purview of research in the area of PV emulators is vivid. Research in off-grid and microgrid systems requires cheap and spacious solar panels that are not available in laboratories and onsite locations at the time of designing and testing. Prediction of nearly same VI characteristics of real solar panel is a vital feature of PV emulator [1–4]. Selection of converter for PV emulator configuration is a significant aspect to be considered while designing [5, 6]. A PV emulator consists of DC supply, a converter, a controller and a PWM scheme for providing gate pulses to the converter's MOSFET. In this paper, buck converter is selected for PV emulator due to its simple circuitry and easy operation [7, 8]. P, PI, PID-based controllers are used by many researchers due to their ease of use [1, 3, 5]. Modifications in these controllers help in achieving efficient and fast control. The model implements PSO-tuned PI and fuzzy logic controllers to obtain characteristics similar to PV solar panel. PSO is used in problems where actual position is not known, but the limits or range are known. This attribute is utilized for PV emulator designing, where open-circuit and short-circuit values are already known. Fuzzy logic proves to be useful when there is robust response having uncertainties like variations in load and weather conditions facsimile to solar panel uncertainties.

2 Buck Converter—Small-Signal Analysis

Small-signal analysis signifies that a nonlinear device is operated over a limited range of voltage or current, where the characteristic curve is linear or approximate to it. In practical sense, it means signal level of a volt or less than a volt [9–11]. The transfer function obtained by small-signal analysis [11] is given below (Fig. 1)

$$\frac{V_o}{d} = \frac{V_{\text{in}}}{LCs^2 + \frac{L}{R}s + 1} \tag{i}$$

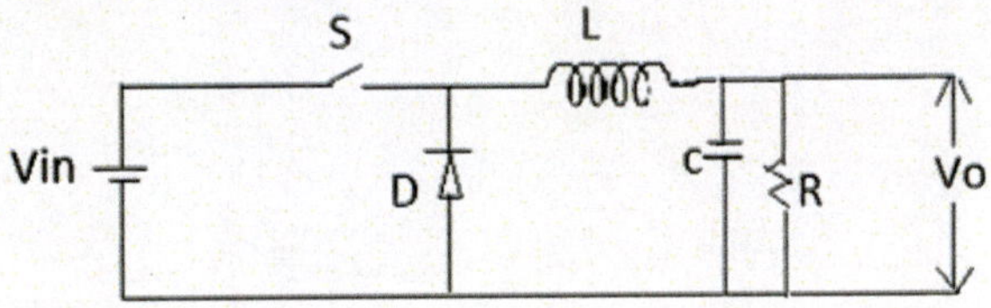

Fig. 1 Buck converter

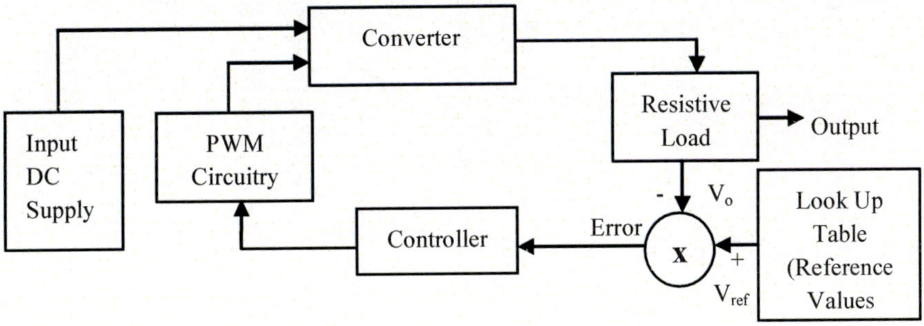

Fig. 2 Block diagram of PV emulator

This transfer function is used for tuning of PI controller using AI. Here, PV emulator is designed for S115 panel with DC power supply of 35 V, a buck converter and close loop. Close loop is formed by a lookup table for reference values, a controller and a PWM circuitry to generate gate pulses for MOSFET in forward path. The block description of PV emulator is shown in Fig. 2.

The prime objective of designing a PV emulator is to save time and money at the time of installation. The values of L and C are calculated using the following formulae

$$L = \frac{V_{\text{in}} d(1-d)}{\Delta I_o f_c} \quad \text{and} \quad C = \frac{(1-d)V_o}{8Lf_c^2 \Delta V_o} \tag{ii}$$

3 Implementation of Artificial Intelligence for PV Emulator Designing

AI is related to field of computer science that appends features of faux ability to understand, think and learn to work better than humans like speech recognition for communication through speaking and listening, natural language processing for reading and writing text, image processing for observing and imagining, pattern recognition by grouping of similar objects, machine learning by providing sufficient data to systems for processing, fuzzy logic for decision making. Here, fuzzy logic and particle swarm optimization are implemented in the controller.

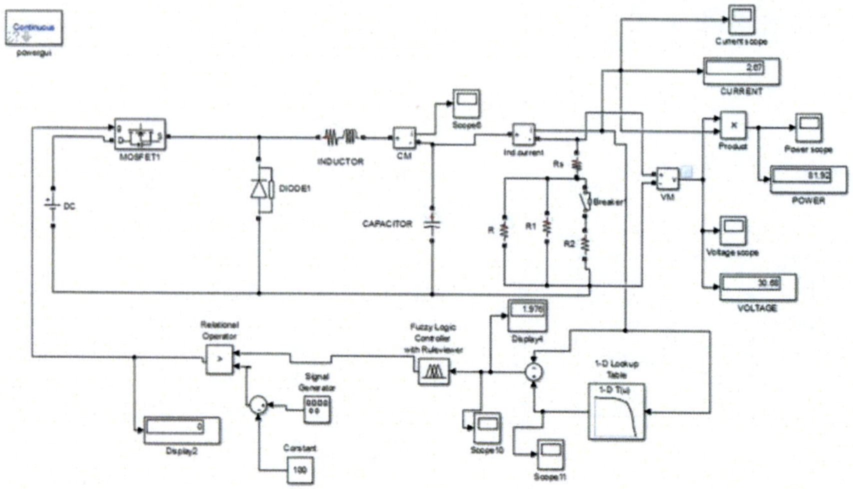

Fig. 3 PV emulator with fuzzy logic controller

3.1 Fuzzy Logic Design

Fuzzy logic is an effective means of representing close but not exact nature of real world. It helps in understanding multi-objective, robust and expert control. It is used for flexible rule-based reasoning, whereas other methods are applied for vision and other sensory processing. It gives good response for even inexactly defined, nonlinear imprecise systems [10]. Expert knowledge and experience play an important role in defining the membership functions, deciding the rules and defuzzification method [12]. In Fig. 3, FL is used for controller design. Input to the controller is an error signal, which is difference in reference and obtained values. Trapezoidal membership functions as input are used. The results obtained by implementation of fuzzy controller and variation of series resistor till 100 Ω (tabulated in Table 1 and plotted in Fig. 4.) have VI characteristics similar to PV panel. The time response plot (Fig. 5) also shows satisfactory results. Load variation and switching at load side help in attaining the required values of voltage and current.

3.2 Particle Swarm Optimization

PSO is a population-based stochastic algorithm which was inspired by social behavior of fish schooling and bird flocking [13]. In case of PV emulator, short-circuit current, open-circuit voltage and maximum power point are available from data sheets. Like GA random population, fitness values and up-gradations are present, but crossover and mutation are not required. Iterations help in locating the best values. It has

Table 1 *V* and *I* values obtained by varying R_s for fuzzy logic controller

S. No.	$R_s(\Omega)$	V (V)	I (A)
1	9	20.49	2.16
2	11	30.68	2.67
3	13	31.15	2.3
4	15	31.5	2
5	20	32.12	1.56
6	25	32.5	1.27
7	30	32.7	1.07
8	40	33.11	0.81
9	70	33.56	0.47
10	100	33.74	0.33

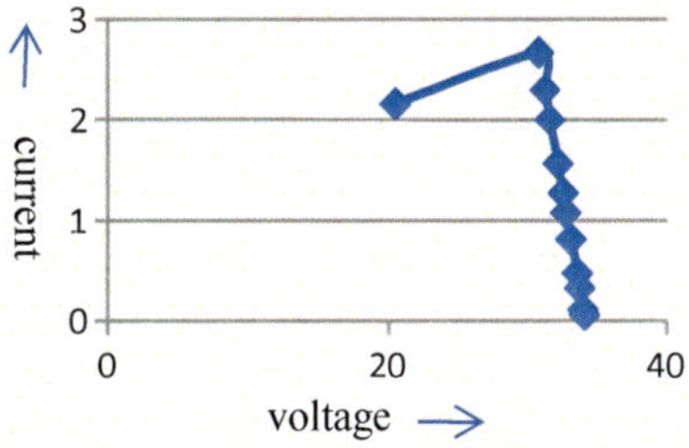

Fig. 4 VI characteristics using fuzzy controller

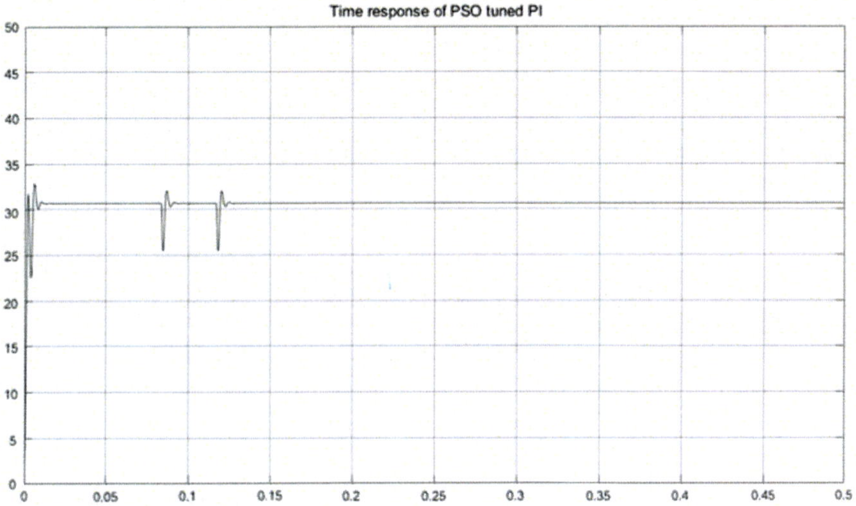

Fig. 5 Time response of FLC

advantages of simple implementation, no requirement of derivations, few algorithm parameters and efficient global search algorithm [14, 15]. Small-signal analysis of buck converter is done to obtain the transfer function on which PSO is implemented to get the best values of k_p and k_i (Fig. 6). After obtaining the best values, the controller gives better results as compared to FLC. The values of V and I are given in Table 2. The plot corresponding to obtained values is shown in Fig. 7, and its time response is shown in Fig. 8.

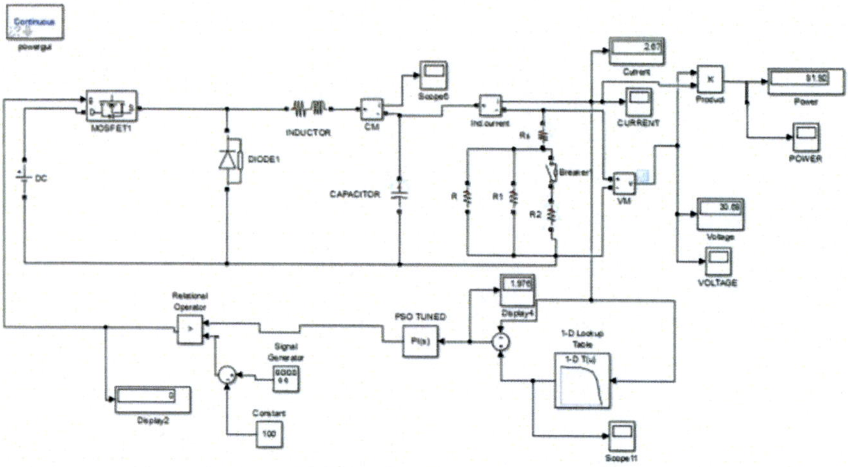

Fig. 6 PV emulator with PSO-tuned PI controller

Table 2 V and I values obtained by varying R_s for PSO tuned PI controller

S. No.	R_s (Ω)	V(V)	I(A)
1	7	20.01	2.67
2	9	24.5	2.58
3	11	30.68	2.67
4	13	31.15	2.3
5	15	31.15	2.03
6	17	31.79	1.81
7	20	32.12	1.56
8	25	32.51	1.27
9	30	32.77	1.07
10	40	33.33	0.81
11	70	33.56	0.47
12	100	33.74	0.33

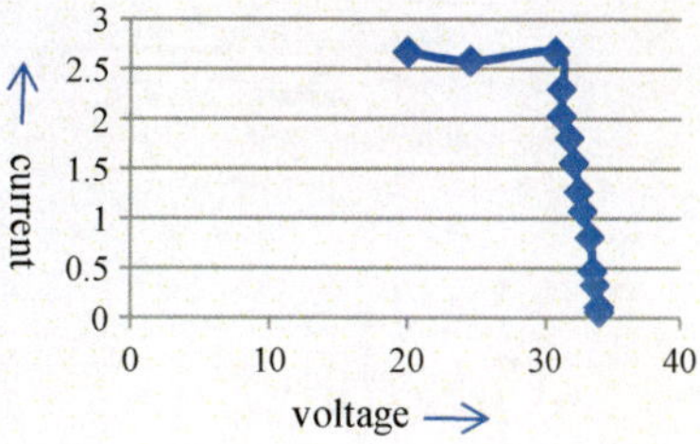

Fig. 7 VI characteristics using PSO-tuned PI controller

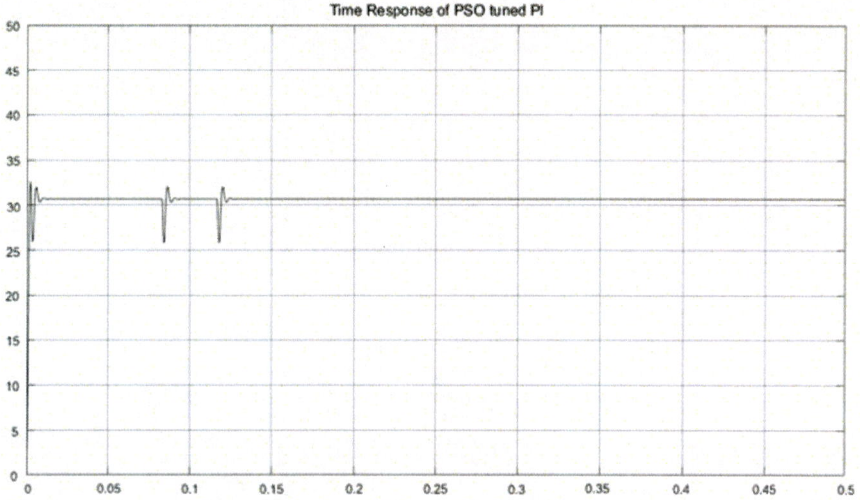

Fig. 8 Time response of PSO-tuned PI

4 Results

The time response at MPP gives the value of settling time for FLC controller as 7.7 ms, and for PSO-tuned PI, the value is 5.6 ms. It implies that PSO-tuned PI has faster response, and the values of ω_n and ξ for FLC and PSO-PI are 0.71, 0.73 and 1.06, 0.67, respectively. These values are used to calculate the transfer function and response of system at MPP. Bode plot and Nyquist plot are obtained using MATLAB, and the values of parameters are shown in Table 3. Figures 9 and 10 show transfer function of PSO-PI model as tf1and transfer function of FLC model as tf2.

Table 3 Stability analysis by bode and Nyquist plot

Controller	Phase margin (degree)	Peak margin (dB)	Delay margin (s)	Frequency (rad/s)	Stable (Yes/No)
PSO-PI	143	0.042	5.28	0.473	Yes
FLC	−180	1.9e−15	Inf.	0	Yes

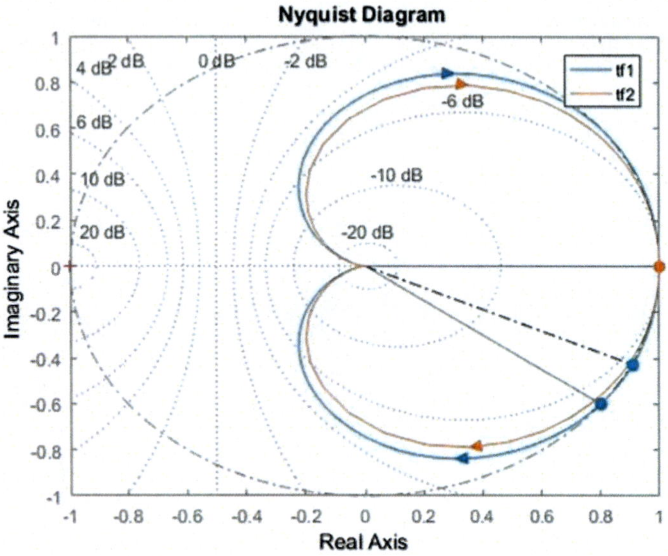

Fig. 9 Nyquist plot of designed model at MPP

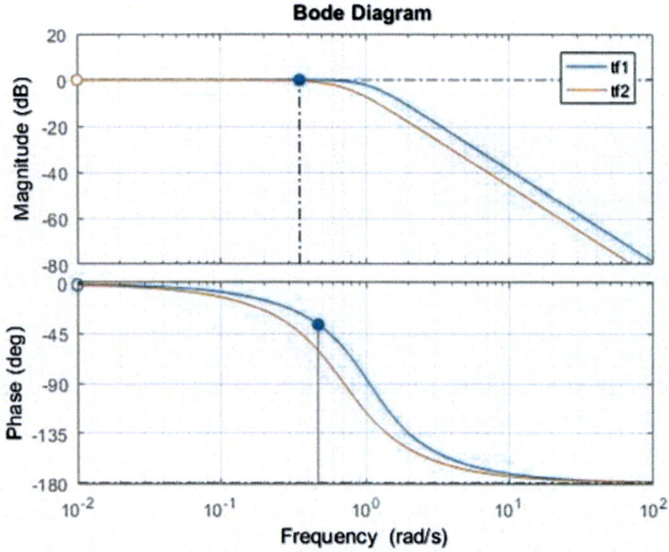

Fig. 10 Bode plot of designed model at MPP

5 Conclusion

The values of voltage and current obtained from designed PV emulator model using AI controllers and resistive load variation (tabulated above) provide required MPP values. Complete stability analysis is done using Bode plot and Nyquist plot to get the graphical representation along with numeric values. It is concluded that the results obtained by PSO-PI are better in terms of time and simplicity. In future, this can be utilized to implement the model for grid or off-grid applications. Other nonlinear control methods like sliding mode control (SMC) and model reference adaptive control (MRAC) can be implemented in future to further understand the nonlinear behavior of the model.

References

1. Iqbal MT, Tariq M, Ahmad MK, Saad Bin Arif M (2016) Modeling, analysis and control of buck converter and Z-source converter for photo voltaic emulator. In: 1st IEEE international conference on power electronics, intelligent control and energy systems (ICPEICES—2016)
2. Chang C (1995) Robust control of DC-DC converters: the buck converter. In: 26th annual IEEE power electronics specialists conference, vol 2, pp 1094–1097
3. Moussa I, Khedher A, Bouallegue A (2019) Design of a low-cost PV emulator applied for PVECS. Electronics 8:232. https://doi.org/10.3390/electronics8020232
4. Sanaullah A, Khan HA (2015) Design and implementation of a low cost Solar Panel emulator. In: Proceedings of the 42nd IEEE photovoltaic specialist conference (PVSC), 14–19 June 2015, New Orleans, LA, USA, pp 1–5
5. Joshi D, Sharma S (2019) PV emulator modeling and design using buck converter. Appl Comput Automation Wirel Syst Electr Eng 553(639):648 Springer
6. Deepak, Pachauri RK, Chauhan YK (2016) Modeling and simulation analysis of PV Fed Cuk, Sepic, Zeta and Luo DC-DC converter. In: 1st IEEE international conference on power electronics, intelligent control and energy systems (ICPEICES)
7. Yusivar F, Farabi MY, Surya Diningrat R, Ananduta WW, Syaifudin Y (2011) Buck-converter photovoltaic simulator. Int J Power Electron Drive Syst (IJPEDS) 1(2):156–167. ISSN 2088-8694
8. Cirrincione M, Di Piazza M, Marsala G, Pucci M, Vitale G (2008) Real time simulation of renewable sources by model based control of DC/DC converters. In: IEEE international conference on industrial electronics ISIE
9. Liu S, Liu PX, Wang X (2016) Stochastic small-signal stability analysis of grid-connected photovoltaic systems. IEEE Trans Ind Electron 63(2):1027–1038
10. Chen L, Xu Y, Liu Y-F, Jin R (2009) Small-signal analysis and simulation of fuzzy controlled buck converter. In: IEEE conference industrial electronics and applications ICIEA, pp 816–820
11. Li Z, Cheng KWE, Hu J (2017) Modeling of basic DC-DC converters. In: IEEE conference power electronic systems and applications-smart mobility, power transfer and security (PESA)
12. Arulmurugan R, Suthanthira Vanitha N (2013) Fuzzy logic controller with maximum power point tracking using creative design of DC to DC buck converter for photovoltaic power system. Int J Inform Technol 116:122
13. Kennedy J, Eberhart R (1995) Particle swarm optimization. In: Proceedings of ICNN'95 international conference on neural networks, pp 1942–1948

14. Ishaque K, Salam Z, Amjad M, Mekhilef S (2012) An improved particle swarm optimization (PSO)—based MPPT for PV with reduced steady-state oscillation. IEEE Trans Power Electron 27(8):3627–3638
15. Sahin E, Ayas SM, Altas IH (2014) A PSO optimized fractional-order PID controller for a PV system with DC-DC boost converter. In: 16th international power electronics and motion control conference and exposition Antalya, 21–24 Sept 2014, Turkey

Controlling Chaos Generated in Predator-Prey Interactions Using Adaptive Hybrid Combination Synchronization

Taqseer Khan and Harindri Chaudhary

Abstract In this research work, we study hybrid combination synchronization (HCS) between chaotic generalized three species Lotka-Volterra (GLV) biological system via adaptive control technique (ACT). Lotka and Volterra discovered the well-known illustrations of primary biological models. This system details the interaction between two species predator and prey. Though it is a very primary model, it has many drawbacks, for example, it avoids several essential characteristics, for instance, interplaying among numerous species of similar ecological community, connectivity with the ecosystem etc. Samardzija and Greller firstly investigated the dynamics and chaotic behavior of GLV biological system in 1988. Subsequently, the area of biological control for numerous biological systems existing in natural habitat has been a significant field for researchers and biologists. We here consider two predators and one prey population present in the system. The adaptive ecological control law in obtaining asymptotic HCS of state variables (SV) of considered system with uncertain parameters has been deduced utilizing Lyapunov stability theory (LST). Further, it is noticed that both anti-synchronization and complete synchronization coexist in HCS. Additionally, numerical simulations using MATLAB are displayed for illustrating the feasibility and efficiency of discussed approach. Remarkably, analytical and computational results agree excellently. The discussed approach is potentially applicable in areas of secure communication, encryption and control theory.

Keywords Chaotic system · Hybrid synchronization · Combination synchronization · Adaptive control · Lotka-Volterra system

T. Khan · H. Chaudhary (✉)
Department of Mathematics, Jamia Millia Islamia, New Delhi, Delhi, India
e-mail: harindri20dbc@gmail.com

T. Khan
e-mail: tkhan4@jmi.ac.in

A. Abraham et al. (eds.), *Proceedings of 3rd International Conference on Computing Informatics and Networks*, Lecture Notes in Networks and Systems 167,
https://doi.org/10.1007/978-981-15-9712-1_38

1 Introduction

Primarily, chaos theory has been characterized as qualitative investigation of deterministic highly complex nonlinear dynamical systems occurring in natural habitat acquiring fundamental properties, for instance, extreme sensitivity for initial conditions, infinite recurrence and boundedness. This high sensitivity for initial conditions is known to be "Butterfly Effect" as described in current literature and is firstly noticed in 1963 by E. N. Lorenz while investigating an atmospheric convention modelling for weather conditions. Most importantly, chaos theory has several variations in applications for many fields of engineering, technology and applied sciences such as biomedical engineering [1], finance models [2], neural networks [3], ecological models [4], oscillations [5], chemical reactions [6], weather models [7], jerk systems [8], circuits [9], encryption [10], etc. Subsequently, chaos synchronization and control have sought significant attention in varied research fields.

It was in 1963, Lorenz [11] first described the widely known chaotic system while studying a weather model for atmospheric convention. Subsequently, Rossler [12] in 1976 identified an algebraically much simpler chaotic system other than Lorenz system. The aforementioned classic systems were eventuated by occurrence of enormous chaotic nonlinear systems, like, Chen system, Cai system, Sprott model, Lu-Chen system, Arnedo system, Tigan system, etc. In recent times, various chaotic systems are also introduced such as Sundarapandian system, Pehlivan model, Vaidyanathan system, Pham system and so on. Interestingly, it has been over three decades since Pecora and Carroll [13] proposed the phenomenon for chaos synchronization in chaotic dynamical systems in 1990. They formulated the notion of chaos synchronization by using master-slave configuration.

Chaos synchronization is described as a procedure of adjusting identical or non-identical chaotic systems in such a way that both depict the same behavior owing to coupling to achieve stability. Till date, various types of synchronized techniques for stabilizing chaotic systems are initiated in synchronization theory, viz., complete [14], lag [15], phase [16], hybrid [17], combination difference [18], combination-combination [19], anti [20], projective [21], function projective [22], hybrid projective [23], modified projective [24] etc. By now, numerous control schemes, for instance, active [25], sliding mode [26, 27], backstepping [28], adaptive [19, 29], feedback [30], impulsive [31] etc. are developed and analyzed in control theory to stabilize the chaos.

Chaos synchronization and control (CSC) for chaotic systems utilizing ACT was firstly advocated in 1989 by Hubler [32]. Mainieri and Rehacek [33] in 1999 introduced the basic criteria of projective synchronization. A detailed study in chaos synchronization for chaotic systems utilizing ACT has been made in [34–38]. Furthermore, in [39] and [40], ACT is discussed to synchronize GLV biological system. Moreover, in [41–43], several control schemes are analyzed for the newly generated chaotic systems in detail.

Considering the aforesaid discussed facts, this manuscript focuses on investigating a hybrid combination synchronization (HCS) in three identical GLV models by using ACT. The manuscript is summarized as: Sect. 2 deals with some preliminaries to be utilized in the upcoming sections. Section 3 consists of some basic structuring properties of considered GLV model. Section 4 investigates ACT along with a parameter estimation update law to attain asymptotic stability globally of chaotic GLV system. Section 5 contains the computer simulations illustrating the effectiveness and feasibility of considered HCS technique. Further, a comparative analysis has been done. Lastly, few conclusions are provided in Sect. 6.

2 Preliminaries

This section deals with the primary notions to achieve combination synchronization in accordance with master-slave composition among three chaotic systems.

Two master systems (MS) and corresponding slave system (SS) are written as:

$$\dot{q_{m1}} = f_1(q_{m1})\beta_1 + F_1(q_{m1}) \tag{1}$$

$$\dot{q_{m2}} = f_2(q_{m2})\beta_2 + F_2(q_{m2}) \tag{2}$$

$$\dot{q_{s1}} = f_3(q_{s1})\beta_3 + F_3(q_{s1}) + w_1 \tag{3}$$

where $q_{m1} = (q_{m11}, q_{m12}, \ldots, q_{m1n})^{\mathrm{T}} \in R^n$, $q_{m2} = (q_{m21}, q_{m22}, \ldots, q_{m2n})^{\mathrm{T}} \in R^n$, $q_{s1} = (q_{s11}, q_{s12}, \ldots, q_{s1n})^{\mathrm{T}} \in R^n$ are state vectors of master and slave systems (1), (2) and (3) respectively, $F_1, F_2, F_3 : R^n \to R^n$ are three nonlinear continuous functions, $\beta_1 = (\beta_{11}, \beta_{12}, \ldots, \beta_{1r_1})^{\mathrm{T}}$ is a $r_1 \times 1$ unknown parameter vector of first master system (1), $\beta_2 = (\beta_{21}, \beta_{22}, \ldots, \beta_{2r_2})^{\mathrm{T}}$ is a $r_2 \times 1$ unknown parameter vector of second master system (2), $\beta_3 = (\beta_{31}, \beta_{32}, \ldots, \beta_{3r_3})^{\mathrm{T}}$ is a $r_3 \times 1$ unknown parameter vector of slave system (3), $f_1 : R^n \to R^{n\times r_1}$, $f_2 : R^n \to R^{n\times r_2}$, $f_3 : R^n \to R^{n\times r_3}$, $w_1 = (w_{11}, w_{12}, \ldots, w_{1n}) \in R^n$ is suitably defined controller.

Definition 1 If there exist constant matrices $A_1, A_2, A_3 \in R^n \times R^n$ and $R \neq 0$ such that:

$$\lim_{t\to\infty}\|E\| = \lim_{t\to\infty}\|(A_1q_{m1} + A_2q_{m2} - A_3q_{s1})\| = 0,$$

then combination of MS (1) and (2) realizes combination synchronization with SS (3) and $\|.\|$ denotes Euclidean norm.

Remark 1 Considering matrices A_1, A_2 and A_3 as scaling matrices. Furthermore, A_1, A_2, A_3 may be expanded as matrices having functions in state variables q_{m1}, q_{m2} and q_{s1} as the entries.

Remark 2 If $A_3 = -I$ and $A_1 = A_2 = -\eta I$, for $\eta = 1$ it converts into complete synchronization and for $\eta = -1$ it reduces to anti-synchronization. Coexistence of anti and complete synchronization composes hybrid combination synchronization (HCS). Therefore, the HCS error attains the form:

$$E = q_{s1} - \eta(q_{m1} + q_{m2}), \tag{4}$$

where $\eta = diag(\eta_1, \eta_2, \ldots, \eta_n)$ with $\eta_i = (-1)^{i+1}$ for $i = 1, 2, 3, \ldots, n$.

3 Description of Chaotic GLV System

Presented by Samardzija et al. [44] firstly in 1988, the discussed GLV system is mentioned as:

$$\begin{cases} \dot{q}_{11} = & q_{11} - q_{11}q_{12} + n_3 q_{11}^2 - n_1 q_{11}^2 q_{13} \\ \dot{q}_{12} = & -q_{12} + q_{11}q_{12} \\ \dot{q}_{13} = & -n_2 q_{13} + n_1 q_{11}^2 q_{13} \end{cases} \tag{5}$$

where $(q_{11}, q_{12}, q_{13})^{\mathrm{T}} \in R^3$ is state variable and n_1, n_2 and n_3 are given positive parameters. Further, in (5), q_{11} denotes the prey population and q_{12}, q_{13} represents the predator populations. For $n_1 = 2.9851$, $n_2 = 3$ and $n_3 = 2$, the system (5) displays chaotic behaviour. In addition, Fig. 1a–d exhibit the phase graphs of (5). Furthermore, detailed theoretical work and numerical simulation outcomes for (5) may be viewed in [44].

The upcoming section describes the HCS scheme for controlling chaos generated in (5) using ACT.

4 Synchronization Methodology

Two master systems (MS) and one slave system (SS) with control functions are defined as:

$$\begin{cases} \dot{q}_{m11} = & q_{m11} - q_{m11}q_{m12} + n_3 q_{m11}^2 - n_1 q_{m11}^2 q_{13} \\ \dot{q}_{m12} = & -q_{m12} + q_{m11}q_{m12} \\ \dot{q}_{m13} = & -n_2 q_{m13} + n_1 q_{m11}^2 q_{m13} \end{cases} \tag{6}$$

$$\begin{cases} \dot{q}_{m21} = & q_{m21} - q_{m21}q_{m22} + n_3 q_{m21}^2 - n_1 q_{m21}^2 q_{m23} \\ \dot{q}_{m22} = & -q_{m22} + q_{m21}q_{m22} \\ \dot{q}_{m23} = & -n_2 q_{m23} + n_1 q_{m21}^2 q_{m23} \end{cases} \tag{7}$$

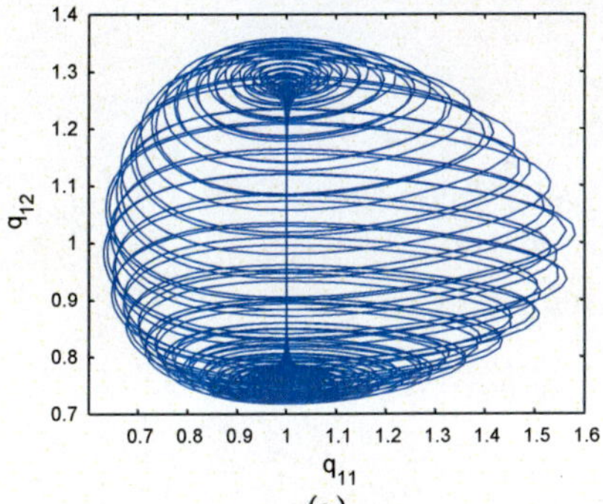

(a)

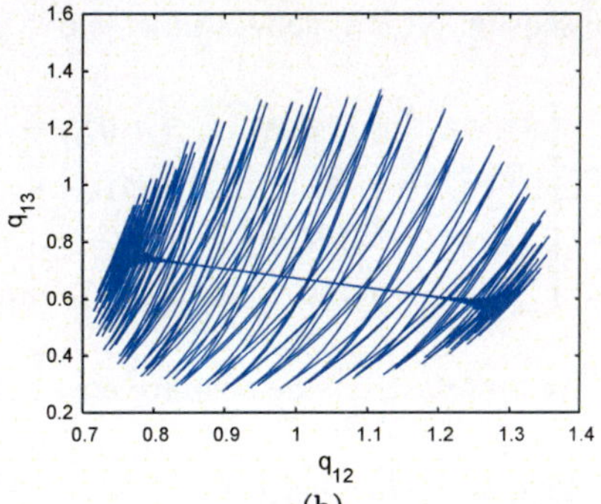

(b)

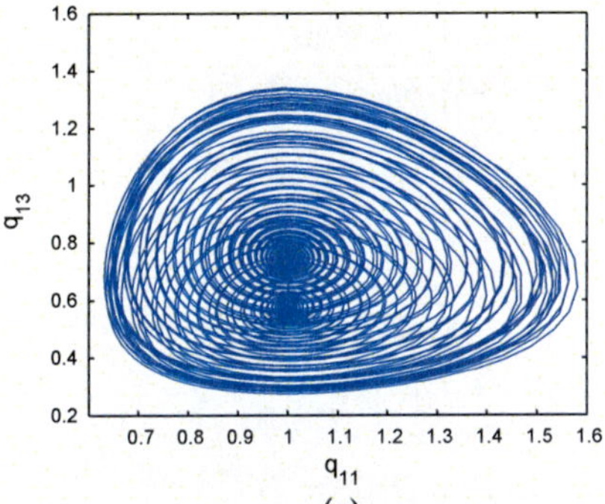

(c)

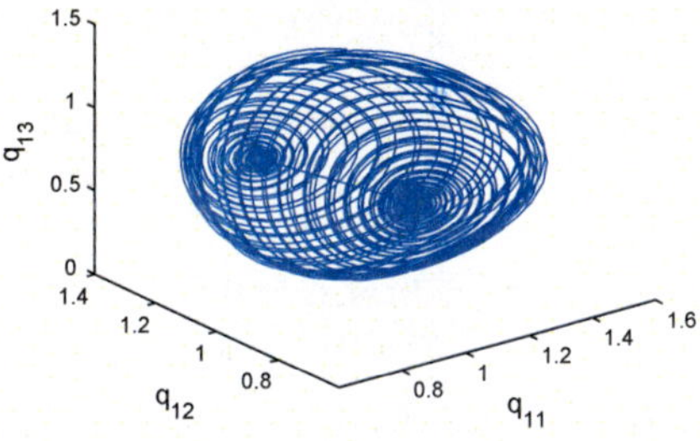

(d) d

Fig. 1 Phase graphs of GLV model in **a** $q_{11} - q_{12}$ plane, **b** $q_{12} - q_{13}$ plane, **c** $q_{11} - q_{13}$ plane, **d** $q_{11} - q_{12} - q_{13}$ space

$$\begin{cases} \dot{q}_{s31} = & q_{s31} - q_{s31}q_{s32} + n_3 q_{s31}^2 - n_1 q_{s31}^2 q_{s33} + w_{11} \\ \dot{q}_{s32} = & -q_{s32} + q_{s31}q_{s32} + w_{12} \\ \dot{q}_{s33} = & -n_2 q_{s33} + n_1 q_{s31}^2 q_{s33} + w_{13}, \end{cases} \tag{8}$$

where w_{11}, w_{12} and w_{13} are adaptive control functions to be drafted so that HCS among systems (6), (7) and (8) will be obtained.

Defining now state errors as

$$\begin{cases} E_{11} = & q_{s31} - (q_{m11} + q_{m21}) \\ E_{21} = & q_{s32} + (q_{m12} + q_{m22}) \\ E_{31} = & q_{s33} - (q_{m13} + q_{m23}) \end{cases} \tag{9}$$

The ultimate aim of the current paper is to formulate control functions w_{1i}, $(i = 1, 2, 3)$ in a manner that resulting state errors provided in (9) satisfy

$$\lim_{t\to\infty} E_{i1}(t) = 0, \quad where \quad (i = 1, 2, 3).$$

Subsequent error dynamics turns into:

$$\begin{cases} \dot{E_{11}} = & E_{11} - q_{s31}q_{s32} + n_3 q_{s31}^2 - n_1 q_{s31}^2 q_{s33} - (-q_{m21}q_{m22} + n_3 q_{m21}^2 \\ & -n_1 q_{m21}^2 q_{m23} + q_{m11}q_{m12} - n_3 q_{m11}^2 + n_1 q_{m11}^2 q_{m13}) + w_{11} \\ \dot{E_{21}} = & -E_{21} + q_{s31}q_{s32} + (-q_{m11}q_{m12} + q_{m21}q_{m22}) + w_{12} \\ \dot{E_{31}} = & -n_2 E_{31} + n_1 q_{s31}^2 q_{s33} - (n_1 q_{m21}^2 q_{m23} - n_1 q_{m11}^2 q_{m13}) + w_{13} \end{cases} \quad (10)$$

We now describe the adaptive control inputs as:

$$\begin{cases} w_{11} = & -E_{11} + q_{s31}q_{s32} - \hat{n_3} q_{s31}^2 + \hat{n_1} q_{s31}^2 q_{s33} - (-q_{m21}q_{m22} + \hat{n_3} q_{m21}^2 \\ & -\hat{n_1} q_{m21}^2 q_{m22} + q_{m11}q_{m12} - \hat{n_3} q_{s31}^2 + \hat{n_1} q_{m11}^2 q_{m13}) - L_1 E_{11} \\ w_{12} = & E_{21} - q_{s31}q_{s32} - (-q_{m11}q_{m12} + q_{m21}q_{m22}) - L_2 E_{21} \\ w_{13} = & \hat{n_2} E_{31} - \hat{n_1} q_{s31}^2 q_{s33} - (\hat{n_1} q_{m21}^2 q_{m23} - \hat{n_1} q_{m11}^2 q_{m13}) - L_3 E_{31} \end{cases} \quad (11)$$

where L_1, L_2 and L_3 are positive gaining constants.
Combining the Eqs. (10) and (11), we have

$$\begin{cases} \dot{E}_{11} = & (n_3 - \hat{n_3}) q_{s31}^2 - (n_1 - \hat{n_1}) q_{s31}^2 q_{s33} - [(l_3 - \hat{l_3}) q_{m21}^2 - (n_1 - \hat{n_1}) q_{m21}^2 q_{m23} \\ & -(n_3 - \hat{n_3}) q_{m11}^2 + (n_1 - \hat{n_1}) q_{m11}^2 q_{m13}] - L_1 E_{11} \\ \dot{E}_{21} = & -L_2 E_{21} \\ \dot{E}_{31} = & -(n_2 - \hat{n_2}) q_{s33} + (n_1 - \hat{n_1}) q_{s31}^2 q_{s33} - [(n_2 - \hat{n_2}) q_{m13} \\ & +(n_1 - \hat{n_1}) q_{m21}^2 q_{m23} - (n_2 - \hat{n_2}) q_{m23} - (n_1 - \hat{n_1}) q_{m11}^2 q_{m13}] - L_3 E_{31} \end{cases} \quad (12)$$

where $\hat{n}_1$, $\hat{n}_2$ and $\hat{n}_3$ are estimating values for unknown parameter n_1, n_2 and n_3 respectively.

Now writing parameter estimation error as:

$$\tilde{n_1} = n_1 - \hat{n}_1, \tilde{n_2} = n_2 - \hat{n}_2, \tilde{n_3} = n_3 - \hat{n}_3, \quad (13)$$

Employing (13), the consequent error dynamics (14) transforms to

$$\begin{cases} \dot{E}_{11} = & \tilde{n}_3 (q_{s31}^2 - q_{m21}^2 + q_{m11}^2) - \tilde{n}_1 (q_{s31}^2 q_{s33} - {q_{m21}}^2 q_{m23} + q_{m11}^2 q_{m13}) - L_1 E_{11} \\ \dot{E}_{21} = & -L_2 E_{21} \\ \dot{E}_{31} = & -\tilde{n}_2 E_{31} + \tilde{n}_1 (q_{s31}^2 q_{s33} - q_{m21}^2 q_{m23} q_{m11}^2 q_{m13}] - L_3 E_{31} \end{cases}$$

The derivative of parameter estimation error (13) can be written as

$$\dot{\tilde{n_1}} = -\dot{\hat{n}}_1, \dot{\tilde{n_2}} = -\dot{\hat{n}}_2, \dot{\tilde{n_3}} = -\dot{\hat{n}}_3. \quad (14)$$

Describing $V(E(t)$, the Lyapunov function, by

$$V(E(t)) = \frac{1}{2}[E_{11}^2 + E_{21}^2 + E_{31}^2 + \tilde{n_1}^2 + \tilde{n_2}^2 + \tilde{n_3}^2], \tag{15}$$

which implies that $V(E(t))$ are positive definite.
Derivation of Lyapunov function $V(E(t))$ is described by

$$\begin{aligned} V(\dot{E}(t)) =& E_{11}\dot{E}_{11} + E_{21}\dot{E}_{21} + E_{31}\dot{E}_{31} - \tilde{n_1}\dot{\hat{n_1}} \\ & - \tilde{n_2}\dot{\hat{n_2}} - \tilde{n_3}\dot{\hat{n_3}}. \end{aligned} \tag{16}$$

Keeping (16) in view, we formalize parameter estimates laws by the following rule:

$$\begin{cases} \dot{\hat{n_1}} = & -(q_{s31}^2 q_{s31} - q_{m21}^2 q_{m23} + q_{m11}^2 q_{m13})E_{11} + (q_{s31}^2 q_{s33} - q_{m21}^2 q_{m21} \\ & +q_{m11}^2 q_{m13})E_{31} + L_4\tilde{n_1} \\ \dot{\hat{n_2}} = & -E_{31} + L_5\tilde{n_1} \\ \dot{\hat{n_3}} = & q_{s31}^2 + q_{m11}^2 - q_{m21}^2 + L_6\tilde{l_3}, \end{cases} \tag{17}$$

where L_4, L_5 and L_6 are gain constants which are positive.

Theorem 1 *The chaotic (6)–(8) systems are asymptotically globally hybrid combination synchronized in each initial states $(q_{m11}(0), q_{m12}(0), q_{m13}(0)) \in R^3$ by adaptive controller (11) and updated parameter law (17).*

Proof The function $V(E(t))$ (Lyapunov) as mentioned in (15) is a positive definite function in R^6. By simplifying Eqs. (16) and (17), we get

$$\begin{aligned} V(\dot{E}(t)) =& - L_1E_{11}^2 - L_2E_{21}^2 - L_3E_{31}^2 - L_4\tilde{n_1}^2 - L_5\tilde{n_2}^2 - L_6\tilde{n_3}^2 \\ & < 0. \end{aligned}$$

Clearly, $\dot{V}(E(t))$ is negative definite.

Thus, in view of LST, we establish that the proposed HCS error $E(t) \to 0$ asymptotically and globally as t tending to infinity for each initial conditions $E(0) \in R^3$ which completes the proof. □

5 Numerical Simulation

We present, in this section, some essential simulation experiments which illustrate the efficiency of discussed HCS scheme using ACT. Parameters of GLV model (5) are selected as $n_1 = 2.9851$, $n_2 = 3$ and $n_3 = 2$ that confirm the chaotic behaviour

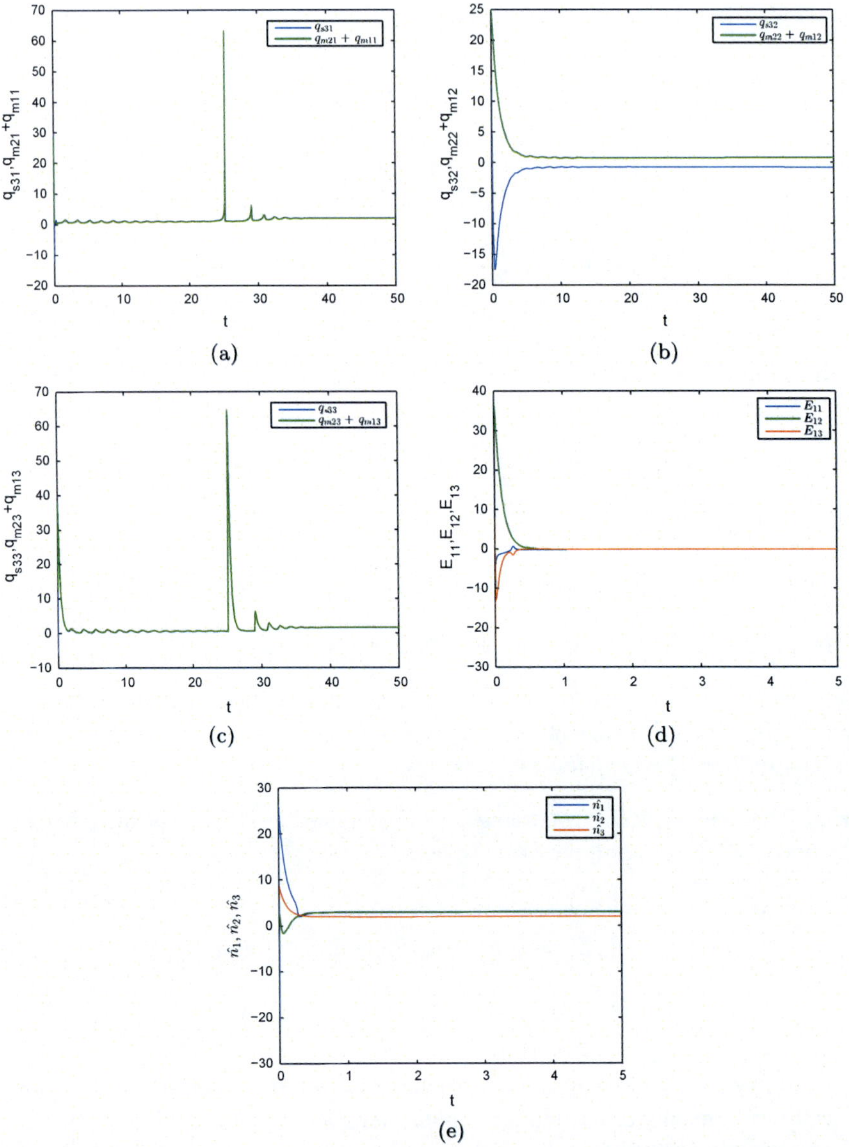

Fig. 2 Phase graphs of HCS trajectory of chaotic GLV system **a** among $q_{s31}(t)$ and $q_{m21}(t) + q_{m11}(t)$, **b** among $q_{s32}(t)$ and $q_{m22}(t) + q_{m12}(t)$, **c** among $q_{s33}(t)$ and $q_{m23}(t) + q_{m13}(t)$, **d** synchronized error, **e** parameter identification

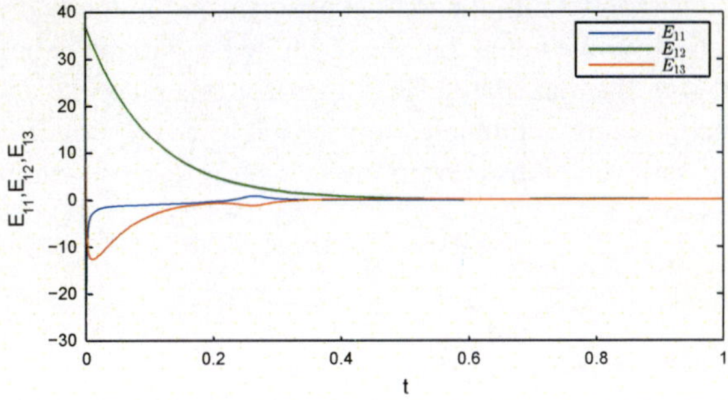

Fig. 3 Phase graph of HCS error

of (5) without controllers. Initial values of two MS and one SS systems (6)–(8) are (27.5, 23.1, 11.4), (1.2, 1.2, 1.2) and (2.9, 12.8, 20.3) respectively. We achieve HCS scheme between systems (6)–(8) by selecting matrix η with $\eta_1 = 1, \eta_2 = -1, \eta_3 = 1$. Here, we have chosen control gains as $L_i = 6$ $(i = 1, 2, 3)$. Simulations depicted through Fig. 2a–c exhibit the synchronized state trajectories for both master (6)–(7) and slave systems (8). Synchronized errors $(E_{11}, E_{21}, E_{31}) = (-25.8, 37.1, 7.7)$ as displayed in Fig. 2d approaches zero as t tending infinity. Moreover, Fig. 2e depicts that estimated values $(\hat{n}_1, \hat{n}_2, \hat{n}_3$ of uncertain parameters converging asymptotically to their respective considered values. Thus, the investigated HCS approach in master and slave systems has been validated computationally.

5.1 *Comparative Survey Among the Proposed HCS Technique and Earlier Published Work*

In [39], hybrid synchronization has been obtained via ACT when conducted in identical GLV system having same parameters. It is noted that synchronization error converging zero at $t = 0.8(approx)$, while on the contrary in our study, the HCS scheme has been realized utilizing ACT, in which synchronization errors converge to zero effectively as displayed in Fig. 3 at $t = 0.5(approx)$. This shows that the proposed HCS approach via ACT is more preferable over earlier publicized researches.

6 Conclusion

In this work, the proposed HCS technique between identical chaotic GLV system via ACT is investigated. By designing suitable control functions based on LST, the considered HCS technique is attained. The particular cases of anti-synchronization, complete synchronization, and chaos control problem are also included. The efficiency of theoretical results is justified by conducting simulations in MATLAB software.

Remarkably, both theoretical analysis and numerical results are in complete compatibility. We noted that the discussed technique is basic yet theoretically rigorous. In fact, the considered HCS approach has potential and advantages as it is applicable in secure communication as well as encryption. In our investigation, time taken by error functions converging to zero is lesser in comparing over earlier publicized researches.

References

1. Provata A, Katsaloulis P, Verganelakis DA (2012) Dynamics of chaotic maps for modelling the multifractal spectrum of human brain diffusion tensor images. Chaos Solitons Fractals 45(2):174–180
2. Tong Xiao-Jun, Zhang Miao, Wang Zhu, Liu Yang, Ma Jing (2015) An image encryption scheme based on a new hyperchaotic finance system. Optik 126(20):2445–2452
3. Bouallegue Kais (2017) A new class of neural networks and its applications. Neurocomputing 249:28–47
4. Sahoo B, Poria S (2014) The chaos and control of a food chain model supplying additional food to top-predator. Chaos Solitons Fractals 58:52–64
5. Ghosh D, Mukherjee A, Das NR, Biswas BN (2018) Generation & control of chaos in a single loop optoelectronic oscillator. Optik 165:275–287
6. Han SK, Kurrer C, Kuramoto Y (1995) Dephasing and bursting in coupled neural oscillators. Phys. Rev. Lett. 75(17):3190
7. Russell FP, Düben PD, Niu X, Luk W, Palmer TN (2017) Exploiting the chaotic behaviour of atmospheric models with reconfigurable architectures. Comput. Phys. Commun. 221:160–173
8. Wang X, Vaidyanathan S, Volos C, Pham V-T, Kapitaniak T (2017) Dynamics, circuit realization, control and synchronization of a hyperchaotic hyperjerk system with coexisting attractors. Nonlinear Dyn. 89(3):1673–1687
9. Shi Z, Hong S, Chen K (2008) Experimental study on tracking the state of analog chua's circuit with particle filter for chaos synchronization. Phys. Lett. A 372(34):5575–5580
10. Wu G-C, Baleanu D, Lin Z-X (2016) Image encryption technique based on fractional chaotic time series. J. Vib. Control 22(8):2092–2099
11. Lorenz EN (1963) Deterministic nonperiodic flow. J. Atmos. Sci. 20(2):130–141
12. Rössler OE (1976) An equation for continuous chaos. Phys. Lett. A 57(5):397–398
13. Pecora LM, Carroll TL (1990) Synchronization in chaotic systems. Phys. Rev. Lett. 64(8):821
14. Singh AK, Yadav VK, Das S (2017) Synchronization between fractional order complex chaotic systems. Int. J. Dyn. Control 5(3):756–770
15. Li C, Liao X (2004) Complete and lag synchronization of hyperchaotic systems using small impulses. Chaos Solitons Fractals 22(4):857–867
16. Ma J, Mi L, Zhou P, Ying X, Hayat T (2017) Phase synchronization between two neurons induced by coupling of electromagnetic field. Appl. Math. Comput. 307:321–328
17. Sudheer KS, Sabir M (2009) Hybrid synchronization of hyperchaotic lu system. Pramana 73(4):781
18. Khan, T., Chaudhary, H.: Estimation and Identifiability of Parameters for Generalized Lotka-Volterra Biological Systems using Adaptive Controlled Combination Difference Anti-synchronization
19. Khan, A. and Chaudhary, H.: Hybrid projective combination–combination synchronization in non-identical hyperchaotic systems using adaptive control. Arab. J. Math. pp. 1–15 (2020)
20. Li G-H, Zhou S-P (2007) Anti-synchronization in different chaotic systems. Chaos Solitons Fractals 32(2):516–520

21. Ding Z, Shen Y (2016) Projective synchronization of nonidentical fractional-order neural networks based on sliding mode controller. Neural Netw. 76:97–105
22. Zhou P, Zhu W (2011) Function projective synchronization for fractional-order chaotic systems. Nonlinear Anal. Real World Appl. 12(2):811–816
23. Xu Z, Guo L, Hu M, Yang Y (2008) Hybrid projective synchronization in a chaotic complex nonlinear system. Math. Comput. Simul. 79(1):449–457
24. Li G-H (2007) Modified projective synchronization of chaotic system. Chaos Solitons Fractals 32(5):1786–1790
25. Delavari H, Mohadeszadeh M (2018) Hybrid complex projective synchronization of complex chaotic systems using active control technique with nonlinearity in the control input. J. Control Eng. Appl. Inf. 20(1):67–74
26. Vaidyanathan S, Mohadeszadeh M (2012) Anti-synchronization of four-wing chaotic systems via sliding mode control. Int. J. Autom. Comput. 9(3):274–279
27. Jahanzaib LS, Trikha P, Chaudhary H, Haider SM et al (2020) Compound synchronization using disturbance observer based adaptive sliding mode control technique. J. Math. Comput. Sci. 10(5):1463–1480
28. Rasappan, S., Vaidyanathan, S.: Synchronization of hyperchaotic liu system via backstepping control with recursive feedback. In: International Conference on Eco-friendly Computing and Communication Systems, pp. 212–221. Springer (2012)
29. Ayub Khan and Muzaffar Ahmad Bhat (2017) Hyper-chaotic analysis and adaptive multi-switching synchronization of a novel asymmetric non-linear dynamical system. Int. J. Dyn. Control 5(4):1211–1221
30. Chen M, Han Z (2003) Controlling and synchronizing chaotic genesio system via nonlinear feedback control. Chaos Solitons Fractals 17(4):709–716
31. Li D, Zhang X (2016) Impulsive synchronization of fractional order chaotic systems with time-delay. Neurocomputing 216:39–44
32. Hubler AW (1989) Adaptive control of chaotic system. Helv. Phys. Acta 62:343–346
33. Mainieri R, Rehacek J (1999) Projective synchronization in three-dimensional chaotic systems. Phys. Rev. Lett. 82(15):3042
34. Liao T-L, Tsai S-H (2000) Adaptive synchronization of chaotic systems and its application to secure communications. Chaos Solitons Fractals 11(9):1387–1396
35. Yassen MT (2003) Adaptive control and synchronization of a modified chua's circuit system. Appl. Math. Comput. 135(1):113–128
36. Li Z, Daolin X (2004) A secure communication scheme using projective chaos synchronization. Chaos Solitons Fractals 22(2):477–481
37. Li S-Y, Yang C-H, Lin C-T, Ko L-W, Chiu T-T (2012) Adaptive synchronization of chaotic systems with unknown parameters via new backstepping strategy. Nonlinear Dyn. 70(3):2129–2143
38. Zhaoyan W, Duan J, Xinchu F (2012) Complex projective synchronization in coupled chaotic complex dynamical systems. Nonlinear Dyn. 69(3):771–779
39. Vaidyanathan S (2016) Hybrid synchronization of the generalized lotka-volterra three-species biological systems via adaptive control. Int. J. PharmTech. Res. 9(1):179–192
40. Vaidyanathan S (2015) Adaptive biological control of generalized lotka-volterra three-species biological system. Int. J. PharmTech. Res. 8(4):622–631
41. Khan A, Bhat MA (2017) Hyperchaotic analysis and adaptive projective synchronization of nonlinear dynamical system. Comput. Math. Model. 28(4):517–530
42. Khan A, Tyagi A (2017) Hybrid projective synchronization between two identical new 4-d hyper-chaotic systems via active control method. Int. J. Nonlinear. Sci 23(3):142–150
43. Khan A, Tyagi A (2017) Analysis and hyper-chaos control of a new 4-d hyper-chaotic system by using optimal and adaptive control design. Int. J. Dyn. Control 5(4):1147–1155
44. Samardzija N, Greller LD (1988) Explosive route to chaos through a fractal torus in a generalized lotka-volterra model. Bull. Math. Biol. 50(5):465–491

Internet Of Things and Environmental Monitoring

IoT for Diabetes: Smart Monitoring and Control

Shweta Taneja and Mohit Chandna

Abstract Nowadays, diabetes has become a very common disease across globe. People with diabetes cost the healthcare system in billions and also result in a great loss to productivity. The root cause for diabetes is still un-identified, and the only way to fight is to either reduce or control the extent. For the same, continuous monitoring of diabetes becomes a core area to be examined. Therefore, in this paper, we have explored the use of IoT devices to monitor and control diabetes. Smart glucometers in form of wearable devices can take a stock of glucose levels in the blood on real-time basis. This data is gathered and streamed to multiple systems for further processing. Data analytics and data engineering methods are used to convert data into useful information which in turn helps us to get insights. In case of abnormal situations (400 < glucose level < 70 mg/dL), notifications in form of alerts are sent out to care givers or doctors. So, this leads to an early or timely intervention of doctor. This timely intervention helps to save cost at both patient and payer level.

Keywords Internet of Things (IoT) · Smart health care · Glucometer (GCM) devices · Chronic disease · Data analytics · Data engineering

1 Introduction

Nowadays, Internet of Things (IoT) has become an important component in our everyday life. IoT enables people and devices to remain connected anywhere and anytime. The IoT contains a network of physical devices containing sensors, software, connectivity, etc., that helps the devices to communicate and exchange data with

S. Taneja (✉)
Department of Computer Science, Bhagwan Parshuram Institute of Technology, GGSIPU, Dwarka, India
e-mail: shwetataneja@bpitindia.com

M. Chandna
ThoughtWorks, Gurgaon, India
e-mail: mohit.chandna@thoughtworks.com

A. Abraham et al. (eds.), *Proceedings of 3rd International Conference on Computing Informatics and Networks*, Lecture Notes in Networks and Systems 167,
https://doi.org/10.1007/978-981-15-9712-1_39

each other on the network. These devices are called as smart devices as they can be controlled remotely. The physical devices can be home appliances, vehicles on the road, smartphones, etc. The terms like smart home, smart city, and smart health are becoming more popular nowadays.

IoT has many applications; among them, smart health is one of the upcoming domains [1, 2].

1.1 IoT in Health Care

IoT will help healthcare industry and bring more value. This value addition is propagated across all major stakeholders and includes beneficiaries, payers, providers and pharmacy. Intelligent system is evolved once billions of devices connect to each other. Continuous share of data by intelligent systems on cloud goes through churning process and results in business insights. These insights can do wonders ranging from improved care coordination to sensing a cardio attack. Increasing costs for payers is the most critical problem in healthcare industry today. One of the effective ways to curb costs is to analyze and examine the existing data (originating from varied data sources). Outcome of data analysis becomes an action area to reduce costs for associated business segments. The significance of IoT in health care is depicted in Fig. 1.

In USA, 90% of annual healthcare expenses are chalked out on chronic and mental health conditions. Among all chronic diseases, diabetes is burdensome both in terms of commonality and associated cost. Complication with diabetes is two-folded. Being

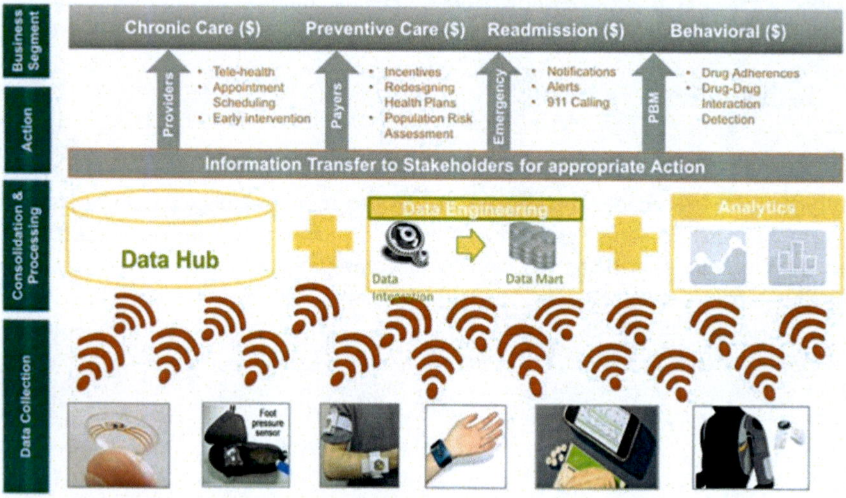

Fig. 1 IoT in health care

a slow poison in itself, it is an originator for multiple problems like cardiovascular arrest, blindness, kidney failure, hypertension, functional disability, etc. Despite huge investments, no major breakthrough has been achieved in defining ways to better support individuals with diabetes. IoT can act as ray of hope in this darkening horizon.

1.2 Motivation

Nowadays, diabetes has become a very common disease in India as well as abroad. Every second or third person is suffering from this disease. It is called as a slow poison. We cannot totally stop it, but we can try to control it. If diabetes is not treated, it can give birth to other diseases. It can affect our heart, kidneys, and the entire body. It can lead to a state that can be hard to control. In this paper, we have shown the usage of IoT devices in the control of diabetes. In case of emergency situations, how IoT devices can help to generate alarms or reminders to doctors or care givers.

This paper is organized as follows: Sect. 2 discusses the work of other researchers done in this area. In Sect. 3, we have explained our proposed work in detail. In the next Sect. 4, benefits and challenges in the field of IoT in health care are given. This is followed by conclusion and future scope in last section.

2 Related Work

IoT has become a buzzword today. It has applications in a variety of domains. Different researchers are doing work on the use of upcoming and latest methods in this field. In the field of health care, use of IoT has been done in different diseases, but very few authors have done research inn cancer care. In [3], authors have shown implementation of IoT in cancer care along with the use of business intelligence tools. This has helped to gain insights into the patient data and further helped in decision making.

Another fruitful domain area of IoT is in care of elderly people. The authors have proposed a fog IoT cloud-based health monitoring system to provide information about routine activities [4]. In another work [5], authors have presented a HOG-SVM-based IoT system for detecting falls in elder people using deep sensors. On detecting a fall, IoT system sends alert messages to family members or doctor.

Another new feature that is being used in IoT systems is the concept of semantics. Some researchers have used semantics in IoT for healthcare domain. In [6], authors have used semantics-based technology to implement smart patient management. This technology is based on the use of agents and incorporates semantic logic that makes the system intelligent. In an another work [7], authors have used semantic Web technology to manage huge and heterogeneous IoT health datasets.

IoT research is a field of dynamics. New developments are being made from across the globe in this area. In [8], authors have used a new concept, Inertial Measurement Units (IMU) in mobiles. They have presented modified sensing algorithms to sense different events like fall, step count, etc. In another paper [11], authors have used Hidden Markov Model to predict the overall health status of patients.

In the field of health care, IoT has also been used to study patient behavior and gain insights from patient data. The authors have used IoT technologies to study the behavior of patients by using methods like gamification, personalization, etc. They took data from Health Promotion Board, Singapore [9]. In [10], authors have proposed a personalized missing data resilient decision-making approach to record health data. They have taken real-time data of 20 pregnant women who were monitored remotely for seven months.

A complete health ecosystem based on IoT has been created by the authors that can monitor and store health data of a patient in [12]. In [13], authors have shown how users can create IoT apps easily on their smartphones. In another work, authors have used IoT device to monitor the blood glucose level [14].

Different researchers have done research of using IoT devices in the field of health care. But, to the best of our knowledge, no one has explored the use of IoT to control diabetes disease. In our work, we have proposed a novel approach of using IoT devices to monitor and control diabetes.

3 Proposed Approach

In this paper, we have proposed a novel approach to control diabetes with the help of IoT devices. The proposed approach is shown in Fig. 2. It has five stages.

Stage 1: In stage 1, we identify chronic patients those who are suffering from chronic diseases like asthma, diabetes, cancer, etc.
Stage 2: In stage 2, we select patients having chronic diabetes and who need urgent care and attention.
Stage 3: In this stage, the data is gathered from a patient through a smart glucometer (GCM) device. This device is a wearable and portable IoT device that works like a smart watch. It continuously measures the glucose level in blood in every hour. This is live data that is gathered in streams.
Stage 4: In stage 4, this live data is stored on the cloud. This stage is a crucial and important stage of our proposed approach. Data analytics and data engineering methods are used in this stage.
The input here is raw data. This data is converted into information or in a reusable format. That is, the values that are below 70 or greater than 400 are captured. So, two things are captured: value and timestamp. After that, machine learning models are built for churning the data and carrying out exploratory data analysis (EDA). This helps us to get insights into the data.

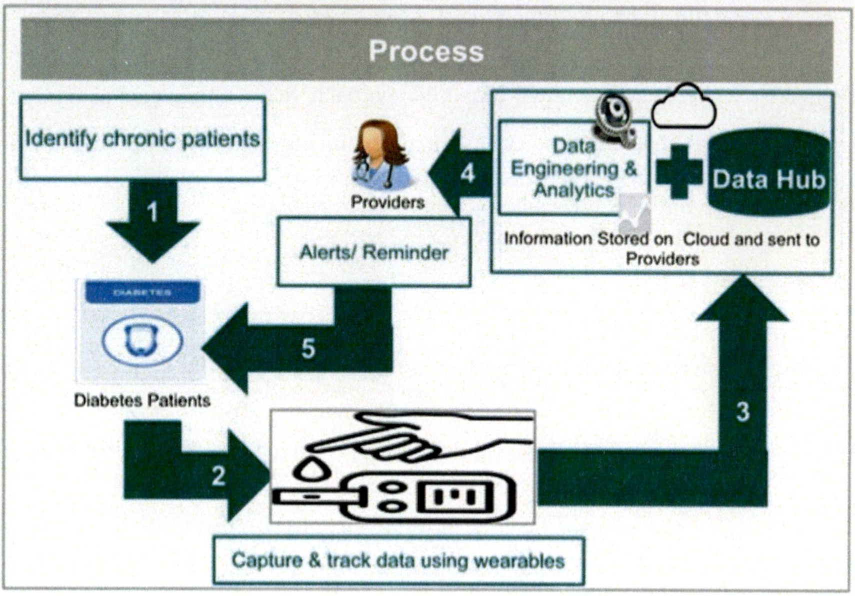

Fig. 2 Monitoring diabetes with IoT devices

Stage 5: This is the last stage, in which the insights obtained in the previous stage are used to generate alerts/reminders to doctors or care givers. The system is designed in such a way that even if there is a single abnormal value (i.e., below 70 mg/dL or greater than 400 mg/dL), it is immediately sent as an alert. The abnormal or alarming conditions are shown in Fig. 3. The doctor may then communicate it to the patient or his family. It is beneficial to those people also who do not take medicines as they are reminded by the doctors if some critical condition arises.

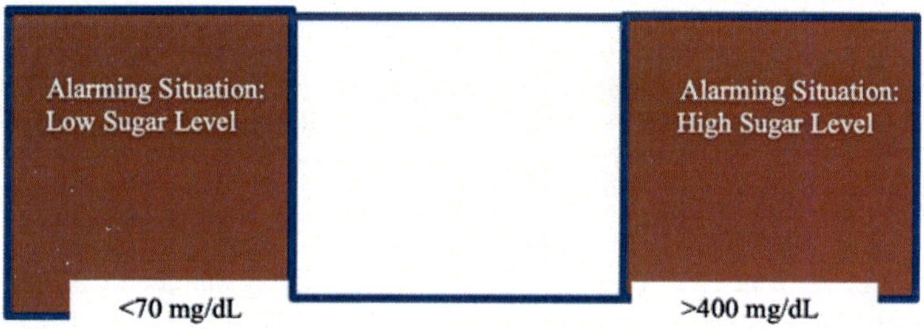

Fig. 3 Abnormal or alarming conditions

3.1 Advantages of the Proposed Approach

There are many advantages of our proposed approach. Some of them are as follows:

- Our proposed approach results in an early or timely intervention by doctors or care givers.
- This leads to optimal utilization of healthcare services.
- It results in reduction in emergency care visit.
- Therefore, it is cost saving.

4 Benefits and Challenging Issues of IoT in Health Care

In the field of health care, IoT has numerous applications like personalized health care, sending alerts to patients, smart sensors, etc. There are many benefits as well as challenges of IoT in health care.

The benefits are as follows:

- Continuous tracking and monitoring;
- Improved accuracy;
- Less overhead;
- Location independence;
- Easy to access.

The challenges are as follows:

- Threat to security;
- Integration of data from multiple devices;
- Handling huge amount of data.

5 Conclusion and Future Scope

In this paper, we have proposed a novel approach of monitoring and controlling chronic diabetes disease using IoT devices. In our approach, smart glucometers are used to collect real-time data. This data is used to build machine learning models and gain insights into it. In case of abnormal conditions, an immediate alert or reminder is sent to doctors or care givers.

In future, the proposed approach may be evaluated and validated with the help of experts or medical practioners. Further research is in progress for identifying optimal techniques for data anonymization (PHI and PII data), data transfer (batch v/s streaming), EDA (exploratory) purpose.

References

1. De Michele R, Furini M (2019) IoT healthcare: benefits, issues and challenges. In: Proceedings of the 5th EAI international conference on smart objects and technologies for social good, pp 160–164
2. Yang P, Xu L (2018) The Internet of Things (IoT): informatics methods for IoT-enabled health care
3. Onasanya A, Elshakankiri M (2017) IoT implementation for Cancer care and business analytics/cloud services in healthcare systems. In: Proceedings of the 10th international conference on utility and cloud computing, pp 205–206
4. Debauche O, Mahmoudi S, Manneback P, Assila A (2019) Fog IoT for health: a new architecture for patients and elderly monitoring. Procedia Comput Sci 160:289–297
5. Kong X, Meng Z, Nojiri N, Iwahori Y, Meng L, Tomiyama H (2019) A HOG-SVM based fall detection IoT system for elderly persons using deep sensor. Procedia Comput Sci 147:276–282
6. Thangaraj M, Ponmalar PP, Sujatha G, Anuradha S (2016) Agent based semantic Internet of Things (IoT) in smart health care. In: Proceedings of the 11th international knowledge management in organizations conference on the changing face of knowledge management impacting society, July 2016, pp 1–9
7. Reda R, Carbonaro A (2018) Design and development of a linked open data-based web portal for sharing IoT health and fitness datasets. In: Proceedings of the 4th EAI international conference on smart objects and technologies for social good, November 2018, pp 43–48
8. Chandel V, Sinharay A, Ahmed N, Ghose A (2016) Exploiting IMU sensors for IoT enabled health monitoring. In: Proceedings of the first workshop on IoT-enabled healthcare and wellness technologies and systems, June 2016, pp 21–22
9. Tan V, Varghese SA (2016) IoT-enabled health promotion. In: Proceedings of the first workshop on IoT-enabled healthcare and wellness technologies and systems, June 2016, pp 17–18
10. Azimi I, Pahikkala T, Rahmani AM, Niela-Vilén H, Axelin A, Liljeberg P (2019) Missing data resilient decision-making for healthcare IoT through personalization: a case study on maternal health. Future Gener Comput Syst 96:297–308
11. Zamanifar A, Nazemi E (2019) An approach for predicting health status in IoT health care. J Netw Comput Appl 134:100–113
12. Zemrane H, Baddi Y, Hasbi A (2019) Improve IoT e-health ecosystem with SDN. In: Proceedings of the 4th international conference on smart city applications, October 2019, pp 1–8
13. Khaled A, Lindquist W, Hela, S (2018) DIY health IoT apps. In: Proceedings of the 16th ACM conference on embedded networked sensor systems, November 2018, pp 406–407
14. Gia TN, Ali M, Dhaou IB, Rahmani AM, Westerlund T, Liljeberg P, Tenhunen H (2017) IoT-based continuous glucose monitoring system: a feasibility study. Procedia Comput Sci 109:327–334

Smart Traffic Light Management System for Emergency Vehicle

Nitin Vohra, Pranjal Kandhari, Abhinav Gupta, Shilpa Gupta, Arpit Shrotiya, and Rohit Dev

Abstract As of right now, there are uncountable individuals ranging from adolescents to elderly irrespective of being adhering to a single age group who become victims of heinous road accidents in India. Giving the claim a more profound relevance, as economic times have mentioned, close to 151,417 individuals were killed in road accidents. Out of which, 69.6% of deaths were caused due to road rage and then due to delay in the arrival of an ambulance to the scene. Of course, a lot of factors play a major role in the circumstances. Our proposed system stresses the time delays caused by redundant, inefficient traffic jams to rectify this problem and improves it with defined logical methods which would be constructively helpful for the society as a whole. The most effective route to the accident is calculated and displayed and then a combined array of servers with a central controller determines the location of the ambulance and changes the traffic lights adaptive to shorten the time delay for the victim as minuscule as possible, with less to none human interference. The implementation of the proposed system is done with keeping the safety of emergency vehicles in mind, giving them a better course of action to follow.

Keywords Emergency vehicle · CCTV cameras · Traffic junction

N. Vohra (✉) · P. Kandhari · A. Gupta · S. Gupta · A. Shrotiya · R. Dev
Bharati Vidyapeeth's College of Engineering, New Delhi, India
e-mail: nitinvohra92@gmail.com

P. Kandhari
e-mail: pranjalkandhari@gmail.com

A. Gupta
e-mail: abhinavgupt99@gmail.com

S. Gupta
e-mail: shilpa.gupta@bharatividyapeeth.edu

A. Shrotiya
e-mail: arpit.shrotriya5945@gmail.com

R. Dev
e-mail: rohitdevs.kmh2@gmail.com

A. Abraham et al. (eds.), *Proceedings of 3rd International Conference on Computing Informatics and Networks*, Lecture Notes in Networks and Systems 167,
https://doi.org/10.1007/978-981-15-9712-1_40

Abbreviations

ANFIS	Adaptive neuro-fuzzy inference system
API	Application programming interface
ATCS	Adaptive traffic control system
CCTV	Closed circuit television
CDS	Cooperative driving strategy
CV	Computer vision
EV	Emergency vehicle
Fig	Figure
GPS	Global positioning system
ILD	Inductive loop detectors
IoT	Internet of Things
MN	Minnesota
R-CNN	Region-based convolutional neural networks
RFID	Radio frequency identification
SVM	Support vector machines
VANETs	Vehicular ad hoc networks

1 Introduction

Vehicle safety at crossroads has been a problem for governments in many major cities around the world to address for too long we cannot even imagine. India is a developing country with a second-largest population in the whole world. Despite being an incredibly fast-growing economy. The country is severely affected by road congestion caused by vehicle overcrowding and slow infrastructure development in space allocated and adhering cost, which further results in causing hindrance in the systematic execution of norms and services. Many different innovative systems have been proposed, but still, older timing circuits are used by today's systems to reflect the responses in an orderly fashion. But, the current systems do not account for different test cases experienced in emergencies resulting in loss of valuable time. In most of today's scenarios, the ambulances are caught up in the traffic jams en route to a near-death victim. As congestion often results in wastage of time and productivity. Driving to the accident scene in a safe manner and quickly is a major challenge for emergency vehicles, particularly, as traffic is increasingly heavy and traffic patterns in modern cities have become more complex. So making both an effective and adaptive system is a task bigger than we can perceive. Thus, the smart traffic light management system takes the computer vision as well as Internet of things (IoT) approach to further implement that by switching traffic lights for emergency vehicles, mainly ambulance on the designated shortest path which is regulated through a nexus of servers and maintaining personnel approach through a custom application to facilitate the proper functioning and creating an effortless experience for patient recovery by the respective vehicles saving many innocent lives in the process. Further, the system can be deployed further for a plethora of emergency vehicles to further regulating the traffic for our country. Thus, making a better solution altogether.

2 Literature Survey

Increasing traffic problems is a huge concern in cities nowadays. Growing population results in the number of vehicles being increased in cities [1]. This results in congestion on roads which increases the traveling time. Mittal and Bhandari [2] have discussed the green wave system, which basically turns all the red lights into green lights, thereby granting permission to emergency vehicles. The green wave system has a drawback that on being suddenly changed due to the alteration of green wave traffic problems can be caused. This leads to the increase in the vehicles in the queue of a green wave which leads to vehicles not being able to reach the green light in time and once the wave is disturbed, its disturbance can cause a major increase in traffic problems. In these situations, a green wave will grow until the line of vehicles becomes too big hence stopping vehicles to reach the green light. This is called as oversaturation [3–5]. Zang et al. [6] studied the utilization and implementation of RFID in VANETs. In the proposed mechanism, RFID is embedded in each vehicle to identify the vehicles with the help of computers till the time the tag is present within the vehicle. Its limitation is it can cover and detect within a brief distance only. Hegde et al. [7] an ambulance system was proposed an automatic route clearance using RFID and GPS-based system. The disadvantage of this system is that it would not work for the ambulance if the starting point and the endpoint of the trip are not known in advance. Guo et al. [8] inspected some factors related to traffic and proposed a mechanism based on a distributed transportation system. In it, a method was proposed to calculate the speed from source to destination. This method will only show the alternative routes or the shortest routes but not the area with traffic congestion. Soni et al. [9] studied numerous technologies and sensors for reducing traffic congestion across the country. The study examined the ultrasonic sensors, adaptive traffic control system (ATCS) and inductive loop detectors (ILD). The technique proposed leads to an increase in traffic on the streets. Kale et al. [10] has proposed a smart ambulance that consists of temperature sensor and heartbeat sensor. The big disadvantage of this is that all ambulances must have more special instruments other than the medical ones. Dubey et al. [11] proposed an image processing technique that uses micro-controller which is installed on the traffic light along with the four cameras that are installed on it. The technique uses OpenCV as a tool for implementation, which leads to the decline in private detection range and quality. Thus, it needs to be trained accurately. Harsha et al. [12] studied the density-based traffic system by using a microwave/millimeter-wave radio detection, ranging and magnetic detector (magnetometer). This technology has varied problems like false detection because of multi-path propagation and magnetic detectors will have to be compelled to be embedded in the pavement which might need multiple detectors to notice smaller vehicles. Anna Merine George et al. [13] used the adaptive neuro-fuzzy inference system (ANFIS) to improve traffic conditions. This study has some disadvantages since it is a simulation-based technique and it cannot be embedded on a real-time basis, where the simulation sometimes occurs as statistics, probability and overview of what is happening in the computer model. Anthony Theodore et al. [14] proposed a system using simulation which has the capability to simulate real-time traffic flow using CDS. The limitation of the model is that it is created only by considering the area of Minneapolis, MN. Wen [15] proposed an expert system using six simulation models for solving the problem of road congestion. The results of the proposed solution proved to be efficient as they observed a sharp drop in average waiting time at traffic signals. Sharma et al. [16] proposed the solution by giving priority to the emergency vehicle using RFID. The limitation of this work that the communication between the controller and the emergency vehicle is not discussed. Karthikeyan [17] offered a solution for detecting emergency vehicles using a ZigBee receiver.

3 Proposed Model

The proposed model of the traffic management system provides a solution to solve the problem of traffic management at the traffic signals to give priority to emergency vehicles so that to save time in the absence of some supervision. It will be done by triggering an alert by the system to turn the traffic lights green when the emergency vehicle is in the range of traffic signals and triggering it back to normal when the emergency vehicle crosses the junction. This system acts as a medium to provide communication between various Internet of things (IoT) devices and emergency vehicles (Fig. 1).

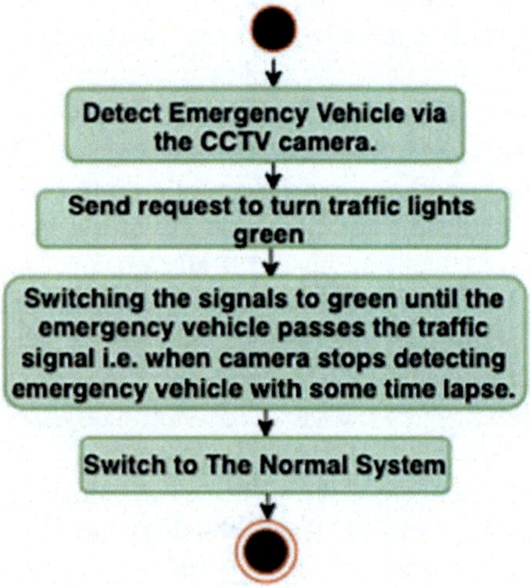

Fig. 1 Flow diagram of proposed model-1

Fig. 2 Flow diagram of proposed model-2

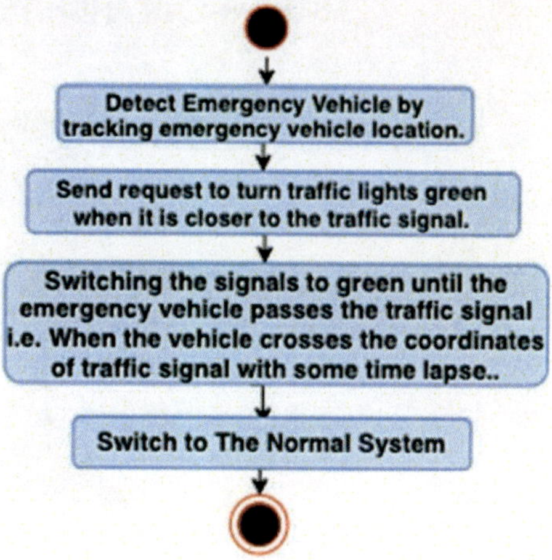

3.1 *Proposed System-1*

This proposed system detects emergency vehicles using CCTV cameras that are placed near the traffic signal. On detection, it triggers the traffic signal and brings the traffic signal lights back to normal scheduling once the emergency vehicle passes the crossing.

Here, two processes are involved (Figs. 3 and 4):

(a) **Emergency Vehicle detection using CCTV cameras.**

This module will be responsible for the detection of ambulances using CCTV cameras which are placed near the traffic signal. When an emergency vehicle comes in the range of the CCTV camera, then by using R-CNN, we will detect the ambulance.

Fig. 3 Ambulance detection using OpenCV

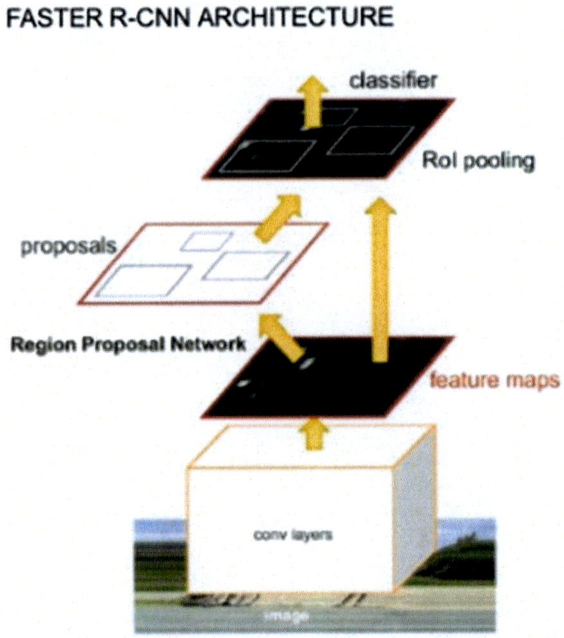

Fig. 4 R-CNN architecture

We have detected an ambulance using R-CNN, or region-based convolutional neural network, which consists of three steps:

(1) The input image is scanned for all the possible objects using selective search, generating more than a thousand region proposals.
(2) After that, we ran a convolutional neural network (CNN) on top of every region proposals found in step 1.
(3) In the last step, the output of each CNN is of feeding it into:

    (a) A support vector machine (SVM) to classify the region.
    (b) Into a module to tighten the bounding box of the object using linear regression if such an object exists. So, in simple terms, we first proposed the regions, then extracted features from them, and then classified those regions based on the features extracted. R-CNN is very intuitive but very slow so to cover up this limitation we have used faster R-CNN that resembled the original R-CNN in many ways but has overcome the limitation of its detection speed through two main augmentations:
        (i) Feature extraction is performed over the image and after the regions is proposed, thus, only running one CNN over the entire image instead of a thousand CNN's over a thousand overlapping regions.
        (ii) The neural network is extended for predictions instead of creating a new model by replacing the SVM with a softmax layer.

(b) **Traffic signal triggering after the detection of the emergency vehicle**

When the ambulance is detected in the CCTV camera placed near the traffic signals, then our central server will send a triggering signal to the micro-controller which is attached to the traffic signals which will then help in switching traffic signals. The traffic signal triggering module is responsible for changing the traffic signals. The function which will help in triggering the traffic signal will take real-time data captured by the CCTV camera as the input and will send a trigger signal as an output to the micro-controller. When the triggering signal is received to the micro-controller, then it will override the traffic signal and switch it to green for making away so that an emergency vehicle can proceed easily. Once an emergency vehicle is no longer visible in the CCTV camera, then the micro-controller turns the traffic signal to its conventional mode after a predefined constant time lapse.

3.2 Proposed System-2

This proposed system provides a real-time location of the emergency vehicle to the centralized server which manages triggering of signals to the traffic signals. The ambulance the driver will constantly use GPS to send its real-time location to the central server (this will be done using the GPS module of the driver's smartphone). Our central server will constantly map the coordinates received from the driver and the coordinates of traffic signals already stored in our database [18]. Once when the distance between the traffic light and the ambulance crosses a threshold, then the central system will trigger a signal to change the signal to green if it is red. This mechanism will help to clear the path of the emergency vehicle at traffic signals at peak time or at junctions where the problem of congestion is very common. The basic activity flow diagram of the system is shown in Fig. 2. Our system is divided into four modules.

(a) **Registration and Authentication of the Emergency Vehicle Driver**

Each emergency vehicle driver must be behaving a smartphone with a properly working GPS module and the provided mobile application installed in it. Also, it is worth noting that the system tracks the locations of the driver and not the emergency vehicle. The major benefit of this is cost-effectiveness as no new GPS module has to be implemented in an emergency vehicle. Therefore, saving the cost of the numerous modules implemented and their recurring maintenance cost. Another aspect of this solution involves concerns regarding taking undue advantage of this system by the drivers and using it for personal use (as the traffic light system is being affected by the driver's location). To resolve this issue, we give an option to the driver to be offline while not driving the emergency vehicle and also we can monitor whether the driver is using the application when not on duty and hence necessary action can be taken against the driver. Another feature that can be added is to connect the driver's attendance database to our system. This will help in the easy attendance of the driver and also managing their working hours. After the authentication is done the driver can access the app using their email ID which can also act as a unique ID for our database (Figs. 5 and 6).

(b) **Detecting the Location of the Emergency Vehicle**

The location of the driver is based on the location of the app of the driver's smartphone location. We receive the location of the driver's application on a real-time database (e.g., firebase). We receive the location of the driver in the form of latitude and longitude coordinates. Additionally, to improve the efficiency of our system, we can also receive the traffic density of the location where the emergency vehicle is presently using Google Maps Application Program Interface (API) (explained in

Fig. 5 Login page of app

Fig. 6 Live location of emergency vehicle

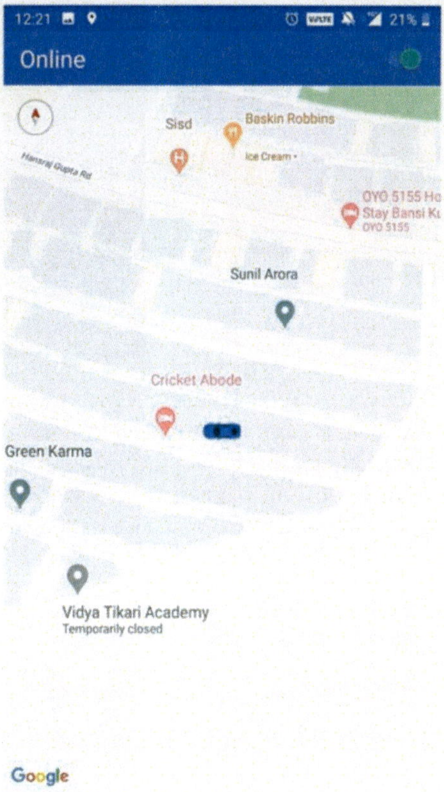

C). Another feature that is included is to show the shortest route to the driver to reach the hospital. Here, one point is worth noting, this system largely depends on the accuracy of the location received by the smartphone application. Hence, a good Internet connection and a properly functional GPS module is a must for the working of this system (Fig. 7).

(c) **Algorithm to Schedule Traffic Signals in Accordance with the Emergency Vehicle.**

The central system (this processing will take place on a cloud-based server) will constantly monitor the real-time location of the vehicle and will be constantly mapping it with the coordinates of the traffic signals. Once an emergency vehicle comes in the vicinity of a traffic light (after crossing a threshold value), we turn the traffic light to green for a specific amount of time. The upper limit to the time has to be specified as in case of heavy traffic, the emergency vehicle might take too long to pass the traffic signal and hence causing traffic blockage on other sides of the crossing. Now, we cannot simply turn on the green signal by taking vicinity as a parameter even if the emergency vehicle is moving away from the traffic light. So

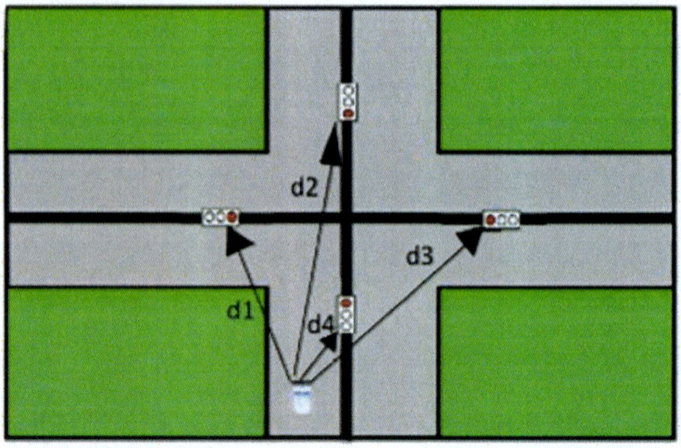

Fig. 7 Traffic light intersection

to handle this case, we also need to store some previous location on the emergency vehicle in order to interpret the direction on the emergency vehicle. We will turn the nearest traffic light to green as shown in Fig. 8. Also, we can use the same directional approach to find if the ambulance has passed the crossing. Once we detect it, we can

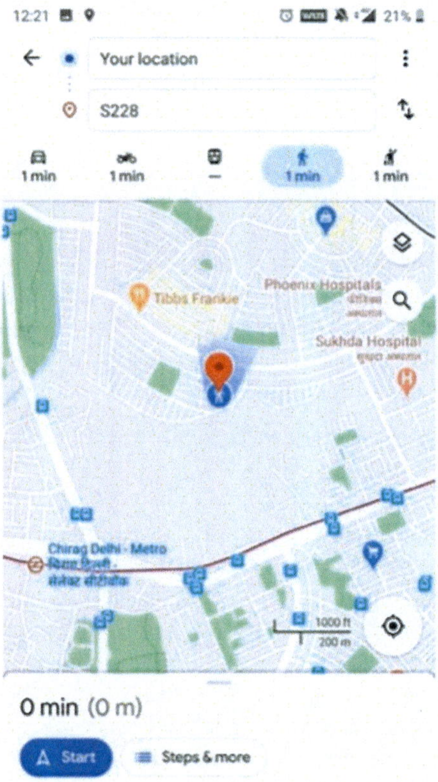

Fig. 8 Optimized path according to traffic

put the traffic light back to its normal schedule. Another noticeable point is that the threshold value will change on different traffic conditions which can be handled by the following approach:

$$\text{Threshold distance} = x * c \tag{1}$$

In the above equation, x is the variable and c is a constant Traffic intensity can be categorized into three levels (which will be provided by Google Maps API): high, medium and low, respectively. In the case of multiple emergency vehicles, we can prioritize them accordingly:

(1) Advanced life support ambulance (max priority)
(2) Basic life support ambulance
(3) Patient transfer ambulance.

Also, if multiple ambulances of the same priority come in the vicinity of traffic lights of the same crossing, then the priority will be given to the ambulance with the shortest distance from the crossing.

(d) **Triggering Mechanism After the Traffic Signal is Scheduled According to the Ambulance.**

When the vehicle crosses a defined stretch the central server relays a trigger signal to the respective micro-controller which is attached to the traffic signals which will help in switching traffic signals. The traffic signal triggering module is bonded for switching the signal conferring with the defined stretch, calculated using the traffic intensity and the live location of the ambulance at that moment of time. The function which will help in triggering the traffic signal will take traffic intensity and the live location of the ambulance as input and will send a trigger signal as an output to the micro-controller according to the threshold distance calculated. When the triggering signal is received by the micro-controller, then it will override the traffic signal and switch it to green for making away so that an emergency vehicle can proceed easily. On the passing of the vehicle across the traffic light and upon crossing a definite stretch, after which the micro-controller turns the traffic signal back to its previous case.

4 Comparison Between the Two Proposed Solutions

See Table 1.

Table 1 Comparison between two proposed solutions

Features	Using CCTV Camera	Using smartphone GPS
Average waiting time	Low, the traffic light is turned green when the EV is detected by CCTV but due to the lower range the system does not work on large traffic jams or when line of sight is blocked	Lower, the traffic light is turned green when the EV crosses the variable threshold distance which can vary according to the traffic intensity at that moment of time
Range	Less, since the detection is dependent on the camera quality and angle	More, since it can vary according to traffic intensity at that moment of time
Dependencies	Range and camera angle	Human error and GPS
Efficiency	Good because it will help EV to pass traffic signal but can fail where traffic intensity is more or line of sight is obstructed	Better as it will help EV to pass and have a variable threshold distance which can vary real time according to traffic intensity
Scheduling as per the priority of emergency	No	Yes, we can prioritize according to type of ambulance

5 Conclusion

In conclusion, we observed that the conventional method was unable to remedy the plight of ambulance getting stuck at traffic signals which often lead to the casualty of patients. So in order to solve this problem, we suggested two solutions to improve the probability of saving a life by eliminating unnecessary time delay at traffic congestion near traffic lights. Considering the dependencies and limitations of the CCTV camera, it does not prove to be a good solution in commonly occurring cases. Also, looking at the huge number of smartphones available in India and good Internet connectivity in cities (which are the major spots for traffic jams), the proposed system 2 (using smartphone's GPS) proves to be a better solution in most cases.

References

1. Kelly B A green wave reprieve. http://gse.cat.org.uk/papers?download=6%3Ablaise-kellya-green-wave-reprieve. Last accessed 2016/07/23
2. Mittal AKR, Bhandari D (2013) A novel approach to implement green wave system and detection of stolen vehicles. In: Proceedings of IEEE conference on advanced computing, pp 1055–1059
3. Zyga L Physics of green waves could make city traffic flow more smoothly. https://phys.org/news/2013-05-physics-green-city-traffic-smoothly.html. Last accessed 2020/06/23
4. Kale SB, Dhok GP (2013) Design of intelligent ambulance and traffic control. Int J Innov Technol Explor Eng (IJITEE) 2(5):2278–3075

5. Sharma S, Pithora A, Gupta G, Goel M, Sinha M (2013) Traffic light priority control for emergency vehicle using RFID. Int J Innov Eng Technol 2(2):363–366
6. Zhang Q, Almulla MA, Boukerche A (2013) An improved scheme for key management of RFID in vehicular adhoc networks. IEEE Latin Am Trans 11(6):1286–1294
7. Hegde R, Sali RR, Indira MS (2013) RFID and GPS based automatic lane clearance system for ambulance. Int J Adv Electr Electron Eng 2(3):102–107
8. Guo C, Li D, Zhang G, Zhai M (2018) Real-time path planning in urban area via VANET-assisted traffic information sharing. IEEE Trans Veh Technol 5635–5649
9. Soni NB, Saraswat J (2017) A review of IoT devices for traffic management system. IEEE Xplore Complaint 2(5):1052–1055
10. Kale SB, Dhok GP (2013) Design of intelligent ambulance and traffic control. Int J Innov Technol Explor Eng (IJITEE) 2(5):211–214. ISSN 2278-3075
11. Dubey A, Akshdeep SR (2017) Implementation of an intelligent traffic control system and real-time traffic statistics broadcasting. In: IEEE international conference on electronics, communication and aerospace technology, vol 2, pp 33–37
12. Naga Harsha J, Nair N, Jacob SM, John Paul J (2018) Density based smart traffic system with real-time data analysis using IoT. In: IEEE international conference on current trends toward converging technologies, pp 1–6
13. George AM, George VI, George MA (2018) IoT: based smart traffic light control system. In: IEEE international conference on control, power, communication and computing technologies (ICCPCCT), pp 148–151
14. Chronopoulos AT, Johnston CM (1998) A real-time traffic simulation system. IEEE Trans Veh Technol 47(1):321–331
15. Wen W (2008) A dynamic and automatic traffic light control expert system for solving the road congestion problem. Exp Syst Appl 34:2370–2381
16. Sharma S, Pithora A, Gupta G, Goel M, Sinha M (2017) Traffic light priority control for emergency vehicle using RFID. Int J Innov Eng Technol 2(2):363–366
17. Karthikeyan V (2017) Intelligent traffic control system for congestion control, emergency vehicle clearance and stolen vehicle detection. Asian J Appl Sci Technol (AJAST) 1(1):122–125
18. George AA, Krishna A, Dias T, Vargheese AS, Divya RS (2017) Golden aid an emergency ambulance system. In: 2017 international conference on networks and advances in computational technologies (NetACT), Thiruvanthapuram, pp 473–476

Performance Evaluation of Rectangular Microstrip Patch Antenna based on Defected Ground Structures

Pawar Umesh Ankush, Abhijyoti Ghosh, Lourembam Lolit Kumar Singh, and Sudipta Chattopadhyay

Abstract A rare and novel comparative study of different defected ground structure integrated rectangular microstrip antenna has been presented in this paper. The defects are placed in such a way to get a clear insight into the concept behind improvements in antenna parameters while compulsorily maintaining the polarization purity by the use of regular defects that are in increasing order of polygons. Different structures of defected ground can give polarization purity of 22–28 dB and gain in the range of at least 8.5–9.1 dBi is suggested herewith. Both the requirements are inescapable being fundamental in nature to the field of polarimetric radars even though various compensating methods are available.

Keywords Rectangular microstrip patch antenna · Polygon · Defected ground structure · Cross-polarization · Bandwidth · Gain

Abbreviations

CP	Co-polarized
DGS	Defected ground structure
DPS	Defected patch surface
MPA	Microstrip patch antenna
PEC	Perfect electric conducting
PMC	Perfect magnetic conducting
RMA	Rectangular microstrip patch antenna
XP	Cross-polarization

P. Umesh Ankush
Indian Army, New Delhi, India

A. Ghosh (✉) · L. Lolit Kumar Singh · S. Chattopadhyay
Department of Electronics and Communication Engineering, Mizoram University, Aizawl, Mizoram, India
e-mail: abhijyoti_engineer@yahoo.co.in

A. Abraham et al. (eds.), *Proceedings of 3rd International Conference on Computing Informatics and Networks*, Lecture Notes in Networks and Systems 167,
https://doi.org/10.1007/978-981-15-9712-1_41

1 Introduction

In present day, continuously evolving wireless communication systems require light weight, confirmable, small size and low cost antennas, and microstrip patch antenna (MPA) is the most suitable candidate for the same. One of the dominating requirement in the modern wireless communication system antenna is larger impedance bandwidth and lower cross-polarization [1, 2]. A conventional rectangular microstrip patch antenna (RMA) resonating in its fundamental TM_{10} mode radiates co-polarized (CP) fields in the broadside direction. Orthogonal radiation called as cross-polarization (XP) fields are also radiated simultaneously along with this co-polarized radiation by the antenna and XP is most prominent in first higher-order orthogonal mode, i.e., TM_{02}. The conventional rectangular microstrip patch antenna has around 2–3% impedance bandwidth [3], however, a broadband system always demands a broad-band antenna/radiator, and hence it's a prime area of research. Various reported techniques to increase impedance bandwidth are L-probe feed inverted EE-H-shaped slotted MPA [4], E [5] and Ψ [6]-shaped patch antenna. Impedance bandwidths reported in those papers are 21.15%, 30%, and 54% respectively. Also, Fractal concept has been applied to various branches of engineering and science including electrodynamics, propagation, and scattering [7–11]. However, the limitations of all above highlighted structures are that they have poor CP-XP isolation (polarization purity). Some significant research results illustrate the investigation on minimizing the radiation of cross-polarized fields by altering the feed structure [12, 13] and ground plane structure [14, 15]. 10–20 dB CP-XP isolation is reported in those articles without any improvement in impedance bandwidth. Recently, defected ground structure (DGS) and defected patch surface (DPS) based on electronic band gap (EBG) theory have been used to improve the XP radiation purity of 10–15 dB is reported in those articles. The use of defected ground structure technique is the most popular and widely acceptable technique to improve polarization purity and other antenna parameters. In fields of not so very exploited polarimetric radars, improvements in antenna parameters while compulsorily maintaining the polarization purity is key to efficient and accurate polarimetric radar.

In the present comparative study between various existing DGS which are particularly in increasing order of polygon (simple RMA without DGS [16], Triangular DGS integrated RMA [17], Pentagon DGS integrated RMA [18], Hexagon DGS integrated RMA [19], and Circular/Ring DGS integrated RMA [17]) has been thoroughly studied with the aim to bring out the concept behind the benefits that they present and implication of the increasing order of polygon in particular.

2 Results and Discussions

2.1 Selection Criteria of DGS

The selection of polygons is practically adequate and covers an aspect of zero sides to infinite sides of the defects placed on the ground plane. The antennas selected for comparison are operating either in X band and Ku band only. The antennas selected for comparison are primarily having good CP-XP isolation (polarization purity), high impedance bandwidth and gain. However, the selections are not random and are motivated with an aim to recommend the concept behind them as opportunity to the field of polarimetric radars.

Also, the comparison is meaningful if the design concepts are on the similar lines. In-depth study of all the structures shows that the consistency in the design concept and method in the analysis of the results. All antennas are probe fed and have linear polarization pattern of radiation. All the antennas have incorporated DGS at the non-radiating sides of the rectangular patch which is the primary source of the cross-polarization radiation [21–24].

The important concept of incorporating defect on the ground plane of antenna structure is being compared and the selection of polygons is in the increasing order. Few dominating key parameters like impedance bandwidth, gain, and polarization purity has been compared. All the structured have been optimized to get the best performance out of each structure.

The bandwidth of the antenna is directly proportional with the different losses provided by the antenna, i.e., as losses increases bandwidth also increases. In each of the structure, studies have been initiated with a small defect size at the appropriate position to improve the performance. Then, the size of the defect increases gradually too get the optimum antenna structure which produces the best performance in terms of impedance bandwidth. The impedance bandwidth may increase with the further increment of the defect size but other parameters like gain and polarization purity may be hampered. Therefore, the optimum defect size has been finalized considering both the input as well as radiation characteristics.

2.2 Results Related to Different Relevant Structures

The selection of the DGSs as mentioned above are very specific and more importantly they are in the increasing order of polygon. Investigation has been performed with similar types of DGS integrated RMA with equal electrical dimension for excitation of similar frequency. Also, the selection of the DGSs are located specifically/nearly centered at a particular reference locations on the ground plain, and the benefits of the same are covered separately. Table 1 shows the details of the all parameters of each antenna used for the comparison.

Table 1 Parameters of the RMA used for investigations (substrate thickness $h = 1.575$ mm)

Name	Antenna structure	L (mm)	W (mm)	Number of sides in DGS (N)	Feed position (offset from center) (mm)
Antenna 1	Conventional RMA	8	12	0	2.4
Antenna 2	RMA with triangular DGS	8	12	3	3.1
Antenna 3	RMA with pentagon dumbelled DGS	8	12	5	4.32
Antenna 4	RMA with hexagon dumbelled DGS	7.1	10.8	6	3.3
Antenna 5	RMA with ring DGS	8	12	∞	3.1

The results obtained from each antenna used for comparison are simulated using [25] and documented in the following sections.

Antenna 1: In Antenna No. 1 that is conventional RMA (without DGS, $N = 0$), the conducting patch and ground plane form perfect electric conducting (PEC) boundary while edges boundaries provides perfect magnetic conducting (PMC) wall [3, 19, 20]. The reflection coefficient profile and the H-plane radiation characteristics are depicted in Fig. 1. Well-known literatures reported that conventional RMA suffers from two common limitations namely narrow bandwidth (typically 3–4%) and low polarization purity (typically 10–12 dB). These two limitations are confirmed from Fig. 1b, c.

Antenna 2: In Antenna No. 2, a pair of triangular-shaped defect (DGS with $N = 3$) has been introduced at the non-radiating sides of patch to effect the cross-polarization radiation of the antenna [17]. The impedance bandwidth achieved through the proposed equilateral triangular-shaped defected ground structure integrated RMA is 17% which is much better as compared to the conventional RMA (Fig. 2b). The polarization purity of 22 dB over a wide angle of ±130° [20] has been achieved through the optimum structure where each side of the equilateral triangular-shaped DGS is 10.4 mm (Fig. 2c).

Antenna 3: Four regular pentagon-shaped defects (DGS with N = 5) have been laid down just below the corners of the RMA in such a way that the center of pentagon coincides the corners of patch [18]. Two slits of 1 mm width are itched out below the non-radiating sides of the patch centrally and these slits connect the pentagon defecting. The structure fails to improve the impedance bandwidth as compared to conventional RMA but significantly improves the polarization purity up to 25 dB over a wide angular range of $\pm 180^0$ when the arm of the regular pentagon is 5 mm [21] (Fig. 3).

Antenna 4: Four hexagonal-shaped defects (DGS with $N = 6$) are itched out at the ground plane in such a way that center of the hexagons matches with the patch

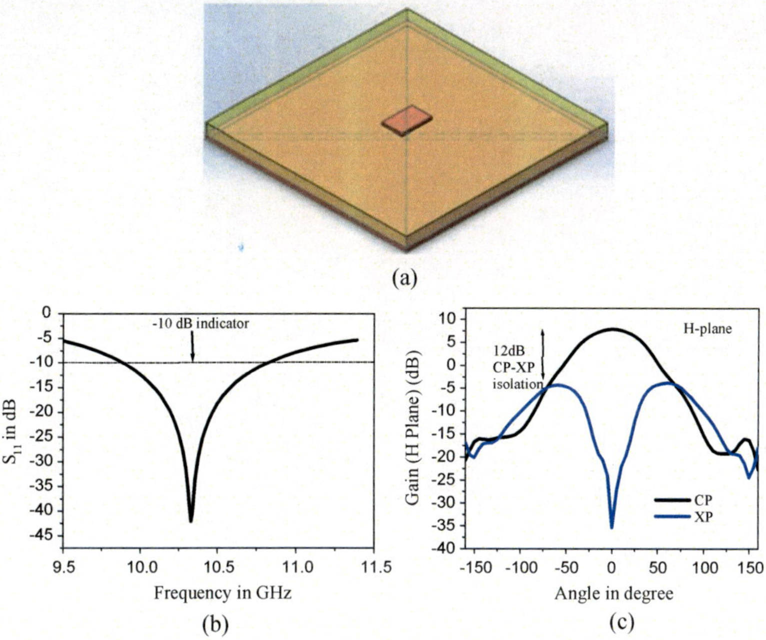

Fig. 1 Conventional RMA **a** 3D view, **b** reflection coefficient profile, **c** H-plane radiation pattern

corners (Fig. 4a) [19]. The fractional bandwidth of hexagonal-shaped DGS integrated RMA with one side of hexagon of 2.9 mm becomes around 58% (Fig. 4b) while the polarization purity varies between 13–22 dB within the wide operating bandwidth which is quite good for such an UWB antenna [22].

Antenna 5: The final structure that has been considered for comparative study is a ring-shaped DGS integrated RMA with external diameter of 6 mm and internal diameter of 2 mm (Fig. 5a). Maximum impedance bandwidth of 47% (Fig. 4b) along with comparable polarization purity has been achieved from the structure [20].

The performance of different antenna referred for the comparative study has been summarized in Table 2.

3 Conclusion

A comparative study between various existing DGSs which are particularly in increasing order of polygon has been methodically studied. The geometry of DGS is a critical factor for the input and radiation properties of the antenna. The bandwidth of an antenna enclosed in a given space can be enhanced only if the antenna exploits the space within it proficiently. In view if the above, it can be concluded that the hexagonal DGS integrated RMA can address all the issues of XP and board

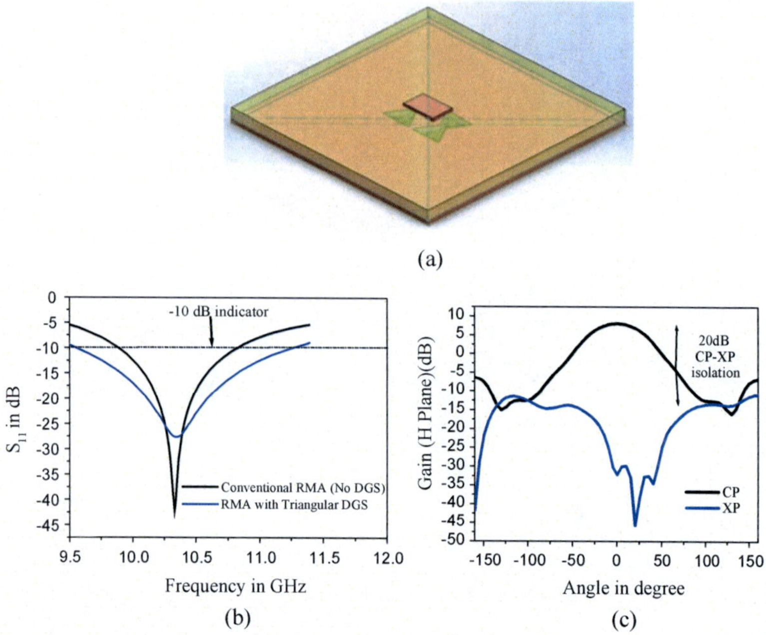

Fig. 2 Equilateral triangular DGS integrated RMA **a** 3D view, **b** reflection coefficient profile, **c** H-plane radiation pattern

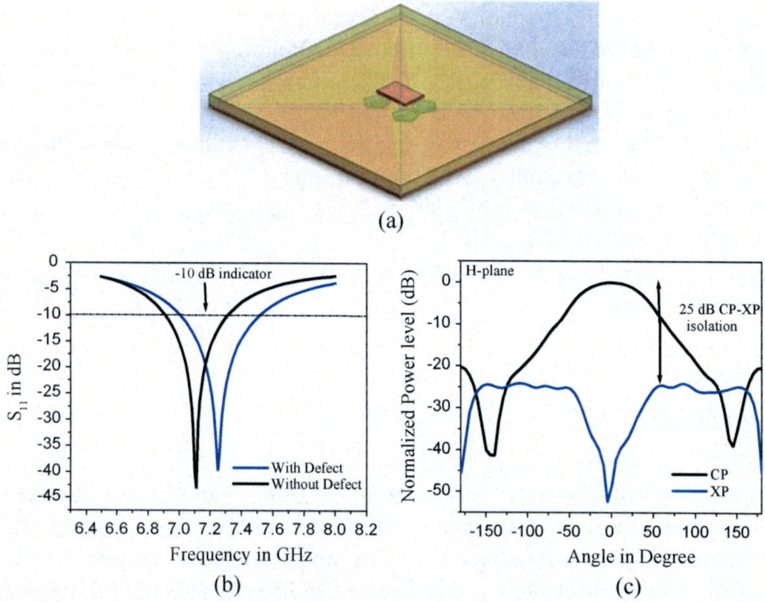

Fig. 3 Regular pentagon dumbelled-shaped DGS integrated RMA **a** 3D view, **b** reflection coefficient profile, **c** H-plane radiation pattern

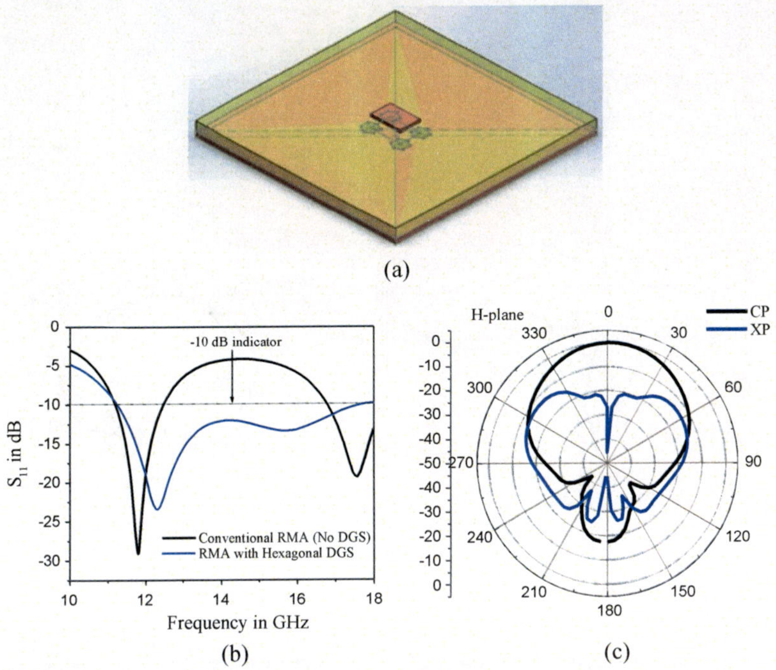

Fig. 4 Hexagon-shaped DGS integrated RMA **a** 3D view, **b** reflection coefficient profile, **c** H-plane radiation pattern

banding along with higher gain of around 9 dBi. Further to it, it offers UWB and these requirements are inescapable being fundamental in nature to the field of polarimetric radars.

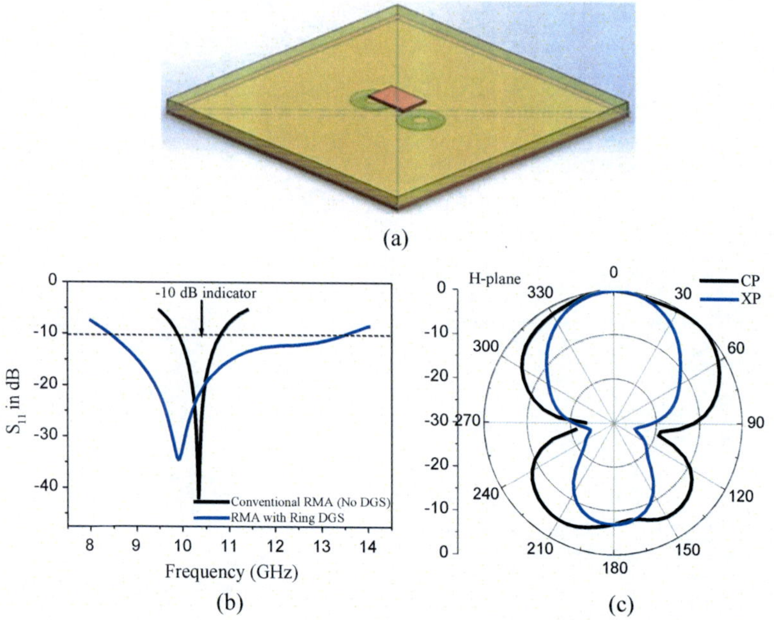

Fig. 5 Ring-shaped DGS integrated RMA **a** 3D view, **b** reflection coefficient profile, **c** H-plane radiation pattern

Table 2 Comparison between input and radiation properties of different antenna structure

Type of DGS	Defect dimension (mm)	Operating band	Resonating freq (GHz)	Variation of gain (dBi)	Impedance BW (%)	Polarization purity (dB)
Conventional RMA	nil	X band	10.33	3–4	3–4	12
RMA with triangular DGS	$S = 10.4$	X band	10.35	7–8.7	17	22
RMA with pentagon dumbelled DGS	$S = 5$	X band	5.6	6.6–7.4	15	28
Hexagonal DGS-integrated RMA	$S = 2.9$	Ku band	12.2	7.5–9.1	58	22.5
RDGS-integrated RMA (ring)	$S_o = 6$ $S_i = 2$	X band	9.86	4.5–5.7	47	22

References

1. Kumar G, Ray KP (2003) Broadband microstrip antennas. Artech House, Norwood, USA
2. Guha D, Antar YMM (2011) Microstrip and printed antennas-new trends, techniques and applications. John Wiley, U.K.
3. Garg R, Bhartia P, Bahl I, Ittipiboon A (2001) Microstrip Antenna design handbook. Artech House, Norwood, USA
4. Islam MT, Shakib MN, Misran N (2009) Design analysis of high gain wideband l-probe fed microstrip patch antenna. Prog Electromag Res 95:397–407
5. Yang F, Zhang X, Ye X, Samii YR (2001) Wide-band E-shaped patch antennas for wireless communications. IEEE Trans Ant And Prop 49(7):1094–1100
6. Sharma SK, Shafai L (2009) Performance of a novel Ψ-shape microstrip patch antenna with wide bandwidth. IEEE Ant Wireless Prop Lett 8:468–471
7. Jaggard DL (1990) On fractal electrodynamics. In: Kritikos HN, Jaggard DL (eds) Recent advances in electromagnetics theory. Springer-Verlag, New York, pp 183–224
8. Jaggard DL (1991) Fractal electrodynamics and modeling. In: Bertoni HL, Felsen LB (eds) Direction in electromagnetics wave modeling. Springer, Boston, pp 235–283
9. Jaggard DL (1995) Fractal electrodynamics: wave interactions with discretely self-similar structures. In: Baum C, Kritikos H (eds) Electromagnetic symmetry. Taylor & Francis, Washington, DC, pp 231–281
10. Werner DH (1955) An overview of fractal electrodynamics research. In: Proceedings of the 11th annual review of progress in applied computational electromagnetics (ACES), vol 2, pp 964–969
11. Jaggard DL (1997) Fractal electrodynamics: from super antennas to superlattices. In: Levy VJ, Lutton E, Tricot C (eds) Fractal in engineering. Springer, London, pp 204–221
12. Petosa A, Ittipiboon A, Gagnon N (1999) Suppression of unwanted probe radiation in wide band probe-fed microstrip patches. Electron Lett 35(5):355–357
13. Li P, Lai HW, Luk KM, Lau KL (2004) A wideband patch antenna with cross-polarization suppression. IEEE Antennas Wireless Propag Lett 3:211–214
14. Wong KL, Tang CL, Chiou JY (2002) Broad-band probe-fed patch antenna with a W-shaped ground plane. IEEE Trans Antennas Propag 50:827–831
15. Hsu WH, Wong KL (2002) Broad-band probe-fed patch antenna with a U-shaped ground plane for cross-polarization reduction. IEEE Trans Antennas Propag 50:352–355
16. Jackson DR, Alexopoulos NG (1991) Simple approximate formulas for input resistance, bandwidth, and efficiency of a resonant rectangular patch. IEEE Trans Antennas Propag 39(3):407–410
17. Ghosh A, Basu B (2019) Triangular slotted ground plane: a key to realize high gain, cross-polarization free microstrip antenna with improved bandwidth. Turk J Electr Eng Comput Sci 27:1559–1570
18. Nath AK, Singh LLK, Chattopadhyay S, Ghosh A (2020) Study of polarization purity of rectangular microstrip antenna integrated with pentagon dumbelled shaped defected ground structure. In: Dawn S, Balas VE, Esposito A, Gope S (eds) International conference on innovations in modern science and technology 2019, LAIS, vol 12. Springer, Cham, pp 484–492
19. Pawar UA, Chakraborty S, Singh LLK, Chattopadhyay S (2018) Application of defected ground structure for augmenting high-gain ultra-wide bandwidth from rectangular microstrip antenna. Electromagnetics 38:123–133
20. Pawar UA, Ghosh A, Singh LLK, Chattopadhyay S (2019) Application of defected ground structure for stable gain with ultrawide bandwidth. In: Bera R, Sarkar S, Singh O, Saikia H (eds) International conference in communication, devices and networking 2018, LNEE, vol 537,. Springer, Singapore, pp 141–149
21. Ghosh D, Ghosh SK, Nandi S, Chakraborty D, Anand R, Raj R, Ghosh A (2014) Physical and quantitative analysis of compact rectangular microstrip antenna with shorted non-radiating edges for reduced cross-polarized radiation using modified cavity model. IEEE Antennas Propag Mag 56(4):61–72

22. Ghosh A, Ghosh D, Chattopadhyay S, Singh LLK (2015) Rectangular microstrip antenna on slot type defected ground for reduced cross polarized radiation. IEEE Antennas Wireless Propag Lett 14:321–324
23. Ghosh A, Chakraborty S, Chattopadhyay S, Nandi A, Basu B (2016) Rectangular microstrip antenna with dumbbell shaped defected ground structure for improved cross polarized radiation in wide elevation angle and its theoretical analysis. IET Microw Antennas Propag 10(1):68–78
24. Chattopadhyay S, Chakraborty S (2018) A physical insight into the influence of dominant mode of rectangular microstrip antenna on its cross-polarization characteristics and its improvement with T-shaped microstrip antenna. IEEE Access 6:3594–3602
25. HFSS High frequency structure simulator, Ver. 14, Ansoft Corp., USA

INDUSTRY 4.0: A Comprehensive Review of Artificial Intelligence, Machine Learning, Big Data and IoT in Psychiatric Health Care

Anoushka Panwar, Neha Malhotra, and Dheeraj Malhotra

Abstract It has been quite well known that modern psychiatric treatments bring along certain side effects and current treatment models are unable to precisely address the complexity of mental illness issues. As a result, there has been a major focus to search and adopt applications of information and communication technology (ICT) as a mode for some additional psychological treatment and alternative diagnose with the help of various technologies. Therefore, the objective of this study is to analyze the technological aspects of using virtual reality, artificial intelligence, machine learning, IoT and big data analytics in the mental healthcare industry. In this review paper, we have accumulated some of the remarkable studies and done a comprehensive analysis of various potential technologies in the field of psychiatric health care and the need for these technologies for improving the quality and accuracy of diagnosis for the patients.

Keywords Industry 4.0 · Artificial intelligence · Machine learning · Big data analytics · IoT · Virtual reality · Mental healthcare

1 Introduction

Computers, robots and artificial intelligence have been in existence from the past few decades, but as a result of the industrial revolution and Internet, new opportunities are being introduced in terms of their use and integration. The field of computer science technology has been brimming up with innovation in research using Industry 4.0 and analyzing new opportunities since the stakes are high. All the major sectors of the society have shown interest in adopting it and now healthcare industry is in avid need for the adoption of Industry 4.0. Generation of fast pace lifestyle demands from healthcare industries to find accurate and precise facilities for their patients' various unmet needs. At times, healthcare industries have faced lawsuits and multiple

A. Panwar (✉) · N. Malhotra · D. Malhotra
Vivekananda Institute of Professional Studies, AU-Block (Outer Ring Road) Pitampura, New Delhi 110034, Delhi, India
e-mail: anoushkapanwar20@gmail.com

A. Abraham et al. (eds.), *Proceedings of 3rd International Conference on Computing Informatics and Networks*, Lecture Notes in Networks and Systems 167,
https://doi.org/10.1007/978-981-15-9712-1_42

backlashes either for not providing proper treatment to the patients or ill-treatment of the patient, altogether due to lack of proper measurement of the patients' condition [1]. Furthermore, mediocre modes of health treatments are somewhat excessive in nature, such as the use of antibiotics and anticholinergics, leading to difficulty in recovery from certain side effects post-treatment.

In order to transform health in the fourth industrial revolution, technologies like the Internet of things (IoT) uses intelligent sensors and gateways to provide an easy lifestyle in many areas. Examples include tracking and analyzing the mental health state of the patients dealing with a type of disorder [2]. A type of tremendous shift is also taking place in the healthcare industry. The keyword used for the change is "Healthcare 4.0." The Industry 4.0 revolution has also walked the steps in the field of healthcare, but the practical adoption is still more so in studies. Information and communications technology (ICT) plays a major role in health care to further generate and prove that there are other alternatives or additional modes which can potentially provide personalized care. Henceforth, technological predictions are taking place more and more in this direction. Technologies like machine learning, big data [3, 4] and IoT aid health care by utilizing the data that is already available in the form of images, clinical records and other electronic health records. For example, a machine learning algorithm known as alternating decision trees can accurately predict patterns in late-life depression and treatment response in patients [5].

However, significant gaps continue to occur in the evidence baseline underlying these technologies. In this context, the emphasis of this study is to give a detailed comparative analysis based on the shortcomings and use-case of various existing technologies in the domain of information and communications technology (ICT).

The rest of the paper is categorized in various sections as follows: Sect. 2 illustrates a detailed category wise literature review of various technologies used in mental health care. In Sect. 3, we have done a comparative analysis with the subject of the paper, finally followed by Sect. 4 which explains the conclusion and future work.

2 Literature Review

The magnitude of an increase in the rate of a fast-paced lifestyle has made the lives of an average human being much complicated if compared from inception to today. Keeping mental health care in focus, recently there has been a major fascination to promote more research and development using smart technologies for Industry 4.0 revolution in health care which has traditionally been slow in the adoption of this revolution. Also, another alternative solution would be finding a path in applying mindfulness-based cognitive technology (MBCT). Also, many organizations have been fascinated by the idea of bringing both of these areas of subjects (mindfulness using smart technologies for Industry 4.0) together into cutting-edge research for the Industry 4.0 revolution. This, in turn, has led to the evolution of several pieces of research and studies in this background.

Table 1 Research categories and sub-categories

Research category	Research sub-category
Review of virtual reality, augmented reality and video games for mental health care	VR and AR gaming applications their implications on young adults and future scope for advancement in mental health care
Review of AI/ML for mental health care	AI/ML-based applications and systems for different mental care problems their shortcomings and design recommendations for the smart health systems
Review of big data analytics and IoT for mental health care	Applications of big data analytics and Internet of things (IoT) system designs for the study, analysis, research and diagnosis of mental healthcare issues

However, for ease of understanding, we have sub-divided this assessment into various categories and their respective sub-categories in Table 1.

The parameters considered for this categorization in Table 1 are as follows:

- Technology adoption by mental healthcare providers.
- Use-case and acceptance for treatment using these technologies by patients.
- Ethical challenges and shortcomings including a lack of research and design of the system.

2.1 Category 1: A Review of Healthcare Studies Using Virtual Reality or Augmented Reality

As the result of hectic lifestyles, the average stress level in a human being has doubled up as compared from past few decades and the lifestyle has become more complicated on an exponential level leading to which healthcare industries are in avid need of finding new alternative methods. Due to which, new technological trends have gained more popularity in health care. Henceforth, we are discussing various studies based on AR/VR for the mental healthcare industry scenario.

2.1.1 Virtual Reality Therapy for Public Speaking Anxiety

Cognitive-behavioral therapy is the largely the common cure for public speaking anxiety disorder. Justas and Audrone et al. describe the architectural and technical decisions formed to design a mobile and cloud-based lightweight system that possibly gets adopted by any psychology clinical center [6]. Furthermore, data gathered from 30 participants are further analyzed taken from those gone through a VRET session for the treatment. The objective of this study is to design a standalone cloud-based VRET system which could be useful for the system as a service (SaaS) solution for people dealing with a public speaking anxiety disorder. The conclusion discussed is that VRET scenarios created a stimulus that is more powerful than that in front of a therapist. Furthermore, this can potentially be introduced as a more universally adaptable virtual reality technique for psychological departments. Scope of the study is further discussed:

- Extract and deliver bio-feedback signal features to the therapist table in real-time.
- Develop a self-treatment online module.
- The system is worn to conduct a complete random clinical test that would consist of the control group.

2.1.2 Virtual Reality Environment and Mindfulness-Based Stress Reduction

This study is based on mindfulness-based stress reduction (MSBR) related to the serious problem known as chronic pain. Tong et al. carried out this conceptual study for the virtual environment system known as "virtual meditative walk" (VMW) to foster patients dealing with chronic pain and help them to direct their attention with ease for mindfulness [7]. The conclusion was a successful discussion on how to design VMW to support and induce learning MBSR techniques. Costa et al. [8] suggested for enhanced immersive environment design to support MSBR in addition to more advancement such as associative nature between three concepts being a presence, meditation depth and perceived restoration about their causal relationship and future scope.

2.2 *Category 2: A Review of Healthcare Studies Using AI/ML*

Although AI/ML is not a new concept these days. The adoption of these technologies is still limited to some major extent. For example, prescribing the right antidepressant medication for acute or chronic depression to the right patient is not an easy job. Hence, advancements in this region have already been formed by taking some explicitly sensible objectives. Machine learning targets health care by utilizing the images, clinical records and other electronic health records (EHR). This has proven out to be revolutionary and helpful for patients' evaluation and treatment. This section discusses some studies targeted specifically for mental health care using artificial intelligence and machine learning.

2.2.1 Smart Mental Healthcare Technological Services

A study conducted by Lee et al. [9] for the prime goal to investigate already applicable smart mental healthcare technologies to deduce an artificial intelligence system model which is to be called as a comprehensive mental healthcare stepped-care model

(CMHCSM) [8]. This particular study encourages using various SmartMentalTech as a mode of self-treatment for users. In a conclusion, it was found out that in South Korea and Japan there are a different set of users who uses a specific type of Smart-MentalTech Furthermore, the study encourages future researches to be conducted in the evidence-based analytical study for the same area.

2.2.2 Mobiles, Wearables and Technologies for Mental Health Care

This comprehensive paper by Luxton and June [10] presented study on artificial intelligence-based applications used by smartphones, wearables or gadgets for mental health care [10]. mHeatlh devices and software applications benefits are discussed followed by AI methods descriptions of mHealth. This study provides design recommendations and ethical considerations for implementations of these smart technologies in mental health care.

2.2.3 Machine Learning in Depression Prediction

This recommended [11] study aims to calculate correct prediction patterns for late-life depression (LLD) diagnosis so that medical care is done more precisely rather than conventional trial and error based on behavioral signs and the symptoms. The machine learning approach known as alternating decision trees predicts a highest accuracy for LLD (87.27%) and treatment response (89.47%). As a result, a combination of multi-modal imaging and/or no-imaging helps in effective prediction for the LLD diagnosis [11].

2.3 Category 3: A Review of Healthcare Studies Using Big Data Analytics and IoT

2.3.1 IoT System to Track Mental Stability in Patients

A paper proposed by Hayati and Suryanegara [2] introduced an IoT-based system design called LoRa system which is short for the long-range radio system can be applied for tracking and monitoring mental stability of the patient. The system is a combination of a tracker device and gateway to be installed on the patient and hospitals and other locations, respectively. A LoRa end-device that has the Dragino LoRa shield wireless, Arduino Uno board, GPS sensor and Wi-Fi module that is deployed in a microcontroller are the constituents of the system. To conclude, parameters considered are feasible based on system efficiency, battery life and adaptability. The future scope is to engineer this design prototype for a real-time system [12].

2.3.2 IoT System for Post-Traumatic Stress Disorder (PTSD)

McWhorter and Brown et al. [12] proposed a comprehensive paper for a wearable monitoring system that is used for tracking sleep patterns, nightmares signs and in the attempt to control them or awaken the patient peacefully [13]. This may help out a patient's lifestyle dealing with post-traumatic stress disorder (PTSD). The conclusion describes that this system is feasible in the long run in an attempt to benefit people suffering from PTSD and reduce their pain, anxiety, depression and hopefully the number of suicide rates.

2.3.3 IoT-Based Affective State Mining

Alam and Abedin et al. [13] proposed an Internet of medical things-based system for emotion detection. Psychophysiological outcomes are acquired via electromyography (EMG), electro-dermal activity (EDA), an electrocardiogram (ECG) medical sensor and calculated with the help of the convolutional neural network (CNN) [14]. This study concludes that this type of state mining method guarantees better results as compared to the previous method.

2.3.4 Big Data Analytics in Predicting Mental Illness

Social media is a goldmine for big data analytics and it can be extremely beneficial for future research conducted as a basis for any type of mental disorder. A study by Thorstad and Wolff [14] investigates the signs present in people's everyday language containing significant phrases or indications to smartly analyze and predict the mental illness for present or future scenario. Social media Web site Reddit is a source for data collection [15]. The conclusion from this study was exactly resulting to deduce various types of psychiatric disorders (ADHD, anxiety, depression, etc.).

3 Comparative Analysis

The following comparative analysis is a comprehensive attempt to describe various studies proposed or applied in the healthcare sector for Industry 4.0. Selected technologies in the comparative analysis such as VR, ML and IoT are in trend in the year 2020 and will surely be used for future developments for the healthcare Industry 4.0.

It has been found from the comparative that there is still a lack of research in these fields but the future scope is definitely possible. Table 2 is a comparative analysis based on various technologies to promote more research in mental health care. These ICT-based wearables, gadgets and sensor systems would aid in more person-centered or personalized experience in the diagnosis of various issues regarding mental health.

The parameters considered for this categorization in Table 2 are as follows:

Table 2 Comparative analysis and shortcomings

References	Type	Primary findings	Shortcomings and real-time use
Šalkevičius and Miškinytė [6]	VR	During the experimental analysis of this study, the system used by patients for VR solution was VR GEAR and by psychologists was WebGL. The objective of this is to create a VR system specifically for the SaaS solution. After the VR session, results have shown that the VRET session is better than a therapist for public speaking anxiety disorder (PTSD)	Cost-ineffectiveness of the software VRET on cloud comes with benefits of user-friendliness and cover a larger diameter in terms of reach as compared to standard VR or just a therapist It can be used as an additional tool for the treatment
Patel and Anderson et al. [11]	ML	An ML-based method, alternating decision trees estimates the highest prediction. Wherein, depression patients ($n = 33$) and non-diagnosed patients ($n = 35$) compared based on cognitive levels. Conclusion from the diagnosis is that the prediction would be more effective if multi-modal imaging is used (with or without a non-imaging)	One of the major shortcomings came in the form of a small sub-sample size for the treatment response prediction The outcomes may help practitioners to understand LLD in a better way rather than going for a conventional trial and error method for the treatment
Alam and Abedin et al. [13]	IoT	During the experimental study, a benchmark dataset (DEAP) calculates the efficiency of the IASM framework by using a neural network (CNN) and a testbed. The Lua language along with Torch 7.0 deep learning framework is applied for developing CNN using a testbed, Lua-JIT is installed after the Torch 7.0. The conclusion comes out to be that, it provides higher accuracy of 87.5% in contrast to the physiology-based recognition methods	Despite having a specific hardware requirement, the system is not 100% accurate to predict State mining can help in proving additional help for early health practitioners who have maybe less experience in psychiatry or those who may need it

(continued)

Table 2 (continued)

References	Type	Primary findings	Shortcomings and real-time use
McWhorter and Brown et al. [12]	IoT	Smart wearable technology like Fitbit charge HR is considered along with the integration of ML-based proposed system which can be deployed in a home automation device carrying HUB like INSTEON that comprises of intelligent outlets, oil diffusers, audio systems, etc. Furthermore, the conclusion of this study provides ongoing comfort and treatment to the users dealing with PTSD	Despite being cheap and easy to deploy, this proposed system can arise some issues like the security of user data and privacy breach due to Bluetooth service as a mode of data transfer The system can be used by healthcare providers as alternative/additional modes for treatment Cheap and easy to use possibly help in preventing any greater concern to the patients

- Type of technologies such as VR, ML and IoT in psychiatric health care.
- Primary findings based on the selected study.
- Shortcomings and uses based on the selected study.

This is how Industry 4.0 is being promoted for psychiatric health care. We found out that IoT could be an efficient approach in the future developments for psychiatric health care. However, there is a huge scope for future advancements to be made and in more common adoption of these technological services by the mental health care with respect to Industry 4.0.

4 Conclusion and Future Work

This study contributes to emphasizing the most dominating challenges faced by psychiatric healthcare industries about how the technological revolution Industry 4.0 can resolve all their concerns. By performing a comparative analysis, we can say that there is an avid demand for health care or psychiatric healthcare industry to adopt these technologies in action. To conclude our study, we have described the use of Industry 4.0 based on its significance to end-users.

1. *Patients*: AI, AR and VR are able to provide comfortable, affordable and convenience rather than going for regular visits to a professional as it would be easily accessible from a smartphone or computer.
2. *Health practitioners and Therapists*: Alternative treatments in the form of virtual reality, faster approach and better measurement and monitoring of the patients' condition through machine learning, IoT and big data analytics.

3. *Research and Development*: Big data analytics and IoT being a goldmine of research and development [16], it would be beneficial for new generation of researchers to further dive into new developments for future. Furthermore, it would open up vast opportunities for the next generation of mental health.
4. *Academics*: Psychology institutional departments and universities will be able to improve the quality and value of academic pool of knowledge and education by introducing new technologies such as big data analytics, IoT and VR based systems to carefully scale up and broaden the horizons to more practical-based knowledge and understanding of the concept.

Finally, we can say that there is a vast potential for research and development in the recently explored area like big data and IoT and the scope of Industry 4.0 revolution for psychiatric health care in high. For a future work of this study, we hope to propose a study aimed at integrating healthcare 4.0 like big data analytics, machine learning and data science for Internet of medical things (IoMT) in analyzing the sleeping patterns of students to analyze any mental health issue caused by lack of sleep and proper rest such as hallucinations, depression, stress, anxiety and post-COVID-19 pandemic.

References

1. Shirer WR, Ryali S, Rykhlevskaia E, Menon V, Greicius MD (2012) Decoding subject-driven cognitive states with whole-brain connectivity patterns. Cerebral Cortex 22(1):158–165
2. Hayati N, Suryanegara M (2017) The IoT LoRa system design for tracking and monitoring patient with mental disorder. In: 2017 IEEE international conference on communication, networks and satellite (COMNETSAT). IEEE, pp 135–139
3. Verma N, Singh J (2017) An intelligent approach to big data analytics for sustainable retail environment using Apriori-MapReduce framework. J Ind Manage Data Syst 117(7):1503–1520
4. Malhotra D, Rishi OP (2018) An intelligent approach to design of E-commerce metasearch and ranking system using next-generation big data analytics. J King Saud Univ Comput Inf Sci 45:42–51
5. Patel MJ (2015) Machine learning approaches for integrating clinical and imaging features in late-life depression classification and response prediction. Int J Geriatr Psychiatry 30(10):1056–1067
6. Šalkevičius J, Miškinytė A (2019) Cloud based virtual reality exposure therapy service for public speaking anxiety. Information 10(2):62
7. Gromala D, Tong X, Choo A, Karamnejad M, Shaw CD (2015) The virtual meditative walk: virtual reality therapy for chronic pain management. In: Proceedings of the 33rd annual ACM conference on human factors in computing .systems, pp 521–524
8. Costa MR, Bergen-Cico D, Grant T, Herrero R, Navarro J, Razza R, Wang Q (2019) Nature inspired scenes for guided mindfulness training: presence, perceived restorativeness and meditation depth. In: International conference on human-computer interaction. Springer, Cham, pp 517–532
9. Lee SKA (2018) Classification of SmartMentalTech services and application for comprehensive mental healthcare stepped-care model (CMHSCM): health psychological approach. Proc Comput Sci 141:302–310
10. Luxton, June (2016) Intelligent mobile, wearable, and ambient technologies for behavioral health care. Artificial intelligence in behavioral and mental health care. Academic Press 137–162

11. Patel MJ, Andreescu C, Price JC, Edelman KL, Reynolds III CF, Aizenstein HJ (2015) Machine learning approaches for integrating clinical and imaging features in late-life depression classification and response prediction. Int J Geriatr Psychiatry 30(10):1056–1067
12. McWhorter J, Brown L, Khansa L (2017) A wearable health monitoring system for posttraumatic stress disorder, In: Biologically inspired cognitive architectures 22:44–50
13. Alam MGR, Abedin SF, Moon SI, Talukder A, Hong CS (2019) Healthcare IoT-based affective state mining using a deep convolutional neural network. IEEE Access 7:75189–75202
14. Thorstad R, Wolff P (2019) Predicting future mental illness from social media: A big-data approach. Behavior Res Methods 51(4):1586–1600
15. Malhotra D, Rishi OP (2017) IMSS: a novel approach to design of adaptive search system using second generation big data analytics. In: Proceedings of international conference on communication and networks. Springer, Singapore, pp 189–196
16. Verma N, Malhotra D, Malhotra M, Singh J (2015) Online libraries website recommendation using semantic web mining and neural computing. Proc Comput Sci 45:42–51

Air Quality Measurement Using Low-Cost Sensors—A Review

Shreevidya Gurudath and K. G. Srinivasa

Abstract Rapid urbanization and a consumption centric economy have created an enormous pressure on the environment. Air, water and soil pollution are a global problem. The affects of pollution are more apparent in developing countries such as China and India, because of various economic and demographic factors. In major cities, such as New Delhi, Beijing, the air pollution reaches hazardous levels, especially during winters. Air quality measurement is the first step toward mitigating the effects of air pollution; hence, there has been an effort to set up air quality measurement stations all over the world. However, the availability of these measurement stations is sparser in developing countries, where the air quality is lower. Hence, there is a need for low-cost air quality measurement devices. The following work presents a brief overview of various low-cost approaches to measuring air quality.

Keywords Air pollution · Low-cost sensors · Air quality index (AQI) · Sensor drift · Sensor calibration

1 Introduction

Air pollution can be defined as the presence of extraneous particles or toxic gases, including those from a biological origin, at concentrations which are deemed to be a health risk to the ecosystem and its inhabitants. Air pollution directly affects the health and quality of life of humans and other organisms sharing the ecosystem. Air pollution asymmetrically affects the poor and vulnerable sections of the population. For example, Chen et al. [1] established a relationship between socio-economic factors and exposure to $PM_{2.5}$ in China using a geographically weighted regression (GWR) modeling.

S. Gurudath (✉)
Ramaiah Institiute of Technology, Bangalore, India
e-mail: shreevidyag@gmail.com

K. G. Srinivasa
NITTTR, Chandigarh, India

A. Abraham et al. (eds.), *Proceedings of 3rd International Conference on Computing Informatics and Networks*, Lecture Notes in Networks and Systems 167,
https://doi.org/10.1007/978-981-15-9712-1_43

1.1 Pollutants

Any substance in the ambient atmosphere which causes ill-effects to humans and environment can be defined as an air pollutant. Pollutants can be in the form of any state of matter, i.e., solid particles, liquid droplets or gaseous. In addition, they may be natural or anthropogenic, i.e., a product of human activity. Different government agencies regulate the levels at which these pollutants are permissible in the ambient air. For example, in the USA, the environmental protection agency (EPA) has set National Ambient Air Quality Standards (NAAQS) which define the permissible limits for each six of the most common pollutants. EPA [2] shows the defined short- and long-term limits for each of the pollutants.

Particulate matter or aerosols, both PM10 and PM25, carbon monoxide, nitrogen dioxide, ground-level ozone and sulfur dioxide are the most common pollutants that are periodically monitored by government agencies and reported in a periodic fashion.

- Traces of *carbon monoxide* occur in the atmosphere naturally and help in the formation of ozone. However, at concentrations of 35 ppm or more is poisonous to humans and animals. The natural concentration of carbon monoxide in air is around 0.2 ppm, and the NAAQS for CO is 35 ppm for an hourly average
- *Nitrogen dioxide* is responsible for photochemical smog and acts as a catalyst for the formation of acid rain.
- *Particulate matter (PM)* also known as aerosols are tiny particles of solid or liquid suspended in air. The NAAQS has multiple levels based on the size of the particulate matter. $PM_{2.5}$ has a primary level at 12.0 $\mu g/m^3$ annual mean and a secondary level at 15.0 $\mu g/m^3$ annual mean, averaged over 3 years. Additionally, it also sets a 24 h average level at 35.0 $\mu g/m^3$. PM_{10} has a limit of 150 $\mu g/m^3$ averaged over 24 h.
- *Ground-level ozone* is a very powerful oxidizing agent, and high concentrations affect our environment. It is also a constituent of smog. Ozone is released into the atmosphere at higher concentrations mainly from the combustion of fossil fuels.
- *Sulfur dioxide* when oxidized forms H_2SO_4 and thus causes acid rains. There are two NAAQS for SO_2. A primary standard of 75 parts per billion (ppb) averaged over one hour and a secondary standard of 0.50 ppm over 3 h.
- *CO_2* is emitted from combustion of carbon-based materials. Earth's atmosphere contains approximately 0.03% of CO_2. It is also a greenhouse gas, hence, contributes toward global temperature rise and climate change.

Additionally, photochemical smog, *ammonia* (NH_3) and volatile organic compounds such as methane and other aromatic compounds such as benzene, toluene and xylene contribute to pollution.

2 Measurement of Particulate Matter (PM_{10} and $PM_{2.5}$)

Suspended particulate matter ($PM_{2.5}$ and PM_{10}) is measured using optical diffraction-based sensors. There are mainly three different types of sensors which are used to measure particulate matter in air. First, there are beta attenuation mass monitoring sensors. These are very large and prohibitively expensive systems which limit their usage scenarios. Secondly, there are sensors which use laser diffraction to measure particulate matter. This is the preferred method in a majority of scientific settings.

Lastly, there are infrared diffraction-based sensors. An infrared emitting diode and a photo-transistor are diagonally arranged in a mixing chamber. The diffraction pattern is then used to determine the concentration of particulate matter in the sample. This class of sensors is low cost and the article mostly concentrates on different sensors belonging this class. Figure 1 shows a schematic representation of the working of IR diffraction sensors.

There are three different sensors widely available and studied. Manikonda et al. [3] and Wang et al. [4] present a laboratory comparison and benchmarking of the below three sensors.

- SHARP GP2Y1010AU0F, SHARP [5]
- SHINYEI PPD42NS, Austin et al. [6]
- Samyoung DSM501A, Samyoung [7].

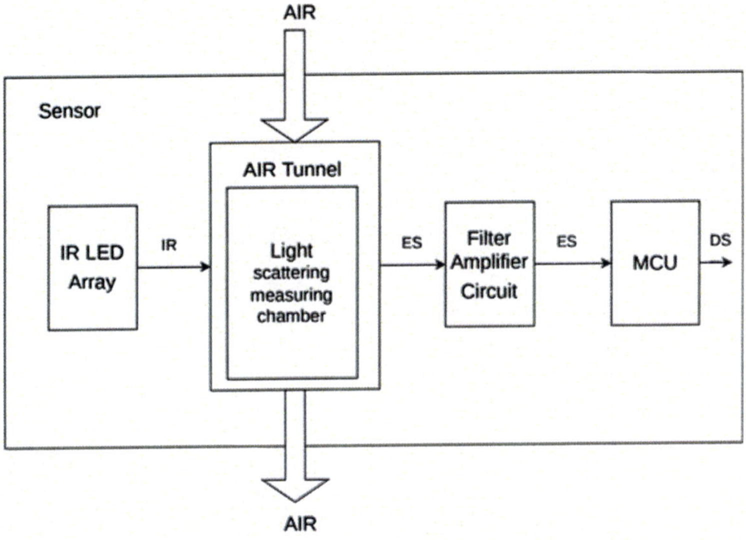

Fig. 1 Schematic representation of IR diffraction sensors for measuring particulate matter in ambient air

The PPD and DSM sensors are quite similar to each other. Both use thermistors to generate heat in order to facilitate a natural convection of air, and hence particles through its measurement chambers. The GP2Y sensor on the other hand has a hole through the center of its body through which the air passes through facilitating detection. Additionally, the PPD and the DSM sensors have a digital out, while the GP2Y outputs analog readings.

Connecting an array of wireless sensors to a server is a simple and low-cost solution to an urban sensor networks. The data can be transmitted via an Internet connection (Wi-Fi for indoor and a mobile data connection for outdoor). Utilizing low-cost air quality sensors and ordinary smartphones, end-users can be more involved and be able to extract, analyze and share local air pollution information. Holstius et al. [8] demonstrated that a low-cost aerosol sensor like the PPD42NS could be used to measure urban $PM_{2.5}$ at a reasonable resolution.

Liu et al. [9] assessed the performance of low-cost SDS011 sensor for a nearly four-month period and concluded that the sensors demonstrate quite high linearity against officially measured concentrations. They also observed that a humidity of 80% or higher, negatively impacted the accuracy of the results. Maricq [10] present a comprehensive comparison of different low-cost sensors against reference readings. They found that GP2Y1010AU0F from SHARP had the most linearity compared to other sensors. Castell et al. [11] evaluated the performance of commercial low-cost sensors (AQMesh v3.5) measuring four gaseous pollutants (NO, NO_2, O_3, CO) and particulate matter (PM_{10} and $PM_{2.5}$).

3 Measurement of Other Gaseous Pollutants

There are a variety of sensors available to detect gases in the ambient air. They differ in their sensing technique, precision and cross-sensitivities. However, low-cost gas sensors belong to three main categories: metal oxide-based sensors (MOS), non-dispersive infrared (NDIR) and electro-chemical sensors.

3.1 Metal Oxide Sensors

Absorption of gases by metal oxides changes its resistance to current. MOS sensors determine the concentration of gases by measuring this change in resistance. Atmospheric oxygen on the surface of the metal oxide reacts with the gases, and the metal oxide is reduced. This increases the flow of electrons in the conduction band and this causes a reversible drop in the resistance. Most available sensors also use a small heater inside with an SnO_2 plate. They are sensitive for a range of gases and are used indoors at room temperature. Figure 2 shows the internal circuit of a metal oxide sensor.

Fig. 2 Circuit diagram of metal oxide-based sensor

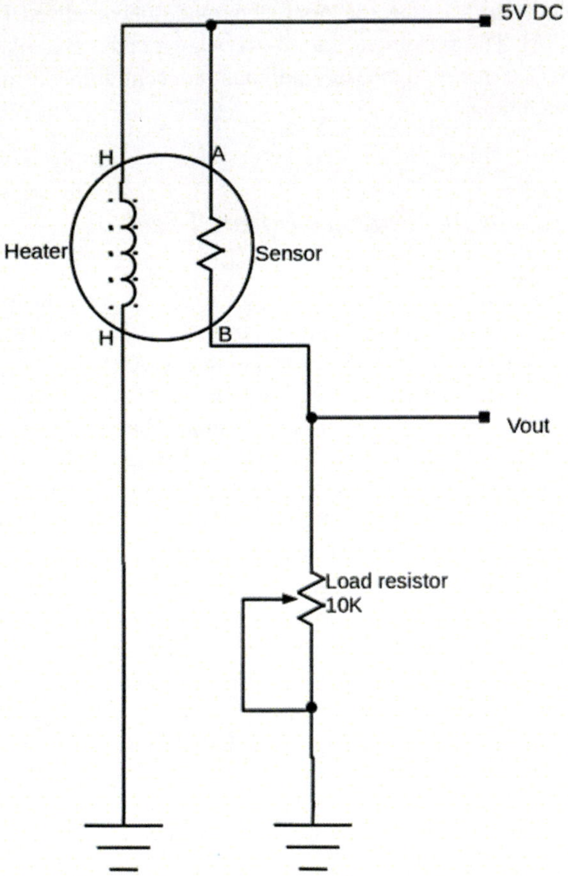

For measuring air quality, an MQ-135 sensor is used which is sensitive to a wide range of gases, including NH_3, NO_x, alcohol, benzene, sulfides, smoke and CO_2. It is low cost and particularly suitable for air quality monitoring applications. The sensor consists of a perspective layer of strontium dioxide (SnO_2) inside aluminum oxide microtubes and a heating element inside a tubular casing. The aluminum oxide tubes function as the measuring electrodes. The end face of the sensor is enclosed by a stainless steel net the connection terminals are on the rear end.

Piedrahita et al. [12] developed air quality monitors (M-Pods) using low-cost, wearable sensors on Arduino platform. A quantification method is developed to provide personal exposure accounting for uncertainties in measurements for CO_2, O_3, NO_2 and CO. M-Pods use metal oxide semiconductor (MO_x) sensors to measure CO,O_3, NO_2 and total VOCs, and non-dispersive infrared gas detectors are used to measure CO_2. An M-Pod collects, analyzes and shares air quality data using a mobile air quality sensing system (MAQS). An MAQS is a personalized mobile sensing system for indoor air quality (IAQ) monitoring. MAQS users carry portable, indoor location tracking sensors that provide personalized IAQ information, Jiang et al. [13]. The system uses an n-gram augmented Bayesian room localization method, CO_2 sensors to measure pollutant concentrations using an air-exchange rate-based technique.

3.2 *Non-dispersive Infrared Sensors*

A non-dispersive infrared sensor works on the principle that different gases absorb different wavelengths of IR radiation. The sensor mainly consists of an Infrared light source, a measurement chamber which holds the air sample, a wavelength filter and an infrared sensor. Air is pumped into the measurement chamber, and infrared source illuminates the gas in the sample chamber. The light passes through the sample chamber then meets the IR sensor. In order to measure the concentration of only the gas of interest, an upstream band pass filter is connected to filter out only specific wavelengths. Figure 3 shows a schematic representation of an NDIR sensor.

The assumption that light of wavelength λ is only absorbed by the gas of interest is rarely true. Absorption spectrums usually overlap, hence, causing cross-sensitivities. This can either be compensated or be avoided by a carefully selecting frequency bands. NDIR sensors have been successfully shown to be able sense close to a 100 different gases. NDIR sensors are non-contact and free of consumption, which makes it the default method in many application areas. Among the common pollutants, CO_2 is usually detected using NDIR-based sensors.

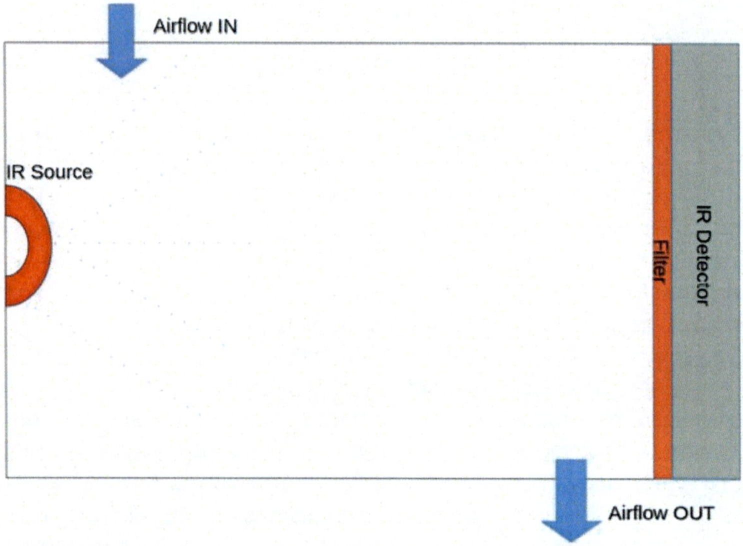

Fig. 3 Schematic representation of an NDIR sensor

Dinh et al. [14] presented a comprehensive review of NDIR-based sensors and also describe the different drawbacks of these sensors. Delebarre et al. [15] developed a system to measure CO_2 concentration for indoor and outdoor environments. They also observed that even though the sensors can accurately estimate concentrations, the sensors were prone to air flow variations. Though NDIR sensors are most widely used to detect CO_2 concentrations, they can also be used to detect other gases. Xing et al. [16] used an NDIR sensor for detection of acetone and ammonia.

3.3 Electro-Chemical Sensors

An electro-chemical sensor is primarily a cell, whose output voltage is directly proportional to the concentration of the entity that is being measured. The concentration of a target gas is measured by the electrolytic reactions which generate a current flow through the sensing circuit. An electro-chemical sensor consists of mainly two electrodes, a "working" electrode and a "counter" electrode. Some sensors have an additional "reference" electrode. The electrodes are suspended in an electrolyte solution. Ambient air interacts with the electrolyte solution through a membrane and a diffusion valve. Figure 4 shows a typical electro-chemical sensor.

Diffusion through the membrane allows the gas being measured to react with the electrolyte solution inside the sensor. There are two kinds of electro-chemical reactions that occur, first, an oxidation reaction, which causes a flow of current from the anode to the cathode, and second, a reduction reaction which causes a flow of

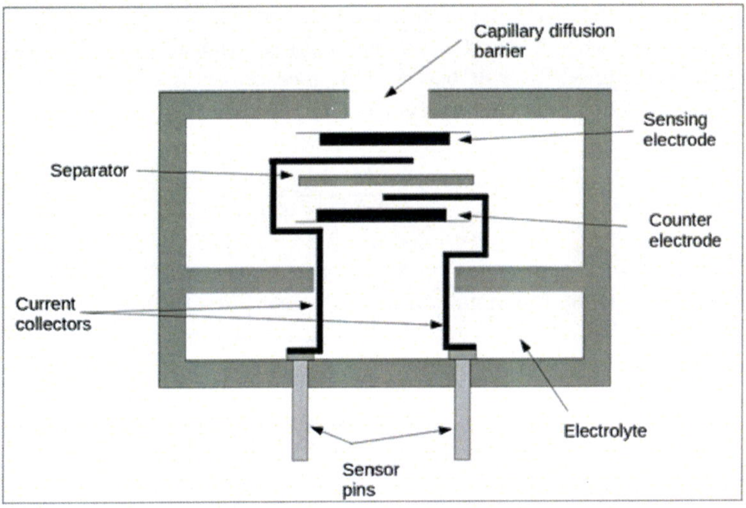

Fig. 4 Typical electro-chemical sensor

current from the cathode to the anode. Measuring the magnitude of this current flow after amplification, filtration gives the concentration of the gas in the air sample.

There have been multiple studies exemplifying the usage of low-cost electro-chemical sensors. For example, Mead et al. [17], Li et al. [18] and Kumar and Hancke [19] all use low-cost electro-chemical sensors to measure pollutant concentration. Afshar-Mohajer et al. [20] presented a comprehensive evaluation of low-cost electro-chemical sensors. The study used sensors for detecting carbon monoxide, nitrogen dioxide and ozone. All the sensors were from Alphasense Ltd. The work also bench-marked CO sensors against oxidative gas sensors. The authors were able to demonstrate the linearity of response w.r.t reference sensors for ambient concentrations of ozone, NO_2 and carbon monoxide.

3.4 Sensor Calibration and Drift

Low-cost sensors such as MQ series of gas sensors and GP2Y1010AU0F particle sensors are calibrated in comparison with a reference sensor. This is usually done either in a laboratory setting or by co-location in the field. Field calibration of sensors by co-location has been shown to be more effective than laboratory calibration, as it can adapt to real-world conditions.

Calibration is performed by comparing the following properties against that of the reference sensor.

- *Linearity of response* (LoR) is assessed using a regression model of the sensor measurements against the reference measurements.
- *Recision of measurements*—Given a pollutant concentration, sensor precision indicates the accuracy of the measurements and also its consistency. This is usually represented using RMS error between the measurements of the sensor under calibration (S) and a reference sensor (R).

$$\mathrm{RMSE} = \frac{2\sqrt{\mathrm{Mean}(S-R)^2}}{\mathrm{Mean}(S+R)}$$

- *Limit of detection* is defined as the lowest concentration that the sensor is capable of detection. It is the minimum concentration at which the sensor signal starts deviating from background static from blank measurements. It can be defined using the Kaiser and Specker method as

$$\mathrm{LOD} = \frac{3\sigma_{blk}}{k}$$

- where σ_{blk} is the standard deviation of static signal, which is obtained by filling the chamber with clean air. k is the slope of the LoR curve.

- *Dependence on composition*—Since light scattering is dependent on the refractive indices of particles, the performance of the sensor depends significantly on particle composition. These differences can be modeled by comparing the measurements against a reference sensor.
- *Sensitivity to particle size*—Similar to composition, sizes of the particles also affect measurements.
- *Relative humidity and temperature influence*—Temperature and relative humidity reduce the accuracy of the sensors. It also affects the distribution of the particles. The effect of relative humidity and temperature is described in little more detail in Sect. 3.5.

Additionally, for the gas sensors, *interference equivalents* need to be calibrated for. This is because equivalent species of gases can cause similar responses. These responses are modeled against the reference sensor. Low-cost sensors are usually unstable and over time prone to saturation and drift. They need to be periodically calibrated with respect to the reference to be able to get reliable measurements.

Spinelle et al. [21] compared performances of different field calibration techniques. The study benchmarked the performance of linear regression, multi-linear regression and supervised learning techniques. Zimmerman et al. [22] proposed a real-time affordable multi-pollutant monitor (RAMP) to resolve spatial heterogeneity in air pollutant concentrations. The study constructed calibration models CO, NO_2, CO_2 and O_3 sensors. The RAMP technique used uni-variate linear regression, multiple linear regression, random forest and other machine learning models to evaluate the calibration. Borrego et al. [23] performed field calibration by using a co-located reference instrument. This technique was used to reduce the considerable inter-nodal variation of O_3 concentrations. The study also suggested a periodical calibration based on the nearby air quality monitoring stations as a suitable method to reduce the deviation in measurements.

A real-time data scanning routine to automate the detection of variation in air quality monitoring data sets was proposed by Alavi-Shoshtari et al. [24]. The technique used a computationally intelligent calibration layer to resolve the problem of heterogeneity of hardware and calibration procedures. Spinelle et al. [25] studied different calibration techniques like orthogonal regression, target diagrams, uncertainty modeling and drift predictions using a cluster of MOX and electro-chemical sensors measuring NO and CO concentrations. They also used miniaturized NDIR sensors for measuring CO_2.

3.5 Effect of Temperature and Relative Humidity

3.5.1 Dust Sensors

Temperature and relative humidity (RH) can affect the concentration and distribution of particulate matter in various ways. Additionally, they have an effect on the sensor output signals too. As RH rises, particles absorb the moisture and increases in size. There is also formation of fog which interferes with the measurements. Sensors without drying facilities at the sample inlets are affected the most because of increase in RH. Jayaratne et al. [26] studied the effect of temperature and relative humidity on the low-cost dust sensors and concluded that the GP2Y and the PPD sensors exhibited fluctuations in the output signal, instead of a linear increase, when the relative humidity reaches 50% or higher.

In the case of sensors which rely on light scattering to measure the mass/concentration of particles, increase in RH causes an over estimation of the particle mass concentrations. This is caused because of absorption of the IR radiation by a film of water. Wang et al. [4] studied this effect in order to determine the influence of humidity and atmospheric fog on the measurements of low-cost particle mass sensors. Hojaiji et al. [27] in their study noted that increase in RH led to inaccurate measurements of dust concentration, however, decrease in RH did not affect the sensor output in a significant manner. The authors theorized that this effect is likely because of water particles interfering with the light trap of the sensors.

3.5.2 Gas Sensors

In metal oxide gas sensors, water absorption at higher levels of RH lowers the sensitivity significantly. Wang et al. [28] also noted that prolonged exposure to humid environments causes permanent deterioration of the sensor sensitivity.

Sohn et al. [29] proposed a model for characterizing sensor response with respect to temperature and relative humidity. The authors define *baseline resistance* R_b as the resistance of the sensor in the absence of any pollutant gases at 20% RH. The sensor response at any RH is a ratio of the measured resistance against baseline.

$$r = \frac{R_\mathrm{h}}{R_\mathrm{b}}$$

where $R_\mathrm{h}(\Omega)$ is the current resistance of the sensor and $R_\mathrm{b}(\Omega)$ is the baseline resistance.

In the case of metal oxide sensors, change in ambient temperature changes the resistance of the metal oxide plate, hence, influencing the concentration measurements. The reason for this is the operating temperature of the sensor is changed. Modeling this change in sensor response to change in temperature is needed to account for temperature variations. However, this is a non-trivial task. Therefore, Sohn et al. [29] described an empirical model to predict resistance variation in a given temperature range.

$$R_\mathrm{T} = R_\mathrm{a} \exp^{(bT)}$$

where $R_T(\Omega)$ is the resistance in dry clean air, T is the ambient temperature and R_a and b are constants.

4 Discussion

Low-cost sensor arrays are a staple for urban air quality monitoring. They are also gaining ground in personal exposure measurements and in building a dense air quality map. Understanding the sensor's capabilities and their drawbacks are essential to effectively utilize these devices.

Numerous studies have been performed on these low-cost sensors evaluating their linearity of response against a reference device, and multiple times have been proved to work. These sensors combined with mobile devices, can act as personal exposure monitor, or be a mode of building a high-density spatio-temporal pollution map.

In the above article, a comprehensive description of various relevant factors associated with low-cost sensors is presented along with previous works from other researchers.

Funding and Ethical Declaration The above work was not funded, and the authors declare that they have no conflict of interest.

References

1. Chen J, Zhou C, Wang S, Hu J (2018) Identifying the socioeconomic determinants of population exposure to particulate matter (Pm 2.5) in China using geographically weighted regression modeling. Environ Pollut 241:494–503
2. EPA (2014) NAAQS Table
3. Manikonda A, Zíková N, Hopke PK, Ferro AR (2016) Laboratory assessment of low-cost Pm monitors. J Aerosol Sci 102:29–40
4. Wang Y, Li J, Jing H, Zhang Q, Jiang J, Biswas P (2015) Laboratory evaluation and calibration of three low-cost particle sensors for particulate matter measurement. Aerosol Sci Technol 49(11):1063–1077
5. SHARP (2006) Compact optical dust. 2006. "GP2Y1010AU0F." SHARP Corporation
6. Austin E, Novosselov I, Seto E, Yost MG (2015) Laboratory evaluation of the Shinyei Ppd42ns low-cost particulate matter sensor. PloS One 10(9):e0137789
7. Samyoung SC (2012) Co, 2012. Dust Sensor Module P/N: DSM501. Samyoung S&C Co. Samyoung S&C Co.
8. Holstius DM, Pillarisetti A, Smith KR, Seto E (2014) Field calibrations of a low-cost aerosol sensor at a regulatory monitoring site in California. Atmospher Measure Techn 7(4):1121–1131
9. Liu H-Y, Schneider P, Haugen R, Vogt M (2019) Performance assessment of a low-cost Pm2. 5 sensor for a near four-month period in Oslo, Norway. Atmosphere 10(2):41
10. Maricq MM (2013) Monitoring motor vehicle pm emissions: an evaluation of three portable low-cost aerosol instruments. Aerosol Sci Technol 47(5):564–573
11. Castell N, Dauge FR, Schneider P, Vogt M, Lerner U, Fishbain B, Broday D, Bartonova A (2017) Can commercial low-cost sensor platforms contribute to air quality monitoring and exposure estimates? Environ Int 99:293–302

12. Piedrahita R, Xiang Y, Masson N, Ortega J, Collier A, Jiang Y, Li K et al (2014) The next generation of low-cost personal air quality sensors for quantitative exposure monitoring. Atmospher Measure Techn 7(10):3325–3336
13. Jiang Y, Li K, Tian L, Piedrahita R, Yun X, Mansata O, Lv Q, Dick RP, Hannigan M, Shang L (2011) MAQS: a personalized mobile sensing system for indoor air quality monitoring. In: Proceedings of 13th international conference on ubiquitous computing. ACM, pp 271–80
14. Dinh T-V, Choi I-Y, Son Y-S, Kim J-C (2016) A review on non-dispersive infrared gas sensors: improvement of sensor detection limit and interference correction. Sens Actuat B Chem 231:529–538
15. Delebarre C, Pujolle T, Cousin G, Domon A, Froux J, Jourdan J (2018) Wireless low cost CO_2 monitoring system design and evaluation using non dispersive infrared sensor. Wirel Sens Netw 10(06):119
16. Xing Y, Urasinska-Wojcik B, Gardner JW (2018) Plasmonic enhanced Cmos non-dispersive infrared gas sensor for acetone and ammonia detection. In: 2018 IEEE international instrumentation and measurement technology conference (I2mtc). IEEE, pp 1–5
17. Mead MI, Popoola OAM, Stewart GB, Landshoff P, Calleja M, Hayes M, Baldovi JJ et al (2013) The use of electrochemical sensors for monitoring urban air quality in low-cost, high-density networks. Atmospher Environ 70:186–203
18. Li H, Mu X, Yang Y, Mason AJ (2014) Low power multimode electrochemical gas sensor array system for wearable health and safety monitoring. IEEE Sens J 14(10):3391–3399
19. Kumar A, Hancke GP (2014) Energy efficient environment monitoring system based on the IEEE 802.15. 4 standard for low cost requirements. IEEE Sens J 14(8):2557–2566
20. Afshar-Mohajer N, Zuidema C, Sousan LH, Tatum M, Rule AM, Thomas G, Peters TM, Koehler K (2018) Evaluation of low-cost electro-chemical sensors for environmental monitoring of Ozone, Nitrogen Dioxide, and Carbon Monoxide. J Occup Enviro Hygiene 15(2):87–98
21. Spinelle L, Gerboles M, Villani MG, Aleixandre M, Bonavitacola F (2015) Field calibration of a cluster of low-cost available sensors for air quality monitoring. Part A: Ozone and Nitrogen Dioxide. Sens Actuat B Chem 215:249–257
22. Zimmerman N, Presto AA, Kumar SPN, Gu J, Hauryliuk A, Robinson ES, Robinson AL, Subramanian R (2018) A machine learning calibration model using random forests to improve sensor performance for lower-cost air quality monitoring. Atmospher Measure Techn 11(1)
23. Borrego C, Costa AM, Ginja J, Amorim M, Coutinho M, Karatzas K, Sioumis T et al (2016) Assessment of air quality microsensors versus reference methods: the Eunetair joint exercise. Atmospher Environ 147:246–63
24. Alavi-Shoshtari M, Salmond JA, Giurcăneanu CD, Miskell G, Weissert L, Williams DE (2018) Automated data scanning for dense networks of low-cost air quality instruments: detection and differentiation of instrumental error and local to regional scale environmental abnormalities. Environ Model Softw 101:34–50
25. Spinelle L, Gerboles M, Villani MG, Aleixandre M, Bonavitacola F (2017) Field calibration of a cluster of low-cost commercially available sensors for air quality monitoring. Part B: NO, CO and CO_2. Sens Actuat B Chem 238:706–715
26. Jayaratne R, Liu X, Thai P, Dunbabin M, Morawska L (2018) The influence of humidity on the performance of a low-cost air particle mass sensor and the effect of atmospheric fog. Atmospher Measure Techn 11(8):4883–4890
27. Hojaiji H, Kalantarian H, Bui AAT, King CE, Sarrafzadeh M (2017) Temperature and humidity calibration of a low-cost wireless dust sensor for real-time monitoring. In: 2017 IEEE sensors applications symposium (Sas). IEEE, pp 1–6
28. Wang C, Yin L, Zhang L, Xiang D, Gao R (2010) Metal oxide gas sensors: sensitivity and influencing factors. Sensors 10(3):2088–2106
29. Sohn JH, Atzeni M, Zeller L, Pioggia G (2008) Characterisation of humidity dependence of a metal oxide semiconductor sensor array using partial least squares. Sens Actuat B Chem 131(1):230–235

Design of IoT-Based Smart Illumination System in Smart Cities

Aditee Mattoo, Kumud Saxena, Somesh Kumar, and Neha Bagwari

Abstract Smart city is a livable, promising and growing concept that is responsible for keeping track of all the assets in real time. These resources are water, air, transport, electricity and solar that is based on pan-city or area-based development as per smart city proposal by the government of India. Smart cities make use of smart independent solutions to provide better hardware infrastructure and reliable services based on the applications of information and communication technology (ICT) for maintaining advance technological interfaces. Among the mentioned yardsticks of the smart cities, the paper focuses on the smart illumination system centered on Internet of things (IoT). However, the conventional structure requires wired and cable connections in bunch, that is, costly and not user interactive. To address the problem, automatic and smart usage of electric systems is analyzed that helps in fault detection, service restoration, easy and economical maintenance and efficient management of resources. This research paper emphasizes on IoT-based method that is used for the implementation of smart illumination system. The paper gives an insight into the connectivity methods, protocols and standards using various sensors to automate lighting system. The proposed system discusses the designing of hardware core components used in the development of the module and the process followed in the software-hardware integration through various network technologies with the dedicated servers. Hence, the proposed smart system may be very useful for managing and controlling the illumination patterns of smart cities of the future.

A. Mattoo (✉) · K. Saxena
Noida Institute of Engineering and Technology, Greater Noida, UP, India
e-mail: additeemattoo6@gmail.com

K. Saxena
e-mail: saxena.kumud@gmail.com

S. Kumar
Moradabad Institute of Technology, Moradabad, UP, India
e-mail: kumarsomesh507@gmail.com

N. Bagwari
Ajay Kumar Garg Engineering College, Ghaziabad, UP, India
e-mail: nghanshala@gmail.com

A. Abraham et al. (eds.), *Proceedings of 3rd International Conference on Computing Informatics and Networks*, Lecture Notes in Networks and Systems 167,
https://doi.org/10.1007/978-981-15-9712-1_44

Keywords Smart cities · Internet of things · Illumination system · Sensors · LUA-programming language · Microchip wireless module

Abbreviations

AC	Alternate Current
AIE	Adaptive Interface Ecosystems
BAN	Body Area Network
CEA	Central Electrical Authority
DC	Direct Current
DRC	Design rule check
EAGLE	Easily Applicable Graphical Layout Editor
ERC	Electrical rule check
ESP8266	Espressif Systems
GSM	Global System for Mobile Communication
GPRS	General Packet Radio Service
GPIO	General purpose input output
GPS	Global positioning system
GUI	Graphical User Interface
IC	Integrated circuits
ICT	Information and Communication Technology
IEEE	Institute of Electrical and Electronics Engineers
IoT	Internet of Things
ITS	Indian Telecom Services
LED	Light Emitting Diode
LDR	Light Dependent Resistor
LoRaWAN	Low-power wide-area network protocol
LTE	Long Term Evolution
LTE-M	Long Term Evolution for Machines
LUA	Programming Language
MW	Mega Watt
MAN	Metropolitan area networks
MCU	Micro controller unit
OC	Opto Coupler
PCB	Printed Circuit Board
QoL	Quality of life
SDK	Software Development Kit
SPIFFS	Serial peripheral interface flash file system
TRIAC	Triode for alternating current
TERI	The Energy Research Institute
TWh	Terawatt-hour
VPS	Visual positioning system
V2I	Vehicle-to-infrastructure

WPAN	Wireless personal area networks
WLAN	Wireless local area network
WAN	Wide Area Network

1 Introduction

As we are moving ahead, lot of enormous energy is consumed in various activities of our day to day life. It can be renewable or non-renewable sources but the excess of their usage is harming our planet. So to overcome the shortcoming, the idea is to build a smart city which plans to utilize the resources in a way to reutilize it or to minimize its usage, thus, smartly handling the resources.

Currently, the maximum emphasis is on saving the consumption of electric energy by the street lights. The smart city concept includes smart energy consideration with preferable usage of renewable energy sources [1]. According to The Energy Research Institute (TERI) and Central Electrical Authority (CEA), the cumulative street lighting loads approximately 4400 MW and that street lighting alone consumes 21 TWh of energy each year. Figure 1 shows the allocation of budget in different projects according to the ministry of urban development where the amount of energy allocation is quite a considerable one [2].

Several studies show that lighting system is controlled (hence reducing the consumption) based on different factors like computing the statistical means of the stream of traffic, differentiated in accordance with the week where a particular day

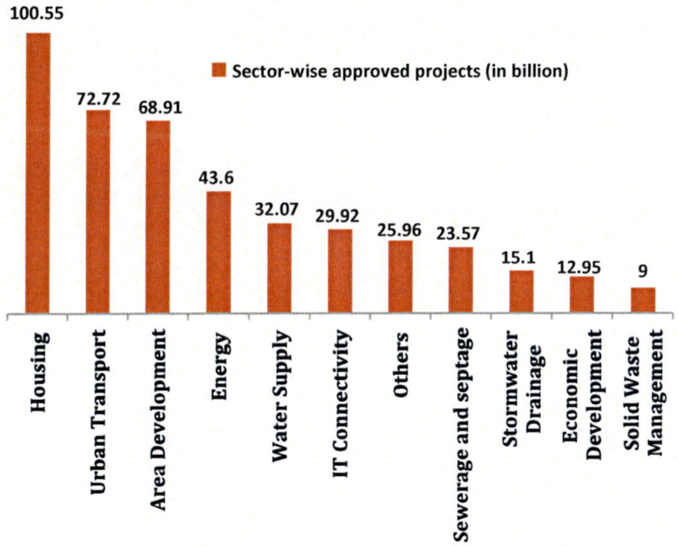

Fig. 1 Smart energy budget in smart city mission

is taken into consideration [3], application of Adaptive Interface Ecosystems (AIE) [4] to implement the controlled lighting system embedded lighting systems plays a vital role and above this IoT tries to interconnect the world. IoT tries to serve better by improving the different logistics for the smart city development [5]. The fundamental proposal of our study is to analyze the usage of things interconnected with one another for controlling the street lights and applying other techniques to find the best suitable way to reduce the consumption of electric energy.

The paper starts with the introduction; the second section covers the use of IoT in smart cities covering the technical aspects of it. Followed by IoT in smart lights in section three, and the fourth section discusses the implementation networking and development boards and finally the conclusion and the future prospects of the proposed model.

2 IoT in Smart Cities

Smart city notion came into being due to increasing rate of growth of population and more and more people shifting to cities for their livelihood. The pressure on the cities has increased a lot mainly on the resources that are to be shared by so many people. Smart city facilitates the management of resources through minimal human interaction [6]. The main characteristics of smart city are sustainability, quality of life (QoL), urbanization and smartness [7]. Sustainability includes structure, capacity, their management, contamination, waste management, social issues, climatic changes, renewable energy, finances and healthiness as its sub-attributes. And to define sustainability, it is the capability of the city, place and thing to strike equilibrium with the above-mentioned aspects while performing city operations [8]. QoL improvement is indicated by the emotional well-being and financial independence of city's population. Urbanization attribute pays attention to the economical, infrastructural, technological and governing aspects of life.

IoT is an enabler of smart city which depends on different communication protocols, different service providers, different network types, standards developed by the standard bodies to provide different kinds of services [9]. IoT offers a variety of applications like insolent transport, smart health, well-being, security, smart four or two wheeler parks, smart managing of waste and energy consumption. As per the application needs, we need to make choice among the available communication protocols, service providers and network types. Below are some of the protocols.

2.1 Communication Protocol

IoT applications depend on different range of communication conventions of transferring packets of data among physical things and back-end servers [10, 11]. Table 1 explains the idea about various applications and wireless technologies used in the

Table 1 Overview of wireless technology

Type of technology	Category of wireless medium	Applications
Short-range wireless technologies	ZigBee network, transmission via Bluetooth and IEEE	Metering, smart health and vehicular communication
Dynamic-dimension technologies	Global system for mobile communication (GSM) and general packet radio service (GPRS), long term evolution (LTE) and advanced	Vehicle-to-infrastructure (V2I), phone-based smart well-being app, smart network and information plus entertainment services
	LTE-M (protocol to enable IoT devices to connect directly to 4G network machines)	Indian Telecom Services (ITS), smart metering, mobile health, logistics, infotainment
	LoRaWAN (low power consumable devices interacts online above wide range wireless connections)	ITS, smart metering, waste management

Table 2 Network types and applications

Type of network	Range	Applications
Capillary IoT networks	Body area network (BANs), wireless personal area networks (WPANs), wireless local area network (WLANs)	Indoor online healthcare services, automation at home level, and in street
Wide range [12]	Mobile data, wide area network (WANs), metropolitan area networks (MANs)	ITS, mobile e-healthcare and waste management

IoT-enabled devices.

2.2 *Network Types*

Table 2 specified below discusses the various network types and their applications.

2.3 *Offered Services*

Various services offered by IoT devices are mentioned in Table 3.

Table 3 IoT-enabled services and their applications

Services	Applications
Smart lighting	The IoT smart modules are deployed in the grid stations, homes and workplaces for consuming energy efficiently
Smart well-being	Automatic devices that are physically connected are helpful in checking various parameters related to human body such as pulse rate, blood pressure, body temperature and level of sugar, and also track the patient's status
Smart traffic IoTs	Surveillance of traffic level using global positioning system (GPS) or visual positioning system (VPS)
Smart waste management	Load level is indicated by waste management smart containers, truck route can be optimized

3 IoT in Smart Lights

Safety, beauty, smooth traffic management and economical power utilization would be important yardsticks [13] for characterizing the future smart cities. Any smart city may be expected to have large industrial complexes, installations, commercial establishments besides complex transportation networks and real-time communication systems that need efficient control and resource management. Among these constituents of the smart cities, there is a scope of optimization of electric power by using automatic illumination and traffic control systems. Smart and energy-efficient systems eliminate the need of manual application of controlling lighting system [14]. Wireless lighting control systems reduce fitting costs, conserves energy and increase flexibility over the manual lighting system.

Smart lighting is an energy-efficient system that is used to regulate the color and intensity of the light according to the user's need and the set regulations. The system automatically controls illumination intensity that depends on the various factors like heat, motion, tenancy, quantity of normal light, etc. The energy consumption can be optimized and energy wastage can be regulated using IoT [15]. IoT-enabled devices are controlled through mobile phones directly. Majority of energy wastage is regularized using IoT, as at the initial stage itself, IoT devices can sense the wastage and take preventive and corrective measures automatically. Each component of the automatic system sends data to the IoT-enabled devices so that energy wastage at any point could be identified and regulated by the IoT. As millions and billions of devices share and process data at their end using electricity and the interconnection between two devices, using Internet and IoT, a large amount of energy will be used by the IoT infrastructure. With the growing popularity of the IoT concept, it is challenging to develop an efficient technology that minimizes the usage of the energy by the IoT, that's why green technology should be explored to yield energy-efficient IoT devices. Different sensors like light sensor, motion sensor, etc., can be used for further improvement in lighting system [16, 17]. The proposed system provides a low-cost solution via wireless network using economical and efficient microcontrollers. The main advantage of the system is that the users can remotely control the system through user-friendly graphical user interface (GUI). Design framework of the proposed system is shown in Fig. 2.

The system comprises of different segments that are interconnected to each other through interface. The system has both centralized and distributed control for handling any emergent situation. The architecture specified in the figure consists of various interfaces like sensor [18], MIC, Bluetooth, high voltage, voice and visual that handles input and output segment of the system through microcontrollers ATmega328 and Wi-Fi module. Line voltage and direct current (DC) conversion supply maintains the power supply of the system. Authorized operator controls the illumination system and view the area illuminated. The system can be used to automatically control lighting system of home, institution or organization [19, 20].

Since the rise of physical interconnected objects is at peak, the concept of wireless networking is used vigorously and the applications of smart lighting lie in various sectors segmented on the basis of lamps and control components. Various sensors, relays and actuators fall under the category of controlled components and are utilized in public sectors, semi or fully government-aided infrastructures, industrial sectors, etc. [21].

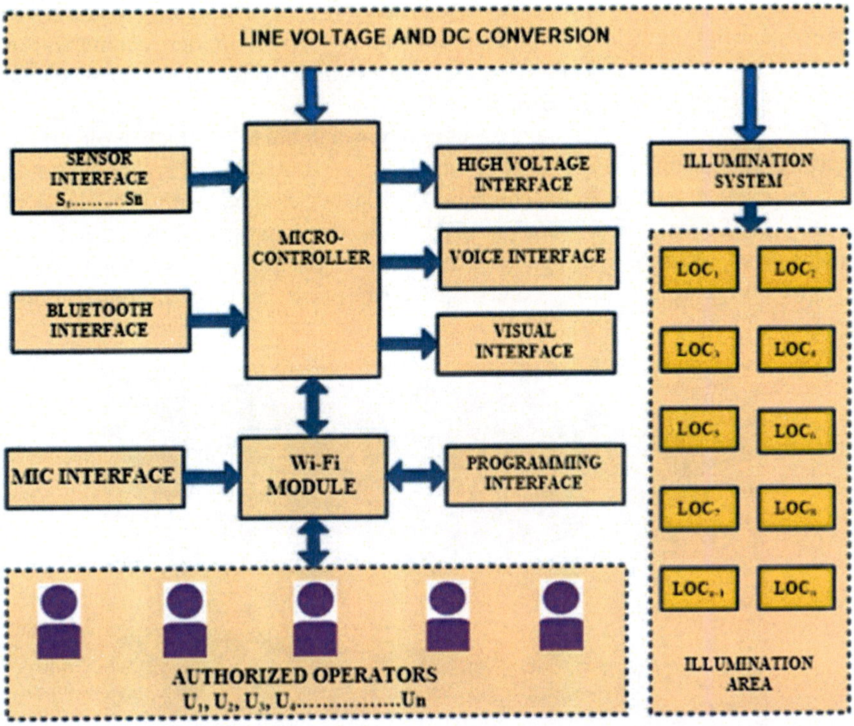

Fig. 2 Proposed system architecture

4 Development of LUA-Based ESP Interface

LUA [22], a programming language interface is used for programming of lights in ESP8266 chip to control the light pattern of the hardware circuit. NodeMCU (micro-controller unit) firmware is applicable for the development of physical configured devices. The IoT modules are based on the Espressif software development kit (SDK) that uses eLUA. NodeMCU LUA utilizes less memory, i.e., saving space and a flash resource in various processors as in ESP8266 that provides a platform to build useful applications. Firstly, communication port for the microprocessor ESP8266 is opened by using appropriate universal service bus port and serial peripheral interface flash file system (SPIFFS) for which a reliable Wi-Fi connection is required. GPIO0 (general purpose input output) pins are set to low at starting for enabling ESP8266 firmware flashing [23]. Figure 3 displays the communication between the ESP8266 chip and port.

EAGLE 7.2.0 professional tool is used for the device's circuit implemented in automatic illumination control system. Easily applicable graphical layout editor, i.e., EAGLE has various inbuilt libraries like Adafruit and Sparkfun, which helps to provide basic electrical components, modules, integrated circuits (IC's) that are needed to make a circuit. Also, EAGLE provides a feature of designing custom user-defined libraries according to the chip, module or microcontroller requirement of the design in the project. In this proposed design model, ESP library for ESP8266 module is designed using the EAGLE tool.

There are two stages and steps to correctly synchronize and realize the circuit interface, i.e., schematic and printed circuit board (PCB) design. Schematic capture is prepared in the first stage. After schematic, PCB layout is created with .brd extension. Routing the air wire connections perfectly is the main concern for the PCB design and then it validates the sync in between schematic view and PCB. Routing between components takes place in the second stage with the placement on the net connections.

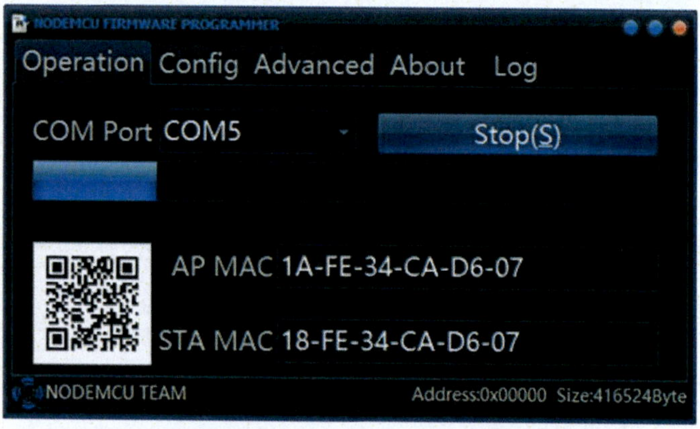

Fig. 3 NodeMCU firmware showing transfer data among ESP8266 and port

Since the auto-routing feature is also applicable, but manual routing is preferable between the two as the net wire connection is done in accordance with the design. PCB layout is displayed in Fig. 4. Error checking through electrical rule check (ERC) and design rule check (DRC) is provision in EAGLE for smooth operation of circuits.

Dimensions of the PCB circuit should be correctly measured as then only correct size board will be prepared with proper drilling and correct positioning of components. Three test cases in different forms of light were passed to the automatic light system Fig. 5 displays the components used in the prototype circuit for control-

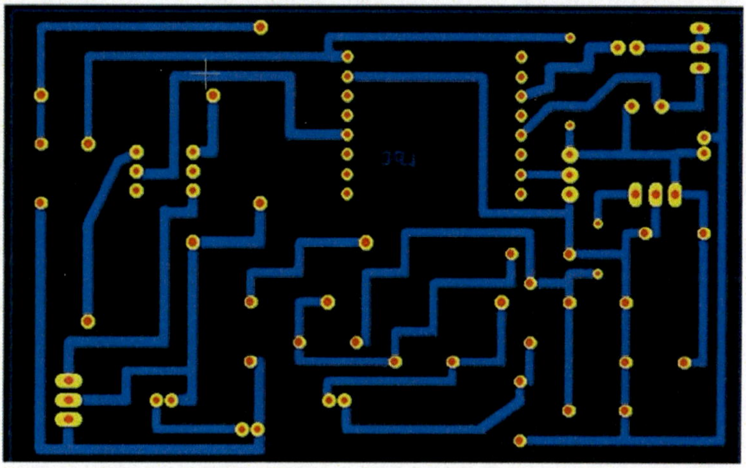

Fig. 4 PCB layout for automatic illumination control system

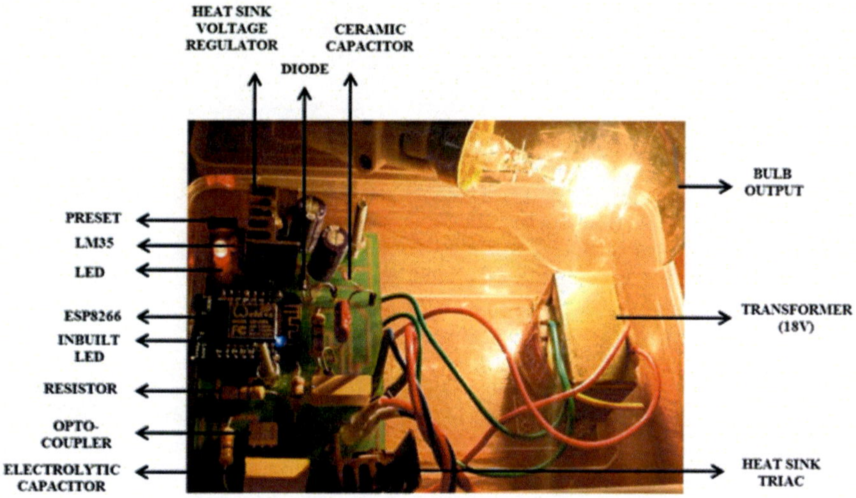

Fig. 5 Prototype model using designed circuitry interface

ling light like inbuilt LED and bulb. Table 4 discusses the components used in the prototype model. Smartphone application developed using android studio is used to operate the illumination system with different activities mapped to it like login and street map [24].

5 Conclusion and Future Scope

There are various significant players for smart lighting like policies of law, energy, electronics and sensor technology development, increase in LED, evolution of wireless technologies, high growth in street lighting system and removal of the incandescent lamps. Also, there are several moderations in this market like incompatible wireless communication solutions, technology costs, lack of awareness among people and organizations. There is also merger with automobile industrial companies Audi and Mercedes-Benz that comprises of automatic lighting system for their luxury cars.

A control system that is been designed using Wi-Fi modules, microcontroller units and interfacing circuits along with a dedicated server can be further used for optimization and security. Automatic light system can be enhanced using nature-inspired algorithms based on several constraints such as available power, illumination area, maintenance cost and priorities set by the establishment under consideration. Block chaining can be used to secure the IoT-enabled devices and exchange data using financial transactions through a decentralized and trustless blockchain.

Table 4 Peripheral components and their description

Peripheral components	Description
Light dependent resistor sensor (LDR)	A photoelectric sensor that detects light emitted by the element and is used in light/dark sensor circuits
PRESET (variable resistor 3296)	Preset is a variable resistor and is connected with LDR to maintain the resistance flow across LDR. One pin from both preset and LDR is connected to digital pin D5 of the ESP module
Light emitting diode sensor (LED)	LED are tiny bulbs and requires less power to light up. In the circuit, it is connected to the digital pin D7 of ESP module. Long lead of LED is anode (+) and short lead called as cathode (−). Red spot in LED is known as flat spot
OptoCoupler (OC) MOC3041 (627Q)	High speed OC is applied for communication and computing purpose. Second pin of OC is connected to digital pin D2 of ESP module. Other pins are connected to resistors and TRIAC
TRIAC (BT139)	TRIAC, triode for alternating current is a three-terminal semiconductor device for controlling current and is a bidirectional device TRIAC symbol carries three terminals, one gate and two anode or main terminal. As TRIAC can easily be heated up while testing the circuit or due to high voltage, so heat sink is used with it to keep it cool and absorb excessive heat
Temperature sensor (LM35)	It measures temperature that provides output voltage in celsius. Temperature is directly proportional to voltage
Voltage regulator (LM7805)	Ic 7805 is used for regulating voltage and is a fixed linear voltage regulator. The difference between value of input and output voltage generates heat. If the difference is high, more heat will generate causing fault
Resistors (330, 360, 470, 39 Ω/1 W, 1 M Ω)	Resistors tend to reduce current flows, adjust signal levels and divide voltage properly
Capacitors (104 J/400 V, 47 NK, 103 (0.01 μF), 1000 μF)	Capacitor stores an electric charge in the form of electrostatic field between its plates and consists of one or more pair of conductors separated by insulator. Therefore, it is called as condenser
Diodes (IN4007)	Diode is a semiconductor device with two terminals, but allows unidirectional flow of current. It blocks current in reverse direction and can convert AC into pulsating DC and therefore called as rectifier

References

1. Mohanty SP, Choppali U, Kougianos E (2016) Everything you wanted to know about smart cities: the internet of things is the backbone. IEEE Consum Electron Mag 5(3):60–70
2. Solanki VK, Díaz VG, Davim JP (2019) Handbook of IoT and big data. CRC Press, Tylor & Francis Group
3. Amini MH, Boroojeni KG et al Sustainable interdependent networks II: from smart power grids to intelligent transportation networks
4. Sanchez AJ, Rodriguez S et al (2019) Adaptive interface ecosystems in smart cities control systems. Future Gener Comput Syst 101:605–620
5. Kauf S (2019) Smart logistics as a basis for the development of the smart city. Transp Res Proc 39:143–149
6. Albino V, Berardi U, Dangelico RM (2015) Smart cities: definitions, dimensions, performance, and initiatives. J Urban Technol 22:3–21
7. Priano FH, Guerra CF (2016) Fully smart cities: analysis of the city to smart city transformation process. In: IEEE international smart cities conference (ISC2), Trento, pp 1–8, 2016. http://dx.doi.org/https://doi.org/10.1109/ISC2.2016.7580745
8. Solanki VK, Venkatesan M, Katiyar S (2018) Conceptual model for smart cities for irrigation and highway lamps using IoT. Int J Interact Multimedia Artif Intell 4(3):28–33
9. Mehmood Y et al (2017) Internet-of-things-based smart cities: recent advances and challenges. IEEE Commun Mag 55(9):16–24
10. Jung C, Kim K, Seo J, Silva BN, Han K (2017) Topology configuration and multihop routing protocol for Bluetooth low energy networks. IEEE Access 5:9587–9598
11. Khan M, Silva BN, Jung C, Han K (2017) A context-Aware smart home control system based on ZigBee sensor network. KSII Trans Internet Inf Syst 11:1057–1069
12. Khan M, Silva BN, Han K (2017) A web of things-based emerging sensor network architecture for smart control systems. Sensors 17:332
13. Lee H-C, Huang H-B (2015) A low-cost and noninvasive system for the measurement and detection of faulty streetlights. IEEE Trans Instrum Measure 64:41019–41031. https://doi.org/10.1109/TIM.2014.2361551,art.no.6963461
14. Petritoli E, Leccese F et al (2019) Smart lighting as basic building block of smart city: an energy performance comparative case study. Measurement 136:466–477
15. Sayali A et al (2016) IoT based street lights for smart city. Int J Res Appl Sci Eng Technol (IJRASET) 4(12)
16. https://www.engineersgarage.com/articles/pressure-sensors-types-working
17. https://www.azosensors.com/article.aspx?ArticleID=339
18. https://thephotographerblog.com/definition-image-sensor/
19. Sharath Patil GS, Rudresh SM, Kallendrachari K, Kiran M (2015) Design and implementation of automatic street light control using sensors and solar panel. Int J Eng Res Appl 5(6):97–100
20. Khan M, Silva BN, Han K (2016) Internet of things based energy aware smart home control system. IEEE Access 4:7556–7566
21. Iromini NA, Nafiu AS, Ajao AO (2015) Automated light control system for offices. Int J Eng Sci Emerg Technol 7(4):701–706
22. Ururahy C, Rodriguez N, Ierusalimschy R (2002) A Lua: flexibility for parallel programming. Comput Lang Syst Struct 28:155–180
23. Liu L (2010) Research on technology of embedded web server application. In: 2nd IEEE international conference on information management and engineering
24. The Firebase website. Available https://firebase.google.com/

Circular Microstrip Antenna with Circular Shorting Posts for Improved Polarization Purity

Jimmy Rinsangzuala, Zonunmawii, Sudipta Chattopadhyay, Lourembam Lolit Kumar Singh, and Abhijyoti Ghosh

Abstract A probe-fed circular microstrip antenna (CMA) with shorting posts has been studied to minimize the cross-polarization (XP) on the broadside direction on the principle H-plane. A number of circular shorting posts, shorting the patch and ground plane on the periphery of the patch have been proposed. Improvement in co-polarized (CP) to cross-polarized (XP) radiation is obtained on increasing the number of the shorting posts. This proposed structure is investigated using two substrates of RT-Duroid and FR4-Epoxy having dielectric constant $\varepsilon_r = 2.33$ and $\varepsilon_r = 4.4$, respectively.

Keywords Circular microstrip antenna · Shorting posts · Co-polarization · Gain · Cross-polarization

Abbreviations

CMA	Circular Microstrip Antenna
CP	Co-polarization
DGS	Defected Ground Structure
DPS	Defected Patch Surface
XP	Cross-Polarization

J. Rinsangzuala · Zonunmawii · S. Chattopadhyay · L. Lolit Kumar Singh · A. Ghosh (✉)
Department of Electronics and Communication Engineering,, Mizoram University, Tanhril, Aizawl, Mizoram 796004, India
e-mail: abhijyoti_engineer@yahoo.co.in

A. Abraham et al. (eds.), *Proceedings of 3rd International Conference on Computing Informatics and Networks*, Lecture Notes in Networks and Systems 167,
https://doi.org/10.1007/978-981-15-9712-1_45

1 Introduction

The microstrip antenna is very common and most utilized antenna in the era of modern communication technology. Microstrip antenna is basically a low profile, tiny, and planar radiating structure. On the other hand, they suffer from poor gain, bandwidth, and efficiency. Such radiators also suffer from poor polarization purity. The XP radiation on the E-plane is often negligible but the XP radiation on the broadside of H-plane is a prominent issue. CP-XP ratio decreases with resonance frequency and the high dielectric substrate (ε_r) [1, 2]. Many scientists and researchers had proposed their activities in reducing the XP for better performances.

Different methods like defected ground structure (DGS), shorting technique, defected patch structure (DPS), and modification of feed are common methods for improvement in XP radiations. Reduction of XP radiation in CMA has been reported by circular DGS, suppressing 5–8 dB, using arc-shaped and circular ring-shaped suppressing 7–12 dB and 5–7 dB, respectively [3–6] with the technique of defected ground structure (DGS). A circular cut defected patch surface (DPS) achieved about 27–28 dB of CP-XP isolation [7]. Earlier, the technique of shorting located symmetrically at the bore-sight of the patch [8, 9] and shorting of non-radiating edges to ground plane for reducing XP radiation using modified cavity model [10] has also been reported.

Apart from these techniques, there are different model of reducing this cross-polarized radiation like dual feeding, probe and aperture hybrid combinations, post-gap or meandering strip [11–14], ground plane with modified shapes like '*W*' and '*U*' shape [15, 16] were reported. The cross-polarization (XP) increases for prove fed design [17]. It also increases with thickness and the permittivity of the substrate [17]. Complex multiple feeding methods have been reported in [18, 19] to reduce XP radiations from single patch element. The inherent narrow banding characteristics of the microstrip patch can be taken care with thick dielectric substrate. Those above models for reducing XP suffer from complexity in feed design and non-planer ground plane.

The placement of shorting post along the non-radiating edges of CMA has been employed in this study for improvement in polarization purity (Fig. 1).

2 Theoretical Insight, Structural Optimization, and Proposed Structure

2.1 *Theoretical Insight*

The available literatures on microstrip antenna mention that a conventional circular microstrip antenna in its dominant mode (TM_{11}) radiates linearly polarized electric fields along the broad side direction due to the fringing fields that resides at the radiating edges of the patch [2, 7]. Nevertheless, a handful amount of fringing fields are

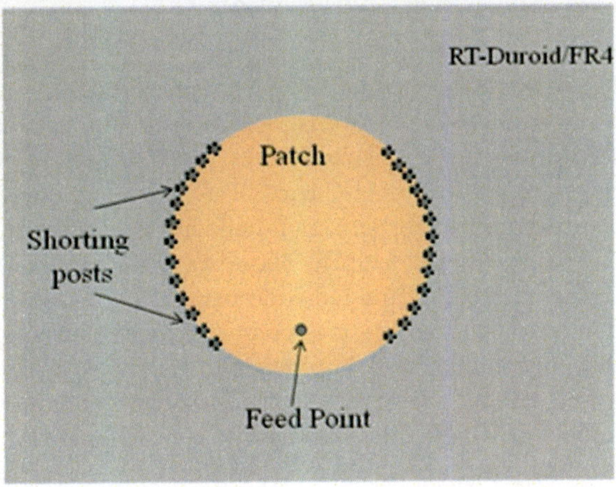

Fig. 1 Top view of the proposed structure with shorting posts at the non-radiating edges

also present in the non-radiating edges of the patch which generates the considerable amount of XP radiation arises from non-radiating sides of CMA as shown in Fig. 2.

The primary sources of the XP radiations in a CMA are the orthogonal resonance fields of week TM_{11} mode and regular excitation of higher-order orthogonal TM_{21} mode [7]. The field distribution in TM_{21} mode is orthogonal to the field distribution of TM_{11} mode is seen from Fig. 1. It also confirms that the primary source of XP radiations is the non-radiating sides, and hence the shorting pins are placed at these regions just to eliminate such radiations without affecting its dominant TM_{11} mode radiations.

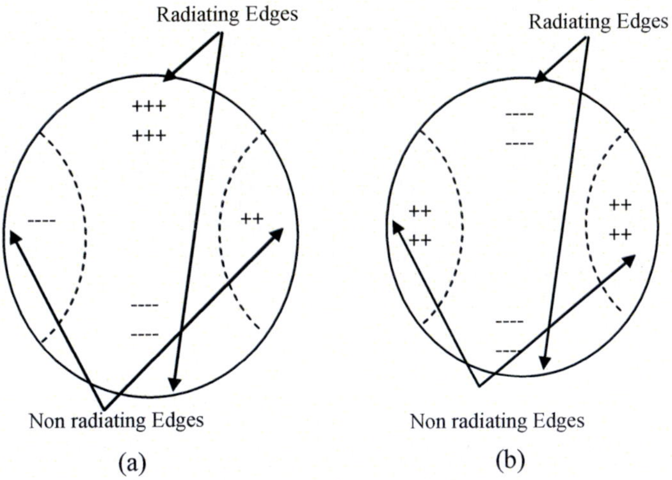

Fig. 2 Electric field distribution on patch **a** fundamental TM_{11} mode, **b** next higher-order TM_{21} mode

2.2 Structural Optimization

Parametric studies to find the best possible proposed antenna geometry has been done using HFSSv14 [20]. At first, a pair of circular patch antennas (conventional) with RT-Duroid ($\varepsilon_r = 2.33$) and FR4 ($\varepsilon_r = 4.4$) substrates have been developed. Both the antennas operate in C band. The parametric study has been started by placing shorting posts of copper at non-radiating sides along the periphery of CMA for both cases (i.e., antenna with RT-Duroid and FR4 substrate).

Figure 3 shows the radiation patterns of the proposed structure with different numbers of shorting posts as compared to the conventional CMA. The figure reveals that polarization purity improves with the increase in number of shorting posts. The lowest H-plane XP level is below −26.5 dB over an angle of ±120° as compared to only 18 dB over ±30° in case of conventional CMA has been achieved with 30 shorting posts, i.e., 15 shorting posts at each non-radiating edges for CMA with RT-Duroid substrate. CMA with FR4 substrate provides the lowest H-plane XP radiation below −31 dB over an angle of ±150° as compared to −21 dB over ±30° in case of conventional CMA has been accomplished with 26 number of shorting posts, i.e., 13 in each non-radiating edges. Further, increase in the number of shorting posts may hamper the co-polarization radiating from the radiating edges of the CMA.

The co-polarized gain pattern of both the antennas (Fig. 4) corroborates that the co-polarization gain is increasing. The highest co-polarization gain of 9 dBi has been achieved with 30 shorting posts with RT-Duroid substrate compared to only 6.5 dBi in case of conventional CMA. In the other hand, the highest co-polarization gain of 5.11 dBi has been achieved with 26 shorting posts with FR4 substrate compared to only 3.8 dBi in case of conventional CMA.

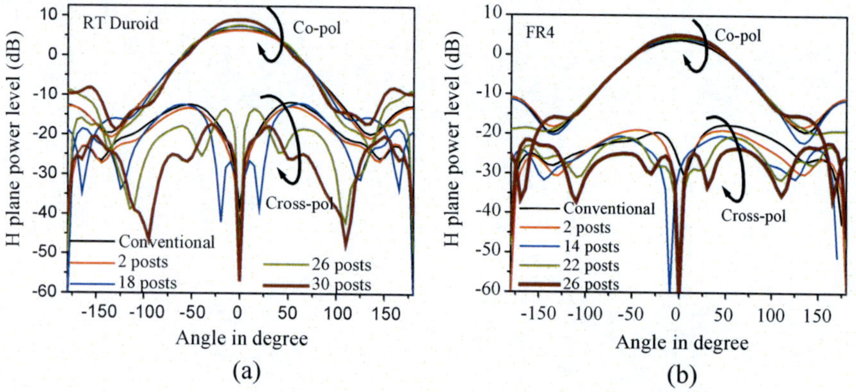

Fig. 3 Comparison of radiation pattern for different numbers of shorting posts **a** RT-Duroid, **b** FR4

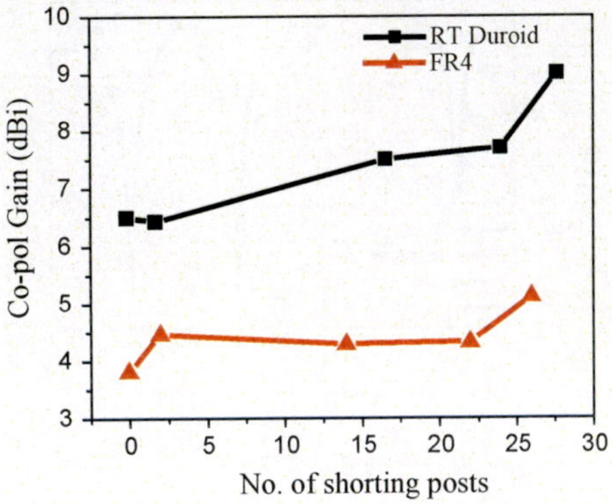

Fig. 4 Co-polar gain variation of investigated structure for different number of posts

Table 1 Details of antenna parameters ($h = 1.575$ mm)

Substrate material	r (mm)	r_p (mm)	Ground plane (mm^2)	f_r (GHz)	ε_r
Fr4	7	0.01	70 × 70	5.6	4.4
RT-Duroid				7.3	2.33

2.3 *Proposed Structure*

Two CMA with radius (r) 7 mm have been designed using RT-Duroid and FR4 substrate ($h = 1.575$ mm) of size 70 × 70 mm^2. Shorting posts of radius 0.5 mm are placed symmetrically on the periphery as is explained in Fig. 1. Firstly, a pair of two posts are placed symmetrically along the y-axis and increased the number of shorting posts as shown in Fig. 1. The gap between each post (r_p) is about 0.01 mm. Table 1 shows the optimum parameter values of the final structures.

3 Results and Discussion

The reflection coefficient (S_{11}) profile for both the antennas with substrate RT-Duroid and FR4-Epoxy is shown in Fig. 5. As the numbers of shorting posts increases the effective size of the circular microstrip antenna increases which results the shift of resonance frequency toward right in the frequency spectrum.

The radiation characteristics of the investigated structures (with shorting posts) using the two substrates (RT-Duroid and FR4-Epoxy) are also depicted in Figs. 6 and 7, respectively. The CP radiation patterns in case of both the antennas do not change much which confirms that the inclusion of shorting posts do not hamper the co-polarization radiation. In the other hand, the polarization purity increases with the increase in the number of shorting posts for both the antennas.

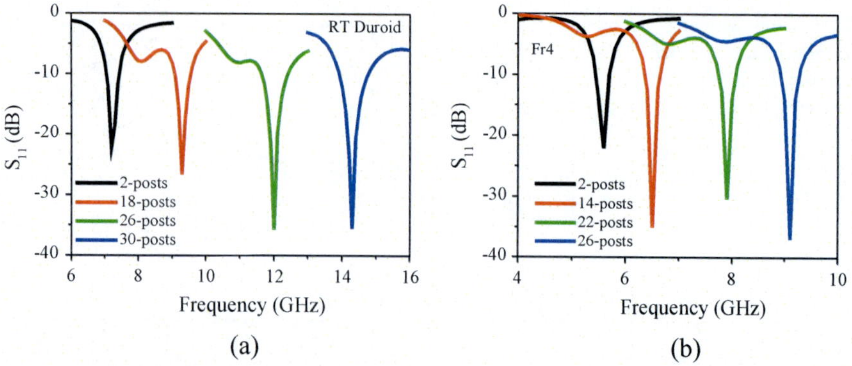

Fig. 5 Simulated reflection coefficient (S_{11}) profile for substrate **a** RT-Duroid, **b** FR4-Epoxy

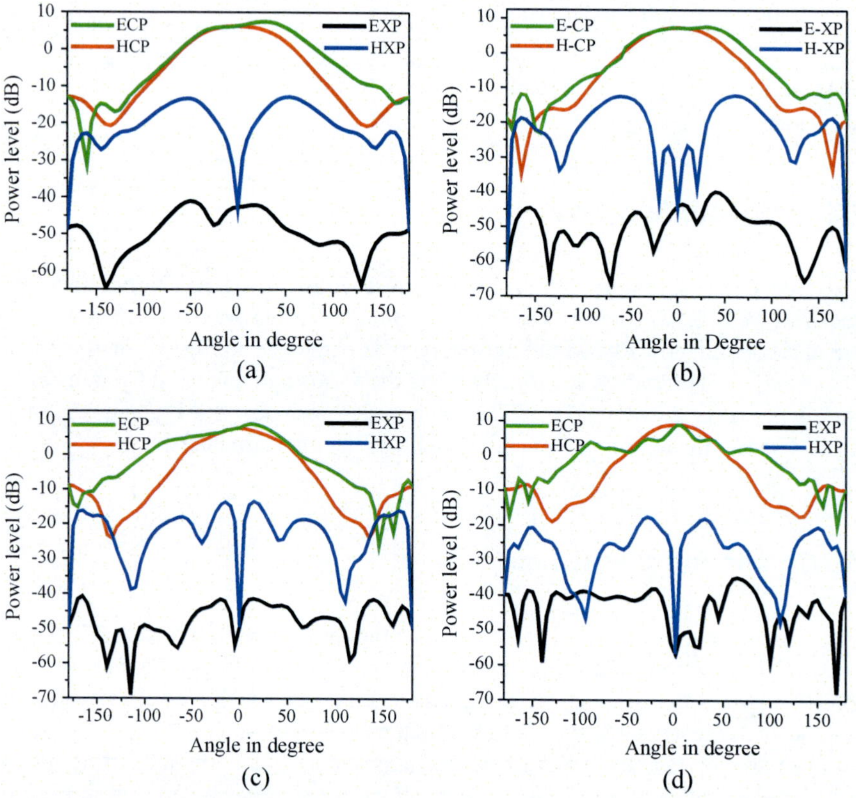

Fig. 6 Radiation characteristic of the CMA with shorting posts using RT-Duroid **a** 2-shorting posts, **b** 18-shorting posts, **c** 26-shorting posts, **d** 30-shorting posts

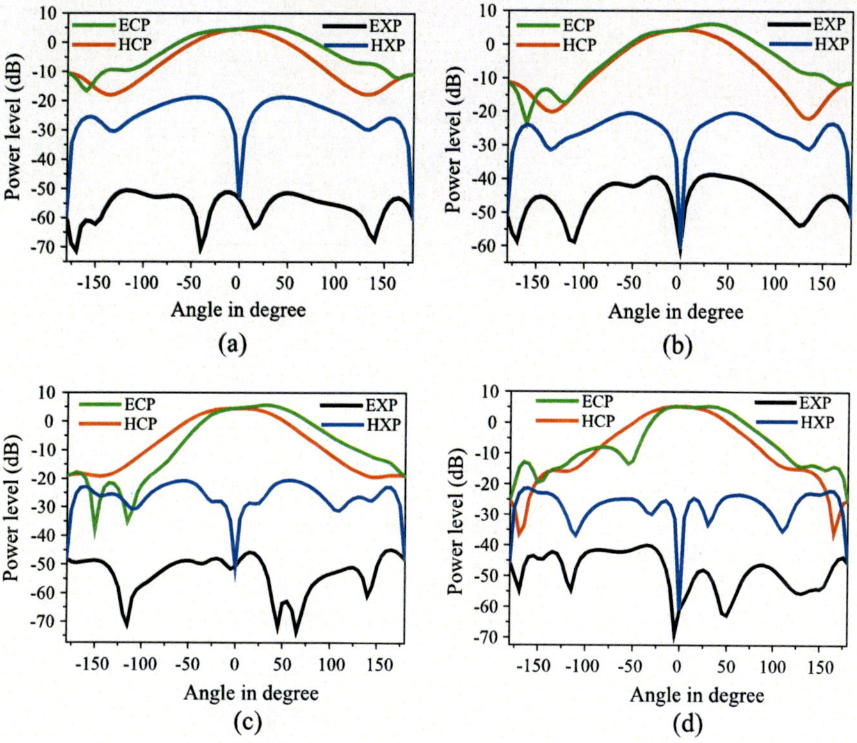

Fig. 7 Radiation pattern for the proposed CMA with shorting posts using FR4-epoxy **a** 2-shorting posts, **b** 14-shorting posts, **c** 22-shorting posts, **d** 26-shorting posts

The summary of simulated results like resonance frequency (f_r), co-polarization gain, polarization purity for the proposed structure with different shorting posts using the two substrates RT-Duroid and FR4-Epoxy are shown in Tables 2 and 3, respectively.

Table 2 Summary of simulated radiation characteristics using RT-Duroid ($\varepsilon_r = 2.33$)

Number of shorting posts	f_r (GHz)	Gain (dBi)	CP-XP isolation (dB)
No (conventional)	7.3	6.5	18.78
2	7.2	6.43	19.25
18	9.3	7.5	20.4
26	12	7.7	21.1
30	14.3	9	26.39

Table 3 Summary of simulated radiation characteristics using FR4-Epoxy ($\varepsilon_r = 4.4$)

Number of shorting posts	f_r (GHz)	Gain (dBi)	CP-XP isolation (dB)
No (conventional)	5.6	3.8	21.8
2	5.6	4.45	23.45
14	6.5	4.28	24.98
22	7.9	4.31	25.33
26	9.1	5.11	30.54

4 Conclusion

Microstrip antenna of circular geometry for different number of shorting posts located at non-radiating sides for improved polarization purity has been investigated. The proposed structure as compared to the conventional microstrip antenna results in an improvement in polarization purity as the number of shorting posts increase. The increasing of shorting posts also results in increasing of the resonance frequency. The change in radiation pattern of the E-plane co-polarized radiation observed using the substrate RT-Duroid when increasing the number of shorting posts can be fixed by using the substrate FR4-Epoxy which has higher dielectric constant compared to RT-Duroid.

References

1. Lee KF, Luk KM, Tam PY (1992) Cross polarization characteristics of circular patch antennas. Electron Lett 28(6):587–589
2. Garg R, Bhartia P, Bahl I, Ittipiboon A (2001) Microstrip antenna design handbook. Artech House, Norwood, USA
3. Guha D, Biswas M, Antar YMM (2005) Microstrip patch antenna with defected ground structure for cross polarization suppression. IEEE Antennas Wirel Propag Lett 4:455–458
4. Guha D, Kumar C, Pal S (2009) Improved cross-polarization characteristics of circular microstrip antenna employing arc-shaped defected ground structure (DGS). IEEE Antennas Wirel Propag Lett 8:1367–1369
5. Guha D, Biswas M, Kumar C (2009) Annular ring shaped DGS to reduce mutual coupling between two microstrip patches. In: Applied electromagnetics conference on AEMC 2009, Kolkata, India
6. Kumar C, Guha D (2009) New defected ground structures (DGSs) to reduce cross-polarized radiation of circular microstrip antennas. In: Applied electromagnetics conference on AEMC 2009, Kolkata, India
7. Ghosh SK, Ghosh D, Chattopadhyay S (2016) Improved polarization purity for circular microstrip antenna with defected patch surface. Int J Microw Wirel Technol 8(1):89–94
8. Samanta S, Reddy PS, Mandal K (2018) Cross-polarization suppression in probe-fed circular patch antenna using two circular clusters of shorting pins. IEEE Trans Antennas Propag 66(6):3177–3182

9. Samanta S, Reddy PS, Mandal K (2017) New dimension in cross polarization reduction of a hexagonal microstrip antenna using two circular substrate integrated cavities. In: Proceedings of IEEE region 10 conference (TENCON), Malaysia, pp 1358–1363
10. Ghosh D, Ghosh SK, Chattopadhyay S, Nandi S, Chakraborty D, Anand R, Raj R, Ghosh A (2014) Physical and quantitative analysis of compact rectangular microstrip antenna with shorted non-radiating edges for reduced cross-polarized radiation using modified cavity model. IEEE Antennas Propag Mag 56(4):61–72
11. Chiou TW, Wong KL (2002) Broad-band dual-polarized single microstrip patch antenna with high isolation and low cross polarization. IEEE Trans Antennas Propag 50(3):399–401
12. Chen ZN, Chia MY (2003) Broad-band suspended probe-fed plate antenna with low cross polarization levels. IEEE Trans Antennas Propag 51(2):345–346
13. Guha D, Siddiqui JY (2002) Simple design of a novel broadband antenna: inverted microstrip patch loaded with a capacitive post. Proc IEEE Antennas Propag Soc Int Symp 2:534–537
14. Li P, Lai HW, Luk KM, Lau KL (2004) A wideband patch antenna with cross-polarization suppression. IEEE Antennas Wirel Propag Lett 3:211–214
15. Wong KL, Tang CL, Chiou JY (2002) Broad-band probe-fed patch antenna with a W-shaped ground plane. IEEE Trans Antennas Propag 50(6):827–831
16. Hsu WH, Wong KL (2002) Broad-band probe-fed patch antenna with a U-shaped ground plane for cross-polarization reduction. IEEE Trans Antennas Propag 50(6):352–355
17. Kishk AA, Shafai L (1986) The effect of various parameters of circular microstrip antennas on their radiation efficiency and the mode excitation. IEEE Trans Antennas Propag 34(8):969–976
18. Petosa A, Ittipiboon A, Gagnon N (1999) Suppression of unwanted probe radiation in wideband probe fed microstrip patches. Electron Lett 35(5):355–357
19. Sim C-YD, Chang C-C, Row J-S (2009) Dual-feed dual-polarized patch antenna with low cross polarization and high isolation. IEEE Trans Antennas Propag 57(10):3405–3409
20. HFSS: High frequency structure simulator, Ver. 13. Ansoft Corp., USA

Computer Applications and Technological Innovations In Social Sciences Health-Care

Design and Development of Mobile Augmented Reality

Mugdha Sharma, Ritivik Vaidya, Ayush Kumar Saxena, Isha Aggarwal, and Jalaj Khurana

Abstract Augmented reality technology has gained attention in entertainment, information and education but its application as a sales and marketing tool has remained largely unexplored. This paper introduces a combination of computer graphics applications such as image and 3D object recognition, in augmented reality and proposes its application in sales and marketing. A mobile application scans the real environment for recognizing the images and real-world 3D objects and superimposes the virtual 3D object over them. Animations are used to explain the product in a detailed and interesting manner which captures user's attention, enhances user experience and increases product awareness and boosts sales. To discover the impact, relevance and scope of augmented reality in marketing, user testing and evaluation in comparison with generic marketing tools such as newspaper advertising, billboards can be conducted. For developing prototype, Vuforia software development kit (SDK) has been used for target management and detection, Blender to create 3D models and Unity 3D Game Engine as the development platform.

Keywords Augmented reality · Image recognition · Object recognition · 3D modeling · Object tracking · Vuforia · Blender · Unity

---

M. Sharma · R. Vaidya · A. K. Saxena · I. Aggarwal (✉) · J. Khurana
Department of Computer Science and Engineering, Bhagwan Parshuram Institute of Technology, New Delhi, India
e-mail: ishaaggarwal1998@gmail.com

M. Sharma
e-mail: Mugdha.sharma145@gmail.com

R. Vaidya
e-mail: vaidyaritivik@gmail.com

A. K. Saxena
e-mail: ayushsaxena92910@gmail.com

J. Khurana
e-mail: jalaj.khurana13@gmail.com

A. Abraham et al. (eds.), *Proceedings of 3rd International Conference on Computing Informatics and Networks*, Lecture Notes in Networks and Systems 167,
https://doi.org/10.1007/978-981-15-9712-1_46

Abbreviations

3D	Three Dimensional
AR	Augmented Reality
AV	Augmented Virtuality
GPS	Global Positioning System
GUI	Graphical User Interface
MAR	Mobile Augmented Reality
MDT	Mobile Data Terminal
MTG	Model Target Generator
PC	Personal Computer
SDK	Software Development Kit

1 Introduction

1.1 Augmented Reality (AR)

Augmented reality is an innovative technology that facilitates blending of ostensible or virtual objects with real-world information. It is an upcoming technology that makes use of display, processor and input devices to perform the following functions:

- Combine real and virtual information: Augmented reality combines digital and virtual information with real-world information seamlessly. Objects are displayed in space, where they are not present in the real world.
- Facilitate user interaction in real time: User interaction is important to capture the user's imagination and attention for prolonged enhancement of user experience.
- Accurate registration of real and virtual objects is necessary to maintain proper interaction and track the locations of objects in 3D space.

1.2 Image Recognition

Image target recognition is the technique where an image is scanned to detect its features such as text, curves in images, designs and logos that can be used to uniquely identify the image. These features can then be recorded and used to detect the image as a target in a real-world scene captured by the camera and augment virtual information like text, images, 3D objects, etc.

1.3 3D Object Recognition

3D object recognition is the technique that involves scanning and recognition of a three-dimensional (**3D**) model of a real-world object and its features such as edges, vertices and contours detected, such that the scanned object can be identified uniquely. These features can then be recorded and used to detect the 3D object as a target in a real-world scene captured by the camera and augment virtual information like text, images, 3D objects, etc.

1.4 Objective

The objectives of using mobile augmented reality (MAR) are that, firstly, augmented reality will help in amplifying the interactivity between the user and the virtual objects in a real world by generating customer-appealing content. It will be helpful in upgrading brand awareness by providing enriched experience to users. And, above all augmented reality will boost the sales of a brand by providing them an opportunity to personalize content virtually. And last but not the least, augmented reality will also help users to acquire information through interactive visualization.

The paper is structured as follows. Section 2 provides the details of mobile augmented reality technology. Section 3 describes the platform and software used. Section 4 discusses the proposed approach, its usefulness and processes involved in creation of database, creation of 3D model and development of an AR application. After which, the result and conclusion are provided in Sects. 5 and 6, respectively.

2 Related Work

Mobile augmented reality (MAR) is a rising technology which overlaps the real scene with virtual data by utilizing the mobile data terminal (MDT) which enables users to have a better understanding of the things present in the real environment and interact with them [1].

Vuforia is a computer vision platform used for AR development and is integrated into newer versions of Unity [2]. There are other AR platforms such as ARKit and ARCore; however, they are only for iOS and Android, respectively. Vuforia works with both iOS and Android, and to some extent, can be used with ARKit and ARCore as needed, therefore, ideal for rapid development as it uses the same code for both operating systems [3].

AR system is a combination of three processes: recognition, tracking, and combining. In recognition, any image, object, human face or body is recognized

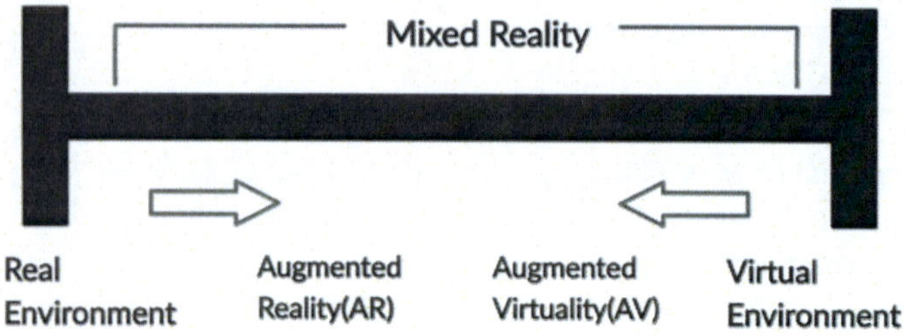

Fig. 1 Milgram's reality–virtuality continuum [5]

on which virtual object is superimposed. In tracking, real-time locating (automatic identification) of the image, object or human face or body takes place and virtual media in the form of text, image, audio, video, 3D model, etc., are superimposed over it [4].

There is a notable difference between the idea of virtual reality and augmented reality which is shown in 'reality–virtuality continuum' given by Paul Milgram and Fumio Kishino's as in Fig. 1.

The real world and a virtual environment are at the two ends of this 'reality–virtuality continuum' and the middle portion is named mixed reality. At the real-world end, augmented reality lies with the concept of computer-generated data superimposing real world. Milgram created the term augmented virtuality (AV) for systems which use figurative languages where some real world is put on virtual objects [4].

AR applications for phones focus on the visual modality. It is based on the concept of augmenting of artificial 3D objects on real-world objects which are identified by camera on the screen. Tracking implies successful scanning of real-world entities for accurate positioning of augmentations over a user's view of the real world [6].

Researchers have defined AR based on its various aspects, in which the data taken from real-world objects and information collected from computer-generated models are imposed on a real world, interact on mobile's screen in real time and display virtual objects properly aligning to real-world coordinates [7].

Apart from that, researchers also gave a broad-ranging view-point, stating augmented reality as a situation in which real-world entities are dynamically superimposed with environment-sensitive virtual information. A less comprehensive definition is also provided, where it is suggested that augmented reality is a system which basically merges computer-generated graphics with visual data seen from a camera. Basically, this idea can be articulated as augmented reality as technology of adding missing information in real life through virtual objects as seen from screen [7].

3 System Design

3.1 Platform

The AR application is compatible with both Android and iOS devices. This research found smartphones and tablets to be an ideal device for AR as it incorporates built-in cameras, accelerometer, GPS and high performing processors, all of these are required to create and execute an AR experience [8].

The application is designed by using Vuforia software development kit (SDK). Vuforia is a library used to build rich AR experiences for mobile devices. It supports multiple target recognition techniques out of which image target recognition and model target (3D object) recognition techniques are used in this application.

3.2 Unity

Unity 3D is used as a development platform. It is a cross-platform game engine which offers an intuitive interface for developers to work and create AR applications [9]. After developing an application on Unity, it can be easily deployed across multiple mobile devices.

3.3 Vuforia

Vuforia is a library used to implement augmented reality on mobile devices. Vuforia analyses images and 3D objects to detect and record features which on detection by the app can be used to project virtual information such as text, image, video or 3D animations on the target images and objects.

Vuforia helps AR developers develop apps because the basic code to implement AR is provided within the library so that the developer can completely focus on the end product that they intend to create and not worry about how to make the system work at a basic level. This library is compatible with Android and iOS devices as well as with Unity Game engine.

Figure 2 shows the architecture of Vuforia where mobile application interacts with the user and initializes the tracker which sets up and detects and tracks targets from the image clicked on the camera. Based on the position of detected targets, rendering information is generated and passed on to the operating system for execution.

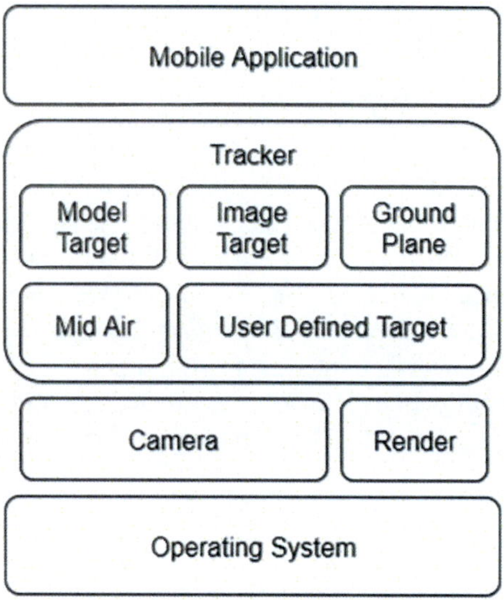

Fig. 2 Vuforia architecture

3.4 *Image Target Recognition*

Image target Recognition is the technique where an image is used to implement AR by detecting and recording its features and textures into a local or cloud database maintained by Vuforia Target Manager. This database is then used to detect images in real-world scenes and augment virtual information like text, images, 3D objects, etc. This can then be used to interact with the users.

3.5 *Model Target Recognition*

Model target recognition is the technique where a 3D object is used to implement AR by detecting and recording its features and design into a local or cloud database built by Vuforia Model Target Generator and maintained by Vuforia Target Manager. Using this database, real-life 3D object can be detected and used to implement augmented reality on mobile devices.

4 Proposed System

This paper proposes a system where a mobile application is used to scan images such as those on brochures or advertisements and augments 3D models of the scanned product over it with animations explaining its features, its specifications and its uses with an intention to generate interest, intrigue and willingness to buy the product [5].

The app shall redirect the user to the sales Web site. Also, the same app can be used to set up the product once it has been bought by scanning the product itself.

The main advantage of augmented reality is that it is developed for mobile devices, tablets, PC, etc., where the hardware is already available. It can give entirely physical experience of interacting with products on screens only. Additional information about the product can also be displayed like comparison between two or more products, discount offers, virtual coupons, etc. It can save a user tons of time. Having able to display personalized content and express things to a user on his mobile screen which is different from standard way of selling makes the brand stand out and is way more engaging.

Figure 3 proposes that on launch, the application shows a splash page and then another page where the user has to choose between image target and model target. If the user chooses an image target, the camera is launched and it starts scanning the environment for the desired images, upon successful detection and registration, the 3D model is augmented as defined by the developer and guides the user through a series of animations describing the product in detailed and interesting manner. Once the journey is completed, the user is guided to the sales Web site. If the user chooses a model target, the camera launches and starts scanning for the desired 3D object in the real world, augments the 3D model upon successful detection and guides the user through the setup and initialization process in an interactive manner. Once the setup is complete and the journey ends, the application terminates.

4.1 Generating Image Targets and Creating Database

Images are uploaded to Vuforia Target Manager for detection and recording of features and textures into a database that can either be locally stored within the application or on Vuforia's cloud platform. This database is then imported into Unity 3D and accessed to develop the various screens of the application [10].

The Vuforia Target Manager is a Web application present on the developer portal of Vuforia. It performs visual evaluation on uploaded targets.

Figure 4 shows the image upload page where the images are uploaded to Vuforia Target Manager to be processed and rated for suitability.

Figure 5 shows the target manager where the processed and rated images are available to be downloaded in the form of databases. The five-star rating suggests that this image target is very good for implementing augmented reality.

4.2 Visual Evaluation of an Image

Figure 6 shows the processed image target and its features (+symbols) detected by the target manager. When most of these are detected in a real scene, the virtual object gets augmented over the detected target and interacts with the user.

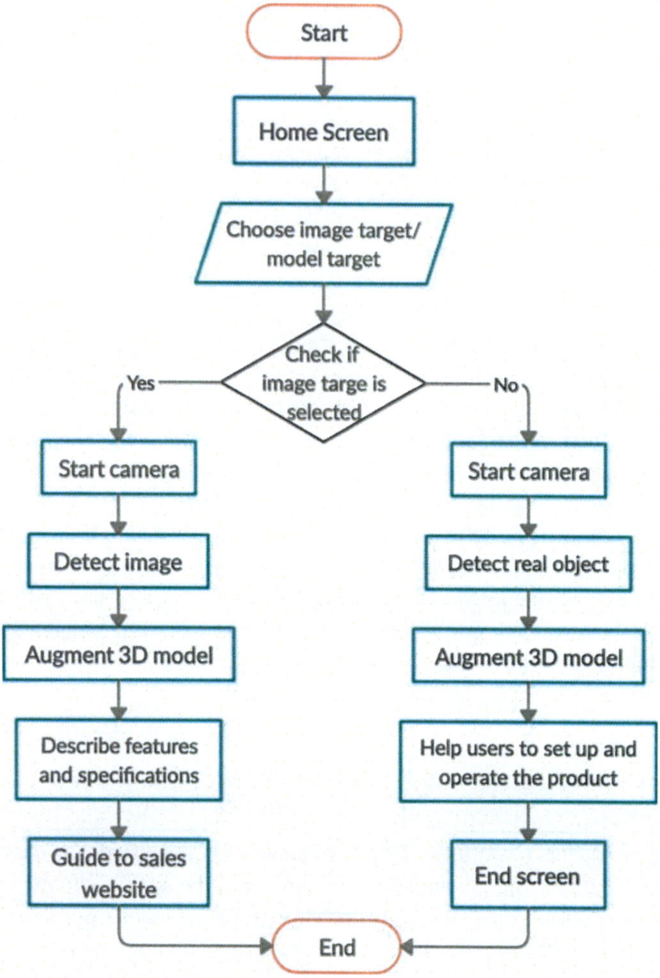

Fig. 3 Application flow

4.3 3D Model Design

3D models and animations were built using Blender 2.8 software that allows developers to design and create 3D models and graphics with user-friendly GUI. It also allows users to create animated films, digital art, visual effects, games, etc. It is free and open-source software. [11]

Figure 7 shows some screenshots of 3D model design where several parts of the model are being created and placed in their right positions to create a realistic model of an air conditioner.

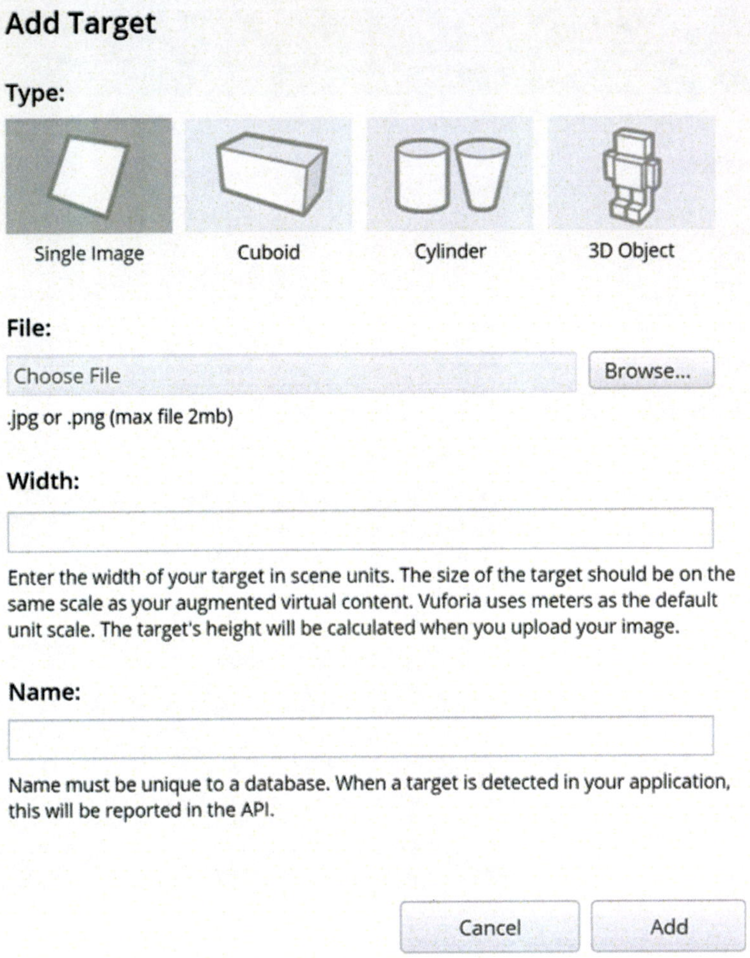

Fig. 4 Image upload page

4.4 *Texturing*

Figure 8 shows the process of adding colors, material and texture to the model, this is called texturing. Texturing is done so that the model looks more realistic to the user, enhancing their experience.

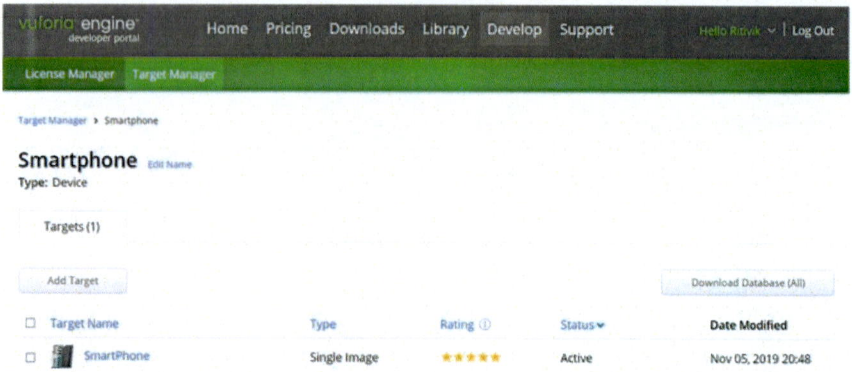

Fig. 5 Vuforia target manager

Fig. 6 Processed image target

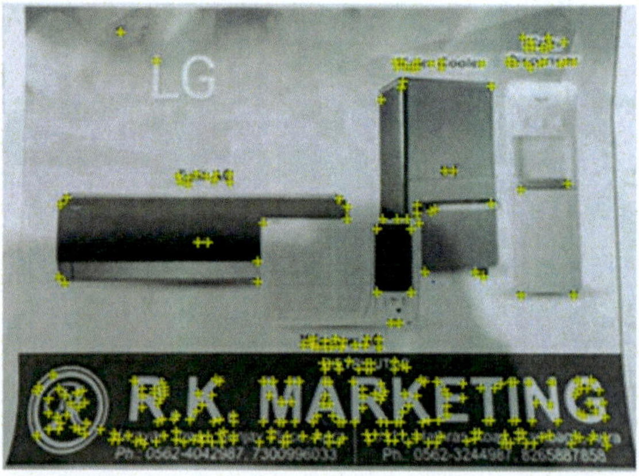

4.5 *Model Target Generation*

Figure 9 shows model target generator which is used to generate guide views of 3D models, which can be imported into the program and be used to implement model target detection.

4.6 *Animation*

Figure 10 proposes creating animations, i.e., moving parts of the model to show and describe each of them separately.

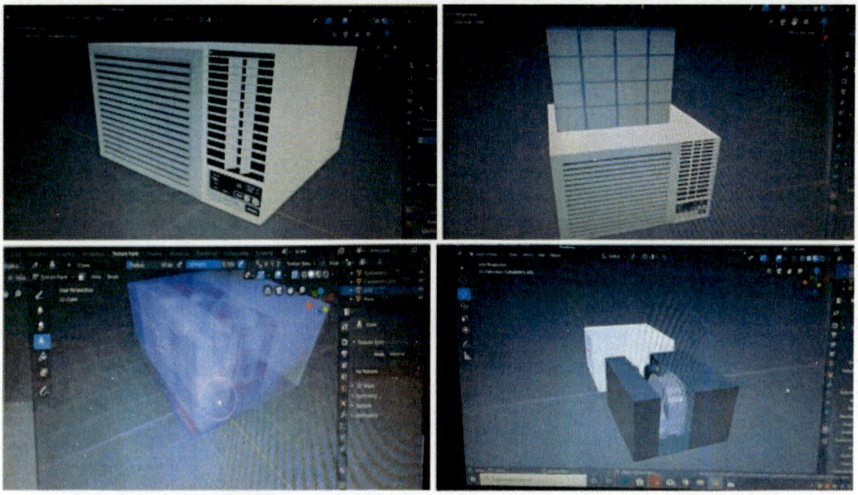

Fig. 7. 3D model design of window air conditioner

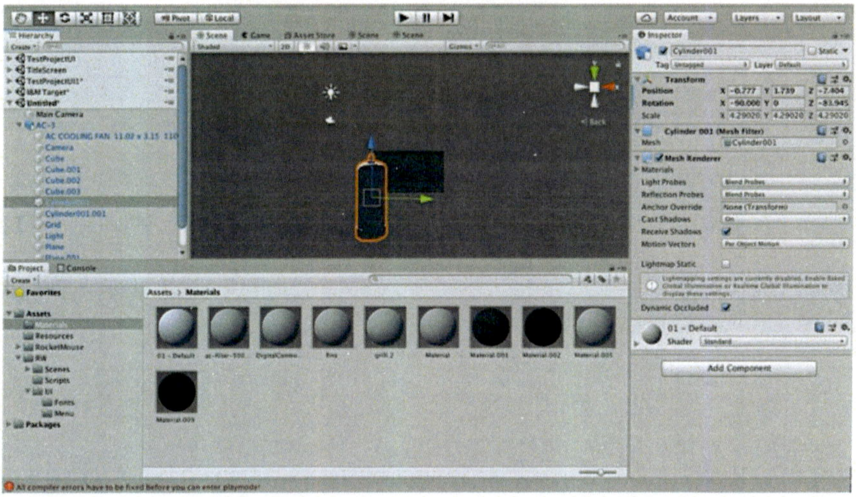

Fig. 8 Texturing

Animations increase the level of user engagement as well as user awareness that can be beneficial when selling or setting up the product. Animations are done using Blender 2.8 Animator Tool to create timed events and record their execution in the object file [12].

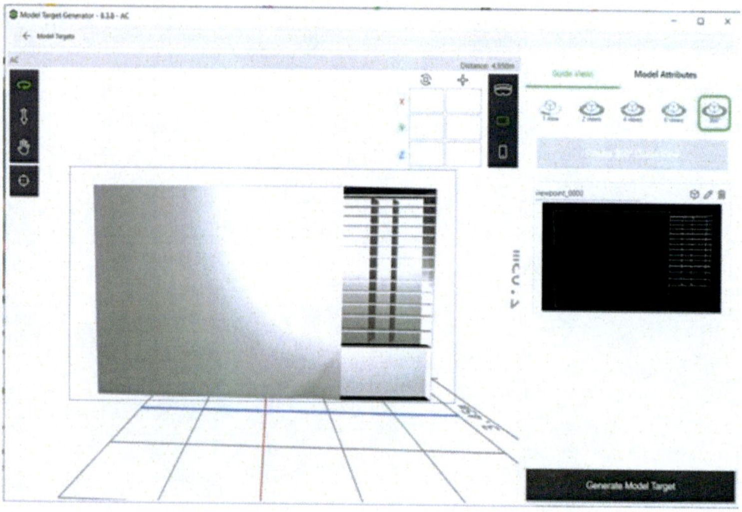

Fig. 9 Model target generator

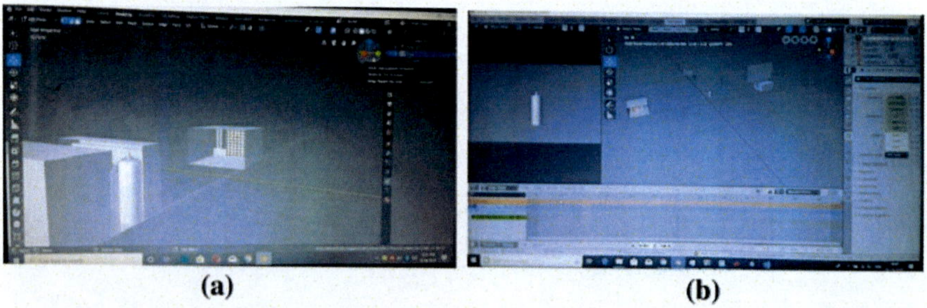

Fig. 10 **a** Animation 1, **b** animation 2

5 Result

The methodology proposed in this paper implements augmented reality in sales and marketing. Firstly, the physical environment is scanned with a smartphone's camera. After that it successfully detects the image targets and real-world objects with the help of Vuforia augmented reality software development kit. And finally, augments the 3D models in the real-world space and aware users about the product specifications which redirects them to the sales Web site to make the purchase.

The practical implementation has been demonstrated by developing and implementing a mobile application that scans an image target and an air conditioner in the real world and provides desired information to the user. It has achieved the following objectives:

- Successfully scans the target and augments 3D model on mobile screen of users.

- Add shopping value and generate more interest among the customers toward buying products online.
- Enhance customer awareness about the product by augmenting 3D models and information on the image or model scanned by the user.
- Engage customers with enhanced interaction capabilities.

6 Conclusion

The proposed system can be used by companies as a marketing tool to interact with customers on their mobile devices which will elevate interest among the users to interact with the advertising campaign of the brand more sincerely. AR gives different digital experiences to users which will enhance bonding between brand and their customers. Here, traditional ways of buying and e-commerce are combined which will lead to quicker purchase and improve the sales of the brand while consumers get to experience a unique and interesting way of interaction with the products and acquire a better understanding of the product and its specifications at the same time from this system.

References

1. Tom Dieck MC, Jung T (2018) A theoretical model of mobile augmented reality acceptance in urban heritage tourism. Curr Issues Tourism 21(2):154–174
2. Jung TH, Lee H, Chung N, Tom Dieck MC (2018) Cross-cultural differences in adopting mobile augmented reality at cultural heritage tourism sites. Int J Contemp Hospital Manage
3. Kico I, Liarokapis F (2019) A mobile augmented reality interface for teaching folk dances. In: 25th ACM symposium on virtual reality software and technology 2019, pp 1–2, Nov 2019
4. Peng F, Zhai J (2017) A mobile augmented reality system for exhibition hall based on Vuforia. In: The 2ND international conference on image, vision and computing (ICIVC). IEEE
5. Rodrigues JMF et al (2017) Adaptive card design UI implementation for an augmented reality museum application. In: International conference on universal access in human-computer interaction. Springer, Cham
6. Xiao C, Lifeng Z (2014) Implementation of mobile augmented reality based on Vuforia and Rawajali. In: IEEE 5th international conference on software engineering and service science. IEEE
7. Muilu T (2019) Technical opportunities and challenges in mobile augmented reality prototyping: CASE: Haaga-Helia UAS Freshman application
8. Amin D, Govilkar S (2015) Comparative study of augmented reality SDKs. Int J Comput Sci Appl 5(1):11–26
9. Kim J-A (2010) A study on the mobile advertisement attitude of smartphone users: focused on mobile advertisements using application. Department of Advertising Design Graduate School of Industry Hongik University
10. Kounavis CD, Kasimati AE, Zamani ED (2012) Enhancing the tourism experience through mobile augmented reality: challenges and prospects. Int J Eng Bus Manage

11. Gerstweiler G, Vonach E, Kaufmann H (2016) HyMoTrack: a mobile AR navigation system for complex indoor environments. Sensors
12. Waruwu AF, Putu Agung Bayupati I, Ketut Gede Darma Putra I (2015) Augmented reality mobile application of Balinese Hindu temples: DewataAR. Int J Comput Netw Inf Secur

A Novel Polycystic Ovarian Syndrome Diagnostic System Using Machine Learning

Rahul Katarya, Aarnav Jindal, Abhinav Duggal, and Abhishek Shah

Abstract Polycystic ovary syndrome is a hormonal disorder that plagues a markedly high population of fertile women around the globe. The percentage of positive cases recorded in countries is alarming ranging from 2.2 to 26% globally. Quick diagnosis and timely medical care can reduce the risk of associated complications like infertility, miscarriage, type 2 diabetes, and heart disease. The current system of diagnosis is time-consuming, laborious, and at the end still prone to errors. In this paper, we present a novel automated diagnostic system for efficient PCOS prognosis using machine learning on clinical data. Feature selection using PSO followed by a modified stacked generalization ensemble learning model is presented. The proposed system achieved a remarkable 90.74% accuracy transcending previously proposed diagnostic systems.

Keywords Polycystic ovarian syndrome · Diagnostic systems · Machine learning · Classification

---

R. Katarya · A. Jindal (✉) · A. Duggal · A. Shah
Department of Computer Science and Engineering, Delhi Technological University, New Delhi, India
e-mail: aarnavjindal1000@gmail.com

R. Katarya
e-mail: rahuldtu@gmail.com

A. Duggal
e-mail: abhinavduggal@gmail.com

A. Shah
e-mail: abhishek.shah24499@gmail.com

A. Abraham et al. (eds.), *Proceedings of 3rd International Conference on Computing Informatics and Networks*, Lecture Notes in Networks and Systems 167,
https://doi.org/10.1007/978-981-15-9712-1_47

1 Introduction

Polycystic ovary syndrome is among the most prevalent endocrine disorders which plagues women. An extensively recognized medical description of PCOS is the association of hyperandrogenism with chronic anovulation in women devoid of characteristic fundamental ailments of the adrenal or pituitary glands [1]. Though there is a lack of consensus on its primal cause, most practitioners agree that anomalous spiked levels of androgen affect the development and creation of eggs by ovaries. Genetic and environmental factors have been linked to androgen production imbalances.

It is thoroughly accepted today that PCOS has a chief effect throughout life on the reproductive, metabolic, and cardiovascular well-being of afflicted women [2]. It may also give way to metabolic syndrome which amps up the possibility of cardiovascular diseases, diabetes, and stroke. Sleep apnea may arise in women with PCOS [3]. The incessant problems of PCOS include but are not limited to depression and anxiety [4], type 2 diabetes, and endometrial and ovarian cancer [5]. It is one of the top causes of infertility [1].

Data mining is the science of exploring datasets to unearth unseen patterns. It lies at the convergence of statistics, machine learning, and database systems. Data mining is now playing a crucial role in the incrementally improving healthcare industry. It is being used extensively for finding possible unknown factors, establishing a relationship between various factors and their contribution to diagnosis and for automating diagnosis [6]. These modern systems are now being embedded in prognosis procedures to improve early detection and aid decision making while simultaneously reducing hospital errors which may prove fatal. Further, these systems are both time and cost-effective, the former being of prime importance. Timely discovery and care can be of utmost importance to avert ovarian failure, cancer, type-2-diabetes, and high blood pressure [7].

We propose an innovative PCOS diagnostic system in this paper. It uses particle swarm optimization for feature selection followed by a modified ensemble learning approach for classification of patient records. The system has shown appreciable classification accuracy.

The following organization describes the remainder of this document: Sect. 2 lists the related work in the field of PCOS diagnostic systems. Section 3 describes the dataset used to train the algorithm. Section 4 details the algorithm and the design of the suggested system. Section 5 shows the experimental results of our suggested system. Lastly, Sect. 6 covers the conclusion of the paper and suggests directions for future research.

2 Related Work

The majority of the early focus in PCOS diagnosis was centered around improving the quality of ultrasound images through preprocessing. A sonograph of the ovary displays blood vessels in the endometrium and stroma along with the follicles which are the objects of interest. The picture is also distorted by noise due to complexities in the procurement method. Thus, preprocessing always precedes the segmentation step to reduce false detections [8].

Efforts in image processing for PCOS diagnosis began with finding follicles and marking their boundary in the images. They utilized rudimentary gray-level thresholding and graph-searching techniques [9] and texture-based object recognition algorithms [10]. They paved the way for future efforts in the field. Focus since their inception has been on contrast enhancement and noise reduction techniques to improve object detection.

A follicle may often appear fragmented in an image leading to false detection of multiple cysts. Region growing algorithm [11] and altered labeled watershed algorithm [12] were introduced to deal with this issue. An adaptive morphological filter was proposed in [12] for denoising of images. Abrupt shifts in pixel values due to speckle noise are suppressed much better due to the above method. The authors also used properties of objects such as roundness and region of existence to improve the classification of follicles. Setiawati [13] employed top-hat transformation to better the contrast in images. It provides contrast stretching for dark as well as bright features to produce an improved image. A cost map methodology is proposed in [14] to differentiate the ovary from its environment and then using dynamic cost-based functions to successfully identify, grow and classify follicles. Swarm optimization-based segmentation [13] and horizontal and vertical scan line thresholding segmentation methods [15] were introduced for improved segmentation of follicles from the surroundings.

Research then shifted focus towards improving the diagnostic accuracy for PCOS. The improved follicle detection capabilities of image processing systems provide reliable input data about the quantity and other characteristics of these follicular cysts. This data along with clinical parameters such as blood level concentration of certain hormones and physiological characteristics like weight, height, age, etc., was then used to develop efficient machine learning diagnostic systems [6].

Statistical analysis and exploration provided insight into the importance of the number of follicles, LH and FSH values, BMI, and cycle duration. They were markedly dissimilar in patients with and without PCOS [16]. Feature selection is an important preprocessing step for machine learning. It helps reduce the humongous volume of data and filter out essential attributes to train the machine learning models while also improving classification accuracy. A neural fuzzy rough set algorithm was proposed in [17] for feature selection. It was then combined with an artificial neural network to achieve high classification accuracy in [18]. An interesting frequent itemset mining approach using the Apriori algorithm was also proposed for classification [19]. A centralized online learning diagnostic system was proposed in which can be connected to numerous infirmaries and clinics [20]. Zhang [21] proposed a system to identify genes important for the diagnosis of PCOS and improving classification accuracy. A competitive neural network was used in [22] for diagnosis.

Table 1 Attributes of the dataset

S. No.	Attribute	S. No.	Attribute	S. No.	Attribute
1	S. No.	15	Pregnant (Y/N)	29	Weight gain (Y/N)
2	Patient file	16	No. of absorptions	30	Hair growth (Y/N)
3	PCOS (Y/N)	17	FSH (mIU/mL)	31	Skin darkening (Y/N)
4	Age (yrs.)	18	LH (mIU/mL)	32	Hair loss (Y/N)
5	Weight (Kg)	19	FSH/LH	33	Pimples (Y/N)
6	Height (Cm)	20	Hip (in.)	34	Fast food (Y/N)
7	BMI	21	Waist (in.)	35	Reg. exercise (Y/N)
8	Blood group	22	Waist: hip ratio	36	BP _Systolic (mmHg)
9	Pulse rate (bpm)	23	TSH (mIU/L)	37	BP_Diastolic (mmHg)
10	RR (breaths/min)	24	AMH (ng/mL)	38	Follicle no. (*L*)
11	Hb (g/dl)	25	PRL (ng/mL)	39	Follicle no. (*R*)
12	Cycle (R/I)	26	Vit D3 (ng/mL)	40	Avg. *F* size (*L*) (mm)
13	Cycle length (days)	27	PRG (ng/mL)	41	Avg. *F* size (*R*) (mm)
14	Marriage status (years)	28	RBS (mg/dl)	42	Endometrium (mm)

3 Dataset

The dataset used for research purposes was a compilation of PCOS patient records obtained from ten hospitals across Kerala, India. It consists of 42 attributes and comprises of 363 negative and 177 positive records, putting the total up to 540 records [23]. The attributes of the dataset are presented in Table 1.

4 Proposed Approach

We present a modified stacked generalization ensemble learning-based diagnostic system. We begin by reducing the complexity of data by feature selection using particle swarm optimization followed by learning using a carefully selected set of models selected for ensemble learning.

4.1 Feature Selection Using PSO

Feature subset selection is an important part of data preprocessing. It helps to reduce the dimensionality of data by identifying irrelevant columns which can be removed from the dataset. Thus, it enhances the classification accuracy of the models while simultaneously reducing the time and memory required for learning [24].

Table 2 Improvement in performance attributes after using PSO feature selection

S. No.	Attribute	Improvement (%)
1	Accuracy	1.50
2	Training time	−48.67
3	Memory required	−47.72

Table 3 Filtered attributes obtained after using PSO feature selection

3, 6, 7, 9, 11, 13, 14, 15, 17, 21, 23, 25, 27, 28, 29, 30, 32, 33, 35, 36, 38, 39

Particle swarm optimization is a population-based search technique. It draws insight from natural organization and behavior of flocks of birds. It is a swarm intelligence-based technique which is useful for multi-modal combinatorial optimization problems.

We used the discrete PSO feature selection algorithm presented in [25] along with SVM as the fitness function. This helped improve the performance of the training process as described in Table 2. It also helped to reduce the initial number of columns of the dataset from 42 to 22 as described in Table 3.

4.2 *Training Stacked Generalization Ensemble Model*

We use six models as part of the ensemble. We use the predicted probability of three models: logistic regression, support vector classifier, and ridge classifier; and three classifier models: random forest classifier, bagging classifier, and gradient boosting classifier. The models were selected on the basis of high individual performance among tested models given in Table 4.

The first three models output a continuous value while the last three output a binary value. The output of the six models is combined with the initial data. The new dataset created is now fed to a neural network for final classification. The architecture of the suggested model is presented in Fig. 1.

Table 4 Performance of models individually

Model	Accuracy (%)
Logistic regression	88.86
Support vector classifier	88.32
Ridge classifier	89.05
Random forest classifier	89.99
Bagging classifier	87.57
Gradient boosting classifier	88.69

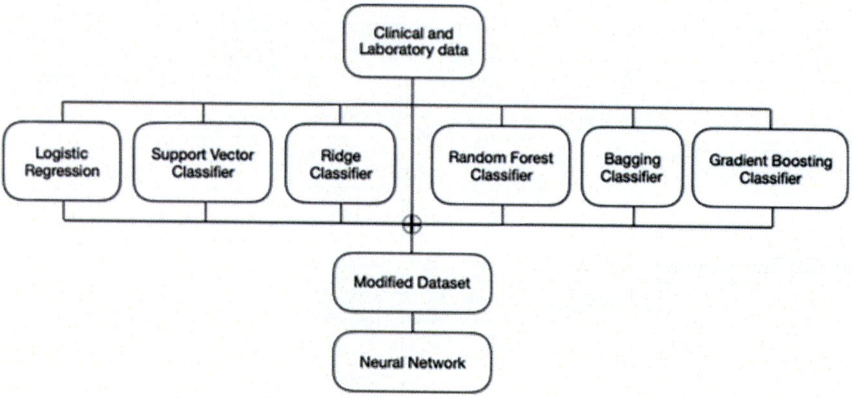

Fig. 1 Architecture of the proposed system

5 Experiments and Results

We compare our proposed system with three other models: ANN + NFRS [17], i-Hope [20], and CNN2018 [22]. True positive (TP), true negative (TN), false positive (FP), and false negative (FN) were evaluated using the dataset described in Sect. 3, which in turn were used to calculate the accuracy, precision, and recall of the system.

Accuracy and sensitivity are both effective measures to judge a model, with accuracy more commonly used in generalized evaluations. But medical diagnostic systems cannot be judged on accuracy alone, where the key to quick treatment is early detection. Thus, these systems need to predict all the positive cases of PCOS with better accuracy. A false positive is significantly more dangerous than a false negative in the case of PCOS. Therefore, along with high accuracy, these systems need to have high sensitivity.

$$\text{Accuracy} = (\text{TN} + \text{TP})/(\text{TN} + \text{FN} + \text{FP} + \text{TP}) \tag{1}$$

$$\text{Recall/Sensitivity} = \text{TP}/(\text{TP} + \text{FN}) \tag{2}$$

Our proposed model was compared with other systems on the basis of both accuracy and sensitivity. The results are shown in Fig. 2. It is clear that the model fares better than the other models in terms of the metrics under consideration.

6 Conclusion and Future Research Direction

A large focus has been given on image enhancement in the automated diagnosis of PCOS to facilitate manual analysis. Data mining and machine learning models

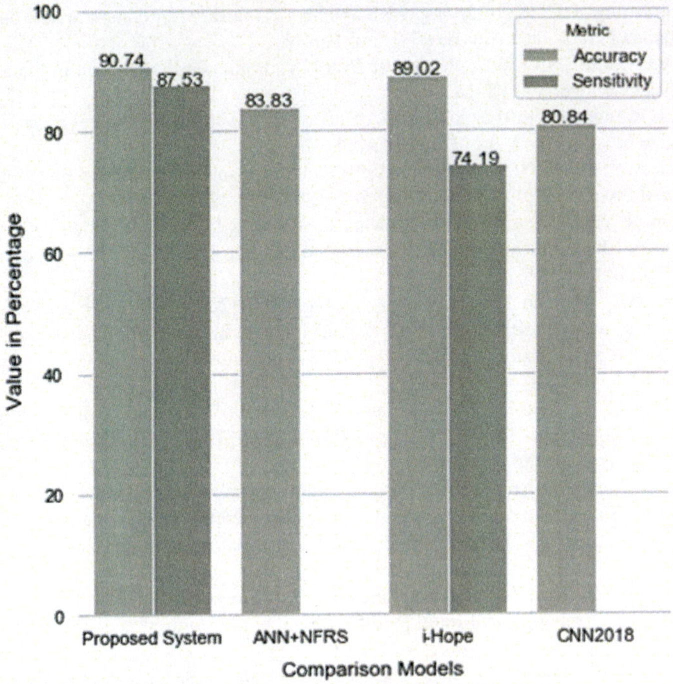

Fig. 2 Comparison of accuracy and sensitivity of the proposed model with other systems

have also been used for diagnosis. It is apparent that a hybrid approach combining enhanced follicle detection, feature extraction, and data mining produces superior results compared to any technique in isolation. Careful data preparation, statistical analysis, and modeling can help in creating superior ANNs with a very high degree of accuracy and sensitivity. Also, the introduction of centralized cloud storage where data from hospitals and clinics could be uploaded regularly can help create an extremely sophisticated online machine learning diagnostic system.

References

1. Franks S (1995) Polycystic ovary syndrome. N Engl J Med 333:853–861. https://doi.org/10.1056/NEJM199509283331307
2. Ehrmann DA (2005) Polycystic ovary syndrome. N Engl J Med 352:1223–1236. https://doi.org/10.1056/NEJMra041536
3. Ehrmann DA (2012) Metabolic dysfunction in PCOS: relationship to obstructive sleep apnea. Steroids 77:290–294. https://doi.org/https://doi.org/10.1016/j.steroids.2011.12.001
4. Cooney LG, Dokras A (2017) Depression and anxiety in polycystic ovary syndrome: etiology and treatment. Curr Psychiatry Rep 19:83. https://doi.org/10.1007/s11920-017-0834-2
5. Dumesic DA, Lobo RA (2013) Cancer risk and PCOS. Steroids 78:782–785. https://doi.org/https://doi.org/10.1016/j.steroids.2013.04.004

6. Vikas B, Sarangi S, Chilla M, Bhargav KS, Anuhya BS (2017) A literature review on the rising phenomenon PCOS. Int J Adv Eng Technol 10:216
7. Chang RJ, Coffler MS (2007) Polycystic ovary syndrome: early detection in the adolescent. Clin Obstet Gynecol 50:178–187
8. Hiremath P, Tegnoor J (2010) Automatic detection of follicles in ultrasound images of ovaries using edge based method. Int J Comput Appl
9. Krivanek A, Sonka M (1998) Ovarian ultrasound image analysis: follicle segmentation. IEEE Trans Med Imaging 17:935–944. https://doi.org/10.1109/42.746626
10. Muzzolini R, Yang Y, Pierson R (1993) Multiresolution texture segmentation with application to diagnostic ultrasound images. IEEE Trans Med Imaging 12:108–123. https://doi.org/https://doi.org/10.1109/42.222674
11. Lawrence MJ, Eramian MG, Pierson RA, Neufeld E (2007) Computer assisted detection of polycystic ovary morphology in ultrasound images. In: Fourth Canadian conference on computer and robot vision (CRV '07), pp 105–112. https://doi.org/https://doi.org/10.1109/CRV.2007.18
12. Deng Y, Wang Y, Chen P (2008) Automated detection of polycystic ovary syndrome from ultrasound images. Conf Proc IEEE Eng Med Biol Soc:4772–4775. https://doi.org/https://doi.org/10.1109/IEMBS.2008.4650280
13. Setiawati E, Tjokorda ABW (2015) Particle swarm optimization on follicles segmentation to support PCOS detection. In: 2015 3rd international conference on information and communication technology (ICoICT), pp 369–374. https://doi.org/https://doi.org/10.1109/ICoICT.2015.7231453
14. Deng Y, Wang Y, Shen Y (2010) An automated diagnostic system of polycystic ovary syndrome based on object growing. Artif Intell Med 51:199–209. https://doi.org/10.1016/j.artmed.2010.10.002
15. Mehrotra P, Chakraborty C, Ghosh Dastidar B, Ghoshdastidar S, Ghoshdastidar K (2011) Automated ovarian follicle recognition for polycystic ovary syndrome. Int Conf Image Inf Process. https://doi.org/https://doi.org/10.1109/ICIIP.2011.6108968
16. Deshpande SS, Wakankar A (2014) Automated detection of polycystic ovarian syndrome using follicle recognition. In: 2014 IEEE international conference on advanced communications, control and computing technologies, pp 1341–1346. https://doi.org/https://doi.org/10.1109/ICACCCT.2014.7019318
17. Meena K, Manimekalai M, Rethinavalli S (2015) Correlation of artificial neural network classification and NFRS attribute filtering algorithm for PCOS data. Int J Res Eng Technol 04:519–524. https://doi.org/https://doi.org/10.15623/ijret.2015.0403087
18. Meena DK, Manimekalai DM, Rethinavalli S (2015) A Novel framework for filtering the PCOS attributes using data mining techniques. Int J Eng Res Technol 4:702–706
19. Vikas B, Anuhya BS, Bhargav KS, Sarangi S, Chilla M (2018) Application of the Apriori algorithm for prediction of polycystic ovarian syndrome (PCOS). In: Bhateja V, Nguyen BL, Nguyen NG, Satapathy SC, Le D-N (eds) Information systems design and intelligent applications. Springer, Singapore, pp 934–944
20. Denny A, Raj A, Ashok A, Ram CM, George R (2019) i-HOPE: detection and prediction system for polycystic ovary syndrome (PCOS) using machine learning techniques. In: TENCON 2019 IEEE region 10 conference (TENCON), pp 673–678. https://doi.org/https://doi.org/10.1109/TENCON.2019.8929674
21. Zhang X-Z, Pang Y-L, Wang X, Li Y-H (2018) Computational characterization and identification of human polycystic ovary syndrome genes. Sci Rep 8:1–7
22. Dewi RM, Wisesty UN (2018) Classification of polycystic ovary based on ultrasound images using competitive neural network. J Phys Conf Ser 971:12005. https://doi.org/https://doi.org/10.1088/1742-6596/971/1/012005
23. Kottarathil P (2020) Polycystic ovary syndrome (PCOS), https://www.kaggle.com/prasoonkottarathil/polycystic-ovary-syndrome-pcos/version/1. Last accessed 10 Jan 2020

24. Masilamani A, Anupriya E, Iyenger NCSN (2010) Enhanced prediction of heart disease with feature subset selection using genetic algorithm. Int J Eng Sci Technol 2
25. Unler A, Murat A (2010) A discrete particle swarm optimization method for feature selection in binary classification problems. Eur J Oper Res 206:528–539. https://doi.org/https://doi.org/10.1016/j.ejor.2010.02.032

Overview of Amalgam Models for Type-2 Diabetes Mellitus

Ravika Rajput, Rakesh Kumar Lenka, Samuel Jacob Chacko, Khan Ghazala Javed, and Abhishek Upadhyay

Abstract Diabetes is reaching in widening proportions in many developing countries in the world due to a lack of health understanding and poor eating habits. Type-2 diabetes (as of late suggested as non-insulin-ward, or adult start) originates from the body's inefficient usage of insulin. Looking over the statistics around 425 million adults, age between 20 and 79 years are suffering from type-2 diabetes; which has caused around 4 million deaths to date. Therefore, it is important to have a machine learning technique that can adequately pinpoint the tendency of type-2 diabetes in the heap of fragments quickly. AI has conjured a lot of enthusiasm among specialists. The proof introduced by a few reports recommends that AI approaches have the potential to yield higher precision in the order of information in contrast with different techniques. This paper reviews five different amalgam classification techniques for the prediction of type-2 diabetes. RF-WFS and XGBoost algorithm give the most elevated precision, affectability, and specificity of the model are 93.75%, 91.79%, and 94.8%, separately in the forecast of the disease.

Keywords Type-2 diabetes · Amalgam models · Machine learning · PIMA indian dataset · Artificial intelligence

R. Rajput (✉) · R. K. Lenka · S. J. Chacko · K. G. Javed · A. Upadhyay
International Institute of Information Technology Bhubaneswar, Bhubaneswar, India
e-mail: a118005@iiit-bh.ac.in

R. K. Lenka
e-mail: rakeshkumar@iiit-bh.ac.in

S. J. Chacko
e-mail: a118007@iiit-bh.ac.in

K. G. Javed
e-mail: a118004@iiit-bh.ac.in

A. Upadhyay
e-mail: a118001@iiit-bh.ac.in

A. Abraham et al. (eds.), *Proceedings of 3rd International Conference on Computing Informatics and Networks*, Lecture Notes in Networks and Systems 167,
https://doi.org/10.1007/978-981-15-9712-1_48

Abbreviations

ADASYN	Adaptive synthetic sampling method
AI	Artificial intelligence
ANN	Artificial neural network
ASSM	Adaptive synthetic sampling method
BMI	Body mass index
DT	Decision tress
FN	False negative
FP	False positive
MLP	Multi layer perceptron
NB	Naïve Bayes
RF	Random forest
SVM	Support vector machine
TN	True negative
TP	True positive

1 Introduction

The early recognition of the illness assumes an extremely urgent job as no solution for diabetes at present exists; however, the illness can go into abatement whenever recognized soon. This methodology can help in approving sort 2 diabetes and in affirming coronary illness with less effort [1]. The proof introduced by a few reports recommends that AI approaches have the potential to yield higher precision in the order of information in contrast with different techniques. This is done to think of a roadmap of ways to deal with order built up procedures of AI in diagnosing maladies of the diabetes.

The endeavors being made are to check how different strategies of AI work [2] in characterizing whether patients having type-2 diabetes or not. This paper gives the outline of five amalgams AI models for the arrangement of the type-2 diabetes nearness in the persistence utilizing the PIMA Indian dataset.

The outline of the paper is as per the following, segment 2 covers the Literature Review, segment 3 gives information on the PIMA Indians informational index and connection among its highlights, segment 4 comprises of five distinctive ensembled models for the classification of the datasets, segment 5 portrays the assessment of the execution of these models, and segment 6 assesses and looks at the exhibitions dependent on numerous parameters of the models. Segment 7 at last finishes up the paper and gives the future extension.

2 Literature Review

Xu and Wang [3] has proposed the half and half model by utilizing RF and XGboost. RF and XGboost ensemble classifiers in information pre-handling have managed missing qualities by placing in the middle. After the pressing the information, XGboost ensembled was applied to perform grouping. Wang and Cao [4] have played out the classification by using NB alongside ADSAY and RF. NB is used to manage the misbalancing and RF performs the classification. Parashar et al. [5] has done the forecast for type-2 diabetes utilizing an amalgamation of K-means and choice tree. K-means deals with outliers in the dataset which is then fed to DT for classification. Ahmed et al. [6] have joined the K-means with SVM for the classification. This algorithm uses the same mechanics of K-means and DT, but the classification is assisted by SVM. Dewangan and Agrawal [7] have joined MLP with Bayesian likelihood for diabetes expectation. This model is a hybrid of MLP combined with Bayesian probability.

3 Materials and Methods

The gathered information was gotten from Kaggle's Web platform PIMA Indians diabetes dataset [8], class variable which is the essential wellspring of Research Center of National Institute of Diabetes and Digestive and Kidney Diseases, RMI Group Leader Applied Physics Laboratory at The Johns Hopkins University. The all-out number of instances is 768, which is totally utilized right now. This dataset involves an aggregate of 268 emphatically tried cases and 500 adversely tried cases, where attributes of dataset are pregnancies—number of pregnancies, glucose—concentration of plasma glucose 2 h in an oral glucose resilience test, blood pressure—diastolic circulatory strain (mm Hg), skin thickness—thickness of the triceps skin overlap in mm, insulin—two-hour serum insulin (mu U/ml), BMI—body mass index, diabetes pedigree function—diabetes family work, age—age in years, and outcome—class variable (0 or 1), i.e., 268 of 768 are 1, the others are 0.

The dataset comprised of an aggregate of 768 examples with eight highlights for every example. The connection between the properties of the dataset is appeared by a relationship network. Figure 1 shows that the class segment 'Outcome' is decidedly associated with every single other parameter. Looking nearer the class segment has a high positive connection with the 'glucose.'

4 Review of the Algorithms

Xu and Wang [3] have proposed a mixture calculation that is *RF X XGBoost* to play out the order on type two diabetes. In the wake of playing out the preparing

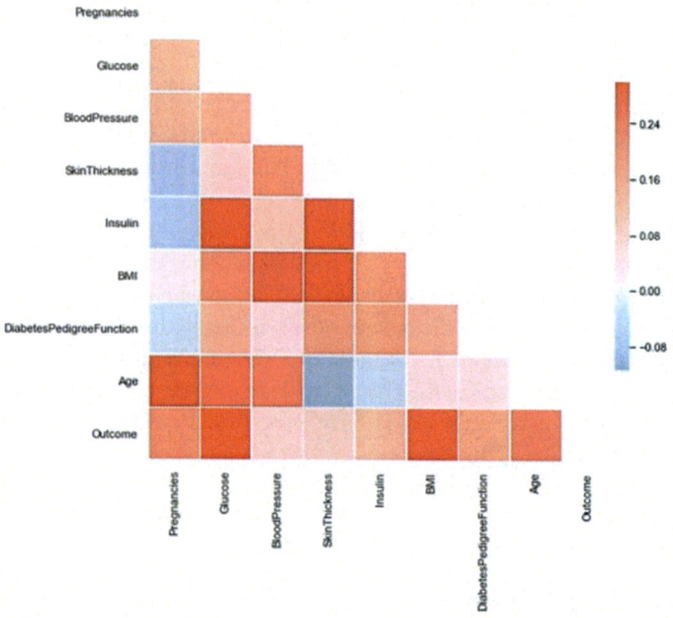

Fig. 1 Correlation graph of PIMA Indian dataset

step, random forest [9] is performed to compute weighted element determination calculation, that is, out of eight traits mulling over just those that contribute most toward the assurance of the event of type-2 diabetes [8]. Right now, though behind actualizing, ideal component determination is proportional down the multifaceted design and to upswing the precision, parity, and extensibility of the model [10].

Naïve Bayes X Adaptive Synthetic Sampling Method X Random Forest classifier [5] has broken the forecast issue into three sections that are missing worth issue [11], class awkwardness, and last expectation. Subsequent to playing out the pre-handling step, the last order is performed utilizing the random forest classifier. RF is a multi-class classifier that expands upon the choice tree. Right now, classifier gains from the highlights independently and gives expectations.

K-Means X Decision tree [4] classifier calculation, to group PIMA India informational collection [8], makes out of four stages that is information pre-preparing, data decrease, classification, and evaluation. In information pre-preparing, as the PIMA India informational collection have missing qualities, to manage them, this calculation proposes to fill the missing an incentive with the mean of the separate property. Information decrease manages the anomalies and the commotion in the information. After spotting the blunders, the broken records are being dispensed with from the informational collection to have immaculate information for order and accomplish the best form of grouping [12].

KSVM [6] composes of the following stages to classify the PIMA India dataset [8]. By acquiring the principle attributes of the training data, we can reduce the magnitude of the dataset which is called dimension reduction. K-means [10] clustering classifies the dataset into two clusters hence providing the sharp two class points to do training on the features of diabetes in K-means of the individual patient which is reviewed to clinch that the features of the samples in a cluster are allied to one another but different from the attributes of the elements existing in other clusters. Further classification is assisted by SVM [6].

The nature of the MLP [7] is the consecutive system that is it is liberated from a circle and goes about as a feedforward system. ANN can proximate the outcomes with the help of performing non-direct capacity on this concealed layer present in the system [11]. Choosing the parameters of the concealed layer and the number of layers is now and again a matter of experimentation and changes with the issue explanation. Sigmoidal capacity goes about as an exchange work comprehensively. The backpropagation calculation is the preparation base for the MLP neural system [7]. To decide the last class in the system, the probabilities are chosen utilizing the Bayesian net.

5 Classification Performance Measurement

Five amalgam ML calculation are tried for the classification of type-2 diabetes informational index.

- True positive (TP)—tests which have a place with positive class and are anticipated with nearness of type-2 diabetes.
- False positive (FP)—tests which have a place with positive class and are anticipated with non-attendance of type-2 diabetes.
- True negative (TN)—tests which have a place with negative class and are anticipated with non-appearance of type-2 diabetes.
- False negative (FN)—tests which have a place with negative class and are anticipated with nearness of type-2 diabetes.

Proportion of execution is given underneath:

Accuracy is characterized by the proportion of the number of tests that are anticipated properly in the dataset to the general size of the dataset.

$$\text{Accuracy} = (\text{TP} + \text{TN})/(\text{TP} + \text{TN} + \text{FP} + \text{FN}) \quad (1)$$

Sensitivity (review or genuine positive) is the proportion of the example anticipated as positive and was having a place with a positive class to the all-out number of positives anticipated.

$$\text{Sensitivity} = (\text{TP})/(\text{TP} + \text{FN}) \quad (2)$$

Specificity (genuine negative) is the proportion of the example anticipated as negative and was having a place with the negative class to the absolute number of unlucky deficiencies of type-2 diabetes anticipated.

$$\text{Specificity} = (\text{TN})/(\text{TN} + \text{FP}) \quad (3)$$

6 Results

See Fig. 2 and Table 1.

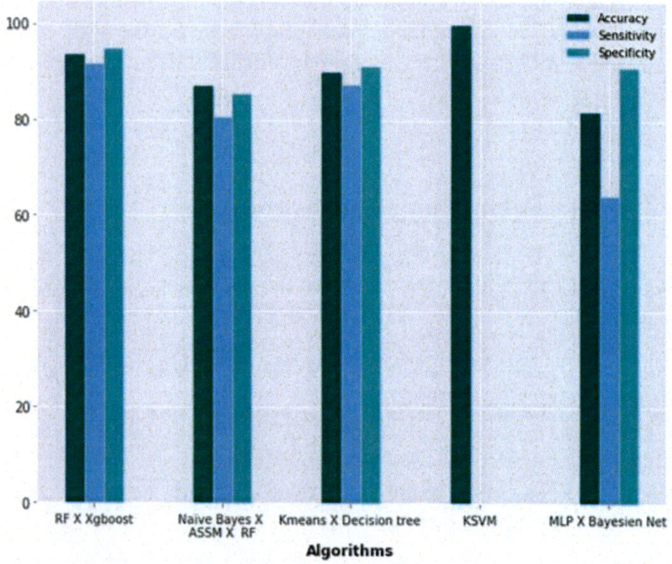

Fig. 2 Performance comparison of the algorithms

Table 1 Comparison between accuracy, specificity, and sensitivity of the algorithms

Algorithm	Accuracy	Specificity	Sensitivity
RF X XGBoost [3]	93.75%	94.8%	91.79%
NB X ADASYN X RF [5]	87.1%	85.4%	80.6%
K-means X DT [4]	90.04%	91.28%	87.2%
KSVM [6]	99.90%	–	–
MLP X Bayesian Net [7]	81.89%	90.90%	64.10%

7 Conclusion and Scope

Random forest X XGBoost [3] classifier show that the best outcomes among all the calculations with a high score of exactness, affectability, and explicitness as 93.75, 91.79, and 94.8% individually. It has managed information pre-handling of missing qualities by placing in the middle. Naive Bayes X ASSM X random forest [5] has separately managed missing worth issue by setting up the mean and has settled the class irregularity issue too; however, it has not upgraded the preparation parameters and the expulsion of the exceptions from the informational collection bringing about least score in assessment among every one of the five calculations. K-means X decision tree [4] has managed missing an incentive by including mean rather than zeros and distinguishes the anomalies by utilizing K-means grouping. Besides, it shows less score of affectabilities demon-starting its powerlessness to manage class irregularity in the informational index. Utilizing the ideal highlights will help in managing the abnormality of the information collection and will help in giving steady outcomes. KSVM [6] execution estimation yields high exactness, yet at the same time cannot be considered as best among all as it does not research the serious issue of awkwardness in the informational index. Moreover, it is just accentuating the precision, not on affectability which is significant with regards to managing imbalanced informational collections like PIMA India. Adjusting the informational index is significant when managing clinical informational collections. This calculation has just played out the element decrease and has not concentrated on clamor in the type of missing worth and exceptions. Random forest and XGboost calculation can be additionally improved by managing the outliers and missing estimations of the informational index.

References

1. WHO. https://www.who.int/diabetes/world-diabetes-day-2018/en/. Last accessed 2020/11/07
2. Malchoff CD (2011) Diagnosis and classification of diabetes mellitus. Diabetes Care 34(Suppl 1):S62–S69
3. Xu Z, Wang Z (2019) A risk prediction model for type 2 diabetes based on weighted feature selection of random forest and XGBoost ENSEMBLE CLASSIfiER. In: Eleventh international conference on advanced computational intelligence (ICACI)
4. Parashar A, Burse K, Rawat K (2014) A comparative approach for Pima Indians diabetes diagnosis using LDA-support vector machine and feed forward neural network. Int J Adv Res Comput Sci Softw Eng 4:378–383
5. Wang Q, Cao W, Guo J, Ren J, Cheng Y, Davis DN (2017) DMP_MI: an effective diabetes mellitus classification algorithm on imbalanced data with missing values. IEEE Access 7:102232–102238
6. Osman AH, Aljahdali HM (2017) Diabetes disease diagnosis method based on feature extraction using K-SVM. Int J Adv Comput Sci Appl (IJACSA) 8(1)
7. Dewangan AK, Agrawal P (2015) Classification of diabetes mellitus using machine learning techniques. Int J Eng Appl Sci (IJEAS) 2. ISSN 2394-3661
8. UCI repository of machine learning databases. University of California at Irvine, Department of Computer Science. http://www.ics.uci.edu/~mlearn/databases/thyroiddisease/newthyroid.data. Last accessed 28 December 2020

9. Sabariah MTMK, Hanifa STA, Sa'Adah MTS (2014) Early detection of type II diabetes mellitus with random forest and classification and regression tree (CART). In: Advanced informatics: concept, theory and application. IEEE, pp 238–242
10. Chen W, Chen S, Zhang H, Wu T (2017) A hybrid prediction model for type 2 diabetes using K-means and decision tree. In: 2017 8th IEEE international conference on software engineering and service science (ICSESS), pp 386–390
11. Vaishnav RL, Patel KM (2015) Analysis of various techniques to handling missing value in data set. Int J Innov Emerg Res Eng 2(2):191–195
12. Anuja KV, Chitra R (2013) Classification of diabetes disease using support vector machine. Int J Eng Res Appl 3:1797–1801

Feature Selection on Public Maternal Healthcare Dataset for Classification

Shelly Gupta, Shailendra Narayan Singh, and Parsid Kumar Jain

Abstract Feature selection is required for effective and efficient data analysis. It is a preprocessing step in data mining which reduces the inputs for analytical task. It is effective in improving the results and increasing the learning accuracy by reducing the data dimensionality and selecting only the relevant variables for modeling. In this paper, we have analyzed the importance of feature selection for classification on maternal health data of Uttar Pradesh for the year 2015–16. In this study, the wrapper method with best first greedy approach is used for features subset selection. The reduced dataset has shown approximately 4.6% increase in the balanced accuracy of the generated classifier over the classifier generated on the high-dimensional original data.

Keywords Feature selection · Data mining · Wrapper model · Classification · Maternal health · National rural health mission (NRHM)

Abbreviations

ANC	Ante natal care
ASHAs	Accredited social health activists
JSSK	Janani Shishu Suraksha Karyakaram
JSY	Janani Suraksha Yojna
MDGs	Millennium development goals
MMR	Maternal mortality rate

S. Gupta (✉) · S. N. Singh
ASET, Amity University, Noida, Uttar Pradesh, India
e-mail: shelly.gupta24@gmail.com

S. N. Singh
e-mail: snsingh36@amity.edu

P. K. Jain
HCMS, Yamunanagar, Haryana, India
e-mail: parsidkumarjain@gmail.com

A. Abraham et al. (eds.), *Proceedings of 3rd International Conference on Computing Informatics and Networks*, Lecture Notes in Networks and Systems 167,
https://doi.org/10.1007/978-981-15-9712-1_49

NPD	Non-priority district
NRHM	National rural health mission
PD	Priority district
SBA	Skilled birth attendant
UNFPA	United nations population fund
UNICEF	United nations international children's emergency fund
WHO	World health organisation

1 Introduction

In this age of information, the organizations have very large volume of data which is increasing day by day. Now, these high-dimensional data repositories have become a big challenge for data mining process. This is a serious problem to many machine learning algorithms with respect to their performance and scalability.

Feature selection is required for effective and efficient data analysis. It is a preprocessing step in data mining which reduces the inputs for analytical task. It is effective in improving the results and increasing the learning accuracy by reducing the data dimensionality and selecting only the relevant variables for modeling.

For this study, the data is collected from NHRM portal of India. The selected dataset is the data about the maternal health of Uttar Pradesh state for year 2015–16. The original dataset has 20 attributes, and their description is given in Table 1.

The purpose of this work is to explore the importance of feature selection on maternal health data for building effective and efficient predictive model using classification methods.

This paper is further organized into six more sections. In Sects. 2 and 3, a brief introduction to wrapper model for feature selection and feature selection for classification is discussed. However, Sect. 4 gives a vision about the current status of maternal health in India. Sections 5 and 6 provide the methodology and experimental results of this paper. At last, in Sect. 7, the conclusion is given on the study.

2 Wrapper Model for Feature Selection

Feature selection is required for effective and efficient data analysis. It is a preprocessing step in data mining which reduces the inputs for analytical task. It is effective in improving the results and increasing the learning accuracy by reducing the data dimensionality and selecting only the relevant variables for modeling [1]. There are mainly two types of features selection models for flat features. One is filter model which selects the features independently without involving any machine learning procedure. This may lead us to select non-optimal features subset as the effect of selected feature subset on the introduced procedure for learning which is totally

Table 1 Attributes description

S. No.	Attribute	Attribute description	Type
1	PD/NPD	District_category_Priority_Non-Priority Districts	Binary
2	P_W_R_ANC	Pregnant_women_Registration_for_Ante Natal Care (ANC)	Range
3	P_W_R_ANC_1ST_TRISEM	Pregnant_women_Registration_for_ANC_within first trimester (within 12 weeks)	Range
4	P_W_R_JSY	Pregnant_women_registeration_under_Janani Suraksha Yojna (JYS)	Range
5	P_W_3ANC_CHECKUPS	Pregnant_women_3_ANC_check ups_Recieved	Range
6	P_W_100IFA_TABS	Pregnant_women_taken_100 IFA tablets	Range
7	D_HOME	Deliveries_at Home_Under_trained or non-trained Skilled Birth Attendent (SBA)	Range
8	NB_VISIT_24 HRS_D_HOME	Newborns_visit_within 24 hours_after Deliveries_at Home	Range
9	JSY_INTENSIVE_D_HOME	mothers_paid_JSY incentive_Deliveries_at Home	Range
10	D_PUBLIC INSTITUTION	Deliveries_at Public Institutions (Including C-Sections)	Range
11	D_PUBLIC INSTITUTION_48 h DISCHARGE	Deliveries_at Public Institutions_Mothers_discharged_under_48 h	Range
12	JSY_INTENSIVE_D_PUBLIC INSTITUTION	mothers_paid_JSY incentive_Deliveries_at Public Institutions	Range
13	ASHA_JSY_INTENSIVE_D_PUBLIC INSTITUTION	Accredited Social Health Activists (ASHAs)_paid_JSY Incentive_Deliveries_at Public Institutions	Range
14	D_PRIVATE INSTITUTION	Deliveries_at Private Institutions (Including C-Sections)	Range
15	JSY_INTENSIVE_D_PRIVATE	mothers_paid_JSY incentive_Deliveries_at Private Institutions	Range

(continued)

Table 1 (continued)

S. No.	Attribute	Attribute description	Type
16	ASHA_JSY_INTENSIVE_D_PRIVATE INSTITUTION	ASHAs_paid_JSY Incentive_Deliveries_at Private Institutions	Range
17	TOTAL_P_W_OC_PUBLIC INSTITUTION	Pregnant_women_Obstetric Complications at Public Institutions	Range
18	PP_CHECKUPS_48 h	Post-Partum_check-up_within_48 h	Range
19	PP_CHECKUPS_48HRS_14 DAYS	Post-Partum_checkup_between_48 h to 14 days	Range
20	PNC_MC	Post-Natal Care_maternal_complications	Range

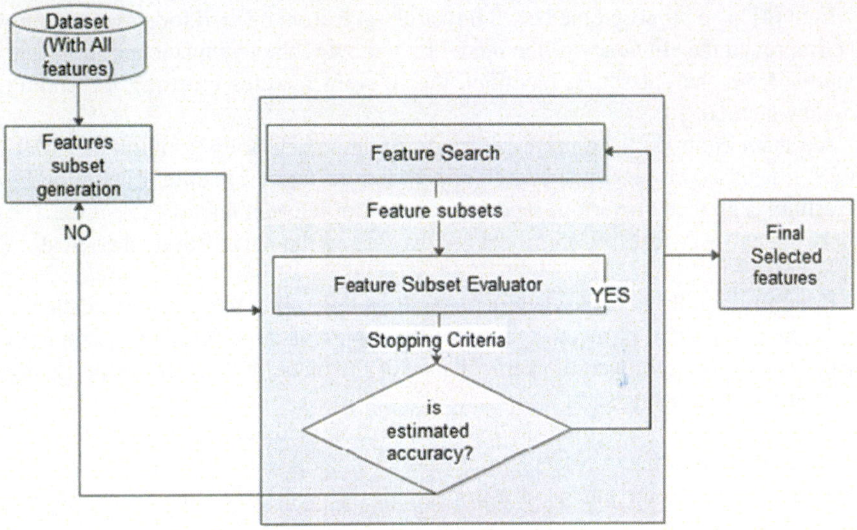

Fig. 1 General wrapper model structure for feature selection

ignored. To overcome this disadvantage, other model is defined, i.e., wrapper model for feature selection. The general structure of the wrapper model is shown in Fig. 1. The wrapper model performs the following steps for optimal features subset selection.

(1) Features subset is searched for evaluation from the all possible subsets for the given data.
(2) The selected features subset is evaluated by the performance of the classifier.
(3) The step 1 and 2 are repeated until the stopping criteria is met, i.e., the estimated accuracy of the classifier.
(4) Finally, the selected features subset is passed for modeling.

3 Feature Selection for Classification

Many research papers related to feature selection and classification of data for data mining are studied to explore the research area.

Novakovic et al. [2] performed a comparative study of various feature selection ranking methods. For this, they have divided the six different ranking methods into two categories, i.e., statistical and entropy-based methods and generated the models using Naïve Bayes and C4.5 decision tree algorithms. In this research, they found that ranking methods give different results for accuracy with different supervised learning methods.

Tiwari and Singh [3] have proposed a new approach based on correlation for feature selection using genetic algorithm. The features get reduced by using this method.

Lutu [4] have proposed the Naïve Bayes classification method for stream mining and improved the efficiency of the model by removing the redundant and irrelevant features from the dataset. It identified the relevant features by using the sliding window method.

Abraham et al. [5] have proposed a novel feature selector algorithm, i.e., CHI-WSS. It is based on greedy approach or best first to select a feature. The proposed algorithm is applied on various medical datasets for relevant features selection. The improvement was observed in model building using the naïve Bayes classification algorithm.

Keerthika and Priya [6] have used Naïve Bayes classifier to classify the data and Naïve feature reduction method to select the inputs of importance. The proposed methodology has improved the performance of the model to be used for designing an intrusion detection system.

Tang and Liu [7] have used the feature selection methods and reduced the data dimensionality for effective data mining on social media data. For feature selection, they have evaluated the effects of user-user and user-post relationships manifested in social media data.

Wang et al. [8] have proposed a family of feature selection methods for online data. They have divided the online feature selection task in two different tasks, one is learning with all attributes, and other is learning with some selected or partial attributes. They have applied the proposed methods on image classification for computer vision problem and in bioinformatics problem for microarray gene expression data for analysis. The results are appreciable for modeling.

From the above discussion, it is very clear that the feature selection and classification have improved the learning in various felids like medical diagnosis, intrusion detection, microarray classification, computer image vision, social media, etc. So, the above discussion gives us the vision that more effective work can also be done in public healthcare data. For this paper, we have chosen the maternal health data.

4 About Maternal Health: India

The health of a woman during her expecting time, baby delivery, and health after the birth of a child is known as maternal health [9], and maternal mortality rate (MMR) is a statistical value to measure the maternal health status of a country or state. It is the count of the maternal deaths of women over 100,000 live births [9]. For a nation, it is important to maintain its maternal health as it is the main health indicator to define the health sector status of that nation. According to a report by WHO, the value for MMR is quite low in developed countries, whereas high in developing countries [9]. According to United Nations Population Fund (UNFPA), there is a decline in MMR from 380 maternal deaths in 1990 to 210 deaths in 2013. The MMR of many countries has reduced to half, but even after such a good statistical value of MMR globally still there are some regions in Africa and Asia where this rate is much higher [10].

The Indian Government has done many efforts to reduce the MMR rate in the country by providing free maternal health services to economically poor mothers and promoting the institutional deliveries via Schemes like Janani Suraksha Yojana (JSY) and Janani Shishu Suraksha Karyakaram (JSSK). As a result, it is found that the MMR has declined from 212 deaths in 2007 to 167 deaths in 2013 which is a remarkable achievement under the Millennium Development Goals (MDGs) [10–12]. However, according to UNICEF, in the year 2017, the recorded MMR of India was 174 [10, 15]. This statistical value has raised question on maternal health status of the women in India again. In our previous work by performing an empirical analysis on public maternal health data OF India, we found that the status of maternal health in Uttar Pradesh state of India among various states is under trouble [13]. This is our motivation to use Uttar Pradesh state data for our work.

5 Methodology

In this paper, the classification, i.e., supervised learning approach along with feature selection is used as the research methodology to classify the given target accurately. The feature selection methods are accountable for the data dimensionality reduction and improving the learning in data mining for effective predictions and knowledge discovery [14, 16]. A general structure of supervised learning approach with feature selection is shown in Fig. 2. The brief description of various steps in supervised learning approach is given below:

Step 1: Data selection and collection: In this step the relevant data is selected and collected from the dataset source. In this work, the public maternal health data of Uttar Pradesh state of India for the year 2015–16 is selected to classify the priority districts and non-priority districts as per the NHRM standards. The data is collected from the NHRM portal of India.

Step 2: Data preprocessing: In this step to upgrade the quality of data, the data is cleaned and transformed. For this study, the data is handled in terms of missing data and to remove the noisy data.

Step 3: Feature selection: In this step, the feature selection is performed to reduce the data dimensionality and to improve the learning of the generated model in terms of accuracy of classifier. For this study, the wrapper model for feature selection is used. Here, the features subset selection is based on the greedy approach, i.e., best first, and the label information for classifier accuracy estimation is provided from the dataset.

Step 4: Classification: In this step, the supervised machine learning algorithm is applied on the reduced dataset to classify the target label. The reduced dataset is firstly divided as training data to generate the model and testing data to validate the model.

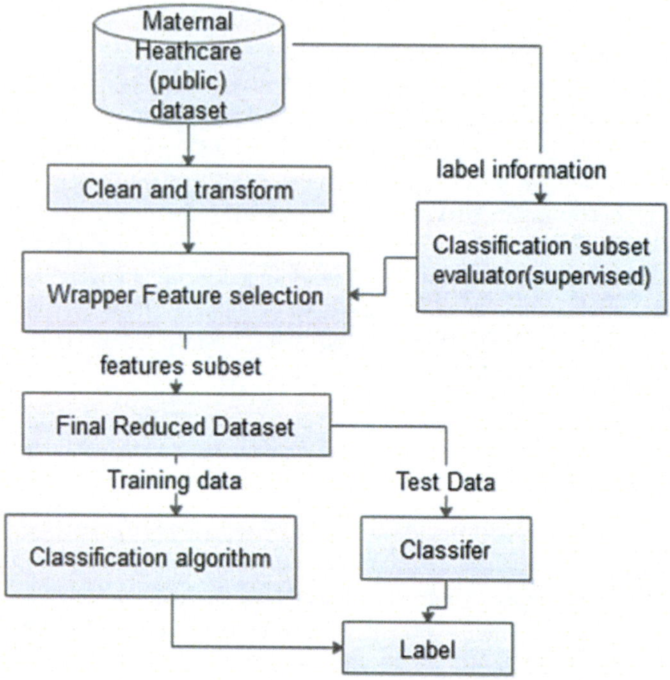

Fig. 2 General structure of supervised learning approach with feature

6 Experimental Results

In this study, the maternal health data of Uttar Pradesh state of India for the year 2015–16 is selected to classify the priority districts and non-priority districts as per the NHRM standards. The data is collected from the NHRM portal of India. Initially, the dataset consists of 20 distinct attributes and 65 instances. The J48 decision tree classification algorithm is applied on the original data in WEKA 3.9 [17], and the classifier has identified the priority and non-priority districts correctly up to 67.69%. The tenfold cross validation is used as test mode. The result is shown in Fig. 3.

In order to improve the accuracy of the model, the wrapper features subset selection is performed. For this purpose, the classifier subset evaluator for J48 classifier has selected the relevant attributes by using the greedy approach, i.e., best first and reduced the original data dimensional from 20 attributes to 6 attributes, as shown in Fig. 4. The detailed description of the selected attributes is given in Table 2.

Now, J48 decision tree classification algorithm is applied on the reduced data in WEKA 3.9, and the classifier has identified the priority and non-priority districts correctly up to 72.3%. The tenfold cross validation is used as test mode. The accuracy gets improved from 67.69 to 72.3% which is a remarkable improvement to validate the importance of the feature selection for maternal health data. The result is shown in Fig. 5.

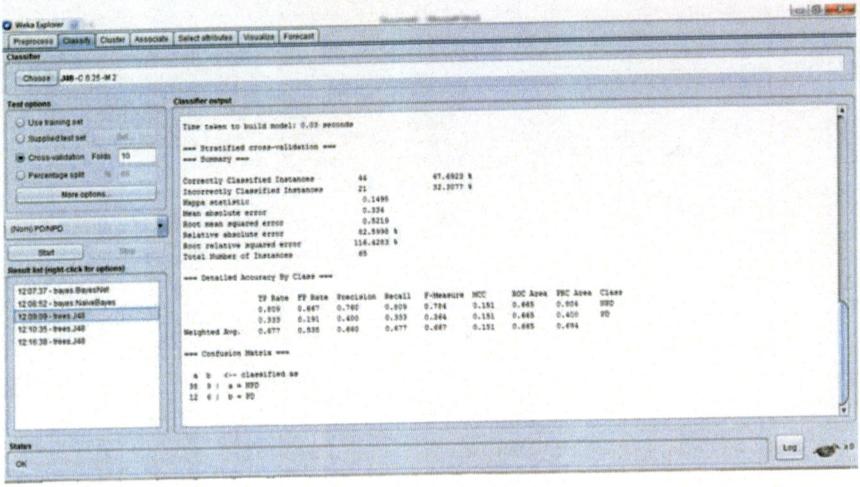

Fig. 3 J48 decision tree classification results on the original dataset

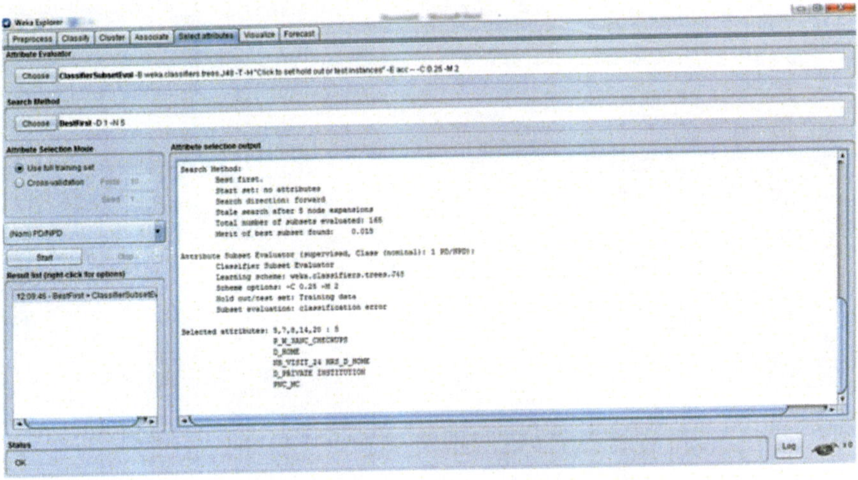

Fig. 4 Feature subset selection using J48 as classifier evaluator

7 Conclusion

In this paper, J48 decision classification technique of data mining is applied on public maternal health data of the Uttar Pradesh state of India for the year 2015–16. Initially, 20 attributes are considered for classification, and then, data is reduced to mainly 6 attributes of high relevance using wrapper method for feature subset selection using greedy approach. This attribute selection approach has improved the balanced accuracy to identify the target correctly approximately by 4.6%. This improvement in

Table 2 Description of the selected attributes

S. No.	Attribute	Attribute description	Type
1	PD/NPD	District_category_Priority_Non-Priority Districts	Binary
2	P_W_3ANC_CHECKUPS	Pregnant_women_3_ANC_check ups_Recieved	Range
3	D_HOME	Deliveries_at Home_Under_trained or non-trained Skilled Birth Attendent (SBA)	Range
4	NB_VISIT_24 HRS_D_HOME	Newborns_visit_within 24 hours_after Deliveries_at Home	Range
5	D_PRIVATE INSTITUTION	Deliveries_at Private Institutions (Including C-Sections)	Range
6	PNC_MC	Post Natal Care_maternal_complications	Range

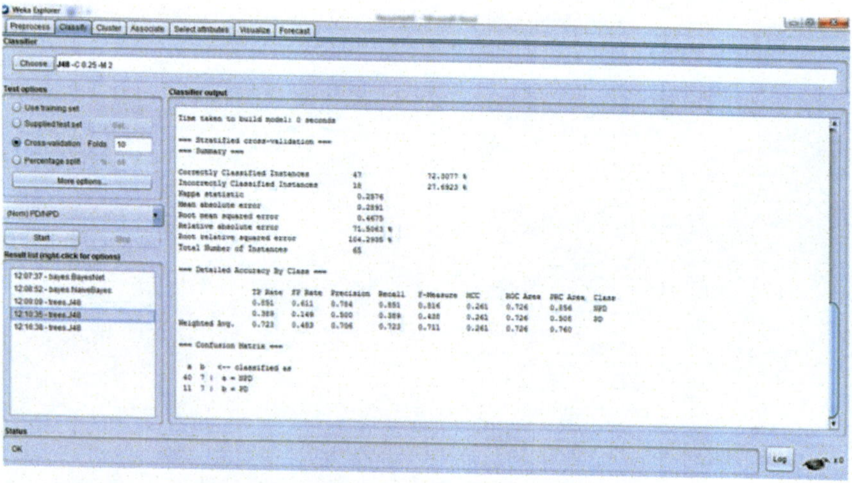

Fig. 5 J48 decision tree classification results on the reduced dataset

accuracy rate is remarkable for building efficient model. It is concluded from this analytical study that by applying appropriate feature selection methods for classification of maternal health, data with given target value can help the health planners to plan maternal health in different districts. Further work can be done on feature selection by using Apriori algorithm for association on given data. This may provide a better planning approach to health planners over small dimensional maternal health data.

References

1. Kohavi R, John GH (1997) Wrappers for feature subset selection. Artif Intell 97(1–2):273–324
2. Novakovic J, Strbac P, Bulatovic D (2011) Toward optimal feature selection using ranking methods and classification algorithms. Yugoslav J Oper Res 21(1):119–135
3. Tiwari R, Singh MP (2010) Correlation-based attribute selection using genetic algorithm. Int J Comput Appl 4(8):28–34
4. Lutu PEN (2013) Fast feature selection for Baive Bayes classification in data stream mining. Proc World Congr Eng 3
5. Abraham R, Simha JB, Iyengar SS (2009) Effective discretization and hybrid feature selection using Naïve Bayesian classifier for medical data mining. Int J Comput Intell Res 5(2):116–129
6. Keerthika G, Priya DS (2015) Feature subset evaluation and classification using Naïve Bayes Classifier. J Netw Commun Emerg Technol 1(1)
7. Tang J, Liu H (2012) Feature selection with linked data in social media. SDM
8. Wang J, Zhao P, Hoi S, Jin R (2014) Online feature selection and its applications. IEEE Trans Knowl Data Eng 26:698–710
9. World Health Organization Maternal Health. http://www.who.int/topics/maternal_health/en/. Last accessed 2019/05/26
10. UNICEF—Maternal Health India. http://www.unicef.org/health/index_maternalhealth.html. Last accessed 2019/05/12
11. Vora KS, Mavalankar DV, Ramani KV, Upadhyay M, Sharma B, Iyengar S, Gupta V, Lyengar K (2009) Maternal health situation in India: a case study. J Health Popul Nutr 27(2):184–201
12. Randive B, Diwan V, de Costa A (2013) India's conditional cash transfer programme (the JSY) to promote institutional birth: is there an association between institutional birth proportion and maternal mortality? PLoS One https://doi.org/10.1371/journal.pone.0067452
13. Gupta S, Singh SN, Kumar D (2016) An empirical analysis of maternal health data: a case study of India. In: 2nd international conference on NGCT. IEEE
14. Kumar V, Minz S (2014) Feature selection: a literature review. Smart Comput Rev 4(3):211–229
15. Health and Family Welfare statistics in India 2015. https://nrhm-mis.nic.in/SitePages/Pub-FW-Statistics2015.aspx. Last accessed 2016/10/16
16. Han J, Kamber M (2001) Data mining: concepts and techniques, 2nd edn. Morgan Kauffmann Publishers, San Francisco
17. WEKA 3.9: a machine learning tool. http://www.cs.waikato.ac.nz/ml/weka/downloading.html. Last accessed 2019/10/24

Investigation of DDoS Attacks Influence on Steering Data in Mobile E-healthcare Sector

Ashu, Rashima Mahajan, and Sherin Zafar

Abstract The most promising upcoming technologies these days are wireless mesh networks (WMN) as it being established with the amalgamation of various wires and wireless networks. This is also known to be one category of mobile adhoc network (MANET) in which various nodes are either stationary or allied with static position. There are many design issues in designing the protocols for MANET routing protocol. Since the applications of such routing protocols vary from remote monitoring of defense-related operations of military, monitoring during crisis situation. Similarly these days due to pandemic situation, remote monitoring is very crucial in e-healthcare sector. Thus, it is important to come up with the best routing protocol for transferring the data from one end to other. An attacker node or device can introduce many malicious activities via internal and external attacks at multiple layers, thus impacting the tasks of important routines which need to be completed in stipulated time. In this, initially, the distributed denial of service (DDoS) attacks are implemented on three most suitable routing protocol on NS-2. Than the most suitable routing protocol among the three were compared, and results were tabularized for two different scenario for the packet loss ratio (PLR) as quality of service (QoS), which aids in providing implications to enhance the most suitable protocols and improve the consequence of instigated DDoS. The performance of adhoc on demand vector (AODV), secured adhoc on demand (SAODV) and hybrid wireless mesh protocol (HWMP) are compared and tabularized. The simulation results show that the HWMP outperformed well than other two routing protocols as to evaluation metrics PLR under the varying number of nodes and is much less vulnerable against DDoS attack dominant in e-healthcare region.

Ashu (✉) · R. Mahajan
MRIIRS, Faridabad 122017, India
e-mail: ashuone@gmail.com

R. Mahajan
e-mail: rashima.fet@mriu.edu.in

S. Zafar
Department of CSE, SEST, Jamia Hamdard, New Delhi 110062, India
e-mail: zafarsherin@gmail.com

A. Abraham et al. (eds.), *Proceedings of 3rd International Conference on Computing Informatics and Networks*, Lecture Notes in Networks and Systems 167,
https://doi.org/10.1007/978-981-15-9712-1_50

Keywords Adhoc on demand vector (AODV) · Secured AODV (SAODV) · Hybrid wireless mesh protocol (HWMP) · Packet loss ratio (PLR) · Adhoc networks for e-healthcare

Abbreviations

ANTHOCNET	Ant based hybrid routing algorithm for mobile and adhoc networks
AODV	Adhoc on demand vector
DDoS	Distributed denial of service
DoS	Denial off service
HCR	Hybrid clustering routing
HWMP	Hybrid wireless mesh protocol
MANET	Mobile adhoc networks
NS-2	Network simulator -2
OLSR	Optimised link state routing protocol
PLR	Packet loss ratio
QoS	Quality of service
SAODV	Secured AODV
SYN	Sync flooding attacks
WMN	Wireless mesh networks
ZRP	Zone routing protocol

1 Introduction

The wireless healthcare system networks are incredibly susceptible to numerous categories of existing attacks because of their intermediate medium, which is universal wireless access and constrained assets. The intended users will be unable to make use of network resources or infrastructure in the presence of such attacks. As the WMN runs on unlicensed 2.4 GHz band, the attacker aims to prevent the service from working efficiently either temporally or indefinitely. To reduce network resources through directing a rush of packets from a malicious node, clients and routers whose number could be one or more, is the focal target of DDoS attack [1]. All the layers of WMN are susceptible to these types of attacks. Disseminated refusal of administration could be one of the consequences, i.e., DDoS assaults can cut down your site and abandon you from reaching to the remainder of the Web for a considerable length of time or days on end. Thirty-three percent of organizations were hit by DDoS assaults in 2017, about twofold the number of assaults in 2016. Eighty-two percent of associations have confronted various DDoS assaults [2]. Such attacks may harm wireless healthcare application network, which can headed toward damage of patient's life. In healthcare system, acute applications such as location tracking, home-assisted monitoring systems are mostly affected by DDoS attacks. In applications related to healthcare sector, there is strong need to send the information from one device to other sensing device which may be far off their radio range. Thus, routing is a very crucial facility for the success of end-to-end announcement. So far, many reactive and hybrid routing protocols have been suggested for networks. MANET has ability of creating network without some setup for an emergency or for a temporary period [3]. Therefore, because of this convergence of MANET and WSN, MANET is appropriate to be integrated with Internet of Things (IoT) for healthcare environment.

1.1 MANET and DDoS Attacks

MANET is one of the broad category of wireless networks, in which nodes communicate among themselves and create a network, without the presence of any infrastructure or centralized administrative support. Innumerable characteristics of MANET had also raised serious concerns for its security and their related challenges. MANET comprehends a number of imperfections which make its prone to various attacks and thus becomes an easy target for attackers [4]. The irrational activity of nodes in MANET makes changes unpredictable in the topology. Scalability is one more apprehension in MANET because of movement of nodes which could access the network as well as leave dependently. The above-mentioned effects make MANET vulnerable to several types of attacks, broadly classified as internal attacks and external attacks. In this research work, the SYN flood attack which belongs to the one of the category of DoS attack is focused [5]. We can categorize the possible attacks in MANET, on the types of sources, second could be the criterion of behavior, and third based on the nodes involved which may lead to diverse kinds of attacks at different layers of MANETS; network layer, transport layer specifically [4]. Among these, the most dangerous category is the denial of service (DoS) attacks. SYN flood attack is one class of DoS assaults. In various application based on cloud services, it is well known that security brings utmost complications specially when dealing with sensitive information belonging to patients. The chief concern as regards to flooding attack is that the malicious flooder node floods the entire network. Flooding attacks main goal is to make higher power consumption in context to battery usage and bandwidth, resulting in degradation in performance of the network. SYN flood: In this attack, malicious flooder node directs the numeral synchronization packets toward the destination node. Hence, the large amount of memory will be consumed through this attack [6].

This category could be caused by manipulating the handshake contrivance of a TCP connection depicted in Fig. 1. An intruder or flooder node propels a series of SYN requests to the destination node or device. Therefore, the node or the device beneath attack is incapable to finish the complete three-way protocol of handshake for establishing a reliable server–client connection.

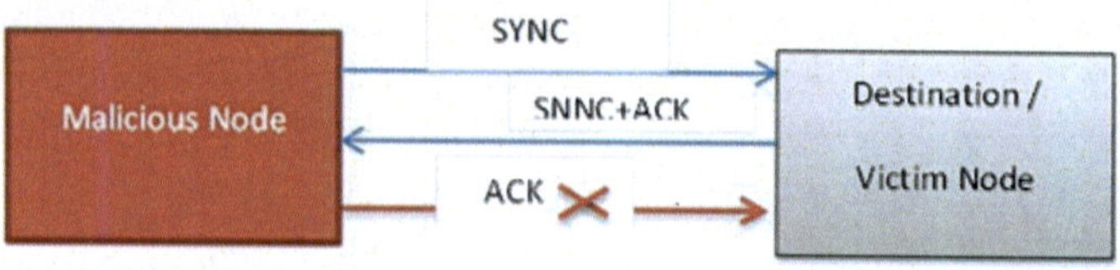

Fig. 1 Schematic description of SYN flood

1.2 E-healthcare

This section elaborates the different layers used in remote monitoring of services. Health care system architecture broadly consists of expert layer comprising physician of patient or healthcare professionals, proficient for allocating procedure for treatment and could approve the recommendations from analysis layer. Secondly layer is comprised of application layer and incorporates both local processing analysis and storage of data as well as complex processing, typically responsible for collecting data and also to integrate the same with external data. This cloud layer is very much susceptible to many kinds of attacks; ultimately resulting in DDoS at each layer [7]. Thirdly private clients' layer consisting of body sensor networks (BSN) layer, which is responsible for collecting the important vital information of physiological variables and then passing on to cloud gateway via routing decision layer. The routing decision layer comprise of IoT devices and sensor controller unit. The IoT devices have a direct link with the local processing, storage, and analysis layer via Wi-Fi, GSM, Ethernet 2G/3G, etc. The smart phone could be an IoT device. These sub-layers are capable, for processing sensor output and retrieving context, integrating with cloud and providing feedback to and from the patient [8, 9].

Introduction section of the research paper has focused on concepts of networks and attacks, further in Sect. 2, relevant analysis has been done on the various researches carried over for tackling security in healthcare, WMN, and MANET. Section 3 depicts broad area that focuses on various protocols used in transmission layer. Section 4 includes the scenario of simulation model used in this research work followed by results based on PLR as quality of service (QoS) parameters for varying scenarios. The discussions and conclusion on the research work conducted are listed in Sect. 6, followed by future and current work of research and references which are listed in the end.

2 Literature Review

Relevant analysis attends to the various investigations carried out in the security field for numerous categories of attacks in different types of complex networks. Studies carried on complex networks for a finding of attacks could be classified as for improving the routing protocol [10]. Thus, assessment of reactive and hybrid routing protocol becomes the foundation of the planned research work. Various researchers have analyzed network performance under black hole attack through NS by taking into considering the most popular AODV protocol. Network conditions were simulated with dissimilar affecting speeds, different quantities of communication, and malicious nodes. The investigational results display several influences of black hole attacks on QoS factors, which endorse the investigation of MANET security measures [11]. A novel structure called pigeon principle and hybrid encryption headed for sensing and overcoming the flaw named watchdog discussed in this study encounters various problems in MANET like: partial dropping, complexity, and ambiguous impacts, which have not been taken care by other existing research schemes. However, there is a need to improve the rate of data transmission that needs to be considered along with complete energy harvesting [12]. In one research work, qualnet simulator 5.0 was used for simulation of ZRP and OLSR for throughput and delay [13]. A review of potential security and protection dangers that can bargain the usefulness of a WSNs medicinal services framework is discussed by the researchers [14]. Researchers in one of the study overviews various parts of IoT-based human services advances and exhibits different social insurance arrange structures and stages that help access to the IoT spine and encourage restorative information transmission and gathering.

3 Routing Protocols

The increased interest of researcher in the network security and routing protocols is majorly due to wide usage of Internet to each and every device. The security provision for Internet depended devices can be assured or recognized in terms of three ways as firstly quality of service (QoS), secured network structure, and security of the end equipment's [15]. End system security is classically achieved by setting up a post-security barrier, and new firewall technology is utilized to expand "defense perimeter" by protecting Internet and intranet for various organizations with a trivial amount of firewall systems. End-to-end security is well suitable to offer confidentiality, authentication, and integrity. Protected QoS presents a numeral of novel security challenges: authentication and authorization of operators who require affluent network assets, mutually to prevent resource holdup and to avoid denial of service due to unapproved traffic, etc. An influence on system arrangement can cause denial of service from the user's viewpoint; however, for network engineer, the invader is taking benefit of the absence of authenticity, integrity, and perhaps privacy [16]. In the upcoming subsection, routing protocols for WMN are analyzed.

3.1 AODV

The most universally used reactive protocol is AODV [17] that produces and maintains paths simply when they are demanded. In comparison with the dynamic source routing, in which a node can cache multiple route for same destination, AODV maintains a routing table, and each node will have a information about the previous and the next node. The dynamic establishment of route table entries, reflecting the involvement of only active nodes for maintaining routing info thus emphasizing that the routing table entry is expired if not used recently. The time stamp is associated with each entry of the routing table, specifying the time at which that entry should be removed from routing table in form of destination sequence number. In case, if information is old, than the destination sequence number is used to check whether the incoming information is fresh or not.

3.2 SAODV

AODV works in an implicit 'trust your neighbor' mode. There are two main security requirements (a) node authentication and (b) message integrity. Secure AODV (SAODV) adds an extension to different AODV packet formats to incorporate digital signature for protecting the non-mutable information and hash chains to protect the mutable information (hop count). SAODV also has two mechanism divided as route discovery and secondly maintenance of the route similar to like that of AODV. The major alteration amid them is about the process of route detection process. SAODV increases the process of directly verifying the destination node by using the exchange of random numbers [18].

3.3 HWMP

Hybrid protocols conglomerate features of proactive and reactive protocols for achieving enhanced results. Such routing protocols fix the overhead routing of proactive processes and postpone reactive processes. In this, networks are separated into zones and appropriate for large networks. Hybrid protocols practice reactive routing, for route discovery scenario. However, for table maintenance process, the proactive protocols are widely utilized by hybrid routing protocols. The kind of hybrid routing protocols that are most widely used for discovering routes are: ZRP, hybrid clustering routing (HCR) and ant-based hybrid routing algorithm for mobile and adhoc networks (ANTHOCNET) [19]. They have the prospective to deliver higher scalability in comparison with pure reactive or proactive protocols. Hybrid procedures can deal with greater scalability in comparison with traditional ones. Since the HWMP emerged because of combination of reactive plus proactive routing protocols, making an easiest way to establish a link with IoT–WMN–MANET based technologies. However, the most severe cons of such protocol is the inability to identify attacker nodes [20].

4 Simulation Results

This research study is implemented on mobility model for evaluating AODV, SAODV, and HWMP routing protocols. The important characteristic of adhoc networks is the frequent changes in incoming and outgoing nodes for sending the information from source to destination. Packet loss ratio (PLR) is the unwavering quality of a correspondence that organizes the packet loss rate which dictates way. This measurement is equivalent to the number of packets not got partitioned by the complete number of packets sent. The estimation is accomplished by utilizing an arrangement number in every packet and checking the missing numbers. Simulation environment setup for mobility-based model with varying number of nodes on three protocols AODV, SAODV, and HWMP is given above in Table 1. The second scenario for comparing the PLR among the three routing protocols is given in Table 3 reflecting the variation in simulation time.

The simulation results for AODV, SAODV, and HWMP are presented by varying number of nodes for two QoS parameters namely PLR and throughput which are presented. Comparison of PLR for three routing protocols is given in Table 2. Figure 2 represents the bar graph, whereas Figs. 3 and 4 are x graphs for PLR of three protocols in absence and in presence of attacks, respectively.

From comparison Table 2, it is prominent that for varying number of nodes scenario, HWMP outperforms both from AODV and SAODV both under the influence and absence of DDoS attack. Packet loss ratio is decreased in case of HWMP.

For 150 nodes, HWMP is 28.99 and 12.99% better than AODV and SAODV, respectively, in absence of attacks. Under the influence of attacks for 150 nodes, HWMP is 29.27 and 7.60% better than AODV and SAODV, respectively. From comparison Table 3, it is prominent that HWMP outperforms from AODV and SAODV both under the influence and absence of DDoS attack.

Table 1 Simulation environment with varying nodes on routing protocols in MANET

Parameters	Value used
No. of nodes	30, 60, 90, 120, 150
Area traffic	1800 × 840 CBR
Simulation time	60 s
No. of connections Traffic rate Speed	20 4 packets/s 20 m/s
Packet size	1024

Table 2 Assessment of PLR with changing number of nodes for routing protocols

Routing protocols	PLR for varying no. of nodes (30, 60, 90, 120, 150)				
	30	60	90	120	150
AODV without attack	47.0218	65.9206	52.0082	60.8259	65.0271
AODV with attack	54.8218	73.351	58.5882	65.1351	69.2239
SAODV without attack	30.569	37.175	40.897	48.993	52.868
SAODV with attack	32.724	38.986	45.08	50.099	52.991
HWMP without attack	27.697	34.965	36.227	41.063	46.174
HWMP with attack	31.355	37.597	38.159	44.167	48.959

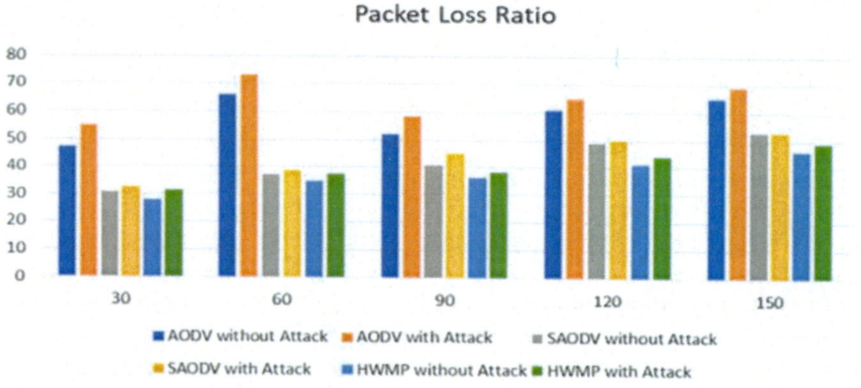

Fig. 2 Comparison of PLR for routing protocols with respect to varying nodes

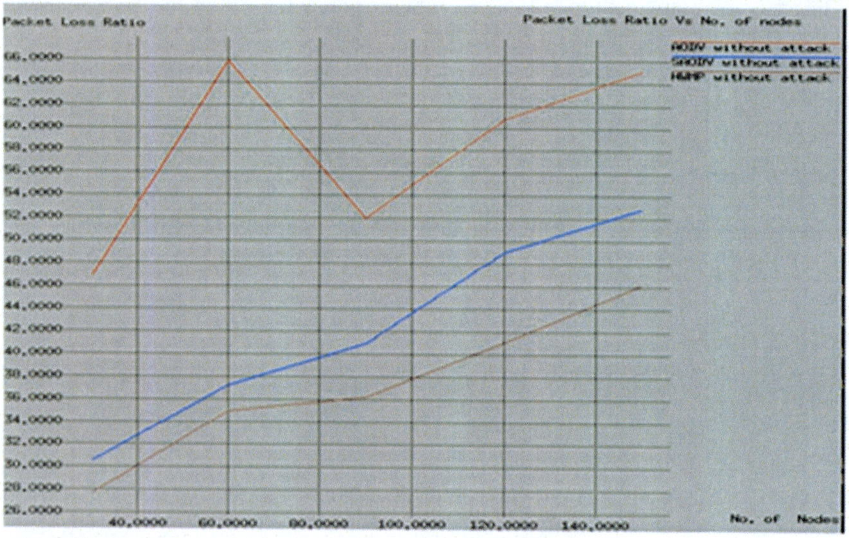

Fig. 3 PLR in absence of attacks for AODV, SAODV, and HWMP

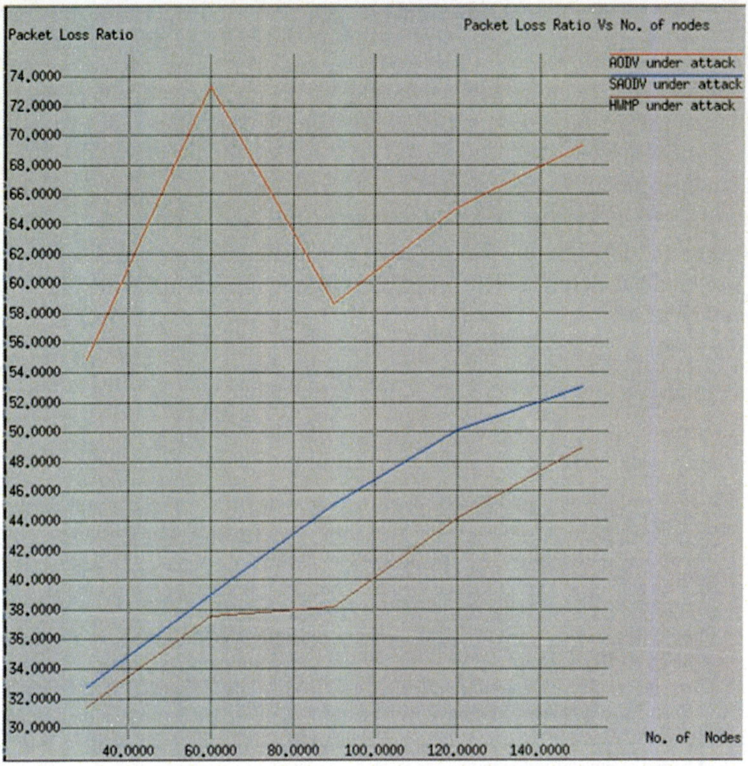

Fig. 4 PLR under influence of DDoS attacks for AODV, SAODV, and HWMP

5 Discussions and Conclusion

For ensuring privacy and security of patients' data during transmission in healthcare sector, MANET could be used as one of the important technologies combined with IoT. For dealing with end-to-end data transmission of patients' sensitive data, MANET protocol plays a vital role for sending the information securely. The effect of DDoS attacks has been shown and compared. The simulation results of NS2 for the PLR as QoS parameter evaluate that HWMP outperforms from the other two traditional routing protocols considered in the simulation. This research work analyzed the performance of three routing protocols AODV, SAODV, and HWMP, respectively, both in absence and under the influence of DDoS attacks; hence, HWMP can be quiet beneficial to be adopted for the healthcare environment The result emphasize that hybrid routing protocol should be preferred in situations where data to be transmitted is very important and any packet loss during transmission is not at all desirable specifically related to healthcare sector.

6 Current and Future Developments

In future, this research study will be considering the measures on other QoS parameters for different scenarios on network simulator 2. The main objective will be to provide enhancement in security of the best suitable routing protocol among the compared three routing protocol, as security is the major concern in all e-healthcare-related data exchange of patients. Therefore, the decentralized form of block chain will be explored for ensuring the utility of legitimated equipment's used for monitoring the vital characteristics of patients, being the main issue in data transferring in e-healthcare.

References

1. Alippi C, D'Alto V, Falchetto M, Pau D, Roveri M (2017) Detecting changes at the sensor level in cyber-physical systems: methodology and technological implementation. In: 2017 international joint conference on neural networks (IJCNN). IEEE, pp 1780–1786
2. Reference. Available from https://www.msspalert.com/cybersecurity-research/kaspersky-lab-study-average-cost-of-enterprise-ddos-attack-totals-2m/
3. Alvarez J, Maag S, Zaidi F (2017) Monitoring dynamic mobile ad-hoc networks: a fully distributed hybrid architecture. In: 31st international conference on advanced information networking and applications (AINA). IEEE, pp 407–414
4. Brill C, Nash T (2017) A comparative analysis of MANET routing protocols through simulation. In: 12th international conference for internet technology and secured transactions (ICITST). IEEE, pp 244–247
5. Degirmencioglu A, Erdogan HT, Mizani MA, Yılmaz O (2016) A classification approach for adaptive mitigation of SYN flood attacks: preventing performance loss due to SYN flood attacks. In: NOMS 2016 IEEE/IFIP network operations and management symposium, pp 1109–1112
6. Džaferović E, Sokol A, Abd Almisreb A, Norzeli SM (2019) DoS and DDoS vulnerability of IoT: a review. Sustain Eng Innov 1(1):43–48. ISSN 2712-0562
7. Islam SR, Kwak D, Kabir MH, Hossain M, Kwak KS (2015) The Internet of Things for health care: a comprehensive survey. IEEE Access 3:678–708
8. Oh S, Cha J, Ji M, Kang H, Kim S, Heo E, Han JS, Kang H, Chae H, Hwang H, Yoo S (2015) Architecture design of healthcare software-as-a-service platform for cloud-based clinical decision support service. Healthcare Inform Res 21(2):102–110
9. Yadav P, Agrawal R (2018) A multi-homing based framework against denial of service open threat signaling in healthcare environment. Int J Control Autom 11(11):1–18
10. Latif R, Abbas H, Assar S (2014) Distributed denial of service (DDoS) attack in cloud-assisted wireless body area networks: a systematic literature review. J Med Syst 38(11):128
11. Zamini M, Hasheminejad SMH (2019) A comprehensive survey of anomaly detection in banking, wireless sensor networks, social networks, and healthcare. Intell Decis Technol 13(2):229–270
12. Petrenko A, Kyslyi R, Pysmennyi I (2018) Designing security of personal data in distributed health care platform. Technol Audit Prod Reserves 4(2(42)):10–15
13. Govindasamy J, Samundiswary P (2017) A comparative study of reactive, proactive and hybrid routing protocol in wireless sensor network under wormhole attack. J Electr Syst Inform Technol

14. Gautam A, Mahajan R, Zafar S (2020) DDoS attacks impact on data transfer in IoT-MANET based E-healthcare for tackling COVID-19. In: International conference on data analystics and management (ICDAM-2020), Jan Wyzykowski University, Poland, 18 June 2020
15. Gautam A, Mahajan R, Zafar S (2019) Implementing blockchain security to prevent DDoS attacks in networks. Int J Secur Appl (IJSIA) 13(4):27–40. NADIA
16. Gautam A, Mahajan R, Zafar S (2020) Repercussions of DDoS attack on MANET based healthcare sector routing protocols performance and ANOVA assessment. Int J Adv Sci Technol 29(05):12157–12177
17. Sharma AK, Trivedi MC (2016) Performance comparison of AODV, ZRP and AODVDR routing protocols in MANET. In: 2016 second international conference on computational intelligence and communication technology (CICT). IEEE, pp 231–236
18. Sengupta A, Sengupta D, Das A(2017) Designing an enhanced ZRP algorithm for MANET and simulation using OPNET. In: 2017 third international conference on research in computational intelligence and communication networks (ICRCICN). IEEE, pp 153–156
19. Walikar GA, Biradar RC (2017) A survey on hybrid routing mechanisms in mobile ad hoc networks. J Netw Comput Appl 1(77):48–63
20. Kaur J, Singh G (2017) Hybrid AODV algorithm for path establishment in MANET using bio inspired techniques. In: international conference on power, control, signals and instrumentation engineering (ICPCSI). IEEE, pp 823–829

Plant Biotic Disease Identification and Classification Based on Leaf Image: A Review

Prabhjot Kaur and Vinay Gautam

Abstract The plant diseases could be the reason of diseases in different parts of a plant such as leaf, root, and stems; however, leaf is one of the most important parts to be observed to identify and detect infection. In literature, different techniques are proposed to identify and classify plant diseases based on texture, shape, and size of leaf image. Each and every technique utilized automated technologies such as artificial image analysis efficiently. This paper summarized the study of various automated tools, techniques, and computation method used in different stages of plant diseases identification and detection with their pros and cons. With the help of discovery of set of aspect, this paper highlights several points which can be considered for future research direction. This review helps others to grab knowledge of automated applications and techniques used to detect plant diseases.

Keywords Biotic diseases · Image · Classification · Identification

1 Introduction

The standard and amount of farming merchandise mainly depends on plants in all countries. There are varied plant lesions that seriously scale back the amount and standard of agriculture products. The detection of plant lesions automatically in leaves gives comparatively reasonable and simpler solutions as it will scale back an oversized work of observation in massive plant fields though several researches use image process technique to ease this tough task [1–5]. A plant is infected with different variety of diseases. It makes the visualization strategies difficult to identify it [6, 7]. Also with the use of medical pesticides, it led to excessive build-up in air, water, soil or even in our bodies. As we know, many humans and animals are

P. Kaur (✉) · V. Gautam
Chitkara University Institute of Engineering and Technology, Chitkara University, Rajpura, Punjab, India
e-mail: prabhjot.k@chitkara.edu.in

V. Gautam
e-mail: vinay.gautam@chitkara.edu.in

A. Abraham et al. (eds.), *Proceedings of 3rd International Conference on Computing Informatics and Networks*, Lecture Notes in Networks and Systems 167,
https://doi.org/10.1007/978-981-15-9712-1_51

dependent on plants and fruits for energy. Over more than 50% of population of Asia and Africa depends on agriculture developments for employment, food security, and earnings [8]. Also the uses of medical attributes come from herbs and leaf of various plants. Annual plant losses for fungal diseases reach around $250 billion and over $650 million are on fungicides [9, 10]. The millions are spent on disease control and to pacify the damage of it. Plant diseases can categorized under two broad categories such as biotic and abiotic as shown in Fig. 1. The detection of diseases experienced by many agricultural experts and plant pathologist using naked eyes may often fail and incorporate which increases complexity that led to wrong solution [11]. The request for prevalent quality and security is main concern. Some of the techniques or methods are fluorescence imaging and chain reaction [12–14].

This paper constitutes different section as follows: Sect. 1.1 introduces plant leaf diseases and symptoms. Leaf diseases detection process is listed in Sect. 1.2. The related work is described in Sect. 2. Finally, Sect. 3 is conclusive section with future aspects.

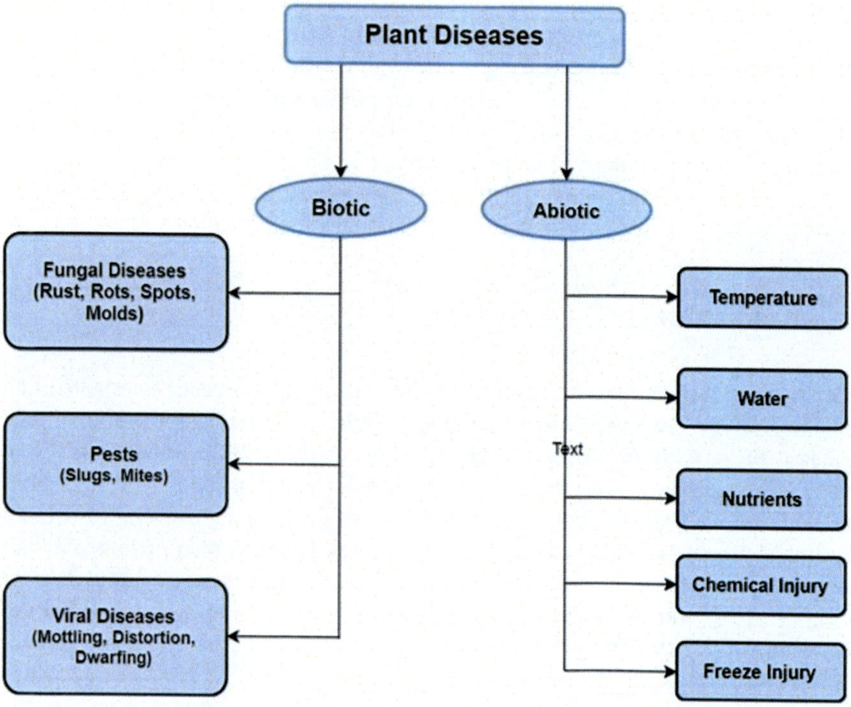

Fig. 1 Types of plant diseases

Table 1 Plant diseases symptoms

Leaf diseases symptoms	Description	Pathogen group
Downy Mildew (*Plasmopara viticola)*	Yellow to white patches, and white to grayish	Fungus
Shot hole	Large holes appear in the leaf	Viruses and water mold
Anthracnose	Lesion becomes covered with pink, gelatinous masses of spores	Fungus
Early blight	Cause leaf surface turn yellow, wither, and die	Rain and fungus
Leaf curl	Turn reddish on leaf area	Fungus

1.1 Plant Leaf Diseases and Symptoms

Some of the plant problems are originated from non-living organisms and some from living organisms. Based on their nature, diseases or damage caused by drought stress, freeze injury, chemical injury, and weather conditions, spring frosts are categories under abiotic or non-infectious. On the other hand, biotic or infectious diseases are caused by fungus, bacteria, and viruses, and are suspected from one plant to another plant as given in Table 1 [14–16].

1.2 Leaf Diseases Detection Process

Different technologies are used for plant diseases detection. Here, the architecture of classification and detection of leaf disease is laid down in Fig. 2. The procedure for detection of diseases in leafs usually includes image formulation, preprocessing stage followed by segmentation stage, and then, identification followed by classification stage.

1.2.1 Image Formulation

The image repair issues are the elimination or reduction of degradations that are included during image processing, e.g., distortion, pixel quality errors, blurring out of focus, and camera movement. Under the diverse background fields and controlled atmospheric condition, the self-collected image datasets are used. In laboratories or a sampling box, images are also acquired for efficient lighting, shading, and brightness regulation [17, 18]. The image sample value is based on the type and direction of the lens. Some experiments uses high precision cameras parallel to the leaf plane with an optical axis, but some used advanced techniques [19–21]. For capturing an image, multispectral CCD camera, Android phones, and hyperspectral imaging system is also used [22, 23]. It is clear that it is easier to process objects obtained under safe

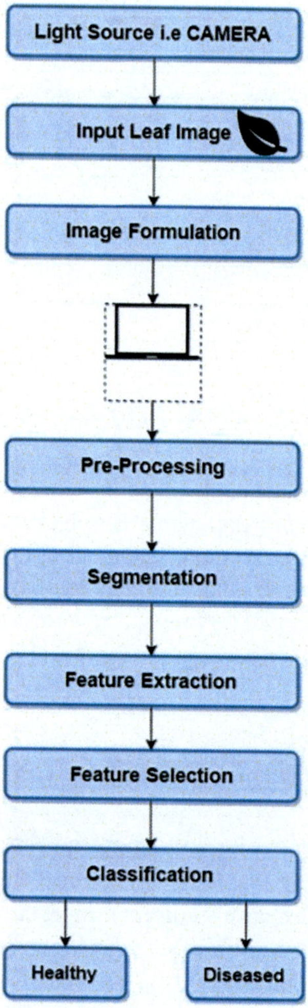

Fig. 2 Architecture of diseases detection system in leaves

environment. Used equipment and techniques provide various details of the image. The quality of a system for detecting plant diseases depends upon the context and capture conditions of the image [6].

1.2.2 Image Preprocessing

In this phase, the intensity of an image is improved for further processing and highlighted the target area to differentiate healthy and diseased area. Algorithms and techniques involved in preprocessing are: color space conversion, contrast limited adaptive histogram equalization (CLAHE) algorithm, smoothing, and cropping. Filters are added for desired improvements after color space transformation, such as improved contrast and brightness. Noise incidence is very common, so median and rank filters are popularly used by these systems. For sharpening, Laplacian filter is used [24, 25]. Recently, the idea of anisotropic diffusion for enhancement is introduced. Some widely used filters include low-pass spatial filters, median neighborhood, and low-pass frequency filters [26].

1.2.3 Segmentation

Segmentation divides picture together with objects of interest into regions with strong correlation. Effectively segmented object features help to easily identify healthy and contaminated specimens, such as the number of peaks in histograms [27]. The region of the contaminated leaf shows significant variations in color which contributes to the spot-based segmentation from its original color [19]. Studies consider clustering such as k-means better than other approaches to segmentation. Some more popular techniques of segmentation are: thresholding process this separates the pixels of image according to their level of intensity, objects from background, and next techniques based on edge detection. Edge detection techniques locate the edges where either the first derivative exceeds a specific inception or the subsequent derivative has no limits. Some more techniques are shown in Fig. 3, and all are unique in comparison with the approach used for segmentation.

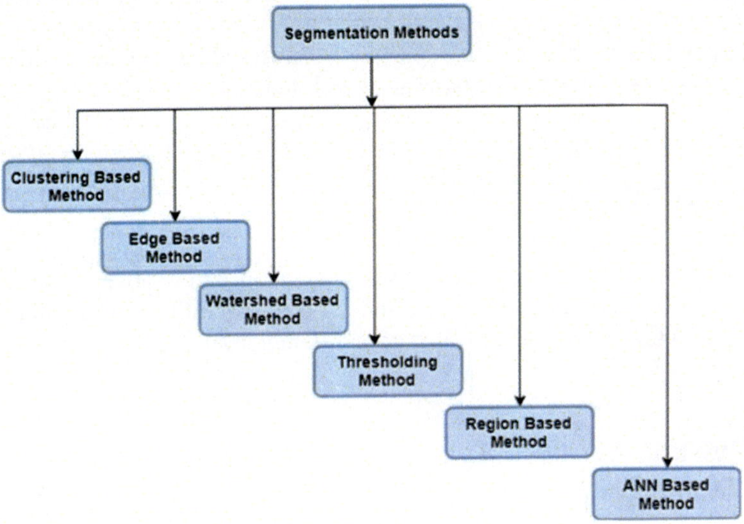

Fig. 3 Segmentation methods

1.2.4 Image Feature Extraction

Images are generally inferred as characteristics of color, texture, and form. The texture can be connected to properties such as contrast, homogeneity, variation, and entropy. Similarly, characteristics are defined for form, roundness, area, excentricity, and concavity [28]. A spatial gray dependency matrix (SGDM) of H image is also used after color space conversion to mine many parameters [26]. Some studies have combined color texture and shape characteristics to detect plant leaf disease type. In order to improve system performance, a combination is found. Some people have removed texture and only dealt with color and shape [19]. Some people have removed all these features, and mean, median, standard deviations are calculated.

1.2.5 Classification

In plant disease detection systems, classification is an essential technique. It is a process of classifying images of plant leaves depending on identified diseases. Scientists have explored a number of methods to identify diseases in multiple cultures through machine learning. The classifier has to distinguish between the plant's healthy and unhealthy image [29]. The methods of machine learning are classified as methods of supervised and unmonitored learning. The training set has inputs for supervised methodologies and the corresponding response values. On the other hand, establish inferences in the training set of missing marked responses. A special class of supervised methods that is semi-supervised uses a mixture of labeled and unlabeled training data. In the domain of plant leaf diseases identification, classification domains are divided into further parts given in Table 2.

2 Related Work

As brought out above, the plant diseases are categories into different types, and its detection process has different stages, and at each stage, different techniques are used. This section covers techniques used at different level in detection process.

2.1 Diseases Detection Analysis with Pre-processing Techniques

Kumbhar et al. [30] presented work using four different classifications of leaves beneath totally unwellness (diseased) conditions. With this, they use SGDM method (spatial gray level dependence matrices). Once color shading of RGB to HSI is done, the H part, i.e. (hue component), is taken into account for additional investigation.

Table 2 Classification domains

Classifi er	**Supervised**	• AdaBoost Method,
		• Neural Networks
		  ◦ Feed Forward Neural Network
		  ◦ Back Propagation Network
		  ◦ Deep Convolution Network
		• K-NN,
		• SVM,
		  ◦ Linear Function
		  ◦ Radial Basis Function

	Unsupervised	• Fuzzy C-means, • LDA, • K-means
	Other Techniques	• Feature Based, • Production Rules, • Fuzzy Logic

The S and I parts are removed because they do not provide any useful information. Ostu method converts green pixels, depending upon the ranging and required threshold. Inside this subsequent stage, the pixels with zero red, blue, and green shading esteem is totally evacuated. After that, diseases classes were categories with features extracted on the basis of color and texture. Ma et al. [31] proposed a technique to recognize diseases in cotton leaf within this three model color measure. Steps followed in this are: capture the image, remove the noise, and use the method to get distinct image, and finally change the color of image. Relative results were given once applying completely different models. The best color model for distinctive leaf pictures is YCbCr model as compared to other models. In case of out of doors situation, leaf shadow and interference of random noise are failed using projected algorithm as verified by the author. Rauf et al. [32] proposed a method based Fourier transforms to identify paddy lesions based on fractal descriptors. Here, the lesions are manually collected and remodeled in HSV color space. Then, after pulling the congestion part, the bar graph exploits (histogram equalization) cut-down of the brightness effects. For closing the classification, from every lesion, the fractal descriptors were obtain and obsessed to PNN classifier. For confirmation and approval, fivefold cross validation is split for testing and training. The author explicit that this technique, if it is combined with alternative options, incorporates a prospective to be used unites of the attainable option, and particularly once, two diseases relatively concerned have identical color.

Tajane et al. [33] using CED algorithm and bar graph exploit proposed a CBIR method (content-based image retrieval) for plant leafs diseases detection. Firstly, the pictures of leafs are remodeled into gray scale pictures, and for the complete image, the color area and features are computed using histogram and also edge histogram is computed. Edge detection algorithm gives effective result to differentiate healthy and diseased plants leafs.

2.2 Thresholding and Clustering Parameters for Diseases Detection and Analysis

Tucker and Chakraborty [34] proposed a method to detect and identify diseases in oat and sunflower leaf. Depending upon the diseases present in leafs, different thresholding techniques were used for segmentation. For implying the lesion part of the images, consecutive points or pixels were associated into clusters using clustering techniques. The author detected better result, but an error occurs due to inappropriate occurring of light at the time of image capturing. Sekulska et al. [35] proposed a lesion detection technique in cucumber and pumpkin leafs. Convert the picture from RGB to HSV color afterward capturing the images. Before grouping, the brighter area is discarded and the fuzzy C-means clustering algorithm is practiced in hue-saturation space. Then, with the interactive phase, these clusters were combined and give two clusters which are final, and from these two groups, one had searched, the diseased area. The author concluded that the good result is proved by first approach and other give different outcomes. Husin et al. [36] discussed the technique for feature recognition of chili leaf diseases. When the plant incited with diseases at the time of applying the chemicals on them. In the first step, preprocessing technique is applied on collected dataset. Then for the better understanding of healthy and unhealthy leafs using color transformation, features are extracted, and then, with the help of above result, the damage area of leaf is differentiated from other area in the groups of colors. The final output the bar graph, i.e., histogram graph is calculated. The given methodology is quick and proficient in detection of plant chili disease. Xiaowei and Zhang et al. [37] proposed an EM segmentation algorithm and superpixel clustering method on cucumber disease leaf images. The superpixel clustering is used to divide image into 150 superpixels after converting the RGB image into LAB color and then segment the image using EM algorithm. The proposed new segmentation algorithm for diseases leaf image detection gives the result that superpixels + EM algorithm which is fastest and best algorithm. The reason is that EM algorithm gives better result after the clustering of pixels from the range of thousands to few hundreds.

2.3 Diseases Detection Using Deep Learning CNN

Baranwal and Arora et al. [38] proposed a deep learning CNN model on apple leaf diseases image samples with four different types of diseases datasets to get accurate results of diseased leafs. The author add new fraction of fine tuning in CNN model by changing the parameters of the layers like dropout value, batch size, training–testing ratio. The overall result accuracy by the model is 98.42% with dropout = 0.2. The accuracy of 98.71, 99.27, 98.70, and 97.3% are achieved by corresponding classes such as Apple Black Rot, Apple Cedar, Apple Rust, Healthy Apple, and Apple Scab, respectively. Liu, Zhang and Yuxiang [39] defined a novel deep CNN approach to identify four varieties of apple leaf diseases. Based on Alex Net, CNN is modeled after removing some fully connected layers, and add some pooling layers and Google Net inception structure to accurately identify apple leaf diseases. The result of diseases detection in apple leafs with accuracy of 97.62% which is far better than other models. Yujian and Edna et al. [40] proposed a model to fine-tune the evaluation of DNN for plant diseases classification. This model includes much different layered architecture. Overall, DenseNets perform best with high accuracy. Therefore, DenseNets achieve a test accuracy score of 99.75% for finding disease identification of plant leaf images. Kumar, Dhingra et al. [41] proposed a computer vision neutrosophic approach on four types of basil leafs for the identification of healthy and disease. Firstly, extract new features with new segmentation approach that combines texture and intensity of leaf image. Than combine feature extraction measure with nine classifiers to get accuracy of each model. The random forest machine learning model gives the best accuracy output of 98.4% for easy identification and detection of diseases leaf image dataset. Amara, Bouaziz et al. [42] proposed an approach to identify two diseases in banana leaf. The deep CNN approach is used for identification and classification of banana lesions. With this approach, farmers can easily identify the type of diseases present in banana at early stage and under any situation.

2.4 *Combined Classifiers to Classify Diseases in Leaf*

Tian et al. [43] presented a model combining three different types of SVM. For these, corresponding SVM classifiers color, features, and shape are extracted, and the part of leaf is segmented using threshold method. As in middle-level and low-level categories, classification is done partially according to the knowledge of plant disease. Later on, the categories extracted from low-level classifiers are then passed to middle-level for extraction. Then, the errors created by color, texture, and shape are cleared by the high-level trained SVM classifier. EI Massi et al. [44] presented serial arrangement approach of two neural network classifiers. In the first step, pictures are remodeled from RGB to LAB color space, and then, using k-means clustering, the diseases part are extracted or segmented. If considering first step, different classes are evaluated, and the diseases with same or neighboring color are evaluated in same class. After this, next class is classified on the foundation of shape and smoothness features. The technique is verifiable on different sections of diseases such as two sections of pest insects and two of fungal lesions. The author halted that the difficulties of single classifiers were solved by serial classifiers division.

2.5 AdaBoost Algorithm for Leaf Diseases Classification

Jagan Mohan et al. [45] proposed a technique for detection and classification of paddy leaf diseases. In the first part, diseased part of leaf of paddy plant is identified and classified using HAAR algorithm and AdaBoost algorithm, respectively, with identification rate 83.33%. In the second part, SIFT method, k-NN, and SVM are used for highlighting the local features of picture and to measured classify rate.

2.6 Detection Using SVM Classifier

Meunkaewjinda et al. [46] described a two-phase technique to identify disease in a grape leaf. In first phase, infected area in an image is identified using an anisotropic diffusion. Afterward, optimization algorithms are used to optimize and to generate group of color in an image. Youwen et al. [47] proposed a method which is used to identify diseases in the cucumber leafs. Here in this, first image regions are segmented using statistic pattern recognition, and then, SVM is used to perform ultimate grouping. Zhang et al. [48] defined singular value decomposition in cucumber leaf for advancing diseases identification levels. Firstly, the watershed algorithm is used for segmentation, then each part is divided into small blocks, and SVD extracts from each block the combination of features of regional specific singular values. After that, vectors are arranged, and their dimensions are modified. Finally, the SVM is used classify the unknown disease in leaf. The researcher concluded that the approach described is producing good results. The drawback of proposed one is that more analytical efforts are required for better recognition.

2.7 Detection Using ANN

Kuo-YI Huang et al. [49] discussed the method for identification of phalaenopsis seeding diseases with color and texture analysis using ANN. The concept of differentiating context (pot), object (leaves), and transforming a factor parameter exponentially to improve the image and determine the quality characteristics of the lesion area for segmenting is done under above conditions, and GLCM was used. The next step is to classify lesions through RGB mean values of lesion area. To identify different lesions, the neural network used is back propagation neural network. The researchers were able to differentiate and categorize the areas of apparent lesions but were unable to analyze the region of infection on the blades. The method of identification and classification may also be beneficial for the production of greenhouse leaves. Santanu Phadikar et al. [50] in this the neural network and zooming algorithm are collectively used to identify lesions in rice plant leaf. Firstly, thresholding method is used to identify infected area from HIS color space. Then, the lesion area of the leaf is generated from trained images using zoom algorithm and SOM. This technique is easy and proficient, resulting in a satisfactory results classification. Hiary et al. [51] proposed a K-means segmentation to identify diseases of crops. In this, K-classes are generated based features the pixels, and using Otsu's thresholding technique, the green pixels were marked. The color matrix are used to extract attributes and to evaluate the texture of the affected leaves. The leaf spots are known as a plant disease marker. Ultimately, artificial neural networks (ANN) are used to classify and evaluate diseases. The proposed method is time and space efficient. Itamar, Adison et al. [52] proposed a technique to identify avocado leaf states such as healthy and unhealthy (nutrient deficient). The technique is divided into two main parts: In the first part, the clusters are generating using k-means. In the next step, for the training of artificial neural network, the histogram is obtained from segmented data of leafs. The author conclude that the histogram gives much information for classification task even after the errors occur in segmentation process the classification gives best results.

3 Conclusion and Future Scope

Motive of the paper is to stress on the concepts, issues, and problems related to discovery and categorization of plant diseases by leaf visualization. Final objective is to minimize the effect of plant diseases on crop productions. That is why it is required to understand the relation between disease cause, its impact, and remedial action. It is a well-known fact that in this world 80% population depends on crops, and crops production can be enhanced by providing fruitful solution to the producer. Therefore, in this paper, we have laid down most of automated methods, tools, and techniques which can be used to analyze, detect, and classify plant diseases based on plant leaf images. This paper outlines pros and cons on all given techniques, and the cause of missing fruitful and effective outcomes for challenging issues. The review concludes that the authors should come-up with more effective and efficient solution and results to recognize diseases in a plant. Points can be considered in future aspects.

(a) In this review, the only parasite diseases are considered, and in future, we are concerned about non-parasite condition which can be the reason of plant diseases.
(b) The existing algorithms can be applied for non-parasite with a common datasets.
(c) Lighting should be considered in detection and recognition techniques of leaf diseases.
(d) App-based solution can be provided to farmer for both parasite and non-parasite conditions.

In this review, different techniques, tools and methods are laid down to identify and categorize plant diseases using automated artificial intelligence. The paper laid down an outline of technical concepts is casted in the current methods related to aim of the paper.

References

1. Bagde S, Patil S, Patil S, Patil P (2015) Artificial neural network based plant leaf disease detection. Int J Comput Sci Mob Comput 4(4):900–905
2. Phadikar S, Sil J, Das AK (2013) Rice diseases classification using feature selection and rule generation techniques. Comput Electron Agric 90:76–85
3. Rastogi A, Arora R, Sharma S v Leaf disease detection and grading using computer vision technology and fuzzy logic. In: IEEE 2nd international conference on signal processing and integrated networks SPIN, pp 500–505
4. Prasad S, Peddoju SK, Ghosh D (2016) Multi-resolution mobile vision system for plant leaf disease diagnosis. Sig Image Video Process 10(2):379–388
5. Khirade SD, Patil AB (2015) Plant disease detection using image processing. In: IEEE international conference on computing communication control and automation (ICCUBEA), pp 768–771
6. Barbedo JGA (2016) A review on the main challenges in automatic plant disease identification based on visible range images. Biosyst Eng 144:52–60
7. Prashar K, Talwar R, Kant C (2015) A review on efficient identification of american cotton leaf diseases through training set. Int J Comput Appl 132(7):32–39
8. Thresh JM (2003) The impact of plant virus diseases in developing countries. Virus and virus like diseases of major crops in developing countries. Springer, Netherlands, pp 1–30
9. González-Fernandez R, Prats E, Jorrín-Novo JV (2010) Proteomics of plant pathogenic fungi. J Biomed Biotechnol 2010:1–36
10. Mahlein AK (2016) Plant disease detection by imaging sensors–parallels and specific demands for precision agriculture and plant phenotyping. Plant Dis 100(2):241–251
11. Akila M, Deepan P (2018) Detection and classification of plant leaf diseases by using deep learning algorithm. Int J Eng Res Technol (IJERT). ISSN 2278-0181
12. Cai HY, Caswell JL, Prescott JF (2014) Non-culture molecular techniques for diagnosis of bacterial disease in animals: a diagnostic laboratory perspective. Vet Pathol 51(2):341–350
13. Eun AJC, Huang L, Chew FT, Li SFY, Wong SM (2002) Detection of two orchid viruses using quartz crystal microbalance (QCM) immunosensors. J Virol Methods 99(1):71–79
14. Dhingra G, Kumar V, Joshi HD (2017) Study of digital image processing techniques for leaf disease detection and classification. Multimed Tools Appl
15. Pujari JD, Yakkundimath R, Byadgi AS (2016) SVM and ANN based classification of plant diseases using feature reduction technique. Int J Interact Multimed Artif Intell 3:6–14
16. Jadhav SB, Patil SB (2015) Grading of soybean leaf disease based on segmented image using k-means clustering. Int J Adv Res Electr Commun Eng 4(6):1816–1822
17. Anthonys G, Wickramarachchi N (2009) An image recognition system for crop disease identification of paddy fields in Sri Lanka. In: IEEE international conference on industrial and information systems ICIIS, Sri Lanka, pp 403–407
18. Pydipati R, Burks TF, Lee WS (2006) Identification of citrus disease using color texture features and discriminant analysis. Comput Electron Agric 52(1):49–59
19. Yao Q, Guan Z, Zhou Y, Tang J, Hu Y, Yang B (2009) Application of support vector machine for detecting rice diseases using shape and color texture features. In: IEEE international conference on engineering computation ICEC, Hong Kong, pp 79–83
20. Huang KY (2007) Application of artificial neural network for detecting Phalaenopsis seedling diseases using color and texture features. Comput Electron Agric 57(1):3–11
21. Oberti R, Marchi M, Tirelli P, Calcante A, Iriti M, Borghese AN (2014) Automatic detection of powdery mildew on grapevine leaves by image analysis: optimal view-angle range to increase the sensitivity. Comput Electron Agric 104:1–8
22. Shrivastava S, Hooda DS (2014) Automatic brown spot and frog eye detection from the image aptured in the field. Am J Intell Syst 4(4):131–134
23. Lu J, Cui D, Jiang H (2013) Discrimination of tomato yellow leaf curl disease using hyperspectral imaging. American Society of Agricultural and Biological Engineers, Kansas City, Missouri, July 21–24, p 1

24. Wang H, Li G, Ma Z, Li X (2012) Application of neural networks to image recognition of plant diseases. In: IEEE international conference on systems and informatics ICSAI, Yantai, pp 2159–2164
25. Youwen T, Tianlai L, Yan N (2008) The recognition of cucumber disease based on image processing and support vector machine. In: IEEE congress on image and signal processing CISP'08, May, vol 2, pp 262–267
26. Sannakki SS, Rajpurohit VS, Nargund VB, Kulkarni P (2013) Diagnosis and classification of grape leaf diseases using neural networks. In: IEEE 4th international conference on computing, communications and networking technologies ICCCNT, Tiruchengode, pp 1–5
27. Phadikar S, Sil J, Das AK (2012) Classification of rice leaf diseases based on morphological changes. Int J Inf Electron Eng 2(3):460–463
28. Gavhale KR, Gawande U (2014) An overview of the research on plant leaves disease detection using image processing techniques. IOSR J Comput Eng 16(1):10–16
29. Mokhtar U, El Bendary N, Hassenian AE, Emary E, Mahmoud MA, Hefny H, Tolba MF (2015) SVM-based detection of tomato leaves diseases. In: Intelligent systems. Springer International Publishing, Berlin, pp 641–652
30. Narvekar PR, Kumbhar MM, Patil SN (2014) Grape leaf diseases detection and analysis using SGDM matrix method. Int J Innov Res Comput Commun Eng 2(3):3365–3372
31. He Q, Ma B, Qu D, Zhang Q, Hou X, Zhao J (2013) Cotton pests and diseases detection based on image processing. Indonesian J Electr Eng Comput Sci 11(6):3445–3450
32. Asfarian A, Herdiyani Y, Rauf A, Mutaqin KH (2013) Paddy diseases identification with texture analysis using fractal descriptors based on Fourier spectrum. In: International conference on computer, control, informatics and its applications (IC3INA), Jakarta, 19–21 November, pp 77–81
33. Tajane V, Janwe NJ (2014) Medicinal plants disease identification using canny edge detection algorithm histogram analysis and CBIR. Int J Adv Res Comput Sci Soft Eng 4(6):530–536
34. Tucker CC, Chakraborty S (1997) Quantitative assessment of lesion characteristics and disease severity using digital image processing. J Phytopathol 145(7):273–278
35. Sekulska-Nalewajko J, Goclawski J (2011) A semi-automatic method for the discrimination of diseased regions in detached leaf images using fuzzy c-means clustering. In: Seventh IEEE international conference on perspective technologies and methods in MEMS design (MEMSTECH), Polyana, 11–14 May, pp 172–175
36. Husin ZB, Shakaff AY, Aziz AH, Farook RB (2012) Feasibility study on plant chili disease detection using image processing techniques. In: Third IEEE international conference on intelligent systems, modeling and simulation (ISMS), Kota Kinabalu, 8–10 February, pp 291–296
37. Zhang S, You Z, Wu X (2019) Plant disease leaf image segmentation based on superpixel clustering and EM algorithm. Neural Comput Appl 31(2):1225–1232
38. Baranwal S, Khandelwal S, Arora A (2019) Deep learning convolutional neural network for apple leaves disease detection. Available at SSRN 3351641
39. Liu B, Zhang Y, He D, Li Y (2018) Identification of apple leaf diseases based on deep convolutional neural networks. Symmetry 10(1):11
40. Too EC, Yujian L, Kwao P, Njuki S, Mosomi ME, Kibet J (2019) Deep pruned nets for efficient image-based plants disease classification. J Intell Fuzzy Syst (Preprint) 1–17
41. Dhingra G, Kumar V, Joshi HD (2019) A novel computer vision based neutrosophic approach for leaf disease identification and classification. Measurement 135:782–794
42. Amara J, Bouaziz B, Algergawy A (2017) A deep learning-based approach for banana leaf diseases classification. In BTW (Workshops), pp 79–88
43. Tian Y, Zhao C, Lu S, Guo X (2012) SVM-based multiple classifier system for recognition of wheat leaf diseases. In: IEEE world automation conference, pp 189–193
44. Massi IE, Saddy YE, Yassa ME, Mammass D, Benazon A (2015) Serial combination of two classifiers for automatic recognition of the damages and symptoms on plant leaf. In: IEEE third world conference on complex systems, pp 1–6

45. Mohan KJ, BalasubramanianM Palanivel S (2016) Detection and recognition of diseases from paddy plant leaf images. Int J Comput Appl 141(12):34–41
46. Meunkaewjinda A, Kumsawat P, Attakitmongcol K, Srikaew A (2008) Grape leaf disease detection from color imagery using hybrid intelligent system. In: Fifth IEEE international conference on electrical engineering/electronics, computer, telecommunications and information technology, pp 513–516
47. Youwen T, Tianlai Li, Yan N (2008) The recognition of cucumber disease based on image processing and support vector machine. In: IEEE conference on image and signal processing, pp 262–267
48. Zhang S, Wang Z (2016) Cucumber disease recognition based on global-local singular value decomposition. Neurocomputing 205:341–348
49. Huang K-Y (2007) Application of artificial neural network for detecting Phalaenopsis seedling diseases using color and texture features. Comput Electron Agric 57:3–11
50. Phadikar S, Sil J (2008) Rice disease identification using pattern recognition techniques. In: 11th international conference on computer and information technology (ICCIT 2008), Khulna, 24–27 December. pp 420–423
51. Al-Hiary H, Bani-Ahmad S, Reyalat M, Braik M, Al-Rahamneh Z (2011) Fast and accurate detection and classification of plant diseases. Int J Comput Appl 17(1):31–38
52. Salazar-Reque IF, Pacheco AG, Rodriguez RY, Lezama JG, Huamán SG (2019) An image processing method to automatically identify Avocado leaf state. In: 2019 XXII symposium on image, signal processing and artificial vision (STSIVA). IEEE, pp 1–5

Rank Correlation for Attribute Selection of Cardiac Patient Data

Richa Sharma and Shailendra Narayan Singh

Abstract Cardiac problem is the major cause of mortality across the whole world, and many factors are responsible for the cause of disease which includes high blood pressure, high cholesterol, inactivity, family history, and many other factors. In this paper, we have taken such factors in consideration to find which is the more responsible attributes for the prediction of cardiovascular disease and for the attribute selection rank correlation method is used. Rank correlation method evaluates the linear correlation between two objects or variables. Aim of the research paper is to use the correlation ranking in medical domain, and for this purpose, cardiac patient data is chosen having nine attributes and 226 instances, dataset collected retrospectively from the SRHC Govt. Hospital, Delhi. Correlation ranking identified the highly correlated attributes responsible for cardiac problem in patients.

Keywords Cardiac disease prognosis · Data preprocessing · Correlation ranking · Statistics · Cardiac health · Heart disease

1 Introduction

Correlation term refers to an association or connection among two variables. Variables co-vary from -1 to $+1$. -1 indicates perfect negative ($-$ve) correlation among two variables, $+1$ indicates perfect positive (+ve) correlation among two variables, whereas 0 signifies no correlation among variables. In statistical terms, correlation can be understood as a method of estimating bilateral linear association among two continuous objects or variables [1, 2]. An association between variables ranges from -1 to $+1$, strong correlation means correlation coefficient comes closer to either $+1$ or -1, and one can say that the variables are highly correlated to each other. If

R. Sharma (✉) · S. N. Singh
ASET, Amity University, Noida, Uttar Pradesh, India
e-mail: richasharma649@gmail.com

S. N. Singh
e-mail: snsingh36@amity.edu

A. Abraham et al. (eds.), *Proceedings of 3rd International Conference on Computing Informatics and Networks*, Lecture Notes in Networks and Systems 167,
https://doi.org/10.1007/978-981-15-9712-1_52

the resultant coefficient is a positive value, variables are directly proportional. This means, increase in the value of one variable will see the increase in value for second variable. If the resultant coefficient is a negative value, variables are inversely proportional. This means that increase in the value of one variable will see the decrease in value for second variable.

Healthcare and technology development contributes widely in quality of healthcare services at affordable cost. Major concern of this work is to contribute the society with quality service by generating quality models or decision support systems which would assist clinical practitioners. In this research work, we majorly focus on attributes or features responsible for cardiac problem, and for the feature selection, rank correlation statistical method is chosen.

WHO reports show 31% of worldwide deaths which means 17.9 million died every year because of cardiovascular disease, and out of those, 85% are because of strokes and heart attacks [3].

2 Related Work

In this section, research work of various researchers in area related to disease prognosis and attribute selection is described:

Schober et al. [4] discussed the correlation measures to guide the researcher for proper use and interpretation of correlation.

Weng et al. [5] has used different neural networks for prediction of disease and examined the performance of individual classifiers used in ensembled classifier and solo classifiers. Ensembled classifier provided the better result as compare to solo classifier with mean difference of 0.0119.

Saxena and Sharma [6] has developed a framework for efficient prediction of disease, dataset considered for the proposed study was taken from UCI machine learning database. For the experimental analysis, knowledge extraction based on evolutionary analysis (KEEL) was considered and achieved the accuracy of 86.7%.

Jabbar et al. [7] presented an effective associative classification algorithm with genetic approach and achieved the accuracy of 88.9%. They compared their work with J4.8, naïve bayes, GNP, and NN.

Lee and Wang [8] presented a fuzzy expert system for diabetic decision support application. They developed a five-layer fuzzy ontology system which contains knowledge layer, group relation layer, group domain layer, personal relation layer, and personal domain layer and further constructed a SDSA: semantic decision support agent. Proposed FDO based fuzzy expert system was implemented by C++ Builder 2007.

3 Research Methodology

Rank correlation is an ordinal association of variables or degree of similarity between two ranks. Some popular rank correlations are as follows:

1. Spearman's
2. Kendall's
3. Goodman and Kruskal's
4. Somers' D.

Spearman's (rho) This method is used to calculate strength of association between two objects or variables. The test is nonparametric and strength of association is denoted by r. A value $r = 1$ signifies perfect positive correlation among the two objects or variables. A value $r = -1$ signifies perfect negative (−) correlation among the two objects or variables.

The meaning of perfect positive correlation is: An increase or decrease in value of one variable will always predict the same directional change in second variable. A famous example of positive correlation is "people with taller height likely to have larger feet and people with shorter height likely to have smaller feet."

The meaning of perfect positive correlation is: An increase or decrease in value of one variable will always predict the opposite directional change in second variable. A few everyday examples of positive correlation are colder the weather, less is the AC cost. If the speed increased, time to reach destination decreases.

The spearman's rho equation is denoted by,

$$rs = 1 - \frac{6\sum D^2}{N3 - N}$$

There must be monotonic association between the variables. Means, either their change should be directly proportional or inversely proportional.

Kendall's (tau) Kendall's tau also is a nonparametric test similar to spearman's rho, but the tau coefficient here returns value of either 0 or 1, where 0 indicates no relationship exists and 1 indicates perfect relationship exists. The test measures the relationship between columns where data is ranked. The oddity of test is that sometimes it can return negative values. For such cases, we usually ignore the '−' sign. The reason is that since the columns are ranked, the negative relationship here do not mean much. So the result of tau test is always an absolute value [9].

There are several versions of tau:

a. Tau A: Tau A test is used for tabular data where given table has equal number of rows/columns.
b. Tau B: B test is used for tables with equal number of rows/columns but can be adjusted for the tied ranks.

c. Tau C: C test is used for tables which do not contain equal number of rows/columns.

The formula to calculate tau coefficient is,

$$\text{Kendall's Tau} = (C - D/C + D)$$

where C = number of concordant pairs and D = number of discordant pairs.

Concordant Pairs When the rank of item A is higher than item B denoted by objects X and Y, the pair item A and item B are said to be concordant.

Discordant Pairs When the rank of item A is higher than item B denoted by object X but as per Y item B is ranked higher than item A, the pair item A and item B are said to be discordant.

Goodman and Kruskal's (gamma) The test mainly performed if your dataset contains outliers. The gamma test tests for the association between the two data points and also tells how strong the strength of that association is. The gamma test results can be:

1: Resultant value of +1 means that there exists a perfect positive correlation among the two variables or data points.
−1: Resultant value of −1 means that there exists a perfect negative (inverse) correlation among two variables or data points.
0: Resultant value of 0 means no relationship exists among the two variables or data points.

The result of gamma test is usually a range between −1 and 1. You can tell the strength of relationship among two variables by looking at how close their result is to 1 or −1. The test can be applied to both continuous and discrete variables. The major goal of the test is ability to predict where the new variables will rank. For example, if a variable has relation of increase/decrease, will the new variable also have the same relationship?

The gamma coefficient is calculated as:

$$\gamma = N_c - N_d/N_c + N_d$$

where N_c = total number of concordant pairs and N_d = total number of discordant pairs.

Yule's Q: Yule's Q is basically a 2*2 version of gamma coefficient which accounts for only concordant and discordant pairs for 2*2 cells. Supposed the cells are denoted as a, b, c, and d, and the Yule's Q formula will be,

$$Q_{(x,y)} = ad - bc/ad + bc$$

Somers' D (Delta) Somers' delta deals with ordinal values. Ordinal values are where variables are ordered. For example: in a sorting order or best to worst or greatest to smallest. The result of this test is measure of agreement between the pairs of the given ordinal values [10]. The result of Somers' delta ranges from −1 to 1 where

−1: All pairs for given ordinal values disagree.
1: All pairs for given ordinal values agree.

Agreement can be told by looking at how close the result is to 1 or −1. Closer the result, better the predictive ability of model.

This agreement defines how well the pairs are connected where connectivity is further dependent upon the concordant and discordant pairs.

Somers' D can be defined in terms of Kendall's tau as:

$$D(Y|X) = \tau(X, Y)/\tau(X, X)$$

where (X, Y) signifies the pair of variables or data points.

4 Result and Discussion

For identification and selection of attributes, WEKA tool is used and out of the eight attributes excluding predicted, 5 attributes are selected for further work which includes gender which specifies that males are more prone to the heart failures, and diabetes and hypertension are in the list on second and third position which specifies patients having diabetic record and suffering from hypertension having high risk of heart-related problem. Low density lipoprotein (LDL) and total cholesterol (TC) are also considered for further prediction of heart disease. For data visualization, kindly refer Fig. 1.

Tenfold cross validation statistical evaluation mode is considered for selection of training data and test data.

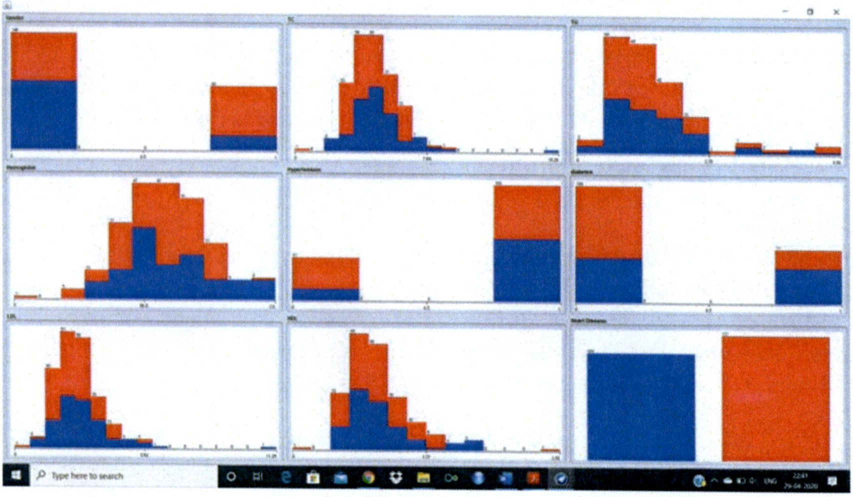

Fig. 1 Data visualization of all nine attributes including the predicted attribute which is depicted in first section

Evaluator: weka.attributeSelection.CorrelationAttribute Eval
Search:
weka. attributeSelection.Ranker -T - 1.7976931348623157E308 -N -1
Relation: Data Instances: 226
Attributes: 9
Gender
TC
TG
Haemoglobin
Hypertension
Diabetes
LDL
HDL
Heart Disease
Evaluation mode: 10-fold cross-validation
=== Attribute selection 10 fold cross- validation (stratified), seed: 1 ===

Average merit	Average rank	Attribute
0.349 ± 0.02	1 ± 0	1 Gender
0.255 ± 0.022	2.1 ± 0.3	6 Diabetes
0.241 ± 0.018	2.9 ± 0.3	5 Hypertension
0.162 ± 0.012	4.1 ± 0.3	7 LDL
0.123 ± 0.014	5.4 ± 0.66	2 TC

(continued)

(continued)

Average merit	Average rank	Attribute
0.118 ± 0.023	5.7 ± 0.78	4 Hemoglobin
0.087 ± 0.022	6.9 ± 0.54	8 HDL
0.036 ± 0.017	7.9 ± 0.3	3 TG

5 Conclusion and Future Scope

In this research work, attributes for further evaluation are selected using correlation ranking method, and it explains the statistical method and results identified patients having problem of diabetes and hypertension are at the higher risk of cardiac problem. Further machine learning algorithms are used for predicting the heart disease. This work helps the medical practioners in decision making.

References

1. Altman DG (1990) Practical statistics for medical research. CRC press
2. Karegowda AG, Manjunath AS, Jayaram MA (2010) Comparative study of attribute selection using gain ratio and correlation based feature selection. Int J Inform Technol Knowl Manag 2(2):271–277
3. Online: http://origin.who.int/cardiovascular_diseases/en/. Last accessed 2020/03/11
4. Schober P, Boer C, Schwarte LA (2018) Correlation coefficients: appropriate use and interpretation. Anesth Analg 126(5):1763–1768
5. Weng CH, Huang TCK, Han RP (2016) Disease prediction with different types of neural network classifiers. Telematics Inform 33(2):277–292
6. Saxena K, Sharma R (2016) Efficient heart disease prediction system. Procedia Comput Sci 85:962–969
7. Jabbar MA, Deekshatulu BL, Chandra P (2013) Heart disease prediction system using associative classification and genetic algorithm. arXiv preprint arXiv:1303.5919
8. Lee CS, Wang MH (2011) A fuzzy expert system for diabetes decision support application. IEEE Trans Syst Man Cybern Part B (Cybernetics) 41(1):139–153
9. Abdi H (2007) The Kendall rank correlation coefficient. Encyclopedia of measurement and statistics. Sage, Thousand Oaks, CA, pp 508–510
10. Newson R (2006) Confidence intervals for rank statistics: Somers' D and extensions. Stat J 6(3):309–334

Breast Cancer Histopathology Image Classification Using Soft Voting Classifier

Deepika Kumar and Usha Batra

Abstract Breast cancer is the reason of mortality rate among women worldwide. Breast cancer survival rate and mortality rate can be improved if it is identified at an initial stage. However, it is a very tough task to classify histopathology images. Therefore, computer-aided detection using various deep learning models is used to categorize the abnormalities in images. In this research, a soft computing classifier for 7-CNN model has been proposed for breast cancer histopathology image classification. The proposed methodology uses the basic CNN with four convolutional layers, basic CNN with five-layer CNN (with data augmentation), VGG 19 transfer learned model (with and without data augmentation), VGG 16 transfer learned model (without data augmentation) and Xception transfer learner model (with and without data augmentation). It uses seven models, and all the seven models have been used to make predictions. The datasets used in the research are hematoxylin–eosin (H&E) for experimentation. The performance of each model has been compared on the basis of accuracy, precision, recall and F1-score. The accuracy has been taken as the main evaluation criteria. The proposed methodology has achieved maximum accuracy of 96.91% on H&E dataset. The accuracy of proposed methodology has been compared with all the transfer learned models.

Keywords Breast cancer · Classification · Ensemble · Histopathology images · Soft Voting · VGG16 · VGG19 · Xception

Abbreviations

CNN Convolutional neural network
DA Data augmentation

D. Kumar (✉) · U. Batra
G D Goenka University, Gurugram, Haryana 122103, India
e-mail: deepika.kumar@bharatividyapeeth.edu

U. Batra
e-mail: dr.ushabatra@gmail.com

A. Abraham et al. (eds.), *Proceedings of 3rd International Conference on Computing Informatics and Networks*, Lecture Notes in Networks and Systems 167,
https://doi.org/10.1007/978-981-15-9712-1_53

DL	Deep learning
Elu	Exponential linear unit
H&E	Hematoxylin-Eosin
ML	Machine learning
ReLu	Rectified linear unit
VGG	Visual geometry group

1 Introduction

Breast cancer is the most common disease identified in females and foremost reason of casualties worldwide [1]. In 2018, World Health Organization released a report that states that breast cancer is affecting approximately two million females each year [2]. The number of breast cancer cases is staggering and constantly rising [3]. Breast cancer survival rate and its outcomes can be improved if it can be identified at an initial stage. Breast cancer screening is the main strategy of its detection. Breast cancer benign and malignant lesions are differentiated by breast cancer screening and biopsies [4, 5]. However, manually assessment of images is a time consuming and laborious task for pathologists because they vary in appearance, texture and structure, and human intervention is also required. Therefore, computer-aided detection (CAD) systems are required for histopathology image classification, whereas disease identification at an early stage plays a very significant role in its diagnosis and treatment. The detection of cancer through histopathological images is considered to be more accurate among all the aforementioned methods, but the final stage of cancer can be determined by visual inspection of the images using a microscope only [6, 7]. The histopathology image-based grading system for prostrate was first introduced by Gleason [8]. The literature shows that deep learning techniques have shown good results in image classification and extracting features efficiently [9]. However, conventional methods identify low-level features, and prior knowledge is also required for feature extraction. The authors are motivated by the efficiency and outperformed results of transfer learning models in the area of biomedical image analysis. Various deep learning models like basic CNN, VGG 16, VGG 19, Xception (ReLu), Xception (Elu) and soft computing classifier have been applied on breast cancer histopathology dataset for breast cancer classification using H&E dataset. Finally, the performance of all models has been compared.

The organization of the paper is as follows: Sect. 2 discusses the literature done in this area. Proposed methodology for breast cancer classification using Soft Voting Classifier is discussed in Sect. 3, whereas results and analysis are explained in Sect. 4 followed with conclusion.

2 Literature Survey

Classification algorithms are widely used in medical image analysis for getting better results in disease prediction. Literature shows that many researchers have worked in areas such as classification, segmentation, detection and prediction [10–13]. A deep learning-based framework has been used for breast cancer classification using breast cytology images. Khan et al. have compared and analyzed the performance of three models GoogLeNet, VGGNet and ResNet [14]. Deep learning models with convolutional layer have been proposed for extracting visual features for breast histology image classification on H&E dataset [15]. Yan et al. have proposed a hybrid neural network model which combines the advantages of both the models and their model achieve an accuracy of 91.3% with multiple classification [16]. A DENSENET-II neural network model for mammogram has been proposed for classification into benign and malignant, and the average accuracy achieved by the proposed model was 94.55% [17]. Another parallel model of CNN and RNN had been introduced for feature extraction from images in [18]. A multiple classification of H&E images using CNN model has been proposed, while SVM classifier was used for extraction of features. The model was capable of attaining precision of 77.8% for four classes and image classification into carcinoma/non-carcinoma with an accuracy of 83.3% [19]. A fine-tuned Google inception-V3 and ResNet50 CNN had been used to identify the histopathology image classification with 97.50% accuracy [20]. Various histopathology images have been classified using CNN containing the residual block. A multilevel classification approach has been suggested by using different deep learning models in [21, 22].

3 Proposed Methodology

The research shows that various machine learning (ML) and deep learning (DL) methodologies have been used for breast cancer image classification [11, 23, 24]. It is a very complex task to classify histopathology images; therefore, deep learning models are used for classification. The Soft Voting Classifier with seven CNN model has been used for the binary classification of images.

3.1 Dataset Description

Different deep learning techniques such as CNN and various transfer learning methods data augmentation, VGG16, VGG19, Xception have been used for the classification. UCSB Bio-Segmentation Benchmark dataset [25] has been used, and it constitutes mount slides of H&E breast cancer images with binary class malignant and benign. It has total 58 patches, out of which 26 patches had benign tumor and 32 patches had malignant tumor.

3.2 Data Preprocessing and One-Hot Encoding

Data preprocessing is the important criteria in machine learning and deep learning. A comma-separated file has been created which was used to keep the class labels corresponding to the labels of the index using the image paths to extract their classes. Later on, the images were resized to 128×128 and interpolated using inter cubic interpolation [26]. The to_categorical () function is used for one-hot encoding the class labels as follows:

- [1. 0.] = 0 for benign breast cancer;
- [0. 1.] = 1 for malignant breast cancer.

3.3 Data Augmentation

Data augmentation is an essential step for the introduction of invariance to the model so that it can scale better rather than amassing the size of the dataset of images. The models use the augmentations are shear, width, zoom and height range of.4 each, rotation with 40°, horizontal and vertical shift, and rescaling is done to 255.

3.4 Model Architecture

The Soft Voting Classifier architecture is based on the seven convolutions-based deep learning models. Soft Voting Classifier takes predictions from seven different models. The predictions of these models are zipped together as a list of tuples. This list of tuples is then multiplied with scaled weights of each model. These scaled weights of the models are obtained by calculating their accuracy on the dataset. The scaling is done such that the sum of all weights (accuracies for each models) comes out to be 1. So, that it is easier for the predictions to be made on this dataset. All the deep learning models have been generated using Matplotlib [27] and Keras [28].

The first model uses four convolutional layers where the first layer introduces the initial convolution operation and the later layers introduce convolutions which are max pooled. The four convolution layers' model outputs (16, 32, 64, 128) feature maps with dropout rate of (0.5, 0.25, 0.25, 0.5), stride of (1, 1, 2, 2) and nonlinear functions in the order (Tanh (Hyperbolic tangent), rectified linear unit (ReLU), exponential linear unit (ELu), ELu), respectively. The features thus obtained are flattened and passed through a layer 144 neurons which are further connected to the output layer of two neurons, passing through a ReLU and dropout of 0.35. The output layer of deep learning model introduces a sigmoid activation (as the output is binary). The sigmoid function is given in Eq. 1:

$$\emptyset(z)_i = \frac{1}{1 + e^{-z}} \tag{1}$$

The second model uses five convolutional layers with data augmentation where the output features had been obtained from (32, 64, 128, 256, 512) feature maps with dropout rates of (0.25, 0.25, 0.25, 0.5, 0.5), stride of 2 and nonlinear functions in the order (Tanh (Hyperbolic tangent) and exponential linear unit (ELu) in next four layers), respectively. The features thus obtained are flattened and passed through a layer 256 hidden layer neurons which are further connected to the hidden layer of 120 neurons, passing through a ReLU and dropout of 0.5. This configuration follows for mapping from 120 hidden layer neurons to two output layer neurons. The output layer of deep learning model introduces a softmax activation which gives class labels' probabilities. This function has been shown in Eq. (2).

$$\sigma(z)_i = \frac{e^{z_i}}{\sum_{j=1}^{K} e^{z_j}} \tag{2}$$

where $i = 1, 2, \ldots, K$ and $z = (z_1, z_2, \ldots, z_K) \in \mathbb{R}_k$.

The third model uses similar augmentation as depicted in model 2. Here, the VGG19 model has been pre-trained on the ImageNet dataset for an input of 128 × 128. The weights obtained from VGG19 (vgg19_input) for this dataset (512 features) have been pooled (Global Average Pooling). These are then passed through four fully connected layers having (256, 128, 64, 32, respectively) neurons which have dropouts of 0.5 and rectified linear unit (ReLU) nonlinear activation. These are then connected to the two-layer output layer which has a sigmoid activation. The fourth model is similar to the third model and only avoids the usage of data augmentation while training of the model. The fifth model does not use data augmentation. Here, the VGG16 model has been pre-trained on the ImageNet dataset for an input of 128 × 128. The weights obtained from VGG16 (vgg16_input) for this dataset (512 features) have been pooled (Global Average Pooling). These are then passed through four fully connected layers having (32, 64, 128, 256, respectively) neurons which have dropouts of (0.15, 0.25, 0.35, 0.45, respectively) and rectified linear unit (ReLU) nonlinear activation. These are then connected to the two-layer output layer which has a softmax activation. The equation for the ReLu and ELu has been shown in Eqs. (3) and (4), respectively.

$$\mathrm{ReLU} = \begin{cases} 0, \text{ for } x < 0 \\ x, \text{ for } x \geq 0 \end{cases} \tag{3}$$

$$R(z) = \begin{Bmatrix} z & Z < 0 \\ \alpha(e^z - 1) & Z \leq 0 \end{Bmatrix} \tag{4}$$

where α had been kept at a value of 1.

The sixth model does not use data augmentation. Here, the Xception model has been pre-trained on the ImageNet dataset for an input of 128 × 128. The weights obtained from Xception (xception_input) for this dataset (2048 features) have been pooled (Global Average Pooling). These are then passed through four fully connected layers having (32, 64, 128, 256, respectively) neurons which have dropouts of (0.2, 0.3, 0.4, 0.5, respectively) and rectified linear unit (ReLU) nonlinear activation. These are then connected to the two-layer output layer which has a softmax activation. The seventh model is similar to the sixth model and only uses data augmentation while training the model. Now, the Soft Voting Classifier model uses a list of tuples (which has a size of 7, corresponding to 7 predictions from 7 models). Here, all the seven models discussed above have been loaded and used to make predictions against 162 images that had been fed to Soft Voting Classifier. The accuracies of these models had then been zipped together to form another tuple of size 7. Both the accuracy tuple and a list of tuples for each of the predictions for all test images had been fed to Soft Voting Classifier. The complete architecture of the Soft Voting Classifier model has been depicted in Fig. 1.

4 Results and Analysis

The detailed architecture of Soft Voting Classifier has been discussed in Sect. 3. It uses basic CNN with four convolutional layers, basic CNN with five-layer CNN (with data augmentation), VGG 19 transfer learned model (with data augmentation), VGG 19 transfer learned model (without data augmentation), VGG 16 transfer learned model (without data augmentation) and Xception transfer learner model (with and without data augmentation). It uses list of tuples which has a size of seven which corresponds to seven predictions from seven models. All the seven models have been loaded and then used to make predictions. The accuracies of these models have been zipped together to form another tuple of size seven. Accuracy, precision, recall and F1-score are taken as the evaluation criteria for various classification algorithms. The formula of various evaluation metric has been shown in Table 1.

The training accuracy curve, loss curve and confusion matrix for all the models have been depicted in Figs. 2a–g, 3a–g and 4a–g, respectively.

Soft Voting Classifier has been inspired by the stacking models [29] which utilizes similar output predictions where weighted Voting Classifier is used for human face recognition. A confusion matrix for this model has been depicted in Fig. 5, and the classification report for the model has been given in Table 2. The accuracy of each model used has been depicted in Table 3.

A detailed accuracy comparison of all the models has been compared which shows that CNN with data augmentation has been achieved the lowest accuracy of 70.37% as compared to other models used for classification. The proposed Soft Voting Classifier using seven CNN models has achieved maximum accuracy of 96.91% on H&E dataset. The accuracy of proposed methodology has been compared with all

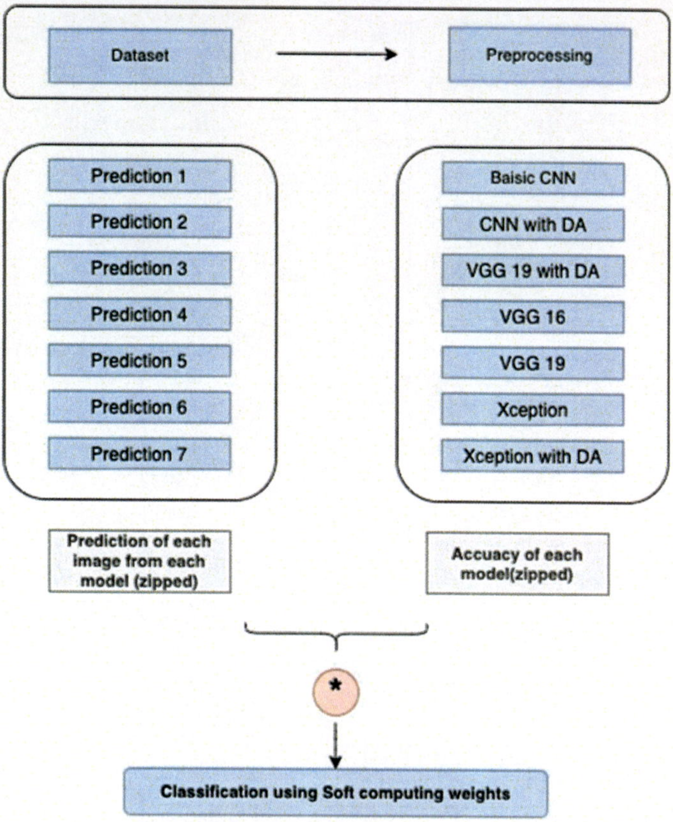

Fig. 1 Architecture of Soft Voting Classifier which zips predictions from each model and bitwise multiplies it with accuracy of that model and then sums the values for each of the seven models for a H&E dataset

Table 1 Description of various evaluation metric used

Evaluation metric	Formula
Accuracy	$\frac{\text{true}_{\text{pos}}+\text{true}_{\text{neg}}}{\text{true}_{\text{pos}}+\text{false}_{\text{pos}}+\text{false}_{\text{neg}}+\text{true}_{\text{neg}}}$
Precision	$\frac{\text{true}_{\text{pos}}}{\text{true}_{\text{pos}}+\text{false}_{\text{pos}}}$
Recall	$\frac{\text{true}_{\text{pos}}}{\text{true}_{\text{pos}}+\text{false}_{\text{neg}}}$
F1 Score	$\frac{2*\text{Precision}*\text{Recall}}{\text{Precision}+\text{Recall}}$

the transfer learned models and proposed methodology outperformed as compared to other models.

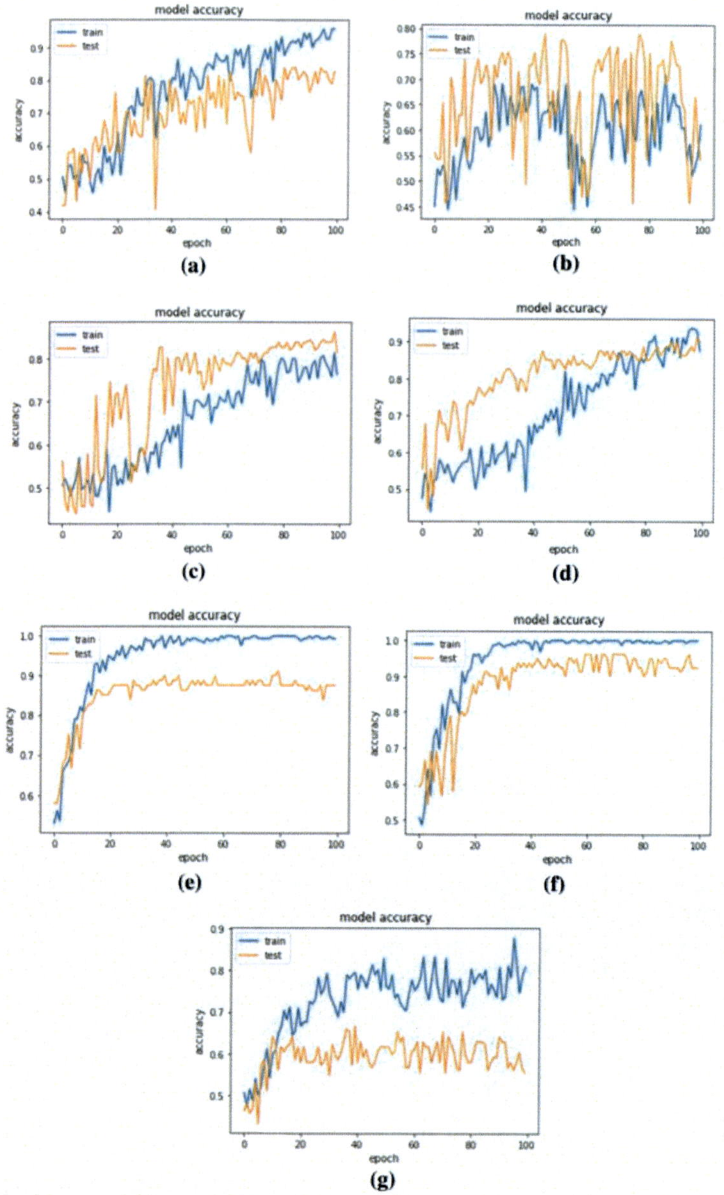

Fig. 2 Training accuracy curve for: **a** (first model for Soft Voting Classifier) Normal CNN without data augmentation. **b** (second model for Soft Voting Classifier) Normal CNN model with data augmentation. **c** (third model for Soft Voting Classifier) VGG19 with data augmentation. **d** (fourth model for Soft Voting Classifier) VGG19 without data augmentation. **e** (fifth model for Soft Voting Classifier) VGG16 without data augmentation. **f** (sixth model for Soft Voting Classifier) Xception without data augmentation. **g** (seventh model for Soft Voting Classifier) Xception with data augmentation

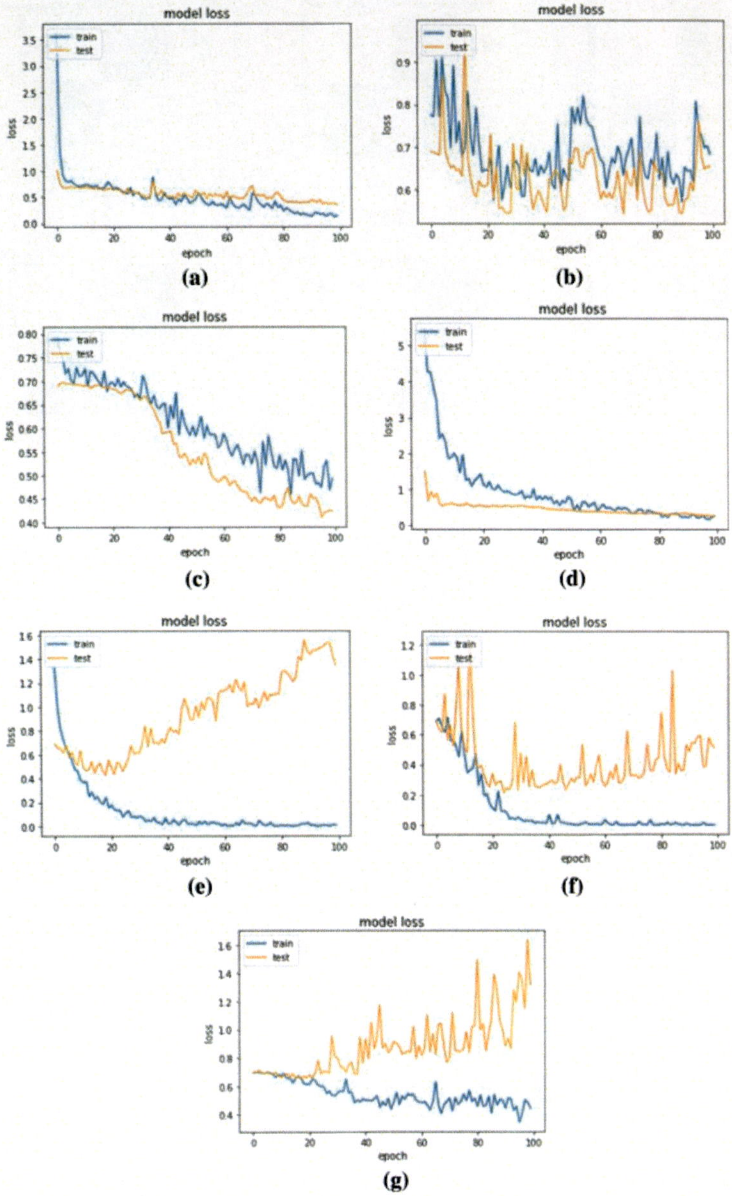

Fig. 3 Training loss curve for: **a** (first model for Soft Voting Classifier) Normal CNN without data augmentation. **b** (second model for Soft Voting Classifier) Normal CNN model with data augmentation. **c** (third model for Soft Voting Classifier) VGG19 with data augmentation. **d** (fourth model for Soft Voting Classifier) VGG19 without data augmentation. **e** (fifth model for Soft Voting Classifier) VGG16 without data augmentation. **f** (sixth model for Soft Voting Classifier) Xception without data augmentation. **g** (seventh model for Soft Voting Classifier) Xception with data augmentation

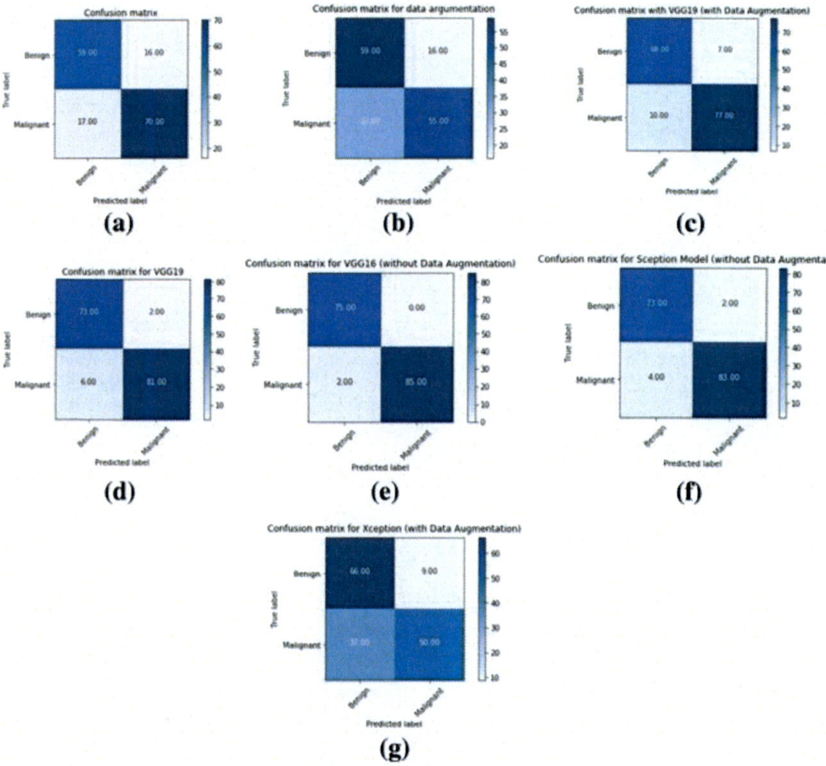

Fig. 4 Confusion matrix for: **a** (first model for Soft Voting Classifier) Normal CNN without data augmentation. **b** (second model for Soft Voting Classifier) Normal CNN model with data augmentation. **c** (third model for Soft Voting Classifier) VGG19 with data augmentation. **d** (fourth model for Soft Voting Classifier) VGG19 without data augmentation. **e** (fifth model for Soft Voting Classifier) VGG16 without data augmentation. **f** (sixth model for Soft Voting Classifier) Xception without data augmentation. **g** (seventh model for Soft Voting Classifier) Xception with data augmentation

5 Conclusion

In this paper, the proposed methodology uses the basic CNN with four and five convolutional layers (with and without data augmentation), VGG 19 transfer learned model (with and without data augmentation), VGG 16 transfer learned model (without data augmentation) and Xception transfer learner model (with and without data augmentation). It uses list of tuples which has a size of seven which corresponds to seven predictions from seven models. All the seven models have been loaded and then used to make predictions. The accuracies of these models have been zipped together to form another tuple of size seven. An accuracy comparison of all the models has been compared with the Soft Voting Classifier. The precision, recall and F1-score of all the transfer learned models have been depicted in Table 2, whereas accuracy of all the models has been depicted in Table 3 utilizing H&E dataset.

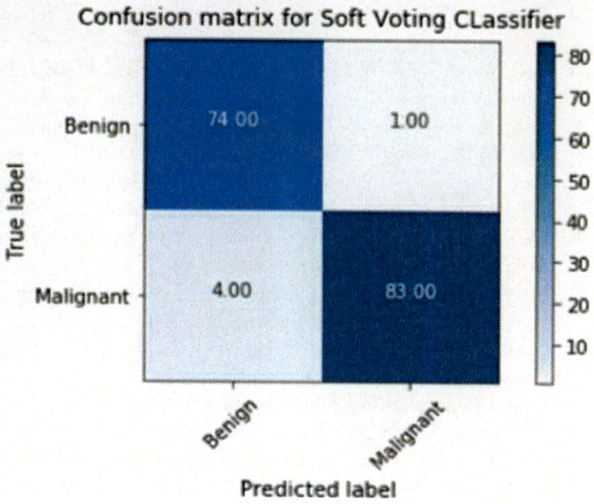

Fig. 5 Confusion matrix for Soft Voting Classifier for H&E dataset

Table 2 Classification report for Soft Voting Classifier for H&E dataset

Class	Precision	Recall	F1-score
0 (Benign)	0.95	0.99	0.97
1 (Malignant)	0.99	0.95	0.97
Avg/total	0.97	0.97	0.97

Table 3 Accuracy for all models used and Soft Voting Classifier for H&E dataset

S. No.	Model	Test accuracy
1.	Basic four-layer CNN	79.63%
2.	Basic five-layer CNN (with data augmentation)	70.37%
3.	VGG 19 transfer learned model (with data augmentation)	89.51%
4.	VGG 19 transfer learned model	95.06%
5.	VGG 16 transfer learned model	95.76%
6.	Xception transfer learned model	96.29%
7.	Xception transfer learned model (with Data Augmentation)	71.61%
8.	Soft Voting Ensemble classifier	96.91%

References

1. Ferlay J, Hery C, Autier P, Sankaranarayanan R (2010) Global burden of breast cancer. In: Breast cancer epidemiology. Springer, pp 119
2. WHO-Breast cancer. https://www.who.int/cancer/prevention/diagnosis-\screening/breast-cancer/en/
3. Takiar R, Nadayil D, Nandakumar A (2010) Projections of number of cancer cases in India (2010–2020) by cancer groups. Asian Pac J Cancer Prev 11(4):1045–1049

4. Kurmi Y, Chaurasia V, Ganesh N (2019) Tumor malignancy detection using histopathology imaging. J Med Imaging Radiat Sci 50(4):514–528
5. Li C, Wang X, Liu W, Latecki LJ, Wang B, Huang J (2019) Weakly supervised mitosis detection in breast histopathology images using concentric loss. Med Image Anal 53:165–178
6. Rubin R et al (eds) (2012) Rubin's pathology clinicopathologic foundations of medicine, 6th edn. Williams and Wilkins, Philadelphia, PA, USA
7. He L, Long LR, Antani S, Thoma GR (2012) Histology image analysis for carcinoma detection and grading. Comput Methods Programs Biomed 107(3):538–556
8. Gleason DF (1992) Histologic grading of prostate cancer: a perspective. Hum Pathol 23(3):273–279
9. Xie J, Liu R, Luttrell J IV, Zhang C (2019) Deep learning based analysis of histopathological images of breast cancer. Frontiers Genet 10:80
10. Pan X, Li L, Yang H, Liu Z, He Y, Li Z, Zhang L et al (2018) Multi-task deep learning for fine-grained classification/grading in breast cancer histopathological images. In: International symposium on artificial intelligence and robotics. Springer, Cham, pp 85–95
11. Rakhlin A, Shvets A, Iglovikov V, Kalinin AA (2018) Deep convolutional neural networks for breast cancer histology image analysis. In: International conference image analysis and recognition. Springer, Cham, pp 737–744
12. Mittal M, Goyal LM, Kaur S, Kaur I, Verma A, Hemanth DJ (2019) Deep learning based enhanced tumor segmentation approach for MR brain images. Appl Soft Comput 78:346–354
13. Herent P, Schmauch B, Jehanno P, Dehaene O, Saillard C, Balleyguier C, Jégou S (2019) Detection and characterization of MRI breast lesions using deep learning. Diagn Interv Imaging 100(4):219–225
14. Khan S, Islam N, Jan Z, Din IU, Rodrigues JJC (2019) A novel deep learning based framework for the detection and classification of breast cancer using transfer learning. Pattern Recogn Lett 125:1–6
15. Vo DM, Nguyen NQ, Lee SW (2019) Classification of breast cancer histology images using incremental boosting convolution networks. Inf Sci 482:123–138
16. Yan R, Ren F, Wang Z, Wang L, Zhang T, Liu Y, Zhang F et al (2020) Breast cancer histopathological image classification using a hybrid deep neural network. Methods 173:52–60
17. Li H, Zhuang S, Li DA, Zhao J, Ma Y (2019) Benign and malignant classification of mammogram images based on deep learning. Biomed Sig Process Control 51:347–354
18. Yao H et al (2019) Parallel structure deep neural network using CNN and RNN with an attention mechanism for breast cancer histology image classification. Cancers 11(12):1901
19. Araújo T et al (2017) Classification of breast cancer histology images using convolutional neural networks. PloS one 12(6)
20. Vesal S et al (2018) Classification of breast cancer histology images using transfer learning. In: International conference image analysis and recognition. Springer, Cham
21. Nahid AA, Kong Y (2018) Histopathological breast-image classification using local and frequency domains by convolutional neural network. Information 9(1):19
22. Han Z, Wei B, Zheng Y, Yin Y, Li K, Li S (2017) Breast cancer multi-classification from histopathological images with structured deep learning model. Sci Rep 7(1):4172
23. Vang YS, Chen Z, Xie X (2018) Deep learning framework for multi-class breast cancer histology image classification. In: International conference image analysis and recognition. Springer, Cham, pp 914–922
24. Golatkar A, Anand D, Sethi A (2018) Classification of breast cancer histology using deep learning. In: International conference image analysis and recognition. Springer, Cham, pp 837–844
25. https://bioimage.ucsb.edu/research/bio-segmentation
26. Sarfraz M (2002) Visualization of positive and convex data by a rational cubic spline interpolation. Inf Sci 146(1–4):239–254
27. Hunter JD (2007) Matplotlib: a 2D graphics environment. Comput Sci Eng 9(3):90

28. François C (2015) keras. Retrieved from https://github.com/fchollet/keras
29. Mu X, Lu J, Watta P, Hassoun MH (2009) Weighted voting-based ensemble classifiers with application to human face recognition and voice recognition. In: 2009 international joint conference on neural networks. IEEE, pp 2168–2171

COVID-19 Pandemic: A Sentiment and Emotional Analysis of Modified Cancellation Policy of Airbnb

Neha Singh, Yash Teotia, Tushar Singh, and Piyush Bhardwaj

Abstract COVID-19 has brought the world to a standstill. Almost every country in the world is facing economic crisis. The most affected sector from COVID-19 is the hospitality and tourism industry. The industry has suffered a loss of billions of dollars. For homestays, Airbnb leads the sector with more than 75% market share in homestays. Forced lockdown in almost all countries leads Airbnb modify its cancellation policy. In this paper, we analyse people's sentiments and emotions for the modified cancellation policy. We have used VADER analysis for sentiment analysis and modified BERT for emotional analysis. The results yield that even though maximum people are positive towards Airbnb, still they are angry with the modified policy. The density of anger is more in European and American continent as compared to other continents.

Keywords COVID-19 · Airbnb · Sentiment analysis · Emotional analysis · VADER · BERT

N. Singh (✉)
Maharaja Agrasen College (affiliated to Guru Gobind Singh Indraprastha University, Delhi 110078), Rohini, Delhi 110096, India
e-mail: join7neha@gmail.com

Y. Teotia · P. Bhardwaj
Bhagwan Parshuram Institute of Technology, Guru Gobind Singh Indraprastha University (affiliated to Delhi 110078), Rohini, Delhi 110089, India
e-mail: yashteotia@gmail.com

P. Bhardwaj
e-mail: piyushb88@gmail.com

T. Singh
Vellore Institute of Technology, Vellore, Tamil Nadu 632014, India
e-mail: tushar12.research@gmail.com

A. Abraham et al. (eds.), *Proceedings of 3rd International Conference on Computing Informatics and Networks*, Lecture Notes in Networks and Systems 167,
https://doi.org/10.1007/978-981-15-9712-1_54

1 Introduction

Due to COVID-19, the world has come to standstill economy around the world is, and one of the worst hit sectors is tourism industry thus in this paper we will see how is Airbnb one of the world's most popular company which is being talked about on the Internet. In this paper, we have extracted tweets from top countries with tourism business from every continent of the world for tweets related to Airbnb. After tweets extraction, they are pre-processed for sentiment as well as emotional analysis. Top 39 countries from around the world that have a booming tourism industry are considered for our research.

A total of 43 thousand tweets were extracted from around the world. All the tweets have been extracted for the month of May. The hashtags used are #airbnb, #airbnbHelpline and #airbnbcutomercare. We have used Valence Aware Dictionary and Sentiment Reasoner (VADER) for sentiment analysis, and an emotion analyser is created by modifying Bidirectional Encoder Representations from Transformers to extract emotion features from the tweets. The created model used here undergoes three main stages, training stage, validation stage and testing stage. The model is trained on the ISEAR data set and undergoes validation to reduce overfitting of data.

Rest of the paper is as follows: Sect. 2 contains a brief literature survey. Section 3 describes the experimental set-up used for analysis. Section 4 provides the analysis of results. Section 5 concludes the paper.

2 Literature Survey

In paper [1], the researcher has tried to mine tweets related to different travel websites from twitter and do the sentiment analysis to tell which is the most popular and loved website of all. Paper [2] tries to mine tweets related to cyber security from different parts of world and used VADER to predict the sentiments of those tweets which tells what are the sentiments of people related to cyber security. In paper [3], the paper author has shown the characteristic feature of people depending on sex, age, etc., and how they react to different social media sites that are Facebook, Instagram, WhatsApp, etc. In paper [4], researcher shows advantages of using different face-APIs such as Face++, IBM Bluemix Visual Recognition, etc., to predict data such as age, race and gender with three data sets. In paper [5], the researcher creates a model to capture emotions from real-time facial features. In paper [6], the author shows how VADER predicts sentiments of tweets from text and how it is best in predicting social media sentiment analysis with respect to different predictors of text. In paper [7], author predicts the emotion from the facial expression just like any other classification algorithm. In paper [8], the author shows the different ways in which machines and humans with the library known as OpenCV can interact with each other and what they can achieve. In paper [9], the author extracts a data set consisting of 400,000 Instagram posts that are from Amsterdam and shows how the city is reassembled on the platform. In paper [10], the author shows how Microsoft and Google face apps operate differently with respect to different images. In paper [8], the author shows the different ways of interaction between machines and humans with the library known as OpenCV. In paper [6], the author shows how VADER predicts sentiments from text and how it is best in social media sentiment analysis. Report [11] is WHO report on coronavirus on 1 April 2020. Report [12] is WHO report on coronavirus 27 March. Report [13] is WHO report on coronavirus 15 March 2020. Report [14] is WHO report on coronavirus 4 March 2020. Report [15] is WHO report on 11 March 2020 declaring coronavirus as a pandemic.

3 Experimental Set-up

3.1 Data Extraction and Cleaning

- Extracting tweets from tweepy with the hashtags of Airbnb and Airbnb Helpline.
- Tweets extracted are for the month of May. All the tweets are extracted with filter of timestamp of tweet being from the month of May.
- Deleting all the tweets that are not in English or does not have a proper place mentioned.
- Cleaning the tweets by removing stopwords and further stemming of tweets as well as lemmatisation.

3.2 Data Set

- Our data set consists of 43 thousand tweets from 39 most popular country in terms of tourism industry
- The data set consists of well-defined information related to date of tweets as well as country from where the tweet has originated

3.3 Sentiment Analysis

- We have used VADER in our data set to predict sentiments from the tweets as shown in paper [6], and we know VADER is best in terms of predicting sentiments from social media. The sentiments that VADER can predict are positive, negative and neutral.
- Vader uses features that are called as lexicals that helps in describing percentage of negativity or positivity present in a sentence

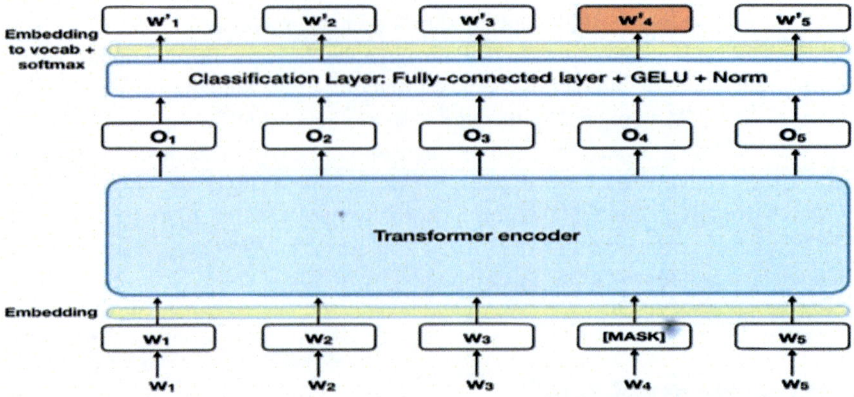

Fig. 1 Block diagram of Bidirectional Encoder Representations from Transformers

3.4 *Emotional Analysis*

- The main endeavour of the model is to predict the emotion labels: joy, anger, sadness, fear for the tweets related to Airbnb's recent cancellation policies (Fig. 1).

We have used modified Bidirectional Encoder Representations from Transformers (BERT) to extract emotion features from the tweets. The created model used here undergoes three main stages: training stage, validation stage and testing stage. The model is trained on the ISEAR data set and undergoes validation to reduce over-fitting of data. The model consists of 12-layer, 768-hidden, 12-heads, 125 million parameters. The model uses a transformer (an attention contraption that identifies and learns contextual relation between the words of the input source). Transformer is basically an encoder–decoder mechanism which reads an input and a decoder that classifies the input and produces predictions. After obtaining a satisfactory training and validation result, the pre-trained model was made to run on the Airbnb data set via transfer learning. The model finally classified all the tweets of the data set, and results were obtained.

3.5 *Graphical Analysis*

- We have used Tableau as well as Python libraries such as matplotlib, seaborn and word cloud to represent the graphs shown in the results section.

4 Results Analysis

- We have used VADER analysis for sentiment analysis and modified BERT for emotional analysis. The tweets are extracted for the month of May 2020. Firstly, analysis of sentiments is done followed by analysis of emotions.

Figure 2 shows the composition of tweets related to different sentiments. There are 19,659 positive tweets, 12,407 neutral tweets and 11,036 negative tweets. Therefore, more than 45% tweets are positive in sentiment.

Figure 3 shows top five countries with positive tweets. USA has the most positive tweets followed by Canada, UK, Netherland and Switzerland.

Figure 4 shows top five countries with neutral tweets. USA has the most neutral tweets followed by Canada, Netherland, UK and Switzerland.

Figure 5 shows top five countries with negative tweets. Once again, USA has the most negative tweets followed by Canada, UK, Netherland and Switzerland.

Figure 6 shows the composition of tweets with respect to sentiments of total number of tweets that is gathered from all the 40 countries. Here the USA is at top with most tweets followed by Canada, Netherland, UK and Switzerland, respectively.

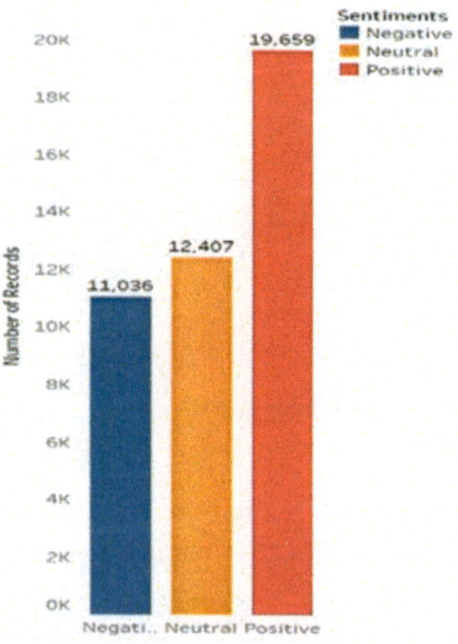

Fig. 2 Sentiment analysis of tweets for the month of May 2020

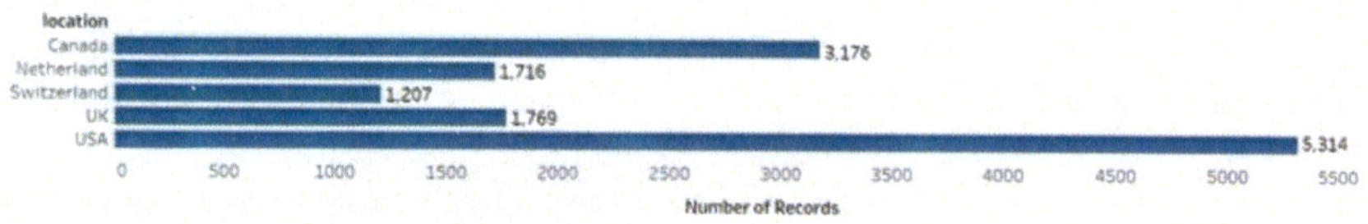

Fig. 3 Top five countries with positive tweets

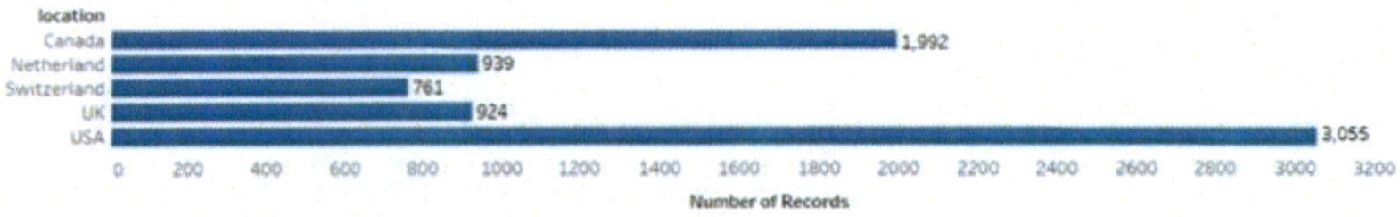

Fig. 4 Top five countries with neutral tweets

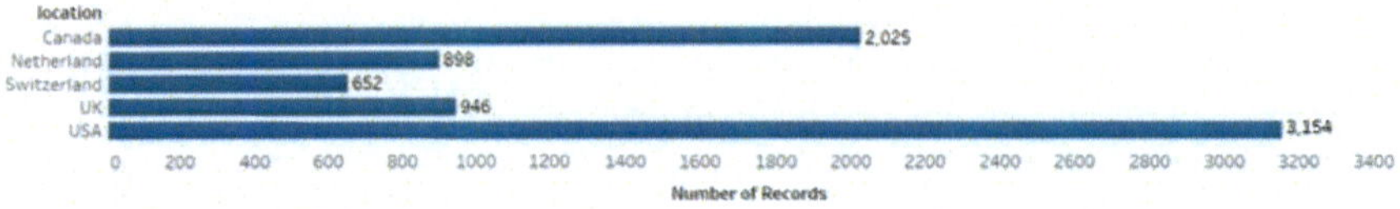

Fig. 5 Top five countries with negative tweets

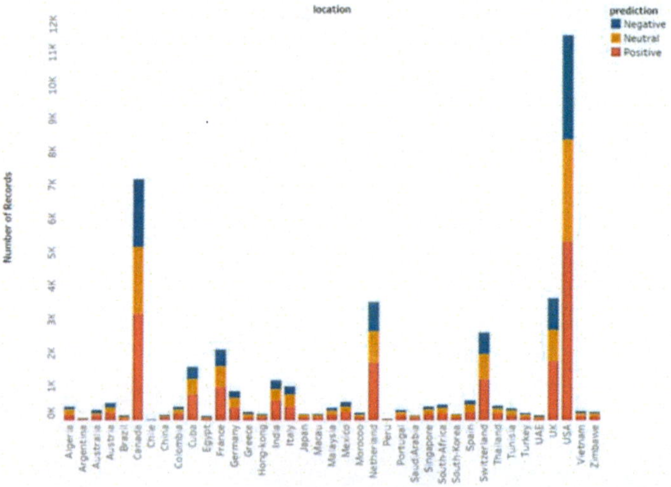

Fig. 6 Country-wise composition of sentiments of tweets

Figure 7 depicts the composition of emotions with respect to 43 thousand Airbnb tweets. More than 39.53% of the tweets contains anger emotion followed by joy, sadness and sear as an emotion.

Figure 8 depicts the composition of emotions of tweets with respect to different countries. Almost all the countries have high percentage of anger emotion followed by joy, sadness and fear.

Figure 9 provides the word cloud of emotion "Fear" in tweets. As it can be seen, Airbnb and coronavirus are most frequently used words associated with fear. A word "return" can be seen used many times. This is because of the modified cancellation policy of Airbnb that people fear whether their money will be returned or not. Another interesting word is "host". Many host fear that they might be a big amount of money because of modified cancellation policy of Airbnb.

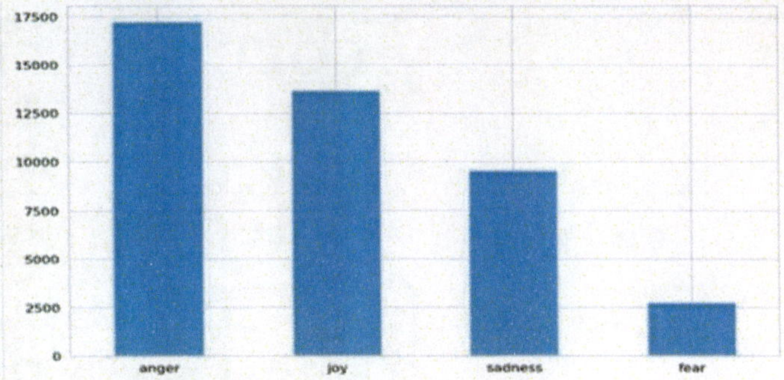

Fig. 7 Emotional analysis of tweets for the month of May 2020

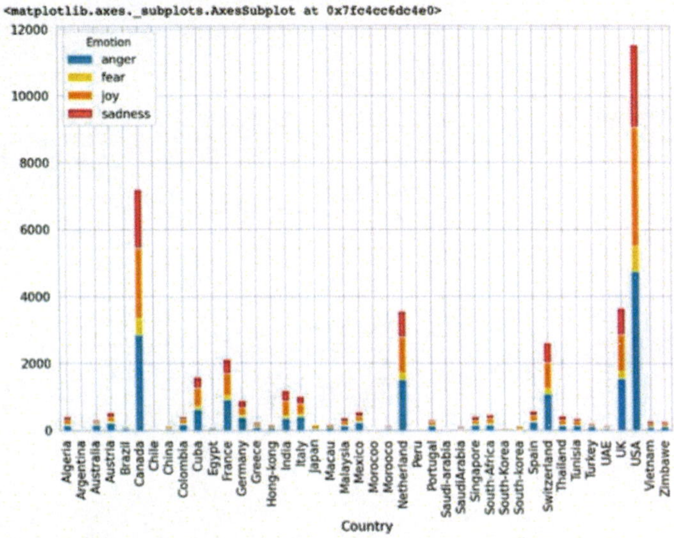

Fig. 8 Country-wise composition of emotions of tweets

Figure 10 provides the word cloud of emotion "Joy" in tweets. As it can be seen, Airbnb, Thank, Founder are most frequently used words associated with joy. A word "new" can be seen used many times. This is because of the new modified cancellation policy of Airbnb that people are happy with after receiving their money. Another interesting word is "experience". Many people seem to have a joyful experience while using the new cancellation policy.

Figure 11 provides the word cloud of emotion "Sadness" in tweets. As it can be seen, Airbnb, host, job are most frequently used words associated with sadness. A word "host" can be seen used many times. This is because of the new modified

Fig. 9 Word cloud of emotion "Fear"

Fig. 10 Word cloud of emotion "Joy"

cancellation policy of Airbnb that many hosts are feeling sad after receiving cancellations at their property. Another interesting word is "decline". Many people seem to be sad after seeing a decline in their income.

Figure 12 provides the word cloud of emotion "Anger" in tweets. As it can be seen "owner", rental, hotels, refund are most frequently used words associated with anger. Almost every owner is angry with the modified cancellation policy as they are the ones who have suffered the most by the new policy.

Figure 13 depicts the density plot of emotion "Anger" across the globe. As it can be seen, the Europeans are very angry with the modified policy followed by Africans and Americans. Asians are least angry with the policy.

Fig. 11 Word cloud of emotion "Sadness"

Fig. 12 Word cloud of emotion "Anger"

Figure 14 depicts the density plot of emotion "Fear" across the globe. As it can be seen, fear emotion has similar density as that of anger. The Europeans are very angry with the modified policy followed by Americans and Africans.

Figure 15 depicts the density plot of emotion "Joy" across the globe. Many Europeans are also happy with the modified policy.

Figure 16 depicts the density plot of emotion "Sadness" across the globe. Many Americans and Europeans are sad with the modified policy of Airbnb.

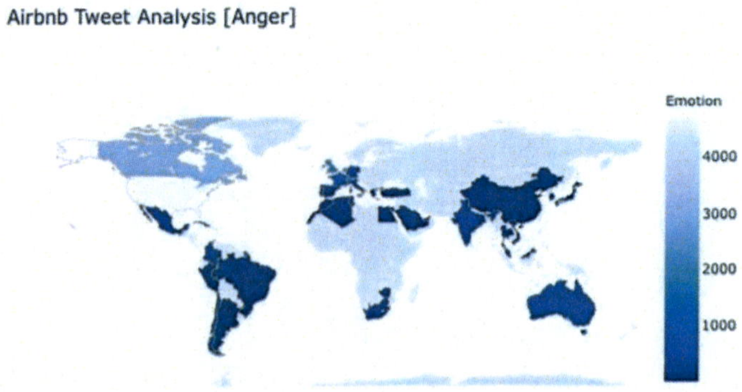

Fig. 13 Density plot of emotion "Anger"

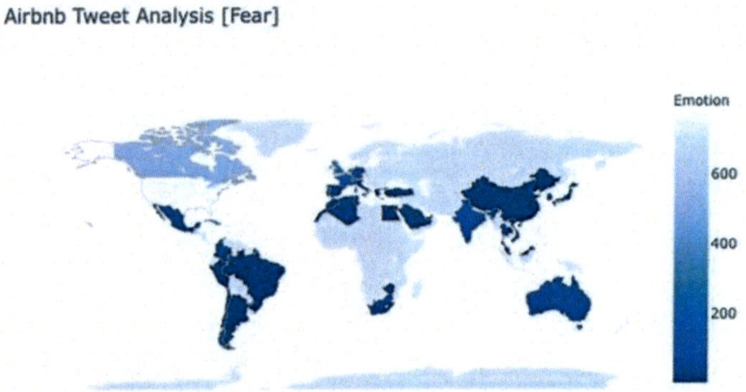

Fig. 14 Density plot of emotion "Fear"

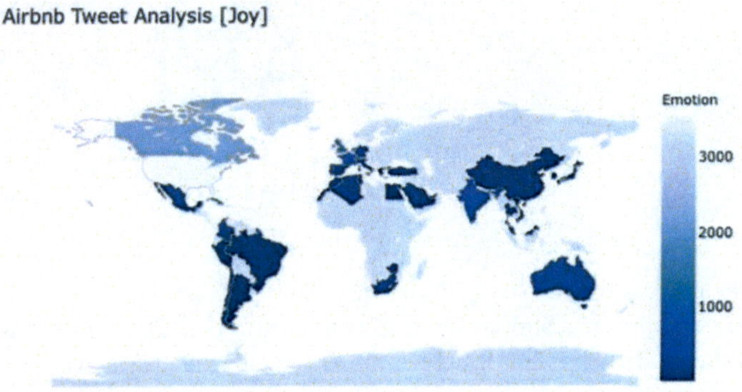

Fig. 15 Density plot of emotion "Joy"

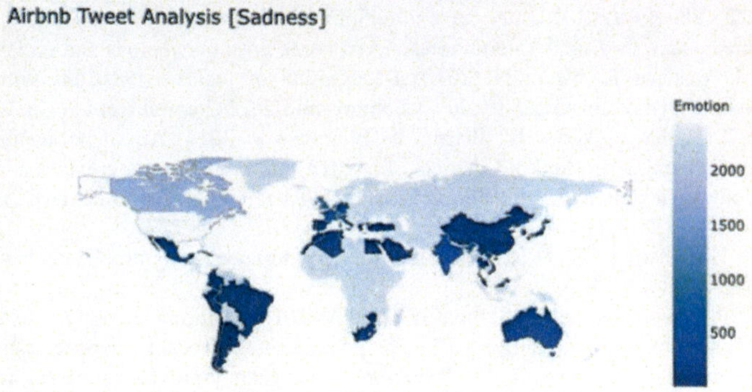

Fig. 16 Density plot of emotion "Sadness"

5 Conclusion

In this research, we have studied the sentiments and emotions of modified cancellation policy of Airbnb due to COVID-19. VADER analysis is used for sentiment analysis, and modified BERT is used for emotional analysis. The analysis yields few points. Firstly, almost half of tweets retrieves positive sentiment, concluding a more satisfied cliental as compared to unsatisfied one. Secondly, that even though maximum people are positive towards Airbnb, still they are angry with the modified policy. The density of anger is more in European and American continent as compared to other continents. Thirdly, few word cloud suggest that owners of rental properties are angry with new cancellation policy, and clients that received full refund are happy with the policy. Lastly, COVID word can be seen in every word cloud indicating the gravity of situation that has arisen due to coronavirus.

References

1. Bhardwaj P, Gautam S, Pahwa P (2017) Opinion mining and sentiment analysis of travel websites through Twitter. Int J Appl Eng Res 12(22):12431–12439
2. Sharma K, Bhasin S, Bharadwaj P (2019, September) A worldwide analysis of cyber security and cyber crime using Twitter. Int J Eng Adv Technol (IJEAT) 8(6S3). ISSN 2249-8958
3. Waterloo SF, Baumgartner SE, Peter J, Valkenburg PM (2018) Norms of online expressions of emotion: comparing Facebook, Twitter, Instagram, and WhatsApp. New Media Soc 20(5):1813–1831
4. Jung SG, An J, Kwak H, Salminen J, Jansen BJ (2018) Assessing the accuracy of four popular face recognition tools for inferring gender, age, and race. In: Twelfth international AAAI conference on web and social media
5. De Silva LC, Hui SC (2003) Real-time facial feature extraction and emotion recognition. In: Fourth international conference on information, communications and signal processing, 2003 and the fourth pacific rim conference on multimedia. Proceedings of the 2003 Joint, vol. 3, pp. 1310–1314. IEEE

6. Hutto CJ, Gilbert E (2014) Vader: a parsimonious rule-based model for sentiment analysis of social media text. In: Eighth international AAAI conference on weblogs and social media
7. Kaur M, Vashisht R, Neeru N (2010) Recognition of facial expressions with principal component analysis and singular value decomposition. Int J Comput Appl 9(12):36–40
8. Sharma T, Kumar S, Yadav N, Sharma K, Bhardwaj P (2017) Air-swipe gesture recognition using OpenCV in Android devices. In: 2017 international conference on algorithms, methodology, models and applications in emerging technologies (ICAMMAET). IEEE, pp 1–6
9. Boy JD, Uitermark J (2017) Reassembling the city through Instagram. Trans Inst Br Geogr 42(4):612–624
10. Khanal SR, Barroso J, Lopes N, Sampaio J, Filipe V (2018) Performance analysis of Microsoft's and Google's emotion recognition API using pose-invariant faces. In: Proceedings of the 8th international conference on software development and technologies for enhancing accessibility and fighting info-exclusion, pp 172–178
11. World Health Organization (2020c) Novel Coronavirus (2019-nCoV). Situation Report 72, 1 April. https://www.who.int/docs/default-source/coronaviruse/situation-reports/20200401-sitrep-72-covid-19.pdf?sfvrsn=3dd8971b_2
12. World Health Organization (2020c) Novel Coronavirus (2019-nCoV). Situation Report 67, 27 March. https://www.who.int/docs/default-source/coronaviruse/situation-reports/20200327-sitrep-67-covid-19.pdf?sfvrsn=b65f68eb_4
13. World Health Organization (2020c) Novel Coronavirus (2019-nCoV). Situation Report 55, 15 March. https://www.who.int/docs/default-source/coronaviruse/situation-reports/20200315-sitrep-55-covid-19.pdf?sfvrsn=33daa5cb_8
14. World Health Organization (2020c) Novel Coronavirus (2019-nCoV). Situation Report 44, 04 March. https://www.who.int/docs/default-source/coronaviruse/situation-reports/20200304-sitrep-44-covid-19.pdf?sfvrsn=93937f92_6
15. WHO Director-General's opening remarks at the media briefing on COVID-19—11 March 2020. https://www.who.int/dg/speeches/detail/who-director-general-s-opening-remarks-at-the-media-briefing-on-covid-19-11-March-2020

Computational Mathematics

Auto Vectorisation Capabilities of the Compilers

Mamta Madan, Pradyumn Nand, and Joy Aneja

Abstract Computer programs today crunch enormous amounts of data and run complex routines to support businesses research. A compiler's job is to convert the source code to the target et language, and it performs optimizations to improve the performance. Vectorisation is one such technique for optimization. The aim of this paper is to determine the auto vectorisation capabilities of four popular compilers—GCC, INTEL C, R and MATLAB and compare them in two groups GCC, Intel C and R, MATLAB. We also investigate the difference in auto vectorisation performances across two architectures—Intel and AMD. The paper not only aims at comparing the compilers but also theoretically and programmatically backing up the investigations and results.

Keywords Optimization · Auto vectorisation · GCC · Intel C · MATLAB · R

1 Introduction

Nowadays, we have huge data and have to run complex routines to support enterprises, research organizations, everyday experiences, etc. A compiler's task is to convert the source code of a program to the target language (often the binary form), but it often also performs various code optimizations [1, 2] to improve the performance of programs. Such performances can improve execution times and memory consumption drastically which help in getting results quickly.

M. Madan (✉)
Vivekananda Institute of Professional Studies, Delhi, India
e-mail: mamta.vips@gmail.com

P. Nand
Makemytrip Pvt Ltd, Gurgaon, India

J. Aneja
Cadence Design Systems, Gurgaon, India

A. Abraham et al. (eds.), *Proceedings of 3rd International Conference on Computing Informatics and Networks*, Lecture Notes in Networks and Systems 167,
https://doi.org/10.1007/978-981-15-9712-1_55

Vectorisation [3] is one of the common program optimization techniques that generalizes operations on scalars to apply transparently to vectors, matrices and higher-dimensional arrays.

Auto vectorisation [4] is a subdomain of parallelization that converts a scalar computer program where a pair of operands are processed one at a time sequentially, to a vectorized version that can compute a single operation on multiple operands at a given time.

Specific compilers have capabilities to convert loops into a set of vector instructions. Auto vectorisation, like other program optimizations, will always preserve the program logic. To parallelize instructions, the compiler's optimizer module has to first interpret dependencies between instructions and re-arrange them if required. Once a dependency graph is created, the module reorganizes the code and vectorizes appropriate statements that can compute on multiple data items in parallel.

2 Auto Vectorisation in R and MATLAB

MATLAB and R support vectorisation through "vectorized" functions rather than direct/raw implementation of vectorized code. These "vectorized" take vectors as arguments and use vector hardware to perform computations on the input arguments. This is considerably different from GCC and intel C, where loop instructions are vectorized due to compiler interpretaion.

In R and MATLAB, loop instructions are not vectorized and hence have very poor performance in cases where the same task can be done through a vectorized function. The two compilers provide vectorisation support only through the library/in-built functions, whereas GCC provides support to vectorize loops/nested loops.

Our analysis is based on the above premise, and we compare and contrast performance of test programs—using loops and using "vectorized" functions.

3 Vectorisation in R

Every element in R is treated as a vector. Therefore, single elements/variables are double-width floating-point atomic vectors of length one. Having this knowledge, it is evident that passing single elements to vectorized functions iteratively through a loop is not the elegant way to use such functions when they can operate on entire vectors, i.e., having

y = VEC_ FUNC (X) // Y and X are vectors is

more suitable than having

for (i)

Y[i] = VEC_FUNC(X[i])

3.1 Difference Between GCC and R Vectorisation

R is not a compiled language like C and hence does not optimize the entire program beforehand, unlike C where during compilation the program binary is organized in an optimal manner for the machine to interpret. Instead R has "vectorized functions" which are actually written in C/C++/FORTRAN and have a small "R wrapper".

Example: (code inspection of "fft" function for R)

```
> fft
function (z, inverse = FALSE)
.Call(C_fft, z, inverse)
<bytecode: 0x7fc261e1b910>
<environment: namespace:stats>
```

Here R passes the data onto a C function called C_fft. The job of R is to interpret the input to this function and make it suitable for the underlying function being called. This is why calling the same function in an iterative manner to process each vector element individually is a bad idea as R will have to repeatedly perform the interpretation and translation of the input and output for the underlying compiled C function which actually uses vectorisation.

Inside the C function, vectors are processed through loops inevitably, which are then optimized using vectorisation. Since this happens on the lower level, it is significantly faster.

3.2 Analysis of Some Popular Vectorized Functions in R

Sample Program Code Non-vectorized code

```
rm(list=ls(all=TRUE)) nums = 1:1000000
cos_novec = function(n) {

ret = rep (NA, length(n)) for (i in seq_along(n)) {
ret[i] = cos(n[i])// calling function for each element

}

return(ret)}

system.time(cos_novec(nums))
```

Vectorized Code

```
rm(list=ls(all=TRUE)) nums = 1:1000000
system.time(cos(nums))// calling function for entire vector
```

Speedup = Time _old/Time_new

NOTE: Min() and Max() have been excluded from the graphs to maintain relative clarity of comparison. The speedups and performance of these two functions are far higher than any of the observed functions (Table 1 and Figs. 1, 2).

Table 1 Trigonometric family—calculating values for natural numbers from 1:106

Function name	Execution Time (non-vectorized)	Execution Time (vectorized code)	Efficiency Improvement
Sin()	2.31 s	0.59 s	39.169
Cos()	2.325 s	0.05 s	43.869
Tan()	2.335	0.10 s	23.35
LOG()	2.393 s	0.047 s	50.914
Exp()	2.328	0.043	54.13
Sqrt()	2.287 s	0.025 s	91.480
Sum()	0.849 s	0.015 s	56.600
Max()	14.044 s	0.016 s	877.750
Min()	14.2555 s	0.017 s	838.529

** for max and min, it is calculated from numbers from 1:107

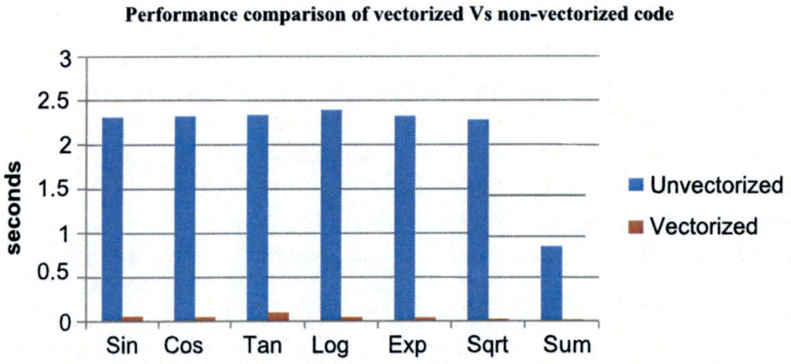

Fig. 1 Comparison for vectorisation of trignometric functions

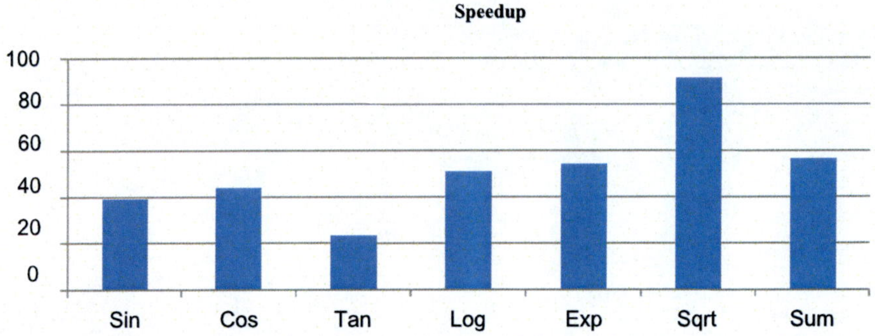

Fig. 2 Efficiency improvement for trigonometric functions

4 Vectorisation in Matlab

Loops are common in most programming languages, they have numerous advantages over normal program as it reduces the job of the programmer of rewriting code every time, but in some languages like MATLAB they are not that fast during runtime, and hence, drastically reduce the performance of the code [5].

4.1 *Difference Between MATLAB and R Vectorisation*

MATLAB Programs [6] Run Very Fast Than R Programs, as R Is not a Compiled Language. Both MATLAB and R Have "Vectorized Functions" Which Are Actually Written in C/C++/FORTRAN and Have Wrappers.

Inside the C function, vectors are processed through loops inevitably, which are then optimized using vectorisation [7]. Since this happens on the lower level, it is significantly faster.

4.2 *Analysis of Some Popular Vectorized Functions in Matlab*

Sample Program:
Non-Vectorized

```
nums=1:1000000;
tic; tsum=0;
for j= 1:1000000
    tsum = tsum + nums(j);
end t1=toc;
```

Vectorized Code

```
tic;
totalSum = sum(nums)
time2 = toc;
```

$$\begin{aligned} Speedup &= \text{Time non - vectorized/time vectorized} \\ &= 0.00446803/0.00095842 = 4.66185479 \end{aligned}$$

which is considerable amount of improvement over the non-vectorized version of the same code (Table 2 and Figs. 3, 4).

NOTE: Min() and Max() have been excluded from the graphs to maintain relative clarity of comparison. The speedups and performance of these two functions are far higher than any of the observed functions (Figs. 5 and 6).

Table 2 Trigonometric family—calculating values for natural numbers from 1:106

Function name	Execution time (non-vectorized)	Execution time (vectorized code)	Speedup
Sin()	0.03077644	0.01443047	2.132739
Cos()	0.03119922	0.01499494	2.08065064
Tan()	0.04237311	0.02195762	1.92976739
LOG()	0.04843214	0.02412971	2.00715788
Exp()	0.01371830	0.00493220	2.78137460
Sqrt()	0.01596391	0.00602118	2.65129345
Sum()	0.00446803	0.00446803	4.66185479
Max()	0.57380843	0.00189523	302.76513346
Min()	0.57500061	0.00188939	304.3317801

** for max and min, it is calculated from numbers from 1:107

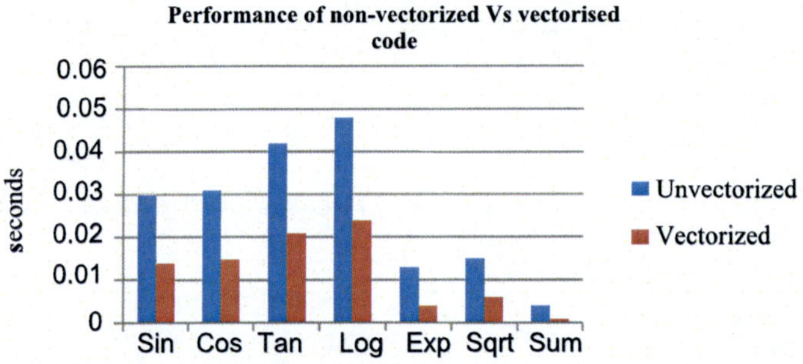

Fig. 3 Vectorisation for trignometric functions in MATLAB

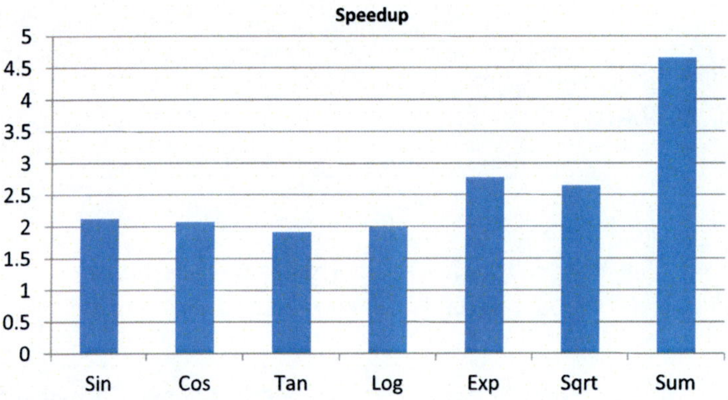

Fig. 4 Efficiency improvement for trigonometric functions in MATLAB

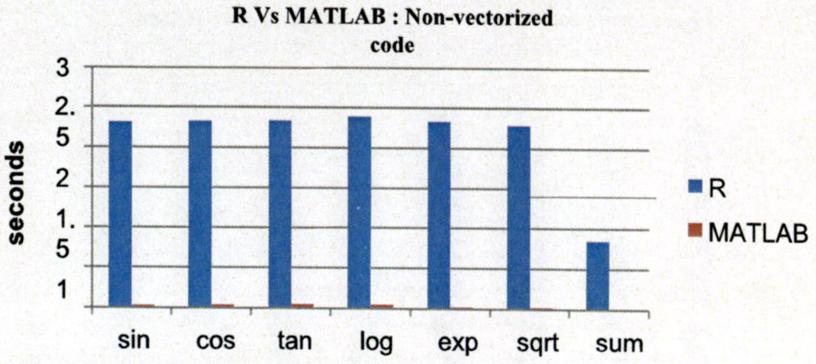

Fig. 5 Comparison of R versus MATLAB on Intel i5 M 450

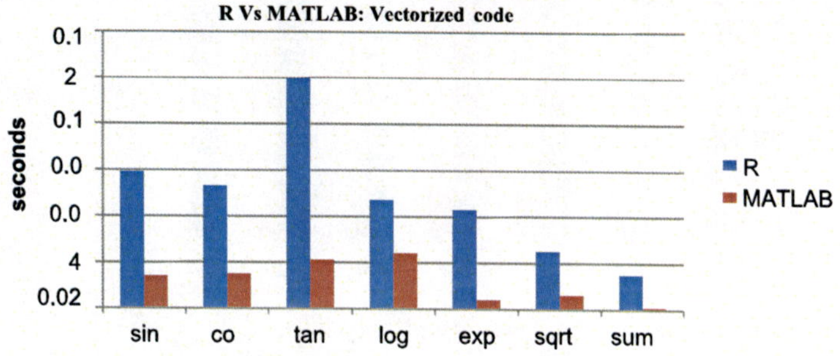

Fig. 6 Efficiency comparison on R and MATLAB

Performance on AMD fx 4300

Performance for *R* test cases (Figs. 7 and 8).

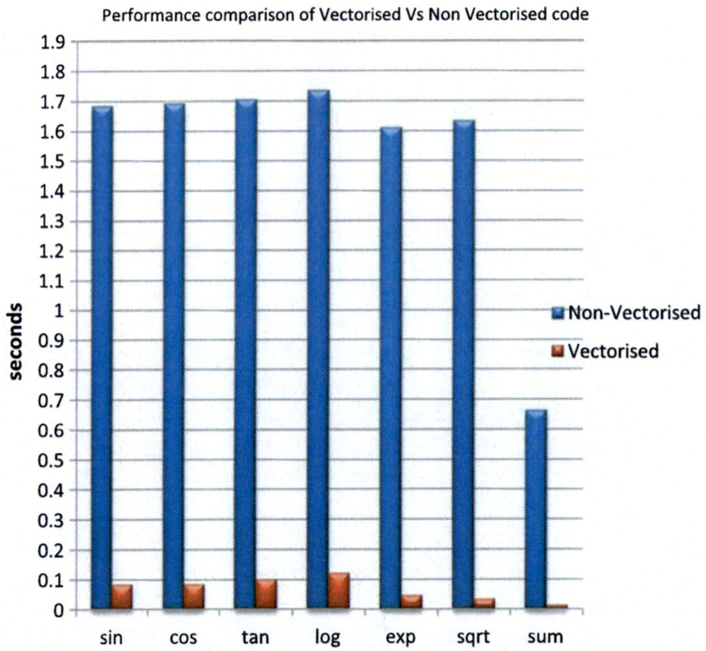

Fig. 7 Vectorizing trigonometric functions on AMD fx 4300

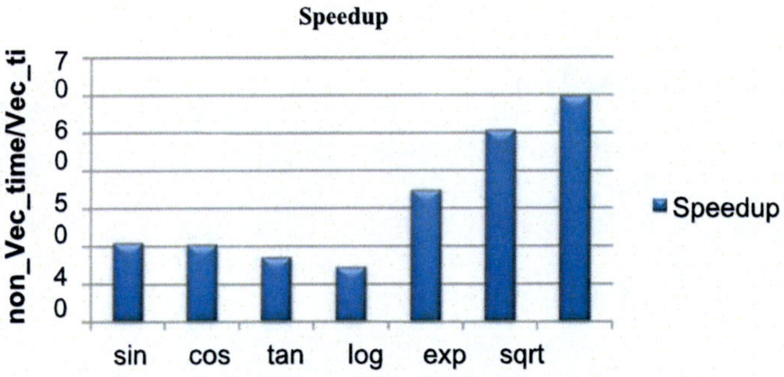

Fig. 8 Efficiency improvement for trigonometric functions

5 Conclusion and Future Work

The preliminary findings of our paper will help programmers make an informed decision about how auto vectorisation will help them and which compiler to choose in case of a choice. The test cases cover a variety of possibilities where auto vectorisation can and cannot work. These decisions can prove important when the programs have to perform large computations as they will help in decreasing the computation time significantly. The results will also help them realize which architectures support auto vectorisation of compilers in a better manner.

References

1. Madan M, Madan R (2013) Optimizing time cost trade off scheduling. Int J Appl Innov Eng Manag (IJAIEM). ISSN 2319-4847
2. Madan M, Madan R (2013) GASolver—a solution to resource constrained project scheduling by genetic algorithm. Int J Adv Comput Sci Appl 4(2) (IJACSA). ISSN 2156-5570(Online)
3. Malekiy S, Gaoz Y, Garzraany MJ, Wongz T, Paduay DA An evaluation of vectorizing compilers
4. AutoVectorization in GCC, Dorit Naishlos. IBM Research Lab in Haifa. dorit@il.ibm.com
5. https://in.mathworks.com/help/matlab/matlab_prog/vectorization.html
6. Kiepas P, Tadonki C, Ancourt C, law Kózlak J (2019) Code optimisations and performance models for MATLAB. MINES ParisTech/PSL University, AGH University of Science and Technology, Poland, 30 January 2019
7. https://insidehpc.com/2018/02/69201/

Author Index

A. Abraham et al. (eds.), *Proceedings of 3rd International Conference on Computing Informatics and Networks*, Lecture Notes in Networks and Systems 167,
https://doi.org/10.1007/978-981-15-9712-1